Silicate Crystal Chemistry

SILICATE CRYSTAL CHEMISTRY

Dana T. Griffen

New York Oxford
OXFORD UNIVERSITY PRESS
1992

Oxford University Press

Oxford New York Toronto
Delhi Bombay Calcutta Madras Karachi
Kuala Lumpur Singapore Hong Kong Tokyo
Nairobi Dar es Salaam Cape Town
Melbourne Auckland

and associated companies in
Berlin Ibadan

Published by Oxford University Press, Inc.,
200 Madison Avenue, New York, New York 10016

Library of Congress Cataloging-in-Publication Data
Griffen, Dana T., 1943–
Silicate crystal chemistry / Dana T. Griffen.
p. cm. Includes bibliographical references and index.
ISBN 0-19-504442-8
1. Silicate minerals. I. Title.
QE389.62.G72 992 549′.6—dc20 91-18669

9 8 7 6 5 4 3 2 1

Printed in the United States of America
on acid-free paper

Preface

In 1974 the Mineralogical Society of America published the first book of a series that was later named *Reviews in Mineralogy*. Most volumes deal comprehensively with a single subject—either a specific mineral group or a particular technique used in the study of minerals—and each has accompanied one of the Society's short courses. In publishing the *Reviews*, which now number over 20, both the Society and the contributing authors have rendered a service of great worth to the professional mineralogical community. Over the same period of time some substantial monographs dealing with individual mineral groups have been written by mineralogists, and these, too, have enriched the literature. Because of their specificity of subject matter, however, none of these is suitable as a textbook for graduate courses in crystal chemistry, and so the gap between such comprehensive professional reference works on specific subjects and the several popular undergraduate mineralogy texts has persisted. This text is intended to help fill that gap.

Because this work is designed to serve as a textbook for a course of one semester or two quarters, it is not a comprehensive reference work. The mineral groups that have been selected for inclusion are not the only ones that could have been included, but they do comprise a set within which the principles important in modern crystal chemistry are well illustrated. Students who become conversant with the material in this text will find themselves capable of reading the scientific literature on tourmaline, epidote, beryl, or any of the other important silicates that have been omitted here.

In general the mineral groups covered have been arranged in decreasing order of tetrahedral polymerization, so that framework silicates (tektosilicates) are addressed first and orthosilicates (nesosilicates) come last. The lone departure from this order is the chapter on amphiboles and nonclassical biopyriboles, which follows the chapter on pyroxenes, rather than coming between micas and pyroxenes; this is done to emphasize the ways in which amphiboles and nonclassical biopyriboles constitute a crystal chemical bridge between sheet silicates and single-chain silicates. We thus begin with the silica polymorphs, and proceed to feldspars, micas, pyroxenes, amphiboles and nonclassical biopyriboles, the aluminum silicate polymorphs, olivines, and garnets.

At the end of each chapter on a particular mineral group is a list of references. These lists are not intended to constitute a comprehensive compendium of available literature, but are rather lists of references from which specific data or interpretations have been taken. Nevertheless, the student who wishes to pursue further study of a mineral discussed in the text will find the references listed a useful basis from which to begin. In this regard, it is recommended that the student striking out on his or her own learn to use *Science Citation Index*, *Mineralogical Abstracts*, and *Chemical Abstracts* to locate studies not specifically cited in this text that have nonetheless added to our understanding.

It is anticipated that students who use this text in a formal class or for independent study will have had an introductory course in mineralogy. Even so, the time available in most undergraduate mineralogy courses and the amount of material to be taught guarantees that many topics that constitute part of the basis for advanced study will be relatively unfamiliar to the student. Such topics include space groups, analytical methods, realistic atomic-bonding models, and phase diagrams. For this reason chapters on these four topics are included as Part II of this text. These chapters are not sufficiently detailed to serve by themselves as texts for these topics, but they provide enough information to make discussions in Part I that depend on such subjects meaningful. They need not be assigned as required reading in a course on silicate crystal chemistry, but students with inadequate foundations in these subjects are encouraged to read them as needed.

No book is written in a vacuum, and many people have directly or indirectly contributed to this text. In general, acknowledgments for specific experimental data, interpretations, or figures are made by way of citations. References are listed alphabetically at the end of each chapter and also numbered sequentially, and cited in the text by number (e.g., [42]); this system, when compared with some others [e.g., the (author, year) format] seems to minimize the disruption of the flow of text. P. R. Buseck, P. E. Champness, H. Hirai, G. Van Tendeloo, and D. R. Veblen kindly provided original prints of photographs reproduced in the text. Joyce Berry, Holland Mills, Dolores Oetting, and Jacqueline Hartt of Oxford University Press have been very supportive and patient, and students who used early manuscript versions of the text in the advanced mineralogy course at Brigham Young University have improved it by their helpful critiques. Paul Ribbe reviewed the "final" version for the publisher and his comments helped make the present version better than that "final" version. For any errors that remain—and there surely are some in a book of this nature—I take sole responsibility.

Aside from the people who have assisted with the technicalities of producing this text, there are those to whom I owe a philosophical debt. Among them are three who deserve special mention. Any who are acquainted with the mineralogical careers of Jerry Gibbs and Paul Ribbe at the Virginia Polytechnic Institute and State University will recognize their influence in this text; it was they who taught me how to start being a crystal chemist. I also owe a great deal to Revell Phillips, who, during my 2 years as a student at Brigham Young University, opened the world of mineralogy to me, and continues to encourage and assist me both as a professional colleague and as a personal friend.

Provo, Utah
June 1991

D. T. G.

Contents

PART I

THE SYSTEMATIC CRYSTAL CHEMISTRY OF SILICATES

1

The Silica Polymorphs

While quartz is one of the most common, important, and well known of the rock-forming minerals, it is only one of about a dozen polymorphic variations of SiO_2. In addition to the crystalline forms, there are also several distinct silica glasses; because glasses have no long-range three-dimensional periodicity in atomic arrangement, these are presumably characterized by the degree and nature of short-range order. Many of these glasses and crystalline compounds do not occur naturally (e.g., keatite), and some probably require trace amounts of "impurities" (atoms other than silicon and oxygen) to exist at all (e.g., melanophlogite). Our discussion will be restricted largely to the most important naturally occurring phases of SiO_2: quartz, tridymite, cristobalite, coesite, and stishovite.

GENERAL OBSERVATIONS

The pressure–temperature (P–T) phase diagram for the pure SiO_2 system (Figure 1-1) provides considerable insight into some of the physical properties of the various polymorphs. In general, it is expected that higher temperatures should result in greater thermal vibration of the atoms, and thus greater effective atomic sizes. Therefore, high-temperature polymorphs should be less dense than low-temperature polymorphs. On the other hand, high pressure should result in closer packing of atoms; high-pressure polymorphs, should, therefore, be more dense than low-pressure polymorphs. Moreover, because electron density governs the velocity of propagation of light, the refractive indices (the ratio of the velocity of light in a vacuum to the velocity of light in the medium in question) should be greater in those polymorphs with greater electron density—and, for constant composition, such as we have here, that means greater crystalline density. Thus refractive index should decrease in a general way from upper left to lower right in the phase diagram.

The plot of mean refractive index versus density for a dozen forms of SiO_2 (Figure 1-2) confirms this prediction. Fulgarites are fused silica glasses formed by lightning, the point plotted for them being an approximate mean refractive index. The other phases are crystalline and possess tetrahedral framework structures, except for stishovite, which has the rutile (TiO_2) structure with the cation (Si in this case) in octa-

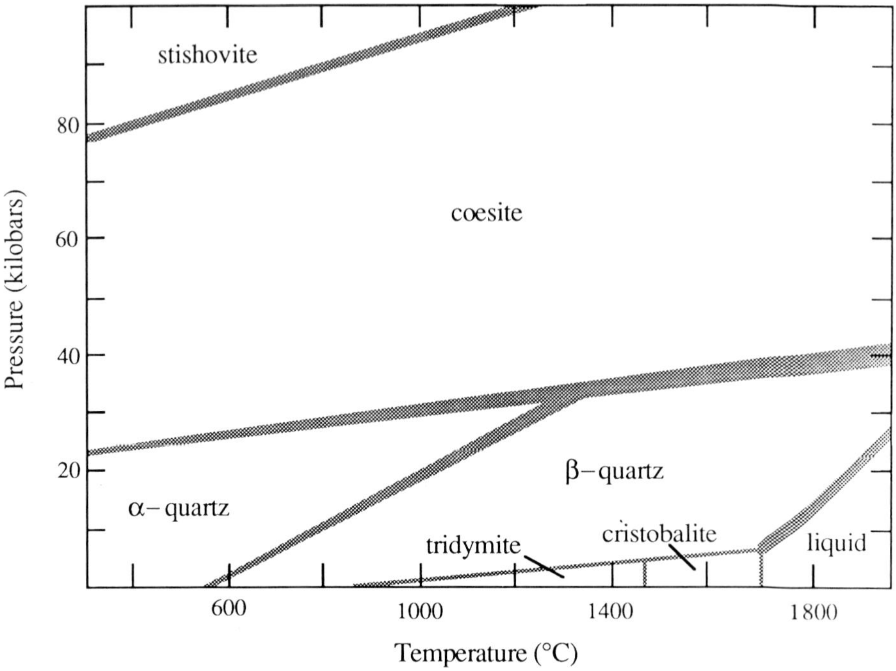

Figure 1-1. Pressure–temperature phase diagram for the SiO_2 polymorphs. Broad phase boundaries indicate, in a semiquantitative fashion, the uncertainties in experimental determinations. (Data from [6;7;32;42]).

hedral coordination. Mohs hardness (H) is seen to increase with density. (It is not true that hardness increases with density for all minerals, because chemical composition plays an important role. Here, composition is not a variable.)

The geologic occurrence of natural silica polymorphs is largely consistent with the foregoing data. Quartz is the most common. Coesite and stishovite originate in very high-pressure environments, such as meteorite impact craters. Tridymite and cristobalite are found in some volcanic rocks and some contact metamorphic rocks; they are also, however, constituents of some opals that form at low temperatures.

Figure 1-1 shows two forms of quartz: β- and α-quartz, often called high and low quartz, respectively. Tridymite and cristobalite also have high- and low-temperature forms, but the low forms are metastable and do not appear on an equilibrium phase diagram. We shall investigate phase transformations among the polymorphs in some detail further on.

CRYSTAL STRUCTURES OF THE SiO_2 POLYMORPHS

As noted previously, the crystalline SiO_2 polymorphs, except for stishovite, are tetrahedral framework structures: every SiO_4 tetrahedron shares each of its corners (that is, each of its oxygen atoms) with one other SiO_4 tetrahedron. Each silicon atom is four-coordinated, and, since there are half as many silicons as oxygens, each oxygen is two-coordinated. In stishovite, each silicon is six-coordinated, while the oxygens are each surrounded by three silicon atoms.

Quartz, tridymite, and cristobalite are known in both low- and high-temperature polymorphs, although, as previously stated, the high-temperature forms of tridymite and cristobalite have no fields of thermodynamic stability. The transformation between high and low polymorphs of the same name (high–low quartz, high–low tridymite, and so forth) is displacive, rapid, and unquenchable. High-temperature forms therefore do not exist below their respective transformation temperatures. The transition from a polymorph of one name to one of another (quartz–tridymite, tridymite–cristobalite, and so forth) is reconstructive and sluggish. Polymorphs such as low tridymite and coesite, which are unstable at surface conditions (Figure 1-1), can and do persist at the surface of the earth because the transformation to the stable form (low quartz) is reconstructive rather than displacive, requiring more energy than is effectively available in the ambient environment. We shall postpone until later in this chapter a detailed discussion of the nature of phase transformations in the silica polymorphs (see p. 15). It will suffice at this point to state that the topological arrangement of silicate tetrahedra for any high–low pair is identical, with the high-temperature modification having higher crystallographic symmetry.

Figure 1-2. Because the velocity of propagation of light is a function of electron density, the correlation between mean refractive index for SiO_2 polymorphs and density is very strong. Hardness is also a function of density.

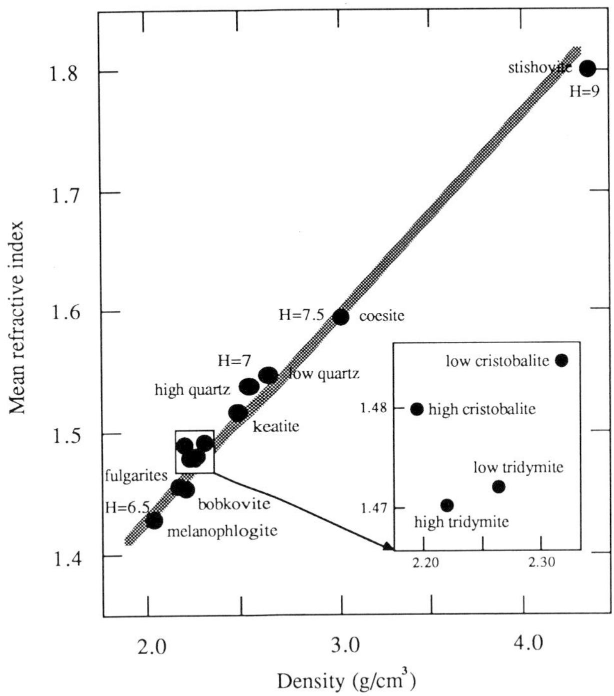

Identical topology means that the tetrahedra are linked together in the same arrangements, although angles within and between tetrahedra (i.e., steric details of the crystal structure) are not the same in the high form as in the low form. For polymorphs with different names, like quartz and coesite, the arrangements of tetrahedra are quite different and thus are said to possess different topologies.

Quartz

The idealized structure of β-quartz, the high-temperature polymorph of quartz, is shown in Figure 1-3. It may be characterized as consisting of double helical chains of tetrahedra running parallel to c. A sixfold screw axis extends parallel to the axes of the chains at the unit cell corners (that is, coaxial with the hexagonal-shaped channels), and a threefold screw axis parallel to c is located at $x = \frac{2}{3}$, $y = \frac{1}{3}$. Diad axes have been omitted for simplicity. The screw hexad may be a 6_2 or a 6_4, requiring the screw triad at $x = \frac{2}{3}$, $y = \frac{1}{3}$ to be a 3_2 or a 3_1, respectively. In the first case the space group is $P6_222$, and in the second it is $P6_422$. These two forms are related by a mirror plane operation, so that the "handedness" of the atomic arrangments is different. $P6_222$ β-quartz is called *left-handed high quartz* because of the 3_2 screw axis it contains, whereas $P6_422$ β-quartz is *right-handed high quartz* because of the 3_1 screw axis. (Pairs of left- and right-handed crystals are known as *enantiomorphs.*

Figure 1-3. The crystal structure of β-quartz. (a) A view down the c axis. (b) A view with the c axis tilted slightly to the right. Stippled tetrahedra comprise one-half of the double helical chain of tetrahedra about the 6_2 axis. (c) A view with the c axis nearly vertical. The stippled tetrahedra here represent the 3_2 screw axis, and show this to be left-handed quartz. [Note: stippled tetrahedra in (b) and (c) are not the same.] (Structural data from [45]).

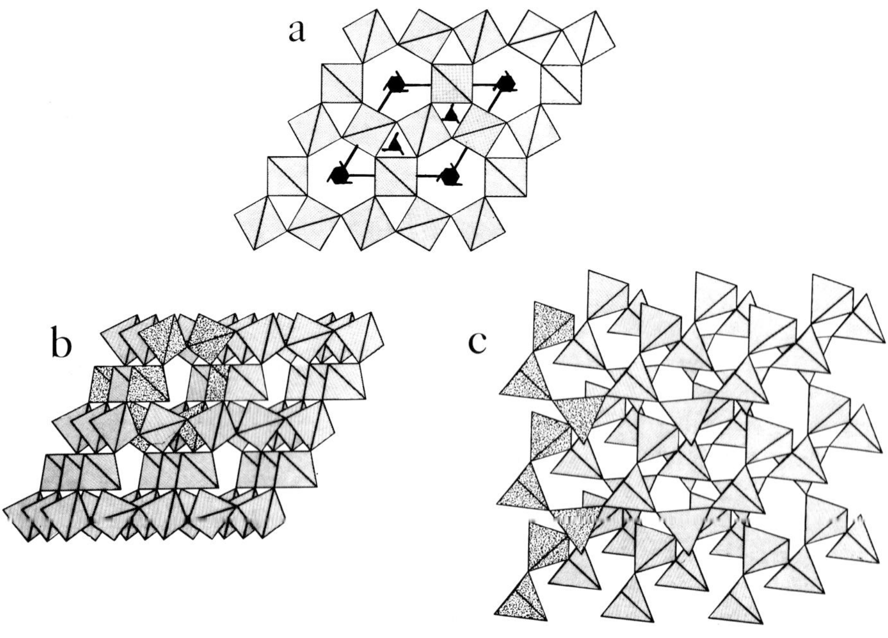

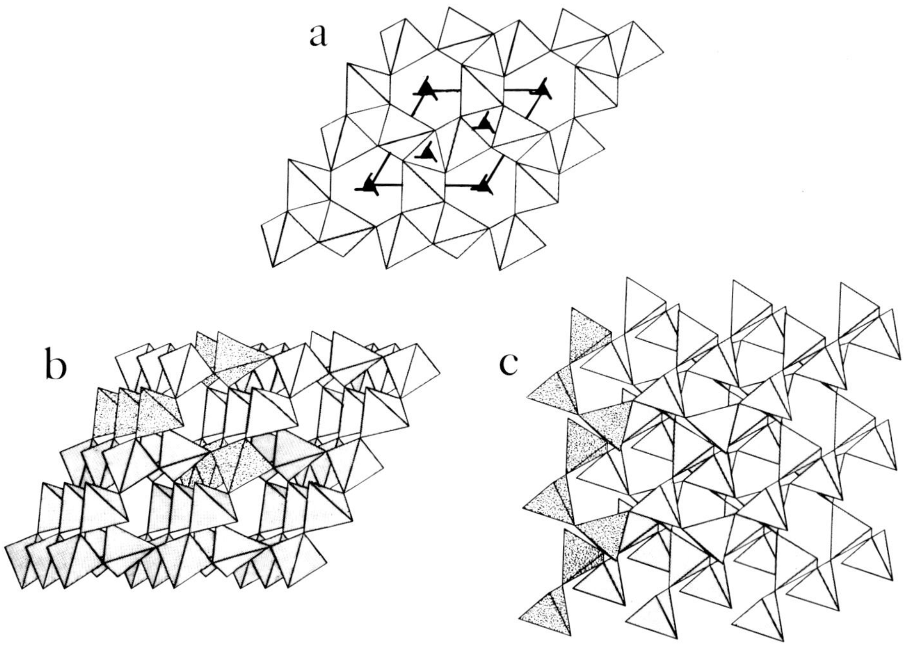

Figure 1-4. The crystal structure of α-quartz. Crystallographic orientations and stippled tetrahedra are the same as in Figure 1-3. Note in (b) that the structural collapse from β- to α-quartz has left only every other tetrahedron in the double helical chain equivalent, reducing the axis from 6_2 to 3_2. (Structural data from [45]).

Enantiomorphism can occur only in the 11 crystal classes that possess no mirrors, centers of symmetry, or rotoinversion axes—that is, in classes 1, 2, 222, 4, 422, 3, 32, 6, 622, 23, and 432.)

In α-quartz (Figure 1-4), the ideal hexagonal helical chains of β-quartz collapse by loss of the diad axis (inherent in the hexad parallel to *c*) to form trigonal helical chains with 3_2 or 3_1 screw axes. Crystals in space group $P6_222$ collapse to yield space group $P3_221$, while those in space group $P6_422$ collapse to give crystals in space group $P3_121$; thus the "handedness" is the same after the transformation as it was before. (Note that in Figure 1-5, however, no handedness is implied.) Because of the loss of the twofold axis parallel to *c* during the transition from either right-handed or left-handed β-quartz (sixfold screw axis) to its α-quartz analog (threefold screw axis), the structural collapse to α-quartz can occur in either of two ways with presumably equal likelihood; these are shown in Figure 1-5. If the sense of tetrahedral rotation that causes the collapse is different for separate regions of the same crystal, then the two kinds of regions (which are related by a twofold twin axis parallel to *c*) meet along irregular interfaces to form a Dauphiné twin (Figure 1-6).

Several observations should be made about Figure 1-6. First, α_1- and α_2-quartz (from Figure 1-5) are different from β-quartz by rotations in different directions; therefore, they cannot meet coherently along a boundary without distortion of the crystal structure. Because the transition from β-quartz to either α_2- or α_2-quartz is

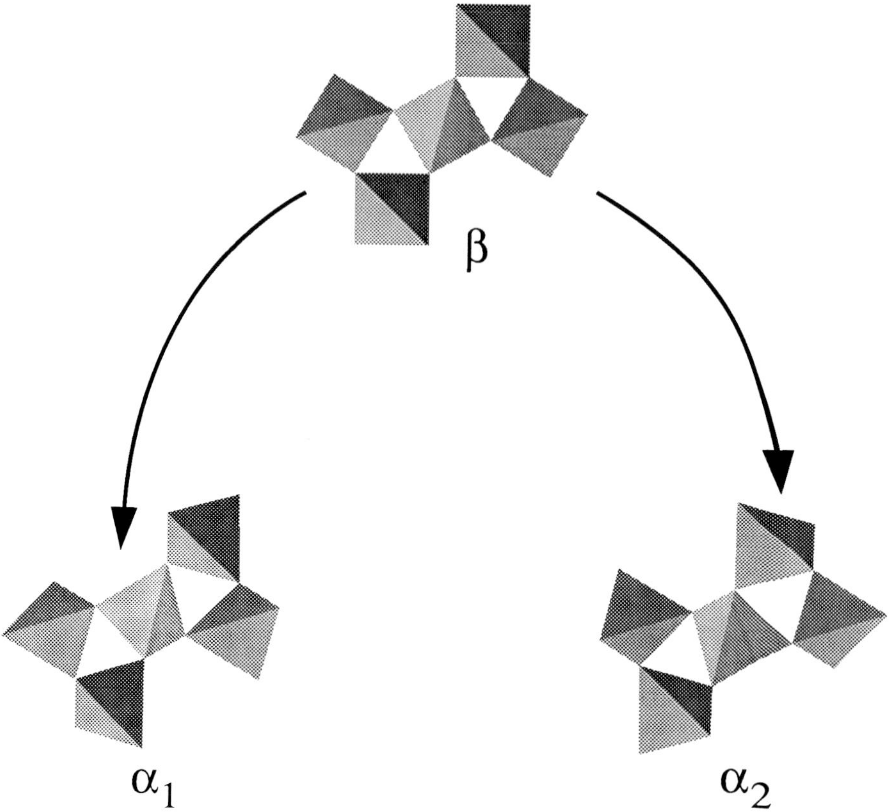

Figure 1-5. The geometry of the β–α collapse, in which the tetrahedra tilt toward the α_1 or α_2 orientations. The structural collapse may proceed in different ways in different parts of the same crystal, in which case a Dauphiné twin results (see Figure 1-6). (After [31]).

displacive, Si–O bonds are not broken across the twin boundary; thus tetrahedra at the boundary have to assume positions intermediate between the α_1 and α_2 configurations, and that intermediate arrangment is the structure of β-quartz. We shall return to these points later when we consider the details of the α–β tranformation (p. 20). (Note that in Figure 1-6 the oxygen positions—that is, the positions of the tetrahedral corners—have been incremented in small steps between their experimentally measured positions in α- and β-quartz, but the lattice parameters have not been varied in making the figure. Of course, the lattice parameters of the two quartz phases *are* different, so that distortions somewhat larger than those portrayed in the figure undoubtedly exist, probably in both the framework and in the tetrahedra themselves. These would not likely be apparent at the scale of Figure 1-6.)

Tridymite

Though it is a tetrahedral framework structure, tridymite may be thought of as consisting of stacked sheets of tetrahedra, with the tetrahedra of any one sheet pointing

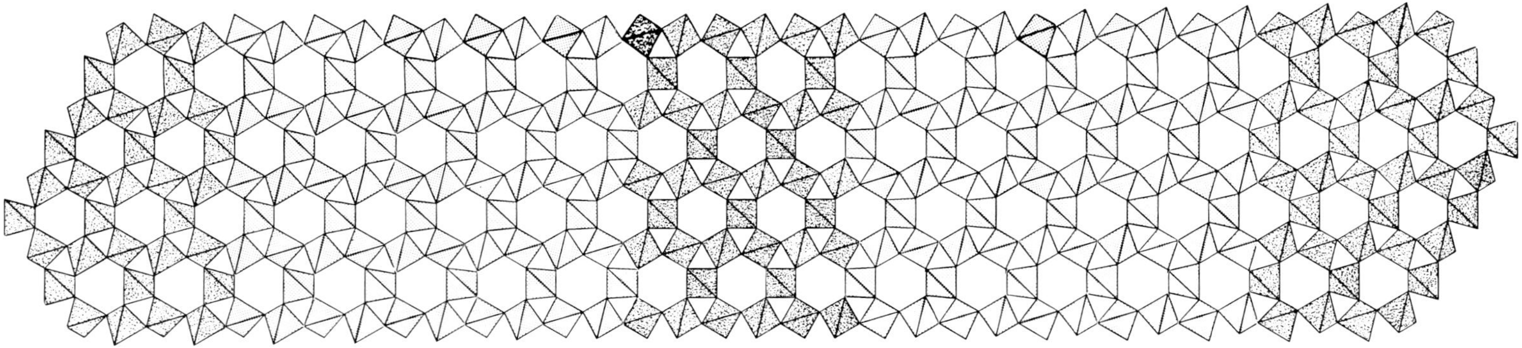

Figure 1-6. Stippled tetrahedra on the left represent α_2-quartz , and those on the right represent α_1-quartz; those in the center represent β-quartz. Between the α and β phases are regions of an "incommensurate phase" (shaded tetrahedra) that is structurally intermediate between the two. In a Dauphiné twin, the β phase would be at the center of the wall between the α_1 and α_2 domains.

Figure 1-7. The "ideal" tridymite tetrahedral sheet, viewed down the hexagonal *c* axis. The unit cell is outlined, and the structure is such that tetrahedra pointing down in this drawing share apical oxygen atoms with tetrahedra pointing up in the next lower sheet, and so on . The structure of high tridymite is represented by this arrangement in only an average sense.

alternately "up" and "down" with respect to the plane of the sheet (Figure 1-7). If such sheets are stacked atop one another so that a mirror plane passes through tetrahedral apices between the sheets, the tridymite topology results. When viewed along the normal to the sheets, the top sheet completely eclipses all the tetrahedra underneath. The arrangement is like hexagonal close packing with atoms missing in the planes of the tetrahedral bases (at the centers of the hexagonal channels) and many more missing in the planes of the tetrahedral apices. The space group is $P6_3/mmc$, and the arrangement is the "ideal" structure of high-temperature polymorph, β-tridymite. We shall discover in the following that the actual structure of β-tridymite is the same as this ideal structure only in an average sense [23;24].

The picture is quite complicated at lower temperatures, and not yet well understood. Among meteoritic and synthetic tridymites, there is an orthorhombic phase, space group $C222_1$, that is related to the hexagonal β form, as suggested by Figure 1-8. A slight distortion of the hexagonal lattice yields a *C*-centered orthorhombic unit cell with analogous lattice points and with essentially the same *b* and *c* dimensions as the hexagonal unit cell, but with a_{orth} approximately equal to $2a_{\text{hex}}$ sin 60°. The transition is displacive. A primitive orthorhombic polymorph (space group $P2_12_12_1$ with an *a* cell dimension of $\sim 3a_{\text{hex}}$ sin 60°) and a monoclinic polymorph (crystallizing in space group *Cc*) have also been reported, as have some that are orthorhombic with $a \approx 2a_{\text{hex}}$ sin 60°, $b \approx 2b_{\text{hex}}$, and $c \approx nc_{\text{hex}}$ (n = 1, 1.5, 2, 5, 6, 10, and 12 so far). Some of these are summarized in Figure 1-9. In addition to these

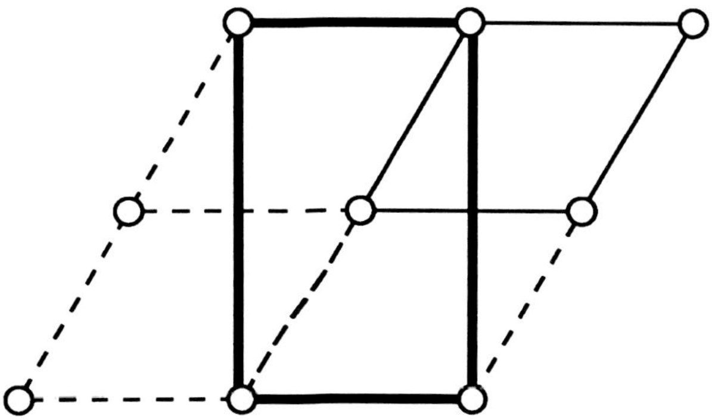

Figure 1-8. Schematic representation of the slight metric distortion that transforms the "ideal" hexagonal lattice of high tridymite to the *C*-centered orthorhombic lattice.

temperature-induced phase transformations, a pressure-induced transition between the monoclinic *Cc* polymorph and a 10-layer orthorhombic form has been observed at room temperature and a pressure of 5 kbar [35]. A terrestrial 10-layer low tridymite with the unconventional triclinic space group *F*1 has also been reported [28]. (Note the use of a face-centered Bravais lattice so as to maintain a

Figure 1-9. Some of the tridymites that have been observed in natural and synthetic specimens at atmospheric pressure. Transformation temperatures are uncertain to one extent or another, and all inversions are not generally observed for any given specimen (for example, see [9]). (Data from [15;23;34;42]).

Temperature (°C)	System	Space group	Cell parameters
400	Hexagonal	$P6_3/mmc$	a=5.05, c=8.28Å
300	Orthorhombic	$C222_1$	a=8.74, b=5.04, c=8.28Å
200			
	Orthorhombic	?	a=105-65, b=5.0, c=8.2Å
	Orthorhombic	$P2_12_12_1$	a=26.65, b=5.02, c=8.15Å
100			
	(Orthorhombic	?	a=17.3, b=10.1, c=8.2Å)
	Monoclinic	Cc	a=18.5, b=5.0, c=25.8, β =118°
0			

triclinic unit cell that is geometrically analogous to unit cells of other tridymites, in order to facilitate comparison among the various polymorphs.)

It is obvious from the silica phase diagram (Figure 1-1) that none of the low-temperature polymorphs of tridymite is stable at the temperatures shown in Figure 1-9. They are therefore metastable, and form only because it takes far less energy to invert from β-tridymite to one of the low-tridymite polymorphs than to the stable silica polymorph, quartz. This is called *alternative metastable behavior*, and will be considered in more detail below (p. 20). It is unlikely that the low-temperature forms of tridymite have any fields of actual thermodynamic stability at all, but the transformation relationships among some of them are quite unclear at present.

Cristobalite

If "tridymite sheets" are stacked so that every other one is rotated 180° about its normal (the high-tridymite c axis), the resulting arrangement is that of "ideal" high cristobalite (β-cristobalite). A view normal to the sheets is shown in Figure 1-10. In this configuration the stacking sequence for oxygen atoms is $\cdots ABCABCA \cdots$; that is, it is like a cubic close packed arrangement with large numbers of anions missing. As a result, the "ideal" unit cell of β-cristobalite is isometric (cubic), and

Figure 1-10. Two adjacent layers of the crystal structure of "ideal" high cristobalite, projected down the [111] crystallographic direction. The tetrahedra of the lower sheet are symmetrically identical to those of the upper sheet. A third sheet would have the same orientation as the top sheet, but would be shifted laterally so that its "up" tetrahedra shared an oxygen with the "down" tetrahedra of the lower sheet; from there on, additional sheets would be eclipsed by one of the top three. Like Figure 1-7 for tridymite, this structure represents only the "ideal" average structure for high cristobalite.

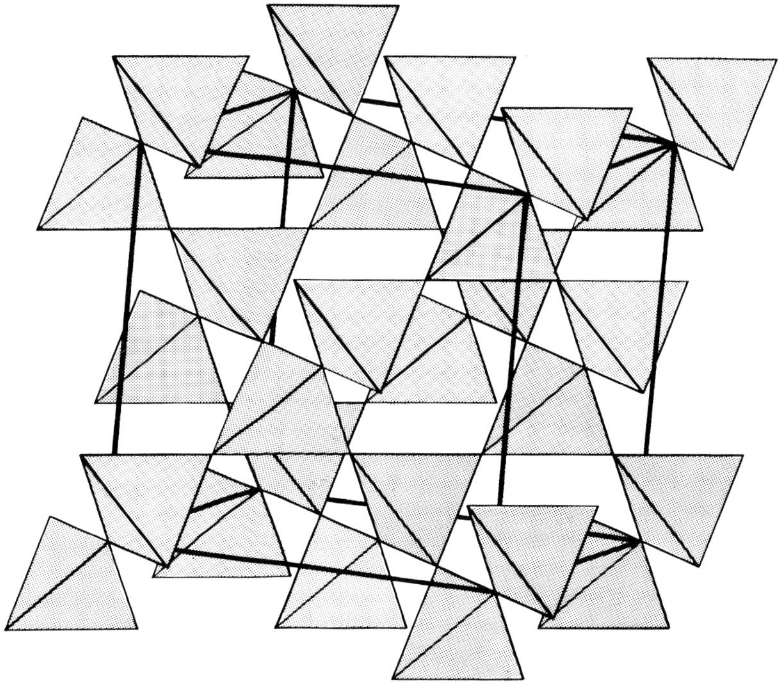

Figure 1-11. The "ideal" high-cristobalite structure, showing it to be isometric.

the "close-packed" planes are seen to be the {111} planes in Figure 1-11. This geometrically ideal model cannot be correct, however, for reasons that will be treated later (p. 26). It turns out that, analogous to the hexagonal structure of high tridymite, that of high cristobalite is isometric (space group $Fd\bar{3}m$) only on the average [29].

The low-temperature polymorph, α-cristobalite, is a displacive distortion of the "ideal" β polymorph, and crystallizes in space group $P4_12_12$ (or the enantiomorph, $P4_32_12$) [14].

Coesite

This high-pressure form of silica crystallizes in only one polymorph, and in space group $C2/c$. Unlike tridymite and cristobalite, the structure is not related to a close-packed arrangement, but shows some similarities to the feldspar framework (p. 40). The silicate tetrahedra are arranged in four-membered rings linked in chains that extend along the [010] direction (i.e., along b). The chains are cross-linked to yield a different kind of four-membered ring parallel to the (010) planes [4].

The mineral coesite is the densest of all the SiO_2 tetrahedral frameworks, and its crystal structure is shown in Figure 1-12. At one time a polymorph of coesite, crystallizing in space group $P2_1/a$, was thought to have been synthesized at pressures of

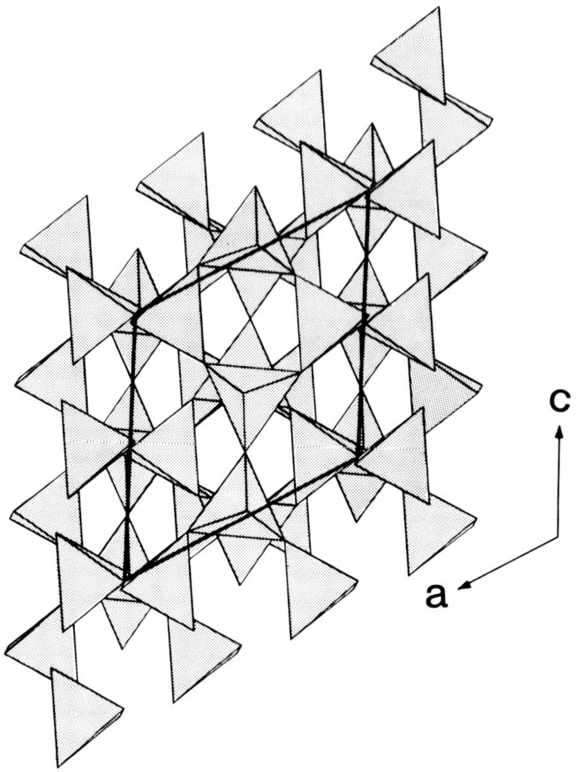

Figure 1-12. The crystal structure of coesite. Note how dense it is compared with the structures of lower pressure polymorphs. (Structural data from [19]).

44 to 49 kbar [26]. Phase transformations between the two polymorphs were never observed, however, and it was subsequently shown that $P2_1/a$ coesite was only $C2/c$ coesite twinned on (100) [8;25;38].

Stishovite

In stishovite, each silicon atom is surrounded by six oxygen atoms, rather than by four as in the other crystalline silica polymorphs. It is therefore not a tetrahedral framework, but has the rutile (TiO_2) structure, and crystallizes in the space group $P4_2/mnm$. The crystal structure is shown in Figure 1-13, where it can be seen that octahedra share both edges and corners to form a relatively dense octahedral network. As expected from Pauling's rules, the shared edges are the shortest in the octahedron, and in fact, are among the shortest O···O separations known (2.29 Å)—well below twice the generally accepted effective ionic radius of oxygen. The O–Si–O angle opposite the shared edge is only 81.3° instead of the ideal 90°, but the Si–O distances involved in this angle are less than the others in the structure by about 3%. This last result is contradictory to the usual expectation (see Figure 3-15) that if such distances do not lengthen to minimize repulsion between cations across the shared edge, they at least will not shorten.

PHASE TRANSFORMATIONS

The transition from one polymorphic form to another is called a *phase transformation* (although phase transition, phase change, and inversion are used synonymously). Phase transformations may be characterized in any of several ways. We shall use the following four ways, to one extent or another, to describe phase changes not only in the silica polymorphs but in other minerals as well:

1. The change from one mineral to another may be described as discontinuous or continuous. *Discontinuous transformations* are those in which a new phase

Figure 1-13. The crystal structure of stishovite, a high-pressure form of SiO_2 in which the Si is in sixfold coordination, and the octahedra share edges. (Structural data from [21]).

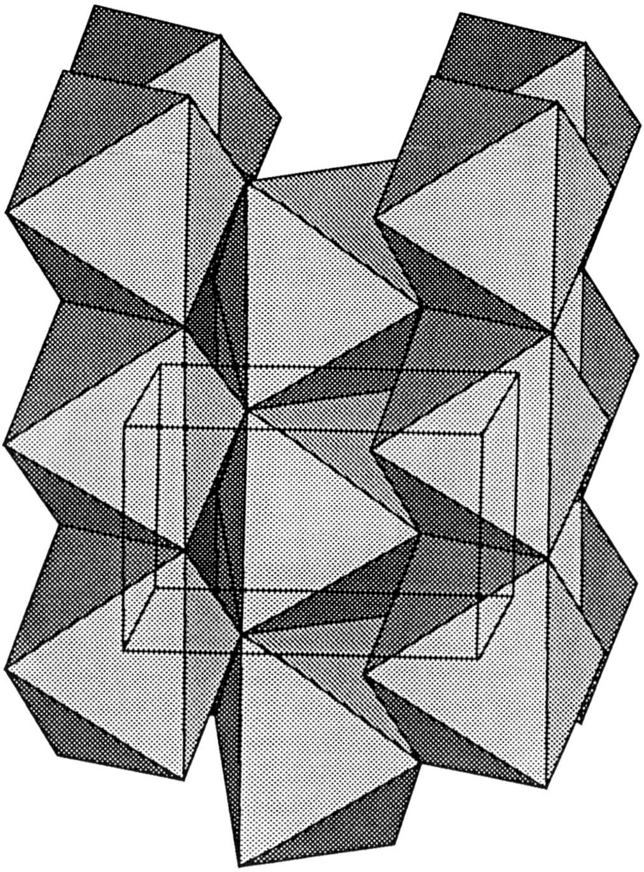

nucleates at discrete locations in the host phase, then grows at the latter's expense. The structure of the host and transformed phases are quite different, and a distinct interface separates the two. The initial stages of transformation involve nucleation of some sort at discrete positions within the host—perhaps dislocations, grain boundaries, or other sorts of imperfections.

A *continuous transformation*, on the other hand, occurs simultaneously throughout the host as a structural fluctuation. The initial stage involves not nucleation, but modulation of the parent structure in response to some external stimulus; the modulated structure is not exactly identical to either the parent or the daughter phase, but is somehow intermediate. Often the modulation can be characterized as a "distortion wave" that displaces atoms small distances, and in somewhat predictable ways, from their normal positions, thus disrupting (if only slightly) the rigid periodicity that is theoretically characteristic of a crystalline solid. Such distortion waves typically have wavelengths of several to many times the unit cell dimensions. Because the strict crystallographic equivalence of otherwise identical atomic positions in adjacent unit cells is lost, such modulated structures are often called *incommensurate* phases.

2. The terms *reconstructive* and *displacive* have already been used in describing polymorphic changes (p. 5). A reconstructive transformation requires the breaking of atomic bonds, whereas a displacive transformation involves only small shifts in atomic positions but leaves atomic bonds intact. The breaking of bonds requires energy, called the *activation energy* for the transformation, and this energy constitutes a barrier to the inversion; unless the activation energy is supplied (usually in the form of heat), the transformation cannot proceed. Thus reconstructive transformations are sluggish, tending to be comparatively slow even when the activation energy is available. Displacive transformations involve only trivial or no activation energy and are therefore rapid; moveover, they are also unquenchable, meaning that the high-temperature form cannot be "frozen in" even by the most rapid quenching from above the transition temperature.
3. From a crystal structural point of view, phase transformations can be categorized as (a) those in which primary coordination is changed, (b) those in which only secondary coordination is changed, and (c) those involving a change in ordering.
   a. A change in primary coordination means that coordination numbers change. For example, in the transformation of coesite to stishovite, the coordination number of silicon goes from 4 to 6, while that of oxygen changes from 2 to 3.
   b. If only secondary coordination changes, then coordination polyhedra of anions about cations remain intact, but their arrangements relative to one another change. The transformation from quartz to coesite, for example, is of this sort, because coordination numbers do not change, but the orientations of second-nearest-neighbor atoms (the second coordination sphere) do.
   c. Changes in ordering may be of either the substitutional or the positional sort. Subsitutional ordering occurs as certain atomic species, usually cations, become completely or partially segregated into certain specific structural sites rather than being randomly distributed over them. Because silicon is the only

cation in any of the SiO_2 polymorphs, substitutional ordering plays no role in the crystal chemistry of these minerals; it is of great significance in most of the other silicates we shall consider, however. Positional ordering occurs as atoms that oscillate between or among different positions become ordered on one set of positions as temperature, or pressure, or both change. We shall see below (p. 20) that the α–β polymorphic transitions in the silica minerals can probably be thought of as positional ordering transformations.

4. Finally, polymorphic transformations may be classified on thermodynamic grounds as first order, second order, third order, and so on. The Gibbs free energy G is a defined function of temperature (T) and two quantities known as entropy (S) and enthalpy (H), such that $G = H - TS$. A system is in equilibrium (that is, it is stable under existing conditions) only when, at the given temperature and pressure, G has the minimum possible value. In other words, G is less than it would be for any other configuration or state of the system. Thus the stable polymorph under particular external conditions is the one with the lowest free energy. If two polymorphs coexist in stable equilibrium, then they must have the same Gibbs free energy ($G_{\text{phase 1}} - G_{\text{phase 2}} \equiv \Delta G = 0$). The nature of the change in slope of the G-versus-T curve at the transformation temperature distinguishes first- and higher-order transformations. If there is a discontinuity in the slope ($\Delta(dG/dT) \neq 0$), the transformation is first order, whereas second-order transformations have continuous G-versus-T curves [$\Delta(dG/dT) = 0$]. Put another way, a transition of order n shows a discontinuity in the nth (and higher) derivatives of free energy with respect to temperature. In Figure 1-14, T_t is the transition temperature for two polymorphs whose free energy curves are labeled φ_1 and φ_2. The stable polymorph is the one with the lower free energy: polymorph ϕ_2 above T_t, and polymorph ϕ_1 below that temperature. Stable portions of free-energy curves are solid, unstable (or metastable) portions are dashed. The first-order transformation shows a discontinuity in the slope of the stable free energy curve, whereas the second-order transformation shows a continuous change in slope. While temperature has been used as the intensive variable for illustration, we could as well have plotted free energy against pressure. Thus, any properties that depend on derivatives of G with respect to T or P will show discontinuities appropriate to the order of the transformation at the transformation temperature (or pressure). For a first-order transition, molar volume, entropy, and chemical potential (all of which depend only on first derivatives of G) are discontinuous at the transformation temperature. For a second-order transition, those three quantities are continuous at the transformation temperature, but heat capacity (C_P), compressibility (β), and thermal expansion (α) are discontinuous, because they depend on second derivatives of G.

Another type of phase transformation that is defined on thermodynamic grounds is the "mixed" or lambda (λ) transition. This is distinguished from a second-order phase change by the behavior of the second deriviatives of G, which tend to infinity at T_t, and hence give curves whose shapes resemble the Greek letter λ. While it is presumably possible for all of the second derivative curves to exhibit the λ shape, it is worth noting here that, in practice, C_P is viewed as the important variable for recognizing λ transitions, and that the other second

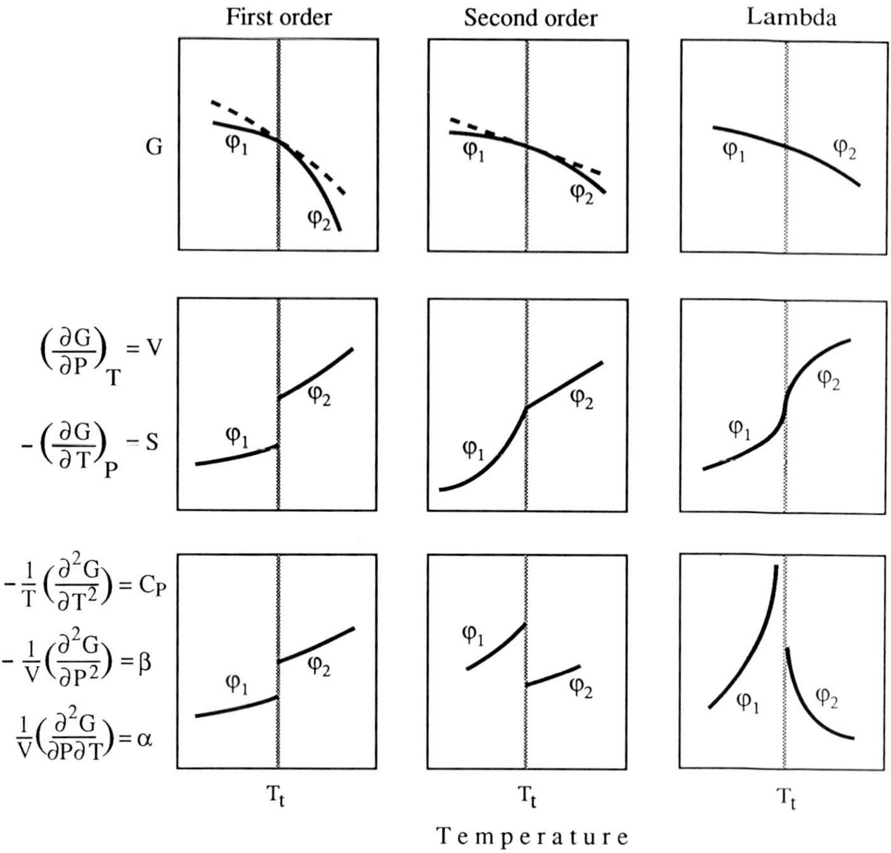

Figure 1-14. Thermodynamic functions versus temperature for three types of transformations. Note that, for first- and second-order transitions, the first and second derivatives, respectively, are the first ones to exhibit discontinuities at the transformation temperatures. Symbols are as follows: *G*, Gibbs free energy; *V*, molar volume; *S*, entropy; C_P, heat capacity; β, compressibility; α, coefficient of thermal expansion. (After [43)].

derivatives may or may not behave similarly. λ transitions are often viewed phenomenologically as near second order, differing only slightly from "classical" second-order phase changes (that is, those represented by the middle column of curves in Figure 1-14); their thermodynamic treatment is, however, quite different.

It is possible to draw some approximate correlations among the different schemes for classifying polymorphic transformations. At one end of the spectrum are discontinuous, reconstructive, primary coordination, and first-order transformations, while at the other end are continuous, displacive, secondary coordination, and second- and higher-order changes. Like any generalization, this simple correspondence does not always hold. Some changes in secondary coordination are also reconstructive, and some changes in primary coordination are displacive. Changes in substitutional ordering are reconstructive and first order; most transitions involving positional order are second order, but some may be first order. Changes in positional ordering may, in principle, be either displacive or reconstructive.

Reconstructive Transformations in the SiO_2 System

The cristobalite–tridymite, tridymite–quartz, cristobalite–quartz, coesite–quartz, and stishovite–coesite transformations are all reconstructive and first order, all are discontinuous, and all but stishovite–coesite involve only rearrangement of tetrahedra (i.e., only a change in secondary coordination). The stishovite–coesite transformation involves a change in primary coordination ($^{[6]}Si$ to $^{[4]}Si$).

The equilibrium phase diagram (Figure 1-1) predicts which polymorph will be the stable one for any pressure and temperature. It does not, however, rule out the possibility that a given phase may persist outside its field of thermodynamic stability. Tridymite, for example, is not stable at ambient surface conditions, but it is nevertheless observed in some volcanic rocks and meteorites at the surface of the Earth. To understand this, consider the schematic G–T curves of Figure 1-15. Tri-

Figure 1-15. Schematic representation of free-energy–temperature curves for silica polymorphs at some low pressure. Black curves indicate phases that are thermodynamically stable under some conditions of P and T, while grey curves indicate phases that are metastable at best. R indicates a reconstructive transformation, D a displacive one. The latter requires very much less energy, and is therefore more likely to occur, leading to alternative metastable behavior and the existence of metastable forms of SiO_2.

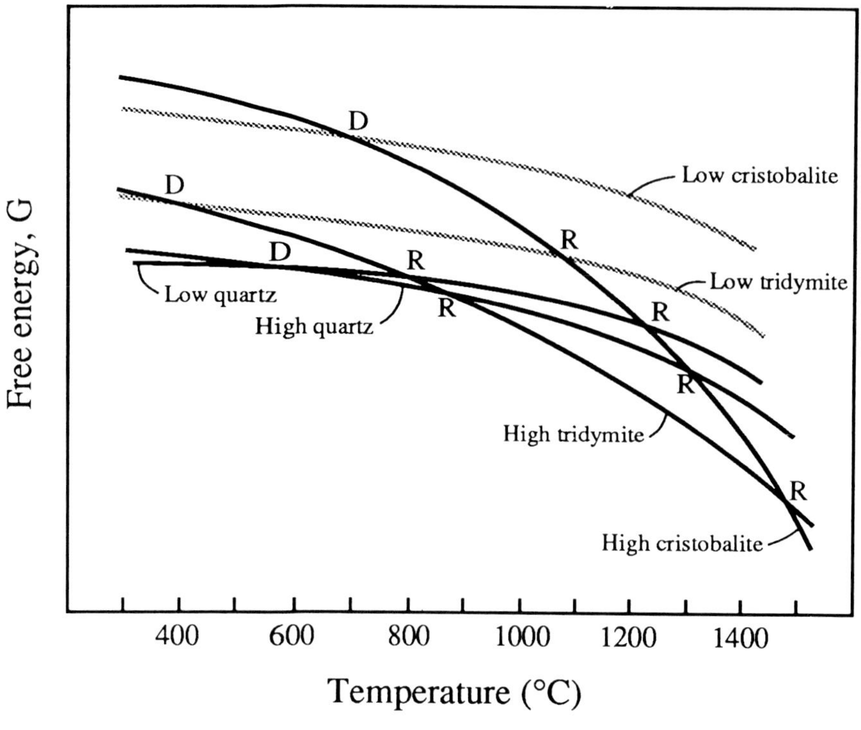

dymite is stable (at atmospheric pressure) from 1470° to 867°C; below that, quartz has a lower free energy and is the stable form. Because the transition from tridymite to quartz is reconstructive, significant activation energy is required for the transition to occur, and this energy must be available at the time the bonds need to be broken and reformed. But unless the cooling rate is very slow, this will not be the case, and tridymite will persist past the transformation temperature. At some lower temperature (see Figure 1-9) the high tridymite will invert to the centered orthorhombic form by a simple displacive transformation that involves little or no activation energy, and thence perhaps to yet lower forms via other displacive mechanisms. The low tridymite thus produced is said to be a metastable phase, and this behavior is called *alternative metastable behavior*. Metastable phases form only on cooling; if quartz is heated, it inverts to high tridymite, never to one of the low-tridymite polymorphs.

Alternative metastable behavior lowers the free energy of the system, but not as much as inversion to the thermodynamically stable phase would have. It is the magnitude of the difference between the free energy of the existing phase (say, high tridymite) and that of the thermodynamically stable phase (say, quartz) that provides the "driving force" for the phase transformation, however, and inversion from high tridymite to low tridymite (instead of to quartz) makes that difference smaller than it would have been had high tridymite persisted. Thus the "driving force" is diminished, making it even less likely that the stable phase (quartz) will form after an alternative metastable transition has occurred. For this reason, metastable phases formed in this way are often called *stranded phases*.

Displacive Transformations in the SiO_2 System

We have already identified the various low-temperature polymorphs that result from displacive inversions from high-temperature forms. Here we shall take a more detailed look at α–β transformations, with emphasis on the one in quartz because it has been the most intensely studied.

Even though it was discovered over a century ago, the α–β inversion in quartz is still not completely understood, and is a matter of continuing investigation. Figure 1-16 shows how some selected physical properties of quartz vary with temperature as the transformation temperature (~573°C at atmospheric pressure) is approached. The Si–O–Si bond angles, molar volume, and thermal expansion appear to change smoothly and monotonically up to 573°C, and then the slopes change abruptly at that temperature. This suggests that the transformation is continuous, and reveals no evidence for a distinct nucleation of the β phase in a matrix of α-quartz. The transformation cannot be stopped and involves shifts in atomic positions that appear to encounter no kinetic barrier, and so it is displacive. From Figure 1-5 it is apparent that the difference between the ideal structure of high quartz and that of low quartz is in the arrangement of tetrahedra; that is, it is a small change in secondary coordination.

Determination of the thermodynamic order of the transformation is not straightforward. Some who have investigated it interpret such curves as are shown in Figure 1-16 as revealing a very small but discontinuous change in properties at the tran-

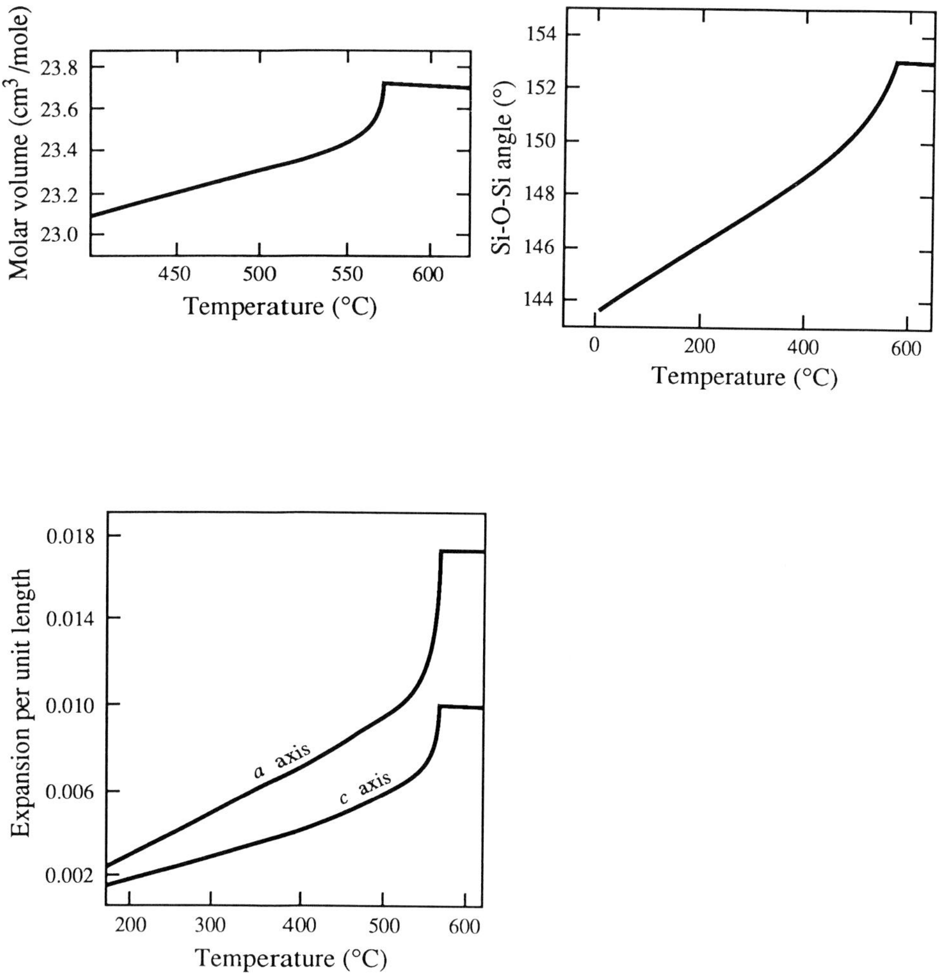

Figure 1-16. Changes in some properties of quartz across the low–high inversion temperature. Note that the temperature scales are not uniform. (Data from [1;18;20]).

sition temperature. Such a change is shown schematically in Figure 1-17. While the transition is completely reversible, many experiments have revealed hysteresis in properties such as coefficients of thermal expansion [5] and birefringence [10], suggesting a first-order phase change. (*Hysteresis*, in this context, means that there is a difference in the transformation temperature depending upon whether it is approached from below or from above.) Even so, the magnitude of the hysteresis effects seems to be dependent on grain size (which affects surface free energy)—with smaller grain size producing greater hysteresis—and thus may reflect sample preparation. Figure 1-18 reveals that heat capacity, C_P, behaves as if the transformation were a λ transition, and an adequate thermodynamic model of the phase change can be constructed by assuming that it is. Although a final first-order "click" in the

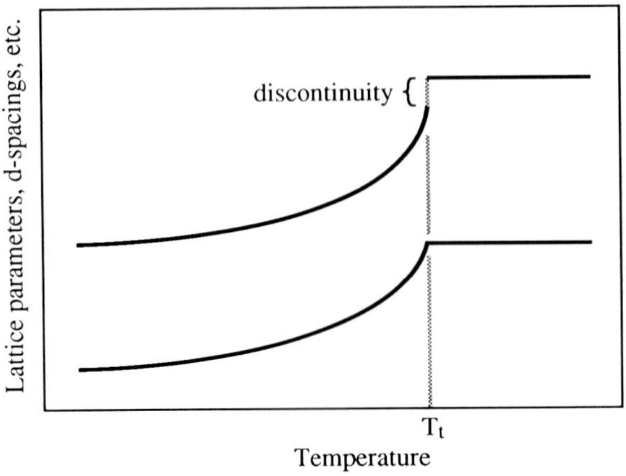

Figure 1-17. Physical properties are expected to exhibit a discontinuity at the transformation temperature for a first-order inversion, but not in a second- or higher-order inversion.

transformation has neither been proved nor conclusively refuted, it is unnecessary to assume it in order to describe the inversion from α- to β-quartz adequately.

It turns out that this exercise in classification really does not provide a definitive answer to the basic question: What actually happens as α-quartz is heated and transforms to β-quartz? There are three possibilities, and the standard x-ray- or neutron-diffraction experiments by which crystal structures are determined cannot discriminate unequivocally among them, because the experiments yield average structures

Figure 1-18. Variation in heat capacity, C_P, for quartz as a function of temperature. Note the resemblance of this curve to the behavior of C_P in a λ transition, Figure 1-14. (After [18].)

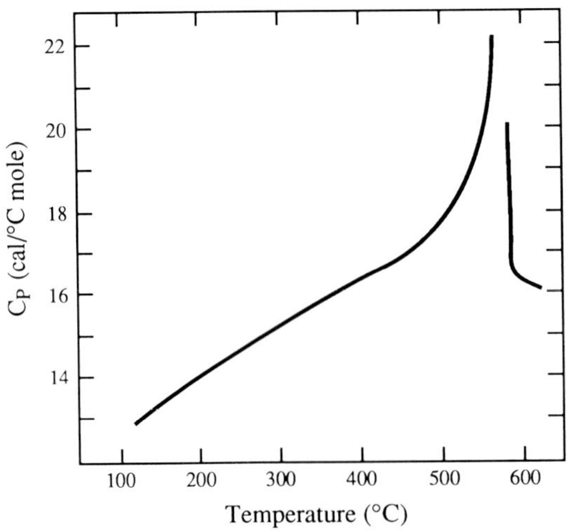

of all the unit cells in the megascopic crystals being irradiated. The three possible models for the β phase are:

1. Above the transition temperature, the structure of each unit cell is truly hexagonal, space group $P6_222$ or $P6_422$, and therefore looks like the unit cell in Figure 1-3. This is what has been tacitly assumed so far in our discussion, because to this point it has made no difference and it is the simplest model.
2. Near the transition temperature, Dauphiné twins form, and they persist above that temperature, but become very small. Thus, above the transition temperature, the crystal consists of static microdomains of each Dauphiné twin individual (α_1 and α_2 orientations of Figures 1-5 and 1-6), and the structure of β-quartz obtained by diffraction experiments is just the average of these two Dauphiné twin orientations taken over the volume of the crystal.
3. Near the transformation temperature, but below it, Dauphiné twin domains form, and the thermal energy causes each domain to oscillate between its two possible orientations. The oscillations are correlated in adjacent domains. As the transformation temperature is exceeded, the oscillations become random and uncorrelated, so that the structure of β-quartz determined by diffraction experiments is both a space and time average of the α_1 and α_2 orientations. Because the oscillations are uncorrelated above the transition temperature, there are no "permanent" regions or microdomains of α-quartz, as there are in model 2 above.

X-ray-diffraction data favor model 1, the ideally hexagonal high-quartz model. Neutron-diffraction crystal-structure refinements, which provide slightly better positional information for the oxygen atoms than do x-ray-diffraction experiments, suggest that the oxygens are disordered over two slightly separated sets of sites, either one of which yields trigonal symmetry, but the combination (superposition) of which is hexagonal. Neutron diffraction data thus favor either model 2 or model 3.

Figure 1-19a is a transmission electron micrograph of quartz viewed down the crystallographic *c* axis. The scale bar shown on the micrograph has a length of 2000 Å. A temperature gradient has been established by the electron beam, so that the temperature increases from the top to the bottom, and is 573°C about half way down the photograph. The uniform area at the top is α-quartz, and that at the bottom is β-quartz. The white triangles all have the same crystallographic orientation and are said to be "in contrast," while the dark triangles are all "out of contrast" because they have some other (but uniform) crystallographic orientation. The light and dark triangles are Dauphiné twin domains of orientations α_1 and α_2. Although the domains look stable enough in this still micrograph, they oscillate rapidly between α_1 and α_2 orientations (light and dark) as the transformation temperature is approached and the domains become finer and finer [12]. The transformation occurs simultaneously everywhere the temperature is appropriate. Above 573°C, the frequency of oscillations are so rapid as to blur the image, yielding an apparently (but not actually) homogenous β-quartz phase in which the sizes of the domains are very small.

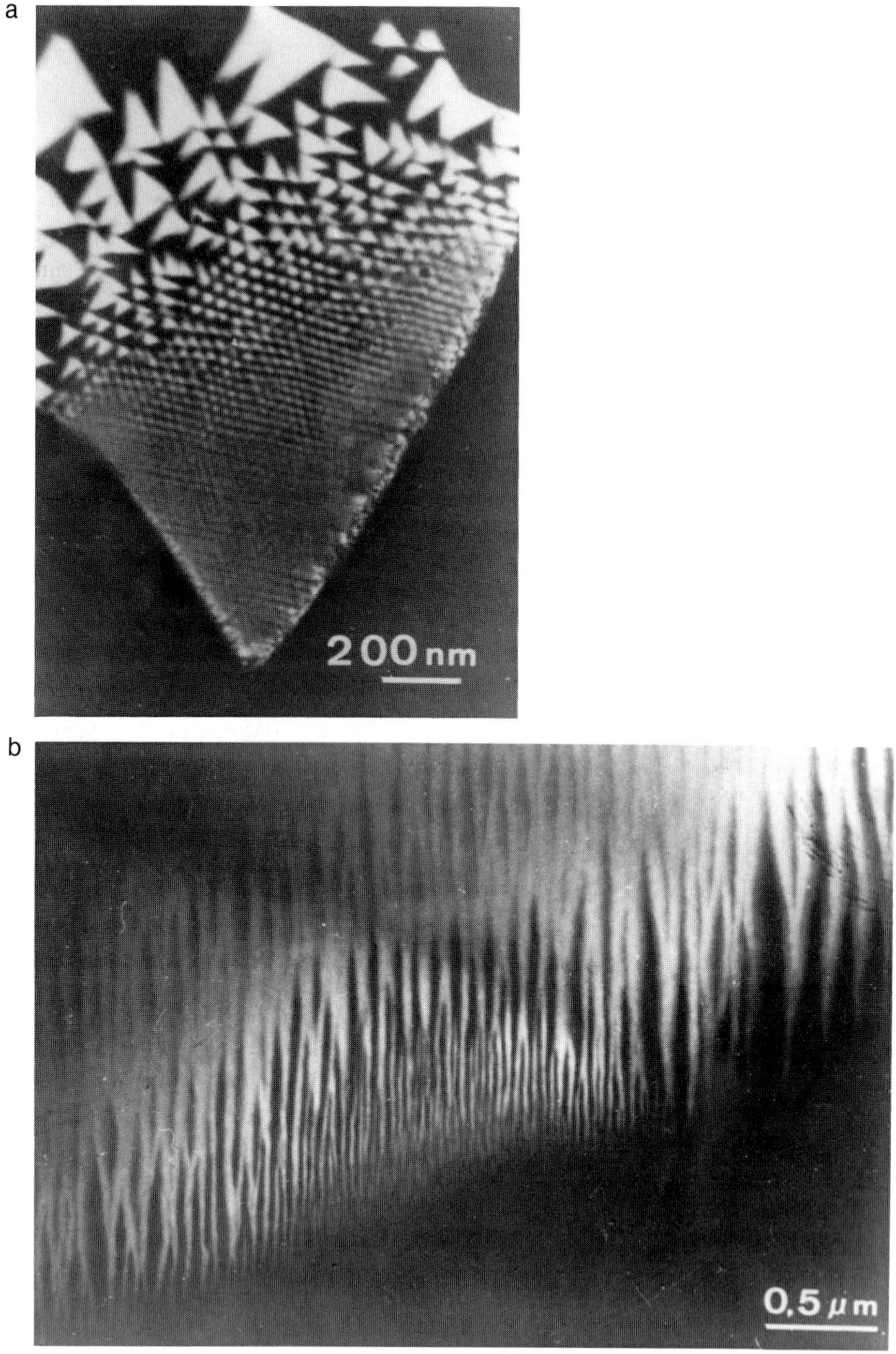

Figure 1-19. Transmission electron micrographs of quartz near the α-β transformation temperature. (a) View down the c crystallographic axis. (b) View normal to the c axis. [Photo from Dolino, G., J. P. Bachheimer, B. Berge, C. M. E. Zeyen, G. Van Tendeloo, J. Van Landuyt, and S. Amelinckx (1984). "Incommensurate Phase of Quartz: III. Study of the Coexistence State Between the Incommensurate and the α-Phases by Neutron Scattering and Electron Microscopy." *Journal de Physique* 45, 901–12.]

Figure 1-19b shows the same phenomenon viewed perpendicular to the crystallographic *c* axis; here the length of the scale bar is 5000 Å. Thus it appears that model 3 is the best for understanding the nature of β-quartz and its transformation from α-quartz: The experimentally determined structure of β-quartz is a space–time average of microdomains oscillating rapidly between the α_1 and α_2 structures.

This model has an interesting consequence. Well below the transformation temperature, the interiors of domains have either the α_1 or α_2 orientation. The domain walls are of finite width, however, and the structure at the center of a domain wall should be the β structure (see Figure 1-6). As the temperature is increased, the domains become smaller, so that the volume occupied by the domain walls becomes a greater proportion of the total volume. As the domain wall structure oscillates, it assumes the β structure in both a time and space average. Thus at some location along the temperature gradient, the physical properties of the specimen change from being predominantly those of α-quartz to being predominantly those of β-quartz. In between there should be a region where the α- and β-quartz volumes are comparable (a coexistence region), and given the right kind of experiment, it should be possible to observe both α-quartz and β-quartz physical properties simultaneously [31]. Refractive indices of both high and low quartz, for example, have been observed simultaneously over a 0.7°C temperature range around the transition temperature.

Yet another apparent complication exists in the α–β-quartz transition, but it derives from what we have already encountered. Figure 1-20 is a plot of heat capacity versus temperature, as is Figure 1-18, but with much higher temperature resolution. The "bump" on the high-temperature side of the λ spike indicates the occur-

Figure 1-20. Heat capacity C_P for quartz versus temperature, with finer temperature resolution than in Figure 1-18. The small bump indicates the existence of an incommensurate phase (see Figure 1-6). (After [13]).

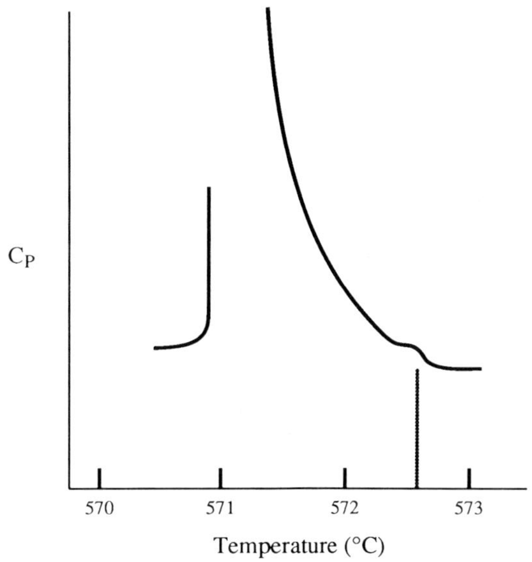

rence of a second-order phase transformation involving a phase that is evidently present as an intermediate step in the transition between α- and β-quartz. Neutron-diffraction experiments [11] have revealed satellite reflections about the Bragg reflections, indicating that the intermediate phase is incommensurate. The satellite reflections are disposed along the three equivalent [100] directions, and, as the temperature drops from that of the incommensurate phase to that of the α phase, the spacing between the Bragg and satellite reflections decreases. This suggests distortion waves or modulations at 120° to one another, whose wavelengths increase with falling temperature. (Remember that decreasing distance in reciprocal space, in which the separations between Bragg and satellite reflections are measured, corresponds to increasing distance in direct space.) The incommensurate phase may thus correspond to the region of the triangular Dauphiné twins seen in Figure 1-19(a), and their twin boundaries, which are themselves certainly incommensurate (see Figure 1-6). Note that the boundary between the incommensurate phase and the β phase is indistinct, consistent with its second-order nature. On the α side of the incommensurate phase, where the λ (or perhaps first-order) phase transformation occurs, the boundary is more well defined, and the triangles are larger—that is, the distortion waves have longer wavelengths.

(In much of the previous discussion, the term *phase* has been used loosely. In the strict thermodynamic sense, a phase must have distinct boundaries, so that it is, in principle, mechanically separable from all adjacent phases. There is no apparent distinct interface between the incommensurate phase and β-quartz, and the use of the term *phase* in this context is not intended to carry any thermodynamic significance.)

The displacive transformations between high and low tridymite and between high and low cristobalite can very likely be explained in much the same way as the α–β transition in quartz. Refinement of the crystal structure of high tridymite at 460°C, assuming the "ideal" $P6_3/mmc$ structure (Figure 1-7), results in large anisotropic temperature factors that indicate unrealistically large thermal motion of the oxygen atoms about lines between silicon atoms [23]. This model also yields unusually short Si–O distances of $\sim$1.55 Å; the Si–O distance in framework silicates is expected to be $\sim 1.61 \pm 0.03$ Å. An alternative model for explaining the same x-ray-diffraction data requires the assumption that the oxygen atoms are statistically divided among six different positions on the circumference of a circle, the plane of which is normal to the Si–Si vector, with $\frac{1}{6}$ of an oxygen at each position, on the average [24]. This corresponds to the superposition of six microdomains of monoclinic symmetry to yield an average hexagonal structure. The domain model gives reasonable steric details (i.e., Si–O and O–O distances, Si–O–Si angles). At present it is not known whether high tridymite is better described as a geometrical average of static microdomains or a time–space average of fluctuating microdomains. It may eventually prove possible to interpret all of the tridymite polymorphic structures on the basis of some sort of domain intergrowth.

In an analogous way, the cubic $Fd\bar{3}m$ "structure" of cristobalite yields unacceptably short bond lengths of 1.54 Å and equally unlikely Si–O–Si angles of 180°. The real structure appears to consist of the average of six microdomains, in which $\frac{1}{6}$ of an oxygen statistically occupies each of six positions on a circle of $\sim$0.45 Å radius, centered on the Si–Si vector [29;36]—remarkably similar to the situation in tridymite. Whether the oxygens are randomly distributed over static positions or fluctuate among positions on a circle is not known. A given microdomain has Si–O distances of ~1.61 Å and Si–O–Si angles of ~148°, which are consistent with values for other silica polymorphs. One treatment of the high-temperature cristobalite structure [29] has assigned the domains to space group *Fdd*2. If this is in fact the case, then the high-temperature structure has lower true symmetry (orthorhombic) than the low-temperature one (tetragonal), and this is generally not expected; in that case, it might be more likely that the tetragonal low-temperature phase is also an average of domains of lower symmetry, but this is only speculation at this point. Equally likely is that the true high-temperature symmetry is tetragonal or higher.

Variations in the α–β Quartz Transformation Temperature

For both natural and synthetic quartz specimens, there is some variability in the temperature at which the displacive transformation occurs. (Some hysteresis is also observed, although, as noted previously, it seems to be connected with particle size in at least some cases.) For example, quartz from rhyolites generally inverts at temperatures below 573°C (at 1 atm pressure), whereas quartz from limestone cavities inverts at generally high temperatures [22]. Figure 1-21 shows that the deviation from 573°C is not large, in any case, for natural specimens. There appears to be a crude inverse relationship between temperature of inversion and temperature of formation for quartz (Figure 1-22), but there are too many exceptions to suggest a general rule.

Figure 1-21. Inversion temperatures of quartz, on both heating and cooling, for specimens obtained from different geologic environments. Isolated outliers have been omitted, so that only the regions in which most specimens fall are shown. (Data from [22]).

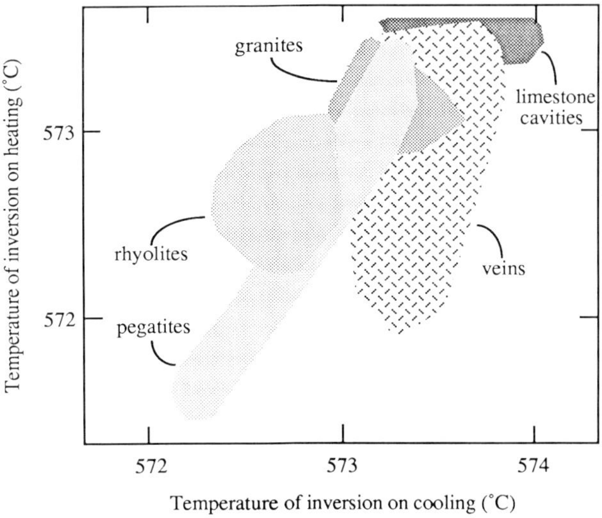

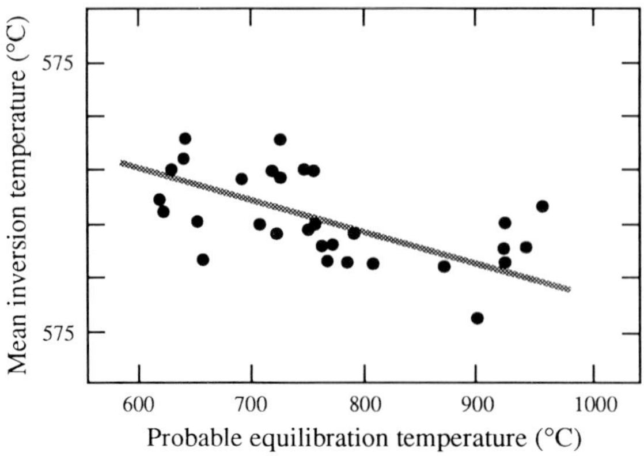

Figure 1-22. Variation in the mean transformation temperature of quartz specimens from different geologic environments as a function of the likely equilibration temperatures. (Data from [18]).

The differences may well have to do with impurities derived from the various geological environments involved. Synthetic quartz grown in the presence of germanium inverts 40°C above the normal temperature [22]. Ge can undoubtedly substitute for Si in quartz, since a quartzlike GeO_2 is known [41]. The Ge–O–Ge angle in GeO_2 is only 130°, as opposed to 144° for the Si–O–Si angle in α-quartz and 151° in β-quartz. Perhaps the bonding forces that cause the Ge–O–Ge angle to be small in GeO_2 also cause early (i.e., high temperature) collapse of β-quartz to the trigonal form if minor Ge is present. Quartz grown in the presence of both lithium and aluminum inverts as much as 120°C below the normal inversion temperature [22]. The Li atoms are located in the hexagonal channels (Figure 1-3), and they

Figure 1-23. Impurities in quartz, in this case Al, have a decided effect on the temperature of inversion. (Data from [18].)

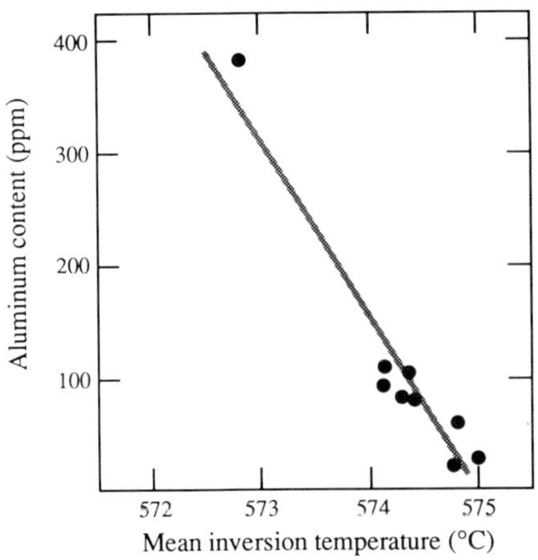

buttress the structure against collapse [30]. (See the discussion of stuffed derivatives to follow.)

In terms of expected substituents in nature, pegmatites are often high in Li and Al; so pegmatitic quartz with low inversion temperatures is not surprising. Figure 1-23 shows that the inversion temperatures of quartz from a suite of volcanic rocks does indeed vary inversely with aluminum content. On the other hand, chemical analyses of carbonate rocks show less Ge than do those of extrusive igneous rocks; so the impurity that causes quartz from limestone cavities to invert at temperatures higher than normal is evidently not germanium.

STUFFED DERIVATIVES OF THE SiO_2 POLYMORPHS

A *derivative structure,* as the name suggests, is one derived from a simpler (and often more highly symmetrical) structure in one of two ways: (1) distortion of the basic structure, or (2) substitution of different atoms into the basic structure. The low-temperature polymorphs of quartz, tridymite, and cristobalite are derivatives of the ideal high-temperature structures by distortion. The space group of a derivative structure must be a subgroup of the space group of the basic structure; that is, the space group of the derivative structure must be derivable from that of the basic structure by loss of specific symmetry elements. For example, $P3_121$ (the space group of right-handed α-quartz) is derived from $P6_422$ (the space group of right-handed β-quartz) by loss of a diad axis; so the structure of α-quartz is a derivative of the ideal structure of β-quartz. Quartz, cristobalite, tridymite, coesite, and stishovite are not derivatives of one another, however, because none crystallizes in a space group that can be derived from that of one of the others by loss of any specific symmetry elements.

A *stuffed derivative* is one in which the substituent atoms of a substitutional derivative have a lower oxidation number (and, therefore, a lower actual charge) than those for which they are substituted, and electrostatic neutrality is maintained by "stuffing" additional cations into available interstices.

Quartz has a fairly dense atomic arrangement compared to tridymite and cristobalite (compare Figures 1-3, 1-7, and 1-10); so the available interstices are not large, and only relatively small atoms may be placed in them. Therefore, only a few stuffed derivatives of quartz are known. Among these is β-eucryptite, $LiAlSiO_4$. A projection of its high-temperature structure is shown in Figure 1-24, and it bears an obvious similarity to the analogous projection of β-quartz (Figure 1-3). Aluminum atoms substitute for half of the silicon atoms, with AlO_4 and SiO_4 tetrahedra in perfect alternation, and lithium atoms (one for every Al) distributed in tetrahedrally coordinated sites in the channels about the 6_2 (or 6_4) axes. (One result of this arrangement is that there are no Si–O–Si or Al–O–Al framework linkages, but only Si–O–Al linkages. This can be rationalized on the basis of the "aluminum avoidance principle," which we shall discuss in the next chapter in connection with feldspars.) The alternation of Si- and Al-containing tetrahedra leads to alternating layers parallel to (0001) that contain exclusively either SiO_4 or AlO_4 tetrahedra, and it also results in doubling of the c axis with respect to quartz; the a and b axes are also doubled owing to the symmetrical inequivalence of chemically identical tetrahedra (i.e., the unit cell contains two distinct Si sites and two distinct Al sites). β-eucryptite is thus a derivative of the quartz structure—often considered a stuffed derivative of high quartz, although it could be viewed as structurally akin to low quartz because of certain structural details it has in common with low quartz, such as the symmetry of individual tetrahedra—and it crystallizes in the high quartz space group, $P6_422$ or $P6_222$ [40]. Another stuffed derivative of β-quartz is $MgAl_2Si_3O_{10}$. Magnesium atoms occupy channel positions, but are surrounded by six, rather than four, oxygen atoms [39].

The crystal structures of the stuffed derivatives of tridymite and cristobalite are also derived from the respective ideal high-temperature structures of these SiO_2 polymorphs. Because the frameworks are relatively more open than that of β-quartz, however, a larger size of "stuffing" atom can be accommodated. The geologically most important of them are nepheline and kalsilite, the room temperature atomic arrangements of which are portrayed in Figures 1-25 and 1-26, respectively. Both are hexagonal at room-temperature, like the "ideal" high tridymite of which they are stuffed derivatives, but the space group is $P6_3$ for both, and the structures are obviously very distorted relative to that of the parent structure (particularly for nepheline—compare Figures 1-25 and 1-7). Above about 1000°C, there is a complete solid solution between nepheline [approximately $(Na_3K)Al_4Si_4O_{16}$ in nature] and kalsilite ($KAlSiO_4$). At atmospheric pressure and 865°C [2], kalsilite undergoes the displacive transformation, from the high to the low form, pictured in Figure 1-

Figure 1-24. The crystal structure of β-eucryptite, viewed down the *c* axis. Tetrahedra containing Si are shaded more lightly than those containing Al. Li atoms are omitted, but occupy positions in the channels. Note that, as in β-quartz, the tetrahedra are not all in the same plane, but form 6_4 or 6_2 screw axes about the channels; Al-containing tetrahedra that appear to share corners do not, because they are at different heights along the screw axes. To avoid confusing overlap of tetrahedra, only those in the lower half of the unit cell are shown. (Structural data from [44]).

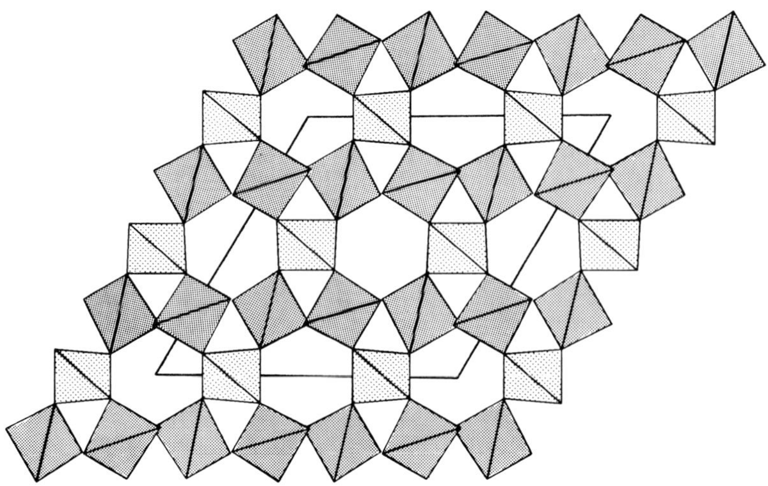

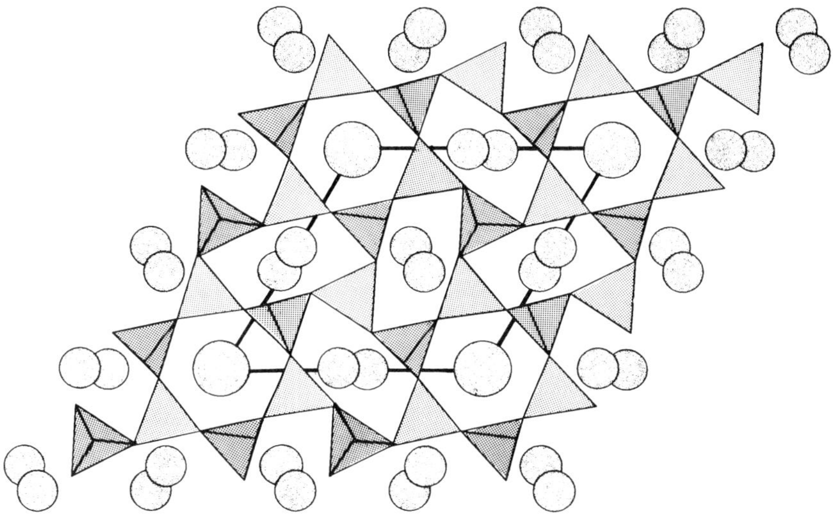

Figure 1-25. The crystal structure of nepheline, a stuffed derivative of "ideal" high tridymite, viewed down the hexagonal *c* axis. Si-rich tetrahedra are shaded more lightly than Al-rich tetrahedra (although Si–Al ordering in nepheline varies with thermal history [17]). K atoms are shown as larger circles, Na atoms as smaller ones. (Structural data from [16]).

26, with the space-group symmetry decreasing from (high-temperature) $P6_3mc$—although this is probably just an average structure comprising several domains—to (low-temperature) $P6_3$. Because this can occur in either of two ways, a domain intergrowth results analogous to Dauphiné twinning in quartz. [The space group of high kalsilite is different from that of "ideal" high tridymite ($P6_3/mmc$) because of ordering of the aluminum and silicon atoms, which alternate as in β-eucryptite. If the aluminum and silicon were disordered over the tetrahedral sites—that is, if there were a random distribution of Al and Si—then tetrahedra would be equivalent on the average, and the space group of high kalsilite would be $P6_3/mmc$.]

ATOMIC BONDING IN THE SILICA POLYMORPHS

Here in the first chapter, it seems appropriate to introduce this topic, because it has assumed great importance in modern mineralogical research. It is not our purpose to review all, or even very many, of the studies aimed at understanding atomic bonding in the minerals discussed, but rather to convey some appreciation of the power and potential of atomic-bonding models for answering some significant questions regarding silicate crystal chemistry. The three models most frequently used, for the different types of bonding situations in which they are applicable, are electrostatic energy calculations, crystal-field theory, and molecular-orbital theory; some theoretical details of each are provided in Chapter 11. Crystal-field theory is

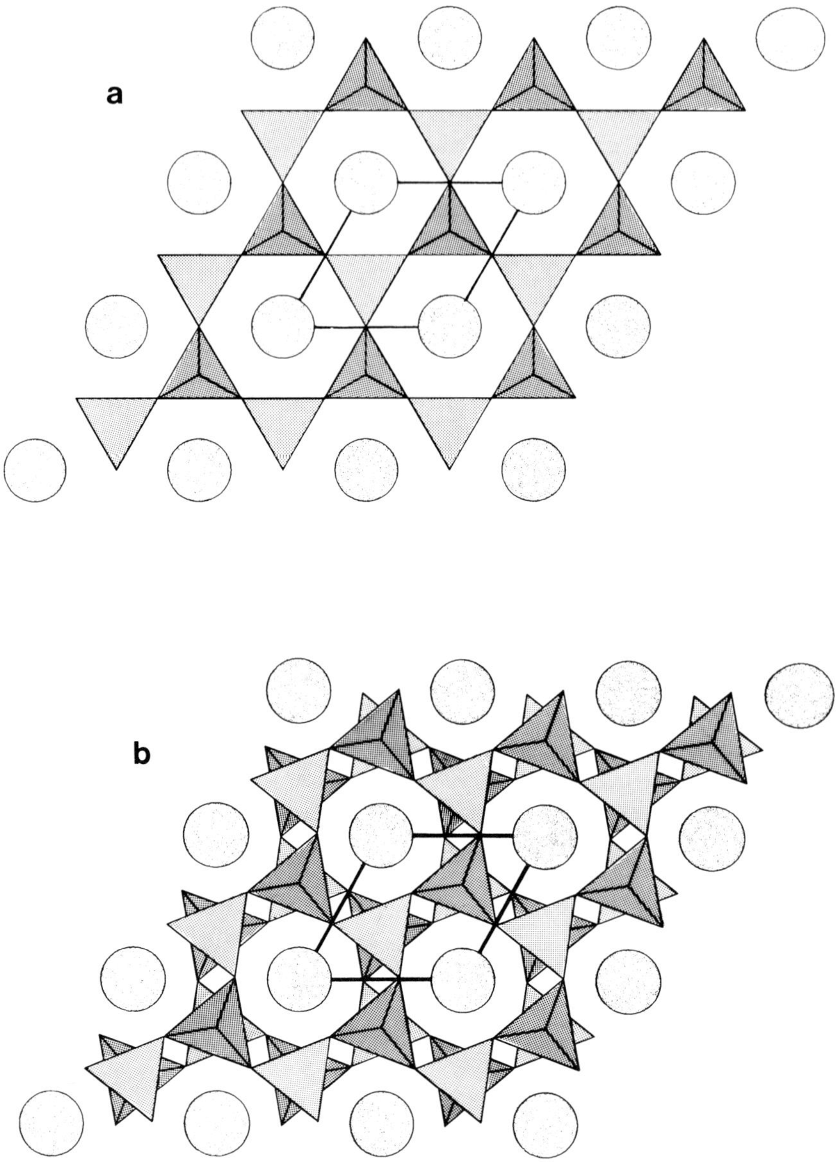

Figure 1-26. The crystal structures of average (a) high and (b) low kalsilite, projected down the hexagonal *c* axis. High kalsilite probably possesses a domain structure like high tridymite. Shading as in Figure 1-25. (Structural data from [3;37]).

inapplicable to the silica polymorphs because they contain no transition metals, but a most interesting application of molecular-orbital theory follows.

Because there is no "molecule" of which any silica polymorph is made up, molecular orbital calculations must be carried out on some small, representative fragment of the structure. Such a fragment (though not the only conceivable one) has the composition $H_6Si_2O_7$, and consists of two tetrahedra that share a common oxygen, with hydrogens bonded to the other six oxygens to effect the charge balance that obtains in the real mineral. In the study under consideration [33], the tetrahedra were constrained to be of ideal geometry, and the bridging oxygen between them was treated as a hinge. For different Si–O bond lengths, ab initio molecular-orbital calculations were used to predict the Si–O–Si angle of minimum energy at the "hinge" oxygen, and from these data, an interatomic force constant for the Si···Si nonbonded interaction was calculated as $\partial^2E/\partial L^2$ (where E is energy and L is the Si···Si distance). The force constant was then used to calculate a value for the bulk modulus of α-quartz. (The bulk modulus is a measure of compressibility, relating the applied pressure to the fractional change in volume due to that pressure.) The calculated value of 397 kbar agrees remarkably well with the experimentally measured value of 393 kbar, indicating that Si···Si repulsion is likely the controlling factor in determining the bulk modulus. On the strength of this perhaps fortuitous but gratifying success, the bulk modulus for α-cristobalite was calculated to be 236 kbar; the fact that it was predicted to be considerably lower than the value for α-quartz seems intuitively satisfying, inasmuch as the framework structure of cristobalite is more open, and thus presumably more deformable, than that of quartz. To this point, no experimentally measured value of the bulk modulus of α-cristobalite is available for comparison with the calculated value.

This example illustrates one of the two major uses to which bonding models have been put in mineralogy: prediction of quantities or trends not yet observed. Prediction of the unknown is, of course, one of the great objectives of all science. More commonly, bonding models have been used for the purpose of rationalizing observed trends and steric (geometrical) details of crystal structures. Several studies of this type have been done for silica polymorphs (for example, [21]). This purpose is important in that it establishes that certain models are (or are not) useful for describing the bonding in given situations, and thus provides insight into the nature of atomic bonding in minerals—eventually leading to the first purpose, prediction. As other silicates are discussed in subsequent chapters, we shall take advantage of selected applications of bonding theory to illustrate its place in modern crystal chemistry.

REFERENCES

[1] Ackermann, R. J., and C. A. Sorrell (1974). "Thermal Expansion and the High-Low Transformation in Quartz. I. High-Temperature X-ray Studies." *Journal of Applied Crystallography* 7, 461–73.

[2] Andou, Y., and A. Kawahara (1982). "The Existence of High–Low Inversion Point of Kalsilite." *Mineralogical Journal* 11, 72–77.

[3] Andou, Y., and A. Kawahara (1984). "The Refinement of the Structure of Synthetic Kalsilite." *Mineralogical Journal* 12, 153–61.

[4] Araki, T., and T. Zoltai (1969). "Refinement of a Coesite Structure." *Zeitschrift für Kristallographie* 129, 381–87.

[5] Bachheimer, J. P. (1980). "An Anomaly in the β Phase Near the α–β Transition of Quartz." *Journal de Physique* 41, L-559–L-561.

[6] Bohlen, S. R., and A. L. Boettcher (1982). "The Quartz–Coesite Transformation: A Precise

Determination and the Effects of Other Components." *Journal of Geophysical Research* 87, 7073–78.
[7] Böhler, R., and J. Arndt (1974). "Direct Determination of the Quartz–Coesite Transition by *in situ* X-ray Measurements." *Contributions to Mineralogy and Petrology* 48, 149–52.
[8] Bourret, A., E. Hinze, and H. D. Hochheimer (1986). "Twin Structure in Coesite Studied by High Resolution Electron Microscopy." *Physics and Chemistry of Minerals* 13, 206–12.
[9] Carpenter, M. A., and M. Wennemer (1985). "Characterization of Synthetic Tridymites by Transmission Electron Microscopy." *American Mineralogist* 70, 517–28.
[10] Dolino, G. (1979). "Elastic Light Scattering and Birefringence Measurements in the Coexistence State of α and β Quartz." *Journal of the Physics and Chemistry of Solids* 40, 121–28.
[11] Dolino, G., J. P. Bachheimer, B. Berge, and C. M. E. Zeyen (1984). "Incommensurate Phase of Quartz: I. Elastic Neutron Scattering." *Journal de Physique* 45, 361–71.
[12] Dolino, G., J. P. Bachheimer, B. Berge, C. M. E. Zeyen, G. Van Tendeloo, J. Van Landuyt, and S. Amelinckx (1984). "Incommensurate Phase of Quartz: III. Study of the Coexistence State Between the Incommensurate and the α-Phases by Neutron Scattering and Electron Microscopy." *Journal de Physique* 45, 901–12.
[13] Dolino, G., J. P. Bachheimer, and C. M. E. Zeyen (1983). "Observation of an Intermediate Phase Near the α-β Transition of Quartz by Heat Capacity and Neutron Scattering Measurements." *Solid State Communications* 45, 295 99.
[14] Dollase, W. A. (1965). "Reinvestigation of the Structure of Low Cristobalite." *Zeitschrift für Kristallographie* 121, 369–77.
[15] Dollase, W. A. (1968). "The Crystal Structure at 220°C of Orthorhombic High Tridymite from the Steinbach Meteorite." *Acta Crystallographica* 23, 617–23.
[16] Dollase, W. A. (1970). "Least-Squares Refinement of the Structure of a Plutonic Nepheline." *Zeitschrift für Kristallographie* 132, 27–44.
[17] Dollase, W. A., and D. R. Peacor (1971). "Si–Al Ordering in Nephelines." *Contributions to Mineralogy and Petrology* 30, 129–34.
[18] Ghiorso, M. S., I. S. E. Carmichael, and L. K. Moret (1979). "Inverted High-Temperature Quartz. Unit Cell Parameters and Properties of the α-β Inversion." *Contributions to Mineralogy and Petrology* 68, 307–23.
[19] Gibbs, G. V., C. T. Prewitt, and K. J. Baldwin (1977). "A Study of the Structural Chemistry of Coesite." *Zeischrift für Kristallographie* 145, 108–23.
[20] Grimm, H., and B. Dorner (1975). "On the Mechanism of the α-β Phase Transformation of Quartz." *Journal of the Physics and Chemistry of Solids* 36, 407–13.
[21] Hill, R. J., M. D. Newton, and G. V. Gibbs (1983). "A Crystal Chemical Study of Stishovite." *Journal of Solid State Chemistry* 47, 185–200.
[22] Keith, M. L., and O. F. Tuttle (1952). "Significance of Variation in the High-Low Inversion of Quartz." *American Journal of Science*, Bowen Volume, 203–80.
[23] Kihara, K. (1978). "Thermal Change in Unit-Cell Dimensions, and a Hexagonal Structure of Tridymite." *Zeitschrift für Kristallographie* 148, 237–53.
[24] Kihara, K. (1980). "On the Split-Atom Model for Hexagonal Tridymite." *Zeitschrift für Kristallographie* 152, 95–101.
[25] Kirfel, A. G., and G. Will (1984). "Ending the '$P2_1/a$ Coesite' Discussion." *Zietschrift für Kristallographie* 167, 287–91.
[26] Kirfel, A., G. Will, and J. Arndt (1979). "A New Phase of Coesite SiO_2. *Zeitschrift für Kristallographie* 149, 315–26.
[27] Klement, W., and L. H. Cohen (1968). "High-Low Quartz Inversion: Thermodynamics of the Lambda Transition." *Journal of Geophysical Research* 73, 2249–59.
[28] Konnert, J. H., and D. E. Appleman (1978). "The Crystal Structure of Low Tridymite." *Acta Crystallographica* B34, 391–408.
[29] Leadbetter, A. J., T. W. Smith, and A. F. Wright (1973). "Structure of High Cristobalite." *Nature Physical Science* 244, 125–26.
[30] Li, C.-T. (1973). "The Role of Lithium in Stabilizing Some High-Temperature Silica Phases." *Zeitschrift für Kristallographie* 138, 216–36.
[31] Liebau, F., and H. Böhm (1982). "On the Co-Existence of Structurally Different Regions in the Low—High-Quartz and Other Displacive Phase Transformations." *Acta Crystallographica* A38, 252–56.
[32] Mirwald, P. W., and H.-J. Massonne (1980). "The Low–High Quartz and Quartz-Coesite Transitions to 40 kbar between 600° and 1600°C and Some Reconnaissance Data on the Effect of $NaAlO_2$ Component on the Low Quartz–Coesite Transition." *Journal of Geophysical Research* 85, 6983–90.

[33] Newton, M. D., M. O'Keefe, and G. V. Gibbs (1980). "Ab Initio Calculation of Interatomic Force Constants in $H_6Si_2O_7$ and the Bulk Modulus of α Quartz and α Cristobalite." *Physics and Chemistry of Minerals* 6, 305–12.
[34] Nukui, A, H. Nakazawa, and M. Akao (1978). "Thermal Changes in Monoclinic Tridymite." *American Mineralogist* 63, 1252–59.
[35] Nukui, A., S. Yamaoka, and H. Nakazawa (1980). "Pressure-Induced Phase Transitons in Tridymite." *American Mineralogist* 65, 1283–86.
[36] Peacor, D. R. (1973). "High-Temperature Single-Crystal Study of the Cristobalite Inversion." *Zeitschrift für Kristallographie* 138, 274–98.
[37] Perrotta, A. J., and J. V. Smith (1965). "The Crystal Structure of Kalsilite, $KAlSiO_4$." *Mineralogical Magazine* 35, 588–95.
[38] Sasaki, S., H.-K. Chen, C. T. Prewitt, and Y. Nakajima (1983). "Re-Examination of '$P2_1/a$ Coesite'." *Zeitschrift für Kristallographie* 164, 67–77.
[39] Schulz, H., W. Hoffmann, and G. M. Muchow (1971). "The Average Structure of $Mg[Al_2Si_3O_{10}]$, a Stuffed Derivative of the High-Quartz Structure." *Zeitschrift für Kristallographie* 134, 1–27.
[40] Schulz, H., and V. Tscherry (1972). "Structural Relations Between the Low- and High-Temperature Forms of β-Eucryptite ($LiAlSiO_4$) and Low and High Quartz. II. High-Temperature Form of β-Eucryptite and High Quartz." *Acta Crystallographica* B28, 2174–77.
[41] Smith, G. S., and P. B. Isaacs (1964). "The Crystal Structure of Quartz-like GeO_2." *Acta Crystallographic* 17, 842–46.
[42] Sosman, R. B. (1965). *The Phases of Silica.* Rutgers University Press, New Brunswick.
[43] Thompson, A. B., and E. H. Perkins (1981). "Lambda Transitions in Minerals." In R. C. Newton, A. Navrotsky, and B. J. Wood, Eds., *Advances in Physical Geochemistry*, Volume 1. Springer-Verlag, New York.
[44] Tscherry, V., H. Schulz, and F. Laves (1972). "Average and Super Structure of β Eucryptite ($LiAlSiO_4$)." *Zeitschrift für Kristallographie* 135, 175–98.
[45] Wright, A. F., and M. S. Lehmann (1981). "The Structure of Quartz at 25 and 590°C Determined by Neutron Diffraction." *Journal of Solid State Chemistry* 36, 371–80.

2
The Feldspars

Feldspar of one composition or another makes up nearly 60% of the crust of the Earth and is a major constituent in more types of rocks than is any other mineral. Fortunately, feldspar crystal chemistry is complicated by such phenomena as exsolution, solid solution, polymorphism, and frequent twinning, so that these minerals record a great deal of information about their conditions of formation and subsequent histories. The interpretation of these phenomena is not always straightforward, but it is at least certain that the crystal chemistry of feldspars holds the key to understanding much about the rocks that contain them.

GENERAL CHEMISTRY

The simplest general chemical formula for feldspar is MT_4O_8. For the most common members of the group, M = K, Na, Ca, or some combination of these three; T = Si and Al. We can consequently write three common end-member formulas as follows:

orthoclase (Or):	$KAlS_3O_8$
albite (Ab):	$NaAlSi_3O_8$
anorthite (An):	$CaAl_2Si_2O_8$.

Throughout this book, and in accepted mineralogical usage, the symbols Or, Ab, and An stand for compositions, not for particular atomic arrangements. Hence, although there is a polymorph of $KAlSi_3O_8$ called orthoclase, the symbol "Or" represents only the chemical composition, and carries no implication as to which polymorph is intended. This ambiguity has occasionally led to suggestions that the polymorph orthoclase be renamed, but the name is thoroughly entrenched in the literature of mineralogy, and the context generally makes the particular meaning obvious. Other polymorphs of $KAlSi_3O_8$ are given different names, and the polymorphs of albite and anorthite are modified by adjectives like "high" and "low." These will all be considered later on.

The An–Ab–Or ternary phase diagram is shown in Figure 2-1, contoured for three temperatures at a water pressure (P_{H_2O}) of 1 kbar. The Or–Ab series constitutes the alkali feldspars, so called because the alkali metals K and Na substitute for each other to yield the continuous spectrum of compositions in the series. Between Ab and Or, there is complete solid solution above about 700°C at P_{H_2O} = 1 kbar. Below this temperature (for example, along the 650°C contour in Figure 2-1), there is a miscibility gap that results in exsolution in alkali feldspars that crystallized above the solvus and then cooled fairly slowly. The binary Or–Ab phase diagram (Figure 2-2) shows the solvus in a semischematic way. The exact geometry of the phase diagram is not yet known, but it must be very much like this.

The ternary diagram suggests complete solid solution between Ab and An at all temperatures, and this compositional range is called the *plagioclase series.* As shown in Figure 2-1, the plagioclases have been given names based on composition within the series. While these are useful for general descriptive purposes, the designation of composition in terms of mole percent An, Ab, and Or is to be preferred. (For instance, $An_{34}Ab_{66}$ and $An_{43}Ab_{53}Or_4$ are both andesines, but the compositional designations obviously provide a great deal more information than does the mineral name.) To the level of resolution of the optical microscope, there appears to be complete solid solution in the plagioclase series. Our perception of the Ab–An series depends to a great extent on the resolution of the technique with which it is observed, however, and we shall turn later to a more detailed consideration of the series.

There is little solid solution between Or and An. The reasons are apparent from elementary crystal chemical arguments. Solid solution between isomorphous minerals tends to be extensive when substituent atoms are of the same charge and are

Figure 2-1. Or–Ab–An ternary diagram, with experimental determinations of the solvus boundary from [69].

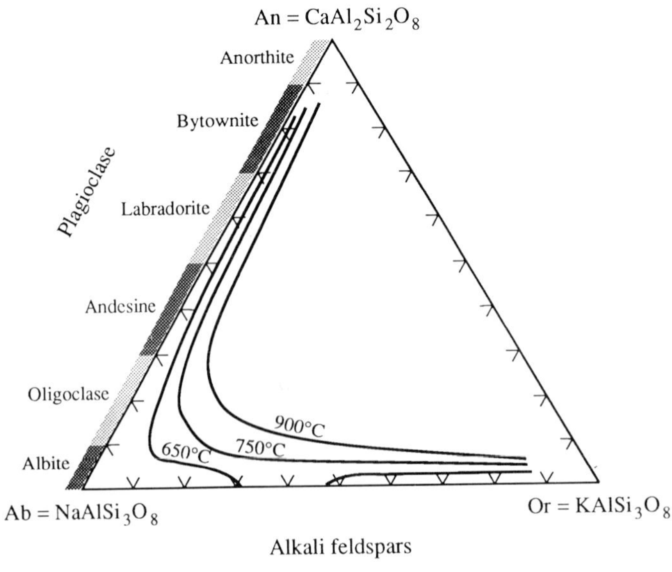

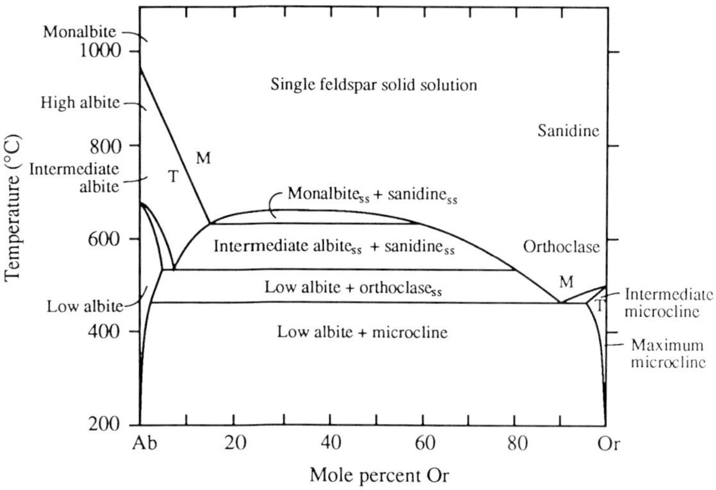

Figure 2-2. Or–Ab binary phase diagram. *M* and *T* stand for monoclinic and triclinic, respectively, and ss is for solid solution. After [70], but note that the monalbite–high-albite transformation is shown as non-first-order after [10].

similar in size to those they replace. In the alkali feldspars, charge is identical in substituent atoms (K^+ and Na^+), but the sizes are different enough (~1.3 Å for K^+ and ~1.0 Å for Na^+, respectively) that solid solution is limited at low temperatures. In the plagioclase series, the sizes of Na and Ca ions are nearly the same, and the coupled substitution NaSi–CaAl is charge balanced. (Again, scale of view affects perception critically, and we shall return to solid solution in the plagioclases on p. 83.) For a hypothetical An–Or series, bóth size and charge are different, and this limits the extent of the solid solution to a few percent Or in An, and vice versa.

Within the three-component region of Figure 2-1, there is restricted solid solution delimited by the isotherms. Outside the isotherms (that is, toward the Or–Ab and Ab–An sides of the triangle) solid solution occurs for the given temperature, whereas inside them (toward the center of the triangle), exsolution occurs. The general formula for such three-component, or ternary, feldspars is $A_{1-x}B_xAl_{1+x}Si_{3-x}O_8$ ($0 \leq x \leq 1$), where A represents some combination of alkali metal atoms (K, Na, or both for natural feldspars, and in terms of Figure 2-1) and B represents an alkaline earth (Ca within the constraints of this diagram). Feldspars in which $x = 0$ constitute the alkali feldspars, while those for which x has any value from 0 to 1, but $K \leq 0.05$ at. %, are the plagioclases. Ternary feldspars, those for which K, Na, and Ca all exceed 0.05 atom per formula unit, have sometimes been referred to as *anorthoclases,* but the definition of this term is so inconsistent among mineralogists that it is probably best to avoid it altogether. *Ternary feldspar* seems adequately descriptive, especially if combined with molar proportions Or, Ab, and An (e.g., $Or_{29}Ab_{65}An_6$).

Three other minerals have the feldspar crystal structure, and thus qualify as uncommon members of the group. Celsian is $BaAl_2Si_2O_8$. Barium atoms are about the same size as potassium atoms, and there is a substantial range of solid solution between celsian (abbreviated Cn) and Or. K-rich intermediate members are called *hyalophanes*. The series has not yet been fully investigated, but present evidence suggests the existence of a miscibility gap. Even less common than celsian are *reedmergnerite* ($NaBSi_3O_8$, a "boron albite" found in the Green River Formation of Utah) and *buddingtonite* ($NH_4AlSi_3O_8 \cdot 0.5H_2O$, an "ammonium orthoclase" with zeolitic water, found in igneous rocks altered by ammonia-bearing hot spring waters in Lake County, California).

Numerous feldspars of unusual compositions have been synthesized. We shall restrict our attention to the natural feldspars of the Or–Ab–An composition range, and celsian.

CRYSTAL STRUCTURE

Feldspars are framework silicates, and it is convenient to envision them as consisting of linked rings made up of four corner-sharing tetrahedra. The rings are interconnected in a way that is unique to feldspars, and, in fact, defines them. In the feldspar four-membered ring (Figure 2-3) two tetrahedra point "up" and two point "down" with respect to the $(\bar{2}01)$ planes. Figure 2-4 shows a partial (and rather stylized) view of the feldspar structure projected onto the (010) plane; the four-membered rings of Figure 2-3 are oriented nearly perpendicular to the **a** crystallographic direction in this view, and they are linked together to form what has come to be called the "double-crankshaft" chain that runs parallel to **a**. The width of the double-crankshaft chain along **c** is the *c* repeat; along **b**, the chains are related by mirrors (in the monoclinic feldspars) or pseudomirrors (in the triclinic feldspars).

The most highly symmetrical of the common feldspars are orthoclase and sanidine, both monoclinic polymorphs of $KAlSi_3O_8$. The Si and Al occupy tetrahedral sites called T_1 and T_2; and the oxygens are designated O_A1, O_A2, O_B, O_C, and O_D. (In triclinic feldspars the nomenclature is related but slightly different.) Figure 2-5 reveals how the four-membered rings are cross-linked to create large voids for the 7- to 9-coordinated *M* cations—K, in the case of orthoclase and sanidine. (Whether the cation is 7- or 9-coordinated depends on what one considers an appropriate bond-length range for K–O bonds with K in irregular coordination, and is thus a

Figure 2-3. A schematic representation of the feldspar four-membered tetrahedral ring as if viewed down the *a* crystallographic axis.

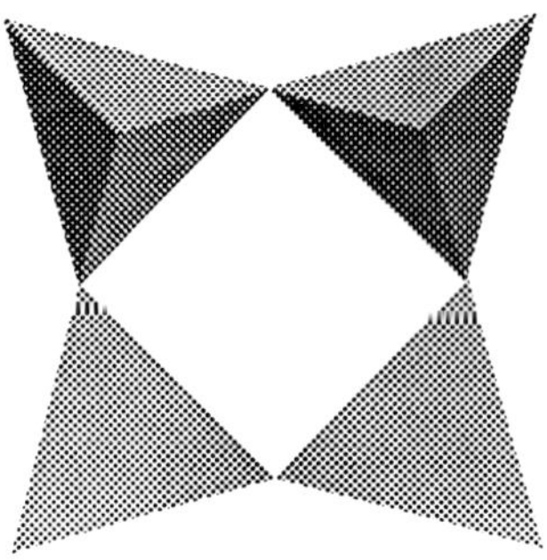

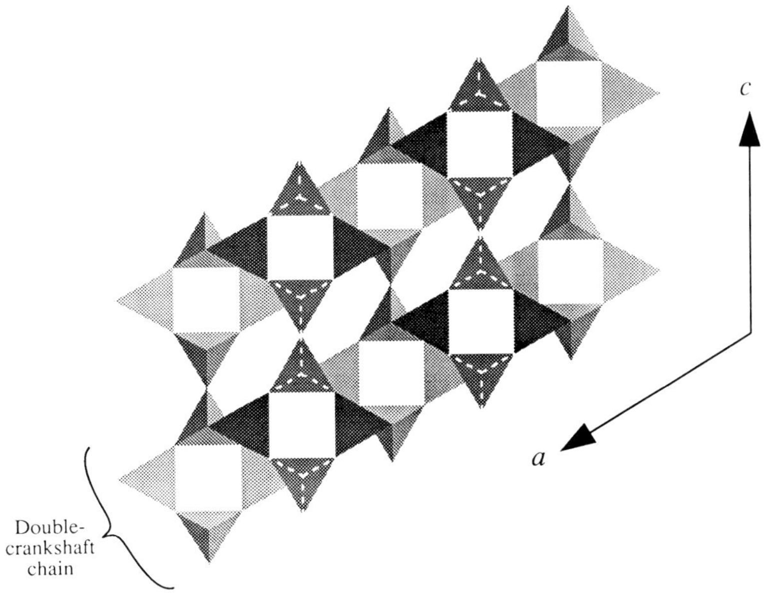

Figure 2-4. The feldspar structure schematically projected onto (010), with the upper layer of tetrahedra shaded darker for clarity. White dashed lines are hidden tetrahedral edges. The four-membered rings shown in Figure 2-3 are seen edge on and nearly perpendicular to *a*. The double-crankshaft chain runs along the *a* axis. (After [51].)

moot question at this point.) Symmetry elements for the monoclinic polymorph are shown, but the basic arrangement is the same for triclinic feldspars, with slight distortions that remove the mirror and the diad. The projection is down **a**, so the double-crankshaft chains are coming out of the page.

All feldspars have similar unit-cell dimensions, except that c may vary. The a cell edge is ~8.5 Å, and b is ~13 Å, but c may be either ~7 Å or ~14 Å. As already noted, the crystal system may be monoclinic or triclinic. If monoclinic, the space group may be $C2/m$ (monoclinic alkali feldspars) or $I2/c$ (celsian). If triclinic, it may be $C\bar{1}$, $I\bar{1}$, or $P\bar{1}$. (Note that the nonstandard nonprimitive triclinic space groups $C\bar{1}$ and $I\bar{1}$ are chosen to facilitate direct comparison of the unit-cell dimensions with the similarly oriented $P\bar{1}$ cell.)

We turn next to the factors that make crystal systems and space groups different, beginning with the alkali feldspars.

TETRAHEDRAL ORDERING AND POLYMORPHISM IN ALKALI FELDSPARS

Potassium Feldspar

In our previous consideration of the SiO_2 polymorphs, we observed that the differences between polymorphs may sometimes be characterized as a change in primary coordination (e.g., stishovite ↔ coesite), and sometimes as a change in secondary

Figure 2-5. A schematic view of the feldspar structure viewed down the *a* axis (as in Figure 2-3), showing how the four-membered rings are cross-linked to create voids for the large atoms such as K, Na, Ca, or Ba. Symmetry elements shown are those of space group $C2/m$, the space group of K-rich alkali feldspars, and is the highest symmetry displayed by members of the feldspar family.

coordination (e.g., coesite ↔ quartz). In examining the feldspars, we find a different type of polymorphism best described as due to a change in the ordering scheme of the cations. We begin by discussing the polymorphism of the K-feldspars, the potassium end of the alkali feldspar solid solution series.

The highest symmetry observed among the K-feldspars is monoclinic, space group $C2/m$. This space group has eight general equivalent positions, and diffraction experiments reveal that the tetrahedrally coordinated cation sites (called T_1 and T_2) are located at general positions (that is, they do not lie on symmetry elements); thus there are eight T_1 sites and eight T_2 sites in the unit cell, and each site in the two sets is equivalent in every way to the other seven sites in its set. The chemical formula for K-feldspar is $KAlSi_3O_8$, and $Z = 4$, so that there are 4 Al and 12 Si atoms that must occupy two sets of eight equivalent positions each. Clearly, the (Al,Si) distribution and the symmetry of this high-symmetry K-feldspar polymorph must be somehow interrelated. There must be some sort of randomness, or disorder, in the (Al,Si) arrangement in order for two sets of chemically different atoms consisting of 4 and 12 members to be distributed so as to yield two sets of eight identical sites each.

Consider Figure 2-6, which is a (001) projection of the feldspar structure from which oxygen atoms have been omitted for clarity; straight lines connect tetrahedrally coordinated atoms that share a common oxygen. The tetrahedral site names T_1O, T_1m, T_2O, and T_2m are standard for alkali feldspars, while T_1Oc, T_1mc, T_2Oc, and T_2mc are their centrosymmetric equivalents. In $C2/m$ feldspars, a mirror plane and two-fold axis make T_1O equivalent to T_1m, and likewise T_2O to T_2m, so that there are only two unique tetrahedral sites (T_1 and T_2, referred to previously), but comparison with polymorphs of lower symmetry is facilitated if we consider all four sites explicitly.

The alkali feldspar polymorph with the greatest degree of cation disorder is called sanidine. In sanidine the distribution of aluminum and silicon atoms among T_1O, T_1m, T_2O, and T_2m is random, and there is consequently a probability of 0.25 of finding an Al atom at any particular tetrahedral site. If we use t_1o, t_1m, t_2o, and t_2m to symbolize the probability of finding Al at T_1O, T_1m, T_2O, and T_2m, respectively, then $t_1o = t_1m = t_2o = t_2m = 0.25$ in sanidine. From Figure 2-6 it is clear that as long as there is complete disorder in the distribution of Al and Si among the four tetrahedral sites, both the mirror plane and the diad must be present, and the space group is $C2/m$. Note that, while there are really only two unique kinds of tetrahedral sites in sanidine—T_1 and T_2—each has 0.25Al + 0.75Si on the average, so that four sites—two of each—are required to account for the one aluminum atom in the formula. Consequently, the probability of finding Al in a T_1 site of sanidine, or any other monoclinic feldspar, is not t_1, but $2t_1$, because $t_1o = t_1m = t_1$ and so $t_1o + t_1m = 2t_1$. For sanidine, then, the probability of finding Al on a T_1 site is $0.25 + 0.25 = 0.50$, and the probability that it is at a T_2 site instead is also 0.50.

Suppose now that $t_1m \neq t_2o$. It is still possible for the space group to be $C2/m$, if $t_1o = t_1m$, and $t_2o = t_2m$. In this case, there is still really only one kind of T_1 site and one kind of T_2 site because the mirror and the diad are still present though the occupancies of T_1 and T_2 are different. Physically, this means that more of the aluminum occupies the T_1 sites than occupies the T_2 sites, and thus, $0.25 < t_1o = t_1m \leq 0.50$, and $0.25 > t_2o = t_2m \geq 0.0$. The polymorph that meets the requirements of partial tetrahedral ordering but monoclinic symmetry is called *orthoclase*, or, by some, low sanidine. (The term *orthoclase*, as used in this context, is a polymorphic modification of $KAlSi_3O_8$, and is not to be confused with the *compositional symbol*, Or.) Orthoclase, then, embraces a range of ordering states consistent with space group $C2/m$. The most ordered orthoclase conceivable would have $t_1o = t_1m = 0.50$ and $t_2o = t_2m = 0.0$, but this degree of order has been reported only in one instance so far [60]; a monoclinic feldspar with this (Al,Si) distribution has been called *fully ordered orthoclase*. In nature the ordering process apparently rarely proceeds as far as fully ordered orthoclase before the next step is encountered, namely preferential segregation of aluminum into T_1O.

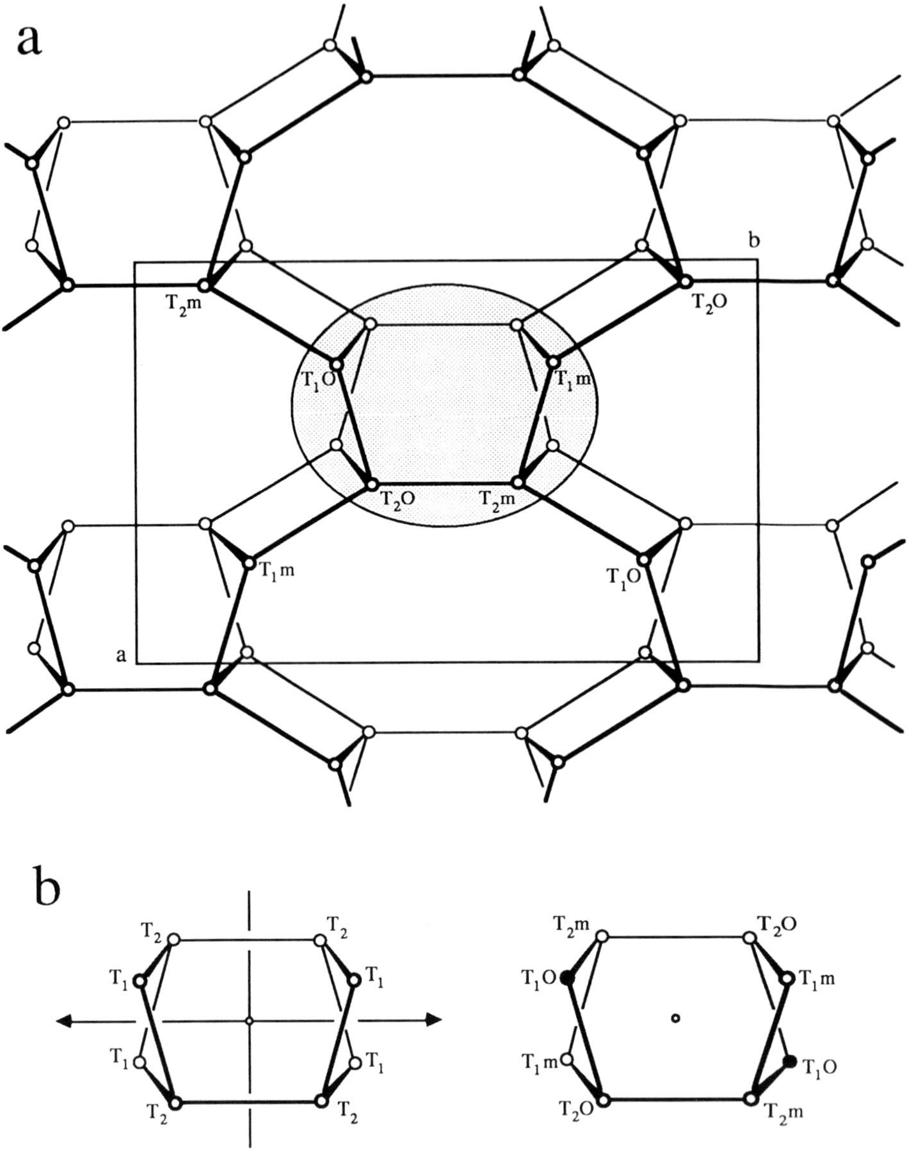

Figure 2-6. (a) A skeletal view of the feldspar structure projected onto (001). Tetrahedral-atom positions are shown as circles, but oxygens are omitted for clarity. (b) The structural unit in the shaded oval of Figure 2-6(a), shown for monoclinic and triclinic polymorphs. In the monoclinic polymorph, the vertical mirror and horizontal diad axis are present, but in the triclinic polymorph only the center of symmetry remains, owing to preferential occupancy of T_1O by aluminum.

As soon as $t_1o > t_1m$, the two T_1 sites become nonequivalent, the diad and mirror are lost, and the symmetry can no longer be monoclinic. The only symmetry element left is the inversion center, and so the space group is $C\bar{1}$. This polymorph is called *microcline* ("small angle"). If ordering proceeds as far as possible, all of the aluminum is found in T_1O (i.e., $t_1o = 1.0, t_1m = t_2o = t_2m = 0.0$), and the mineral is referred to as *maximum* (or low) *microcline.* (Note, again, the choice of an unconventional centered triclinic space group. Another unit cell could have been chosen that is primitive, to give $P\bar{1}$ as the space group, but there is an obvious advantage in retaining the same unit cell, slightly distorted, as is used for the monoclinic polymorphs so that the effects of factors that cause unit-cell distortions can be compared directly.)

The various degrees of (Al,Si) order are called *structural states,* and sanidine and maximum microcline represent two extreme and unique structural states. Orthoclase and microcline include the range of structural states found between the extremes (and the possibility exists that they may be kinetically stranded rather than stable forms), and these two polymorph names convey no specific information about degree of tetrahedral order other than the general ranges. When specific information needs to be conveyed, it is conventionally expressed as values for the probabilities t_1o, t_1m, t_2o, and t_2m.

The concept of electrostatic bond strength and the application of Pauling's second rule can be used to rationalize the tendency of aluminum to order into the T_1 sites. In all feldspars, the atoms in the T_1 sites are bonded to oxygens O_A1, O_B, O_C, and O_D, whereas T_2 atoms are bonded to oxygens O_A2, O_B, O_C, and O_D. Since the only difference lies in the type of O_A atom, we consider the sum of electrostatic bond strengths reaching O_A1 and O_A2. Figure 2-7 represents the cation coordination about these two oxygens. Numbers adjacent to each bond are the electrostatic bond strengths, assuming the *M* cations to be nine-coordinated and the tetrahedral atoms to be disordered (that is, to have an average oxidation number of 3.75 due to average occupancy by 0.25Al+0.75Si, as in sanidine). If ΣS is the sum of the Pauling electrostatic bond strengths received by an anion, then for sanidine $\Sigma S(O_A1) = 2.097$ and $\Sigma S(O_A2) = 1.986$. Thus O_A1 is slightly overbonded, and O_A2 is slightly underbonded if the tetrahedral cations are disordered, and a lower average oxida-

Figure 2-7. Pauling electrostatic bond strengths for O_A1 (left) and O_A2 (right) in sanidine [fully disordered (Al,Si) distribution]. Large spheres represent K, small ones the tetrahedrally coordinated atoms, and intermediate-sized ones the oxygens. O_A1 is overbonded, and O_A2 is underbonded.

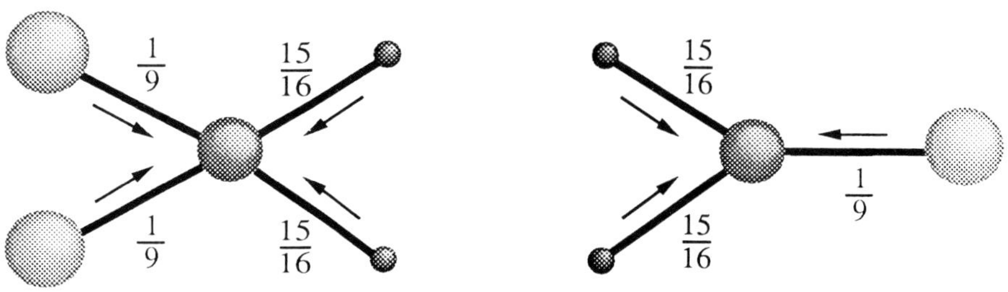

tion number at T_1 and a higher one at T_2 will improve the overall charge balance. For any structural state from fully ordered orthoclase to maximum microcline (i.e., all Al in T_1 sites), $\Sigma S(O_A1) = 1.972$ and $\Sigma S(O_A2) = 2.111$, so that this degree of ordering has over-corrected the imbalance. There is obviously a partially ordered orthoclase for which $\Sigma S(O_A1)$ is equal to $\Sigma S(O_A2)$, and some simple calculations reveal the following: For monoclinic alkali feldspars with structural states between $2t_1 = 0.57$ and 0.89, both O_A1 and O_A2 are overbonded, and O_A1 is formally charge-balanced if $2t_1 = 0.89$ (see the section on ordering paths); the amount of aluminum in the combined T_1 sites that will make ΣS *equal* at both oxygens is 0.72 atoms, yielding $\Sigma S(O_A1) = \Sigma S(O_A2) = 2.04$. Thus it turns out that there is no (Al,Si) distribution that will result in an ideal distribution of Pauling bond strengths (i.e., $\Sigma S = 2.0$) at both oxygen atoms simultaneously because the bond strength sum at both oxygens combined is greater than 4.0; but the distribution $2t_1 = 0.72$, $2t_2 = 0.28$ divides the excess bond strength evenly. So these simple arguments provide some rationale for an ordering "driving force" from sanidine to a lower monoclinic structural state, but they do not explain the tendency of feldspars to order beyond monoclinic symmetry.

A word here about the use of Pauling bond strengths seems advisable. While the deviations of ΣS from 2.00 for O_A1 and O_A2 are not large in the monoclinic feldspars we have just discussed, and certainly fit within the reasonable limits of Pauling's second rule, the charge imbalances they indicate should not necessarily be taken at face value. In the first place, the choice of 9 as the coordination number of K is perhaps the best choice, but it is not the only choice for this large and irregular polyhedron; the use of other coordination numbers would yield different values for electrostatic bond strengths. In the second place, Pauling bond strengths do not take variations in bond lengths into account, and it is obvious that shorter bonds of a given type must be stronger than longer ones. To overcome this limitation, one may calculate bond strengths using empirical relationships that have been developed between bond lengths and bond strengths (see, e.g., [10] or [21]). We shall not do so here because it would add nothing to this discussion, except to show that, indeed, such an approach yields bond-strength sums closer to the ideal values than do Pauling bond strength sums. Note, however, Figure 2-8. In it *M*–O bond lengths of seven monoclinic potassic feldspars are plotted as a function of t_1, the latter being calculated by a method that requires *T*–O distances from a crystal structure refinement [45]. As more aluminum occupies T_1O, the M–O_A1 bond length decreases and the M–O_A2 bond length increases. This is exactly what we would expect based upon the Pauling bond-strength sums, in which bond lengths are assumed to be constant: As O_A1 goes from overbonded to underbonded in the formal Pauling sense, the *M*–O distance shortens to compensate for the decrease in formal bond strength; the opposite occurs for O_A2, which experiences an increase in formal bond strength. (These feldspars have compositions ranging from Or_{85} to Or_{90}, so that the bonds are essentially K–O bonds, and the trends shown are not significantly influenced by compositional variation.) These data suggest that the Pauling bond strengths do a qualitatively reasonable job of characterizing charge balance in feldspars, and also illustrate how bond lengths adjust to provide optimum distribution of charge.

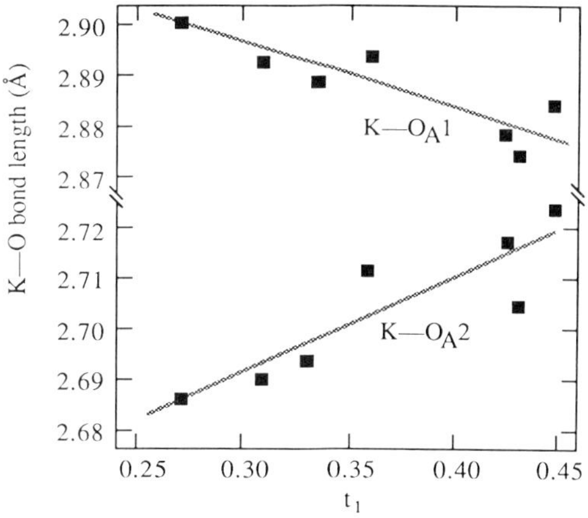

Figure 2-8. Plot of K–O_A1 and K–O_A2 bond lengths as functions of t_1, the probability of finding Al in the T_1 site of monoclinic feldspars, for seven feldspars. Bond lengths respond to increasing (Al,Si) order by decreasing as the oxygen atom becomes less overbonded in the Pauling sense (K–O_A1) or increasing as it becomes less underbonded in the Pauling sense (K–O_A2).

The asymmetric unit (that is, the symmetrically unique portion) of a given unit cell of a particular alkali feldspar ideally contains one aluminum atom and three silicon atoms; the aluminum is, of course, in only one site, a "local T_1O" site that may not be consistent from one asymmetric unit to another. Different "choices" of aluminum sites in contiguous asymmetric units would result in lattice strain, and it is conceivable that the need to relieve local stresses (i.e., to minimize strain energy) produced by differences in ordering of adjacent unit cells may drive the monoclinic–triclinic inversion. At any rate, the process is slow. Ordering at the tetrahedral sites involves the breaking of atomic bonds and moving of atoms, and there are substantial kinetic barriers to overcome in accomplishing this. Experimental data show that ordering takes place much more rapidly in the presence of water than in dry systems. Perhaps the protons facilitate local charge balance while cations are trading places. Regardless of the ordering mechanism, sanidine in volcanic rocks and orthoclase in plutonic rocks persist indefinitely under ambient conditions. Laboratory experiments, hindered by transformation kinetics, have not yet produced triclinic ordering in an originally monoclinic K-feldspar, although sanidines have been ordered to orthoclases. In some rocks orthoclase and microcline even coexist, and there is at present no convincing evidence that maximum microcline is the stable low-temperature polymorph, although most mineralogists and petrologists think of it that way at the intuitive level, and Figure 2-2 suggests that it may well be so.

If the reasons for ordering of Al into T_1O are currently obscure, the consequences at least are evident. In a monoclinic alkali feldspar, there are two equivalent T_1 sites, and neither is really T_1O or T_1m. Accepting on the empirical evidence some driving force for ordering Al into one of the T_1 sites, either one would qualify as the recipient. The site that, on an average throughout the crystal, ends up with excess Al will then, and only then, be identifiable as T_1O, and the other will be T_1m. Because the transformation is very sluggish, the best the monoclinic crystal can do is to establish very small domains in which there is partial "triclinic" order. The domains may well be ordered differently from one another, however, so that adjacent domains will have "chosen" T_1O differently, and this leads to distortion waves, or modulations at the submicroscopic level (Figure 2-9). The structure in this condition is still monoclinic on the scale of optical and x-ray investigation.

The two equally likely choices for T_1O result, at the macroscopic level, in the two types of twins shown in Figure 2-10(a). If, in the modulated structure, microdomains can grow to macroscopic sizes, the resulting texture is a mixture of albite twinning [180° rotation about **b*** with (010) composition plane] and pericline twinning [180° rotation about **b** with (*h*0*l*) composition plane]—the familiar microcline cross-hatched twins [Figure 2-10(b)]. If the twin domains are smaller than the limits of optical resolution but larger than several unit cells, then the mineral will appear to have monoclinic optical properties but will be triclinic to x-ray diffraction.

In the electron micrograph of Figure 2-11, the cross-hatched area is not microcline twinning, but the modulated structure of a monoclinic feldspar that has developed further than that in Figure 2-9. In the modulated structure there are no twin planes, as such, and no twin axes, but only the mutually orthogonal distortion waves due to local Al ordering. The homogeneous area in the middle of the micro-

Figure 2-9. Structural modulations caused by incipient triclinic domains in a sensibly monoclinic alkali feldspar. [Photo from J. D. C. McConnell (1965). "Electron Optical Study of Effects Associated with Partial Inversion of a Silicate Phase." *Philosophical Magazine* 11,1289–1301.]

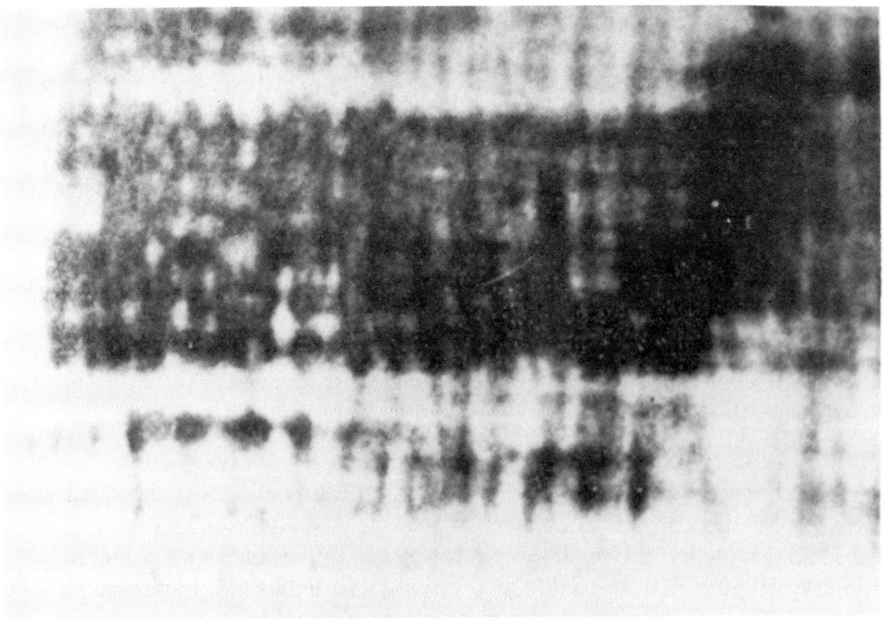

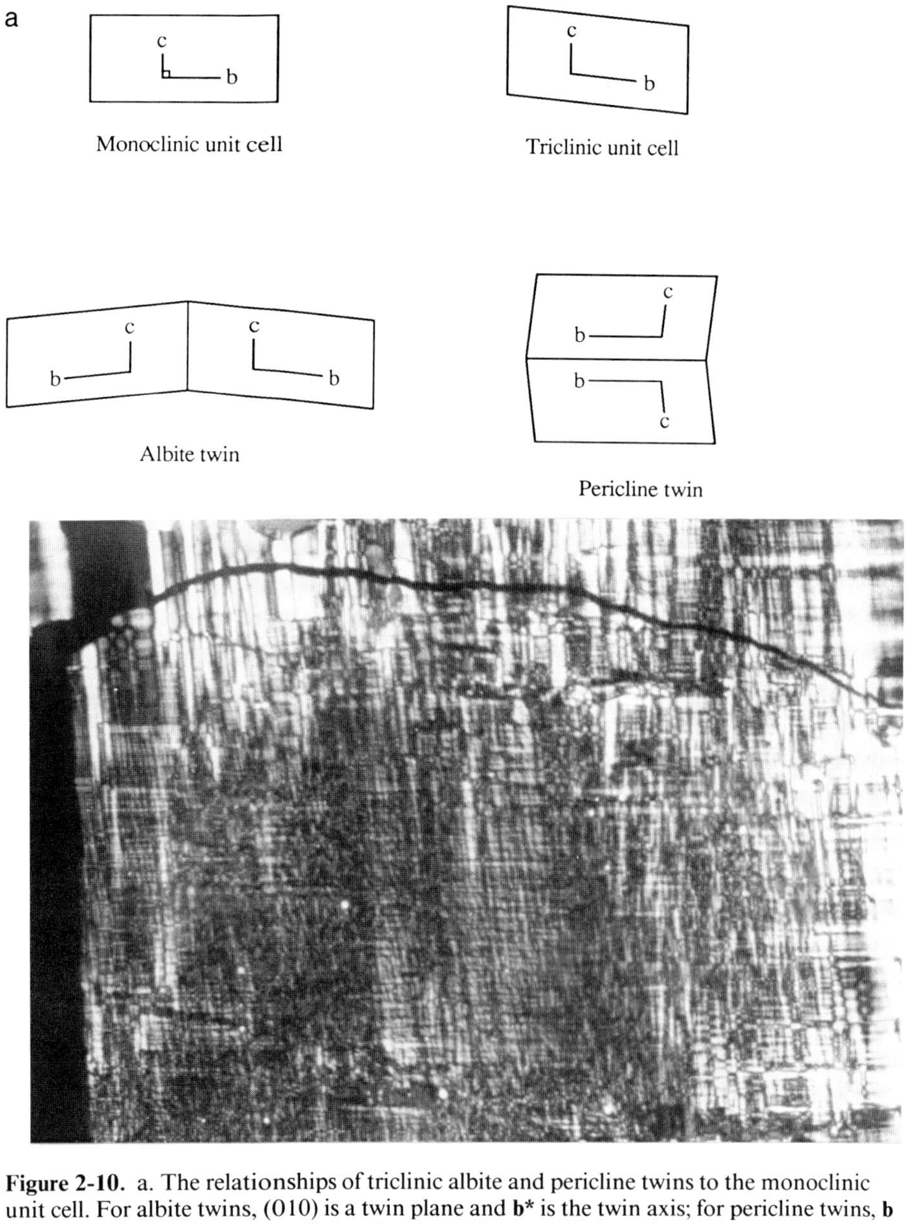

Figure 2-10. a. The relationships of triclinic albite and pericline twins to the monoclinic unit cell. For albite twins, (010) is a twin plane and **b*** is the twin axis; for pericline twins, **b** is a twin axis. b. Micrograph of cross-hatched twinning in microcline.

graph is a region of triclinic feldspar in which one of the alternative T_1 sites has "won out" and become T_1O.

Consider now the implications of cross-hatched twinning. If a microcline grew originally as triclinic, then the growth process would dictate which site became T_1O, and it would be uniform throughout the crystal. Only if an originally monoclinic

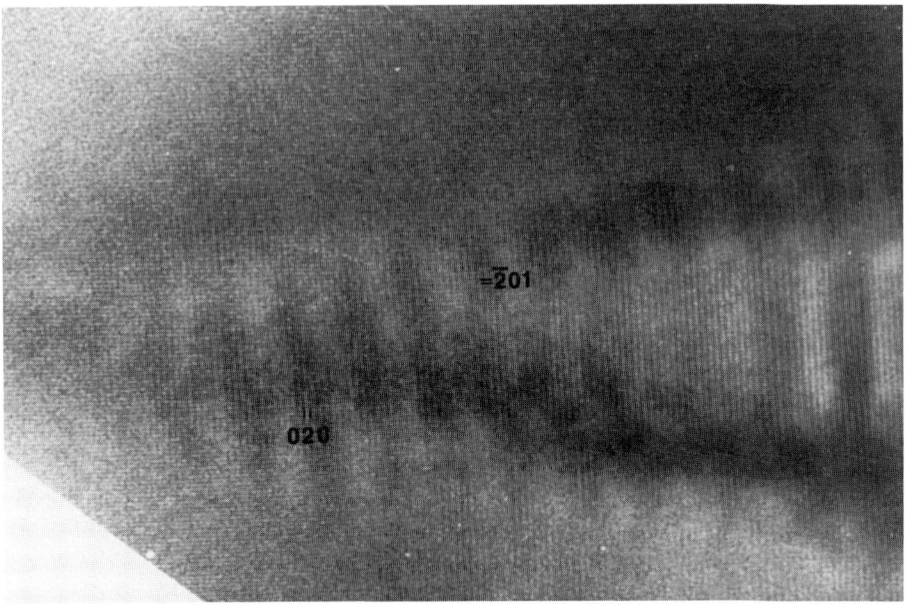

Figure 2-11. The modulated structure of a sensibly monoclinic alkali feldspar that has developed incipient triclinic domains to a greater extent than that seen in Figure 2-9. The homogeneous region is microcline that has nucleated and grown from the orthoclase by preferential growth of one of the alternative triclinic domain orientations. The length of the scale bar is 0.2 μm. [Photo from P. R. Buseck and J. M. Crowley (1983). "Modulated and Intergrowth Structures in Minerals and Electron Microscope Methods for their Study." *American Mineralogist* 68, 18–40 (reproduced there from reference [22]).]

feldspar inverted to triclinic symmetry, with the resultant random and nonuniform "choice" of T_1O, would cross-hatched twinning develop. Moreover, because a cooling feldspar surrounded by other solid minerals in a rock is already physically constrained to the geometrical shape it has developed by then, the inversion *must* produce such twins in order physically to accommodate the new lattice geometry in the rigid volume of the crystal. Consequently, one can say that the presence of both albite and pericline twinning in microcline is evidence that the mineral was once monoclinic. If the feldspar is triclinic but shows no cross-hatched twinning, then it must have crystallized initially as triclinic. The phase diagram of Figure 2-2 suggests that initial triclinicity would require the temperature of formation to be something under ~500°C at low pressure, and it would probably not be much higher under typical geologic pressures. The combination of albite and pericline twins is often called *M*-twinning (*M* for monoclinic), or *inversion twinning*, whereas twins that can form in initially triclinic feldspars (for instance, either albite or pericline twins alone) are called *T-twins*. (For an alternative point of view in which cross-hatched twinning is regarded as a growth feature rather than a phase transformation feature, see [1;2].)

While it has been implied that temperature of formation and cooling rate are the factors that determine which K-feldspar polymorph will persist at room conditions, there are complications that must be admitted, and the situation is not entirely understood. It has already been indicated (p. 47) that H_2O is important in facilitating (Al,Si) migration; in the absence of water, ordering probably will not proceed. Given the presence of an aqueous fluid, its composition influences the rate of ordering; alkaline fluids [those with high $(Na_2O+K_2O)/Al_2O_3$ ratios] accelerate the ordering process in laboratory experiments [49], and may be important even on a geologic time scale. Bulk composition of the rocks hosting the feldspar may influence ordering kinetics. Some K-feldspars, known as *adularias*, grow at the comparatively low temperatures of hydrothermal activity, yet show a surprising range of structural states from disordered sanidine to ordered microcline, sometimes in different growth sectors of the same crystals [2]; the degree of ordering at such low temperatures of growth may be linked to growth rates—high growth rates may exceed ordering rates and preclude ordering at low temperatures—or it may be a function of the two-dimensional symmetry presented by the surfaces of different growth sectors. The complexity of the kinetics of ordering is indicated by a schematic "route map" (Figure 2-12). Beginning with sanidine, maximum microcline may be reached by different routes depending at least on temperature and cooling rate, and probably on some other factors. Different routes imply different ordering paths (that is, different ways of going from $t_1o=t_1m=t_2o=t_2m=0.25$ to $t_1o=1.0$, $t_1m=t_2o=t_2m=0.0$), and we shall return to the question of ordering paths later. For now it may prove helpful just to illustrate the diversity of factors suspected of influencing the ordering kinetics, using observations on natural feldspars,.

1. In the Tauern Window of the Eastern Alps [6], the metamorphic rocks containing maximum microclines are separated from those containing intermediate

Figure 2-12. A schematic "route map" showing various ways in which potassic alkali feldspars can order (or fail to do so) as temperature is decreased. The dashed arrow indicates a hypothetical, and as yet unobserved, one-step ordering path from sanidine to maximum microcline. Modified from [61].

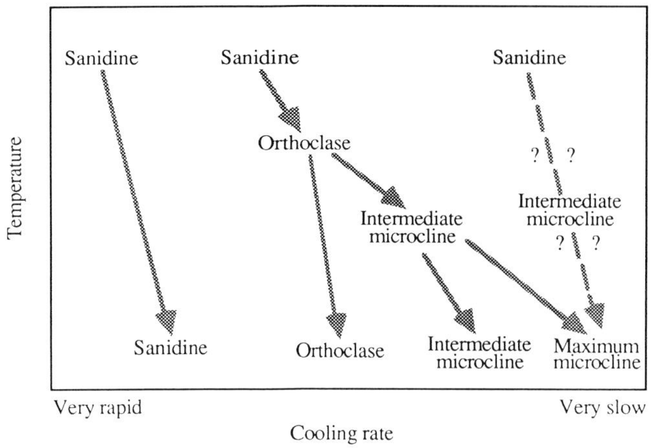

microclines or orthoclases by a rather narrow zone very near the 500°C isotherm, as determined by mineral equilibria and oxygen isotopic data. In this case, it appears that temperature of metamorphism may have been the major factor influencing structural state. (See also [18].)

2. In a suite of dike rocks from the Bohemian Massif [56], comprising the differentiation series minette–alkaline minette–alkaline microsyenite–microgranite, minettes contain orthoclase, alkaline minettes contain both orthoclase and microcline, and alkaline microsyenites and microgranites contain only microcline. The dikes likely had similar cooling rates. As differentiation proceeded in the parent magma, the temperature of crystallization decreased, the alkalinity increased owing to a deficiency of aluminum, and the amount of residual water increased. Although it is likely that final crystallization of the most alkaline rocks occurred at temperatures below that of the monoclinic–triclinic inversion for K-rich feldspars, there is abundant field and experimental evidence that this alone is not always a sufficient condition for inversion in igneous rocks. It is probable that the increasing alkalinity of progressively water-rich magmas influenced the structural states that resulted.
3. In the St. George pluton of New Brunswick [16], orthoclase occurs in the eastern end and microcline in the western end of the igneous body. Textural evidence of an original fluid phase (such as miarolitic cavities and pegmatite pods) is more abundant in the eastern end, which seems to contradict our usual assumption that a fluid phase promotes ordering. Compositions of the K-rich phases of the exsolved feldspars indicates that both ends of the pluton equilibrated to the same subsolvus temperature. Two factors appear to bear on the differences in structural states: (a) The orthoclases are more potassium-rich than the microclines, although only by ~5 wt %, and (b) associated igneous rocks that likely prolonged the heating of the pluton are on the west end. This latter factor constitutes a difference not in temperature of formation, but in cooling *rate.* Within the two ends of the pluton there is slight variation in structural states of feldspars, with a systematic trend for greater ordering in rocks with bulk compositions closer to the granite minimum melt composition; this suggests that degree of differentiation of host igneous rocks (i.e., bulk composition) affects feldspar structural states, as in example (2) above.
4. Feldspars from the Kameruka granodiorite in southeastern Australia [22] are all orthoclases or relatively disordered intermediate microclines, except for those that occur in faults; these are maximum microclines. Transmission electron microscopy reveals that the intermediate microclines consist of domains only 20 to 30 unit cells across and twinned according to the albite law in order to mimic the monoclinic shape of the less ordered precursors. Given the domain sizes and orientations, it turns out that the amount of strain relieved by (Al,Si) ordering is approximately equal to that developed by forcing the triclinic domains to retain their original monoclinic shape as ordering proceeds. Shearing along the faults evidently removed the mechanical constraints imposed by the surrounding solid rock, and allowed the release of strain so that ordering to maximum microcline could occur. In this case, then, neither temperature, nor cooling rate, nor fluids, nor any of the other factors cited previously appear to

have played the key role, although temperatures and cooling rates had to be such that ordering could occur once strain was relieved.

Sodium Feldspar

The ordering process in albite, $NaAlSi_3O_8$, differs from that in the high-sanidine–maximum-microcline series. Above ~950°C (with considerable variation from one specimen to another, as we shall see), it has an (Al,Si) distribution that is probably completely disordered and the structure is monoclinic. (There has been some speculation, based on heating studies at the potassium end of the alkali feldspar series and on crystal structure refinements of albites at high temperature, that the disorder in this monoclinic albite may not be complete, but that instead $2t_1 > 2t_2$.) Around this temperature it undergoes a simple displacive transformation that alters the symmetry to triclinic, space group $C\bar{1}$. The reason for the transformation is to be found in the size of the sodium atom. At high temperature the thermal motion of the Na atom renders it effectively about as large as K in sanidine, and the structure is "propped open" to maintain monoclinic symmetry. As the temperature decreases, so does the effective size of Na, and around 950°C the framework collapses about it to become triclinic (Figure 2-13). At this temperature the (Al,Si) distribution is still almost completely random, so even *disordered* albite is triclinic at temperatures below the displacive transformation.

Experimental evidence indicates that, as temperature decreases, aluminum migrates equally from T_1m, T_2O, and T_2m to occupy T_1O, with Al completely ordered on T_1O around 600–650°C. In contrast with the displacive transformation just described, this one takes place by solid-state diffusion. In this case the symmetry is already triclinic and, unlike in the monoclinic potassium feldspar, there is a geometric distinction between T_1O and T_1m (and between T_2O and T_2m) even when the silicon and aluminum are randomly distributed. The triclinic geometry results in a skewed "cage" for the sodium atom (Figure 2-14), with more Na–O bonds to oxygens surrounding T_1O than to those about other tetrahedral sites. Using simple

Figure 2-13. The arrangements of oxygen atoms about K in sanidine (left) and Na in albite (right). The Na atom is depicted as occupying multiple positions within the collapsed and skewed oxygen "cage." After [63].

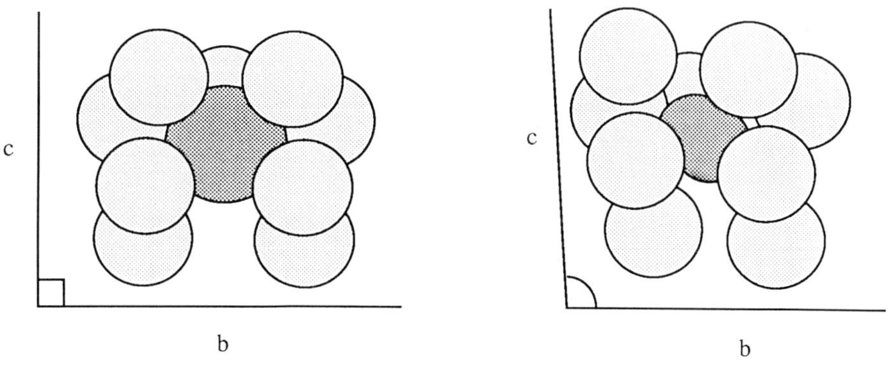

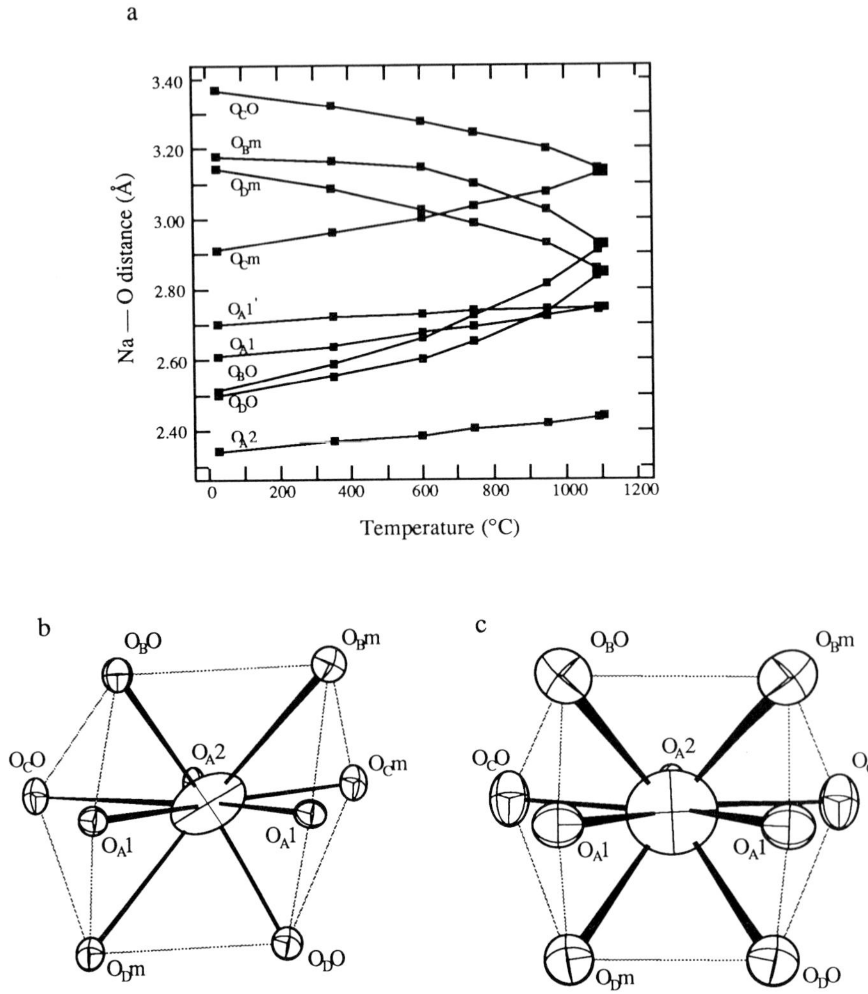

Figure 2-14. The geometry of the Na site in albite. (a) Trends in Na–O distances with changing temperature. (After [59].) (b) The Na site at room temperature. (After [83].) (c) The Na site in high albite at 1040°C. (After [83].)

arguments based on electrostatic bond strengths again, T_1O is found to be overbonded, and the best charge balance occurs in a nearly fully ordered low albite. More sophisticated electrostatic energy calculations [9] predict, in fact, that fully ordered low albite with aluminum in T_1O is more stable than one with aluminum in any other site.

A system of nomenclature has evolved in order to facilitate discussion of these various possible combinations of symmetry and structural state. The *monoclinic* $NaAlSi_3O_8$ that exists only at temperatures above the displacive transformation is called *monalbite.* Because the displacive transformation takes place at a temperature higher than or equal to that of the ordering (or diffusive) transformation

[42;43], high-temperature *triclinic* $NaAlSi_3O_8$ is essentially completely disordered, and is called *analbite;* on heating it inverts to monalbite, but it evidently has no field of thermodynamic stability and has so far been produced only by quenching of monalbite and Na-exchange of sanidine. With the onset of the diffusive transformation the (Al,Si) distribution also becomes triclinic and the feldspar is called *high albite;* the maximum amount of disorder attainable in high albite is not known, but high albites do not invert directly to monalbite on heating. Low-temperature, fully ordered $NaAlSi_3O_8$ is called *low albite.* At structural states intermediate between $t_1o = 0.25$ and $t_1o = 1.0$, the structural states are of the form $t_1o > t_1m = t_2o = t_2m$, and we have intermediate albites. The transition from high to low albite is continuous and higher than first order.

High temperature crystal structure refinements of albites have revealed some complications that are not apparent from the above discussion. Figure 2-14 shows the change in Na–O bond lengths with temperature, and a comparison of the geometry of the sodium "cage" in room temperature ordered and high temperature disordered albite. Note that monoclinic symmetry is not attained until 1105°C for the synthetic albite in Figure 2-14(a). In Figure 2-14(b), a natural albite from Tiburon, California, is not quite monoclinic at 1040°C. Another natural specimen from Amelia, Virginia, was definitely monoclinic by 980°C. The inversion temperature for the displacive transformation may depend on prior thermal history, or slight undetected deviations from complete (Al,Si) disorder, or some other unknown factors.

Aside from the general shape of the sodium "cage," the apparent shape of the sodium atom itself is interesting. Note, in Figure 2-14, how anisotropic the shape of the displacement ellipsoid is. The "thermal ellipsoid" is an ellipsoidal surface that is usually interpreted to depict the magnitudes and directions of thermal vibrations, and therefore the size and shape of the volume occupied, at some level of probability, by an atom over a time average. Extreme anisotropy of "thermal" ellipsoids can be interpreted, however, either (1) as an indication of actual thermal motion, or (2) as the superposition of two or more equilibrium atomic positions, with their associated (smaller) thermal ellipsoids, and which are occupied differently in different domains of the same crystal. (Obviously, one sodium atom can be in only one location at a time in a given unit cell, but remember that a crystal structure refinement using x-ray- or neutron-diffraction methods averages the positions of all the atoms in each of the large number of unit cells being irradiated.) For low albite examined at different temperatures, the isotropic temperature factor for Na—a numerical measure of thermal vibration—extrapolates to very near zero at 0K [71;82]; this indicates that the thermal ellipsoid represents real thermal motion, since there should be none at absolute zero. For high albite, it does not extrapolate to zero at 0K, and thus represents slight but significant variation in the Na position throughout the crystal [59;83]. Similar conclusions for both low and high albite are supported by structure–energy calculations [58]. For high albite, which owing to the rapidity of ordering in this feldspar exists only at high temperature or as a quenched phase, the random (Al,Si) distribution makes many kinds of crystallographically distinct distributions possible—all that is required is that they all average to yield the correct 1:3 ratio of Al to Si (and that an empirical notion called the "aluminum avoidance rule" be obeyed; see p. 78). It is likely that the particular geometry of Al and Si sites about any particular Na "cage" determines the specific Na location within it. Electrostatic "lattice-energy" calculations have confirmed that the potential well for the Na site is broad and elongated along the longest direction of the thermal ellipsoid, so that a well-defined equilibrium position for Na does not exist in albite [9;58]. The lack of positional disorder for sodium in *low* albite is evidently because the strictly ordered (Al,Si) arrangement does not result in variations in the electrostatic field from one Na "cage" to another, and so all Na atoms are in geometrically identical positions.

Although intermediate albites were defined previously, albites found in nature are either the low or high polymorphs at temperatures below the monalbite–analbite displacive transformation. Intermediate albites have been synthesized, however, and appear, from results of experiments in which the degree of order was reversed, to be stable minerals [29]. Moreover, the ordering transformation occurs over a range of temperatures and is second or higher order. In these experiments, which were carried out without water but at a pressure of 18 kbar, disordering proceeded rapidly until it was 80–85% complete, and then it slowed by an order of magnitude. *If* the process on cooling is the reverse of that observed on heating (which, we shall see, is not the case for *potassium* feldspars), then it should be relatively difficult to coax a cooling albite out of its essentially disordered high structural state, but once that is achieved, ordering should proceed rapidly to completion. The weight of theoretical and experimental evidence suggests that ordering is accomplished to a substantial degree by 775 to 700°C, with the driving force arising from the collapse of the tetrahedral framework as temperature decreases [59;68]. This implies that ordering is driven by factors more significant at high temperatures than at low ones, and that ordering and disordering may not be simply mirror-image processes for albite. In fact, structure–energy calculations indicate that the driving force for ordering in albite increases as the temperature is lowered to about 800°C and then remains essentially constant with further decrease in temperature [58].

Alkali Feldspars of Intermediate Compositions

We have seen that completely disordered K-feldspar is monoclinic at any temperature, whereas Na-feldspar with any degree of order is triclinic at temperatures below the monalbite–analbite transition temperature, the difference arising from the sizes of the respective M cations. Obviously then, for any given temperature, there must be some composition at which the mean size of the M cation is just small enough to permit the collapse to the triclinic structure at that temperature, whereas a slightly more K-rich feldspar under the same conditions would be monoclinic. That composition turns out to depend on the structural state of the feldspar series in question. Figure 2-15 shows composition, N_{Or}, as a function of (1-cos ϕ) for four metrically triclinic alkali feldspar series that have monoclinic (Al,Si) distributions. (Such disordered triclinic feldspars, of which analbite is a good example, are said to be *topochemically* monoclinic, but *geometrically* triclinic. Some authors would use "topologically" rather than "geometrically," but topology refers to the connectivity

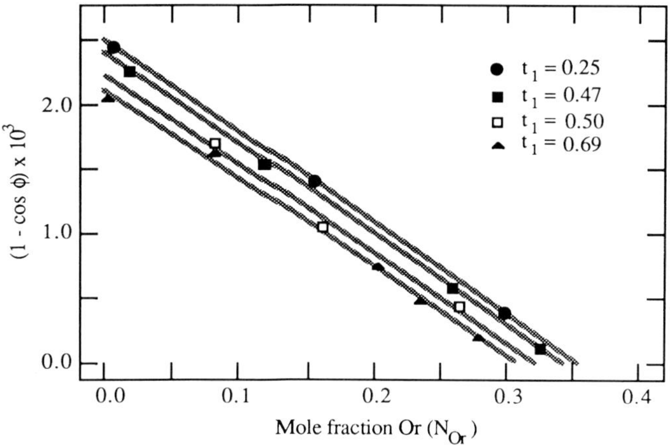

Figure 2-15. Variation in $(1 - \cos\phi)$ with composition for four alkali exchange series of topochemically monoclinic alkali feldspars. ϕ is the angle between **b** and **b***, and $(1 - \cos\phi)$ is therefore zero for monoclinic feldspars. Extrapolation of the lines to $(1 - \cos\phi)$ values of 0.0 yields the compositions at which each series undergoes the monoclinic–triclinic structural collapse at room temperature. (After [39].)

of structural elements, and all feldspars have identical framework topologies.) Here ϕ is the angle between the **b** and **b*** directions. [**b*** is the direction perpendicular to the (010) lattice planes, so that ϕ is zero for monoclinic minerals and nonzero for triclinic ones. The function (1-cos ϕ) is used because it varies linearly with composition for alkali feldspars.] Each series has a constant, topochemically monoclinic (Al,Si) distribution, and all four were produced from different starting materials by the technique of alkali exchange; in this technique, compositions, but not structural states, are changed by immersing the feldspars in baths of molten KCl–NaCl that have [K]/[Na] ratios identical to those desired in the final feldspar products. Alkali exchange is driven by compositional gradients, and the volume of the bath is so large compared to that of the crystalline solid that the final alkali composition of the crystals is essentially identical to the starting alkali composition of the bath. The temperature required for alkali exchange is well below that required to produce disorder of Si and Al. The figure shows that extrapolation to a (1-cos ϕ) value of zero (i.e., monoclinic symmetry) yields different compositions at which the triclinic-monoclinic inversion occurs at room temperature, and that the composition so obtained depends on structural state: More ordered feldspars require more sodic compositions in order for the displacive symmetry inversion to proceed from monoclinic to triclinic. This is presumably because the ordering process, in which the sizes of T_1 and T_2 change, allows small structural adjustments to be made around the M site so that, as ordering proceeds, the M site can tolerate a progressively smaller average atom without collapse.

Figure 2-16 shows how temperature and pressure of inversion vary as functions of composition. The results are as expected: For a given composition that is triclinic at room temperature, there is some temperature at which thermal motion expands the M site, tilting the (Al,Si) tetrahedra in the process, and producing monoclinic

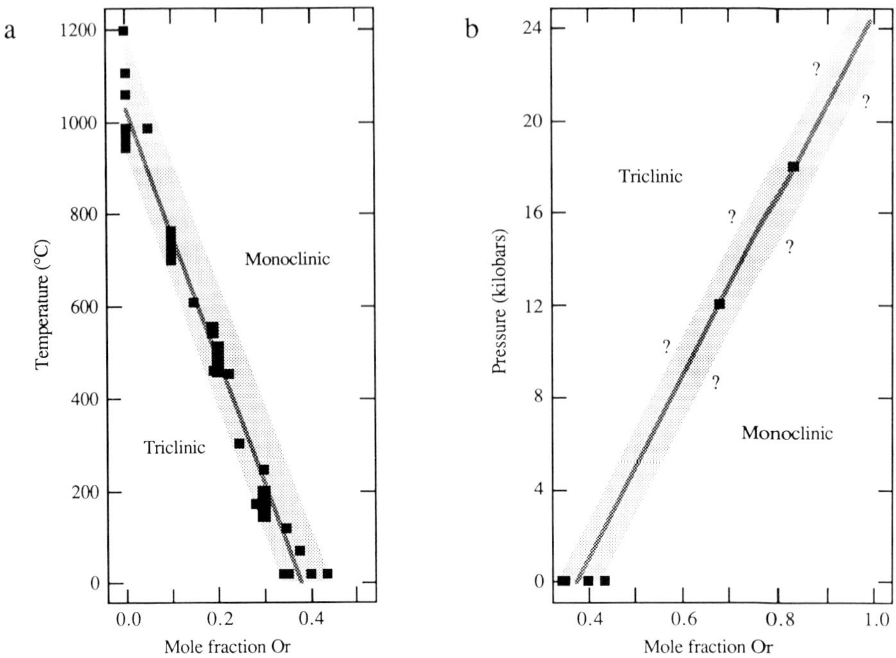

Figure 2-16. Variation in the monoclinic–triclinic inversion with (a) temperature (at 1 atm) and (b) pressure (at room temperature). (After [52].)

symmetry; for a given composition that is monoclinic at atmospheric pressure, there is some pressure at which the structure is collapsed to triclinic symmetry. Probable variations due to structural state, as discussed in the preceding paragraph, are indicated by the widths of the shaded band. The limited number of experimental investigations into the thermodynamics of the monoclinic–triclinic inversion suggest that there are no changes in molar volume, coefficients of thermal expansion, or heat capacity at the transformation temperature. This would require the transformation to be not only displacive, but, according to Figure 1-14 and the discussion on pages 17–18, also higher than second order.

Figure 2-17 shows the transformation geometry in *P-T-X* space (*X* being the customary symbol for composition). Each structural state would transform across a two-dimensional surface, but the addition of the surfaces for all different structural states yields a slab of finite thickness through which the monoclinic–triclinic inversion takes place. Little work has been done at high pressures, and virtually none at simultaneously high pressures and temperatures; so the thickness of the slab is largely schematic. Ignoring these effects of structural state, the composition at which the transformation occurs for any given temperature and pressure is given by [52]

$$N_{\mathrm{Or}} = 0.474 - 0.361[T(\mathrm{K})/1000] + 0.0252P(\mathrm{kbar}).$$

The figure reinforces the observations made in Chapter 1 about the effects of temperature and pressure on crystal structure: High temperature favors high symmetry and low density, whereas high pressure favors low symmetry and high density.

There is strong evidence that potassium and sodium order into distinct atomic positions in alkali feldspars of intermediate composition. In the structure refinement of an Or_{58} crystal, for example [23], the best fit of calculated-to-observed x-ray intensity data was found when a so-called "split-atom model" was used for the alkali metal site (rather like the model for hexagonal tridymite, p. 26). In this model, the alkali metal cavity was split into one site occupied by 0.50K and one occupied by 0.42Na + 0.08K. This adds up, as required, to a total of 1.0 atom in the cavity, and what it really means is that there is one alkali metal site occupied by K alone and one occupied by 0.84Na + 0.16K on the average, and that the structure refinement superimposes these two sites; it makes each alkali metal cavity the average of two different cavities—one K-rich and the other Na-rich—and requires that the alkali metal atoms be ordered in the structure. The K-rich site was nearly isotropic and centered in the cavity, while the Na-rich site was highly anisotropic, a characteristic of Na-rich sites that we have already observed in the albite structures. Interestingly, the vector between the two split-atom sites is essentially normal to the $(60\bar{1})$ and $(80\bar{1})$ planes, which are the planes of coherency in exsolved alkali feldspars, as discussed in the following. It is tempting to suggest that such ordering is a necessary premonitory step to exsolution in alkali feldspars of intermediate, or perhaps any, composition. Studies of six less potassic intermediate alkali feldspars

Figure 2-17. The monoclinic–triclinic inversion in pressure–temperature–composition space. Feldspar of any one structural state would presumably invert across a surface, but, because feldspar of each different structural state inverts at a different composition (see Figure 2-15), the surface becomes a slab of finite thickness. The thickness as shown here is intended to be only schematic. (Modified from [36].)

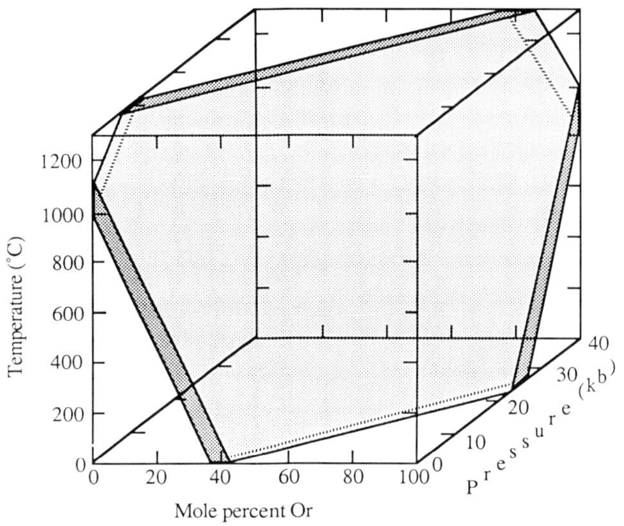

yielded essentially similar results [19;35]. In addition, results of Raman spectroscopic investigations also support ordering at the alkali metal site [67].

ORDERING PATHS IN ALKALI FELDSPARS

The way in which aluminum content changes in each tetrahedral site as an alkali feldspar orders from the most disordered to the most ordered state is called its *ordering path.* For albites, the simplest ordering path is straightforward: The aluminum, which is statistically distributed among all four tetrahedral sites in high albite, migrates equally from T_1m, T_2O, and T_2m into T_1O. Alternatively, if $2t_1 > 2t_2$ in monalbite (p. 53), then aluminum migrates essentially uniformly from T_1m, T_2O, and T_2m into T_1O until some temperature is reached at which a first order transformation to low albite takes place. Figure 2-18 illustrates both of these possibilities.

The ordering path in K-rich alkali feldspars is shown in Figure 2-19, in which mean T–O bond lengths may be interpreted as directly reflecting Al occupancy. The figure shows the mean bond length of each tetrahedral site plotted against the mean bond length of the T_1O site. An increase in aluminum occupancy causes mean bond length to increase, whereas an increase in silicon occupancy causes it to decrease. Unlike the ordering path of albite, this is distinctly a two-step phenomenon. The first step involves the migration of aluminum from the T_2 sites to the T_1

Figure 2-18. One-step ordering paths for albite. (a) Path assuming a non-first-order transition from intermediate to low albite. (After [73].) (b) Path assuming a first-order transition from intermediate to low albite. (After [45].)

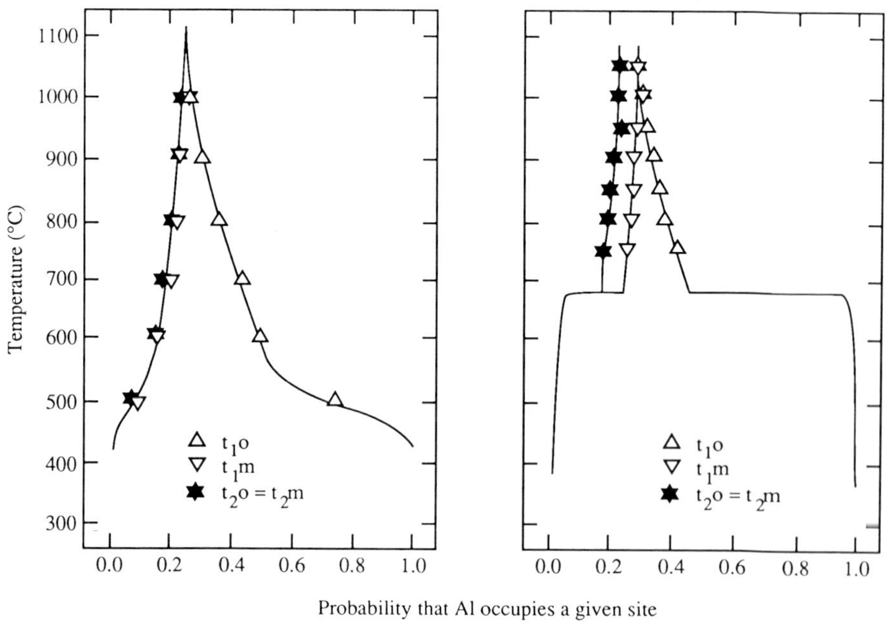

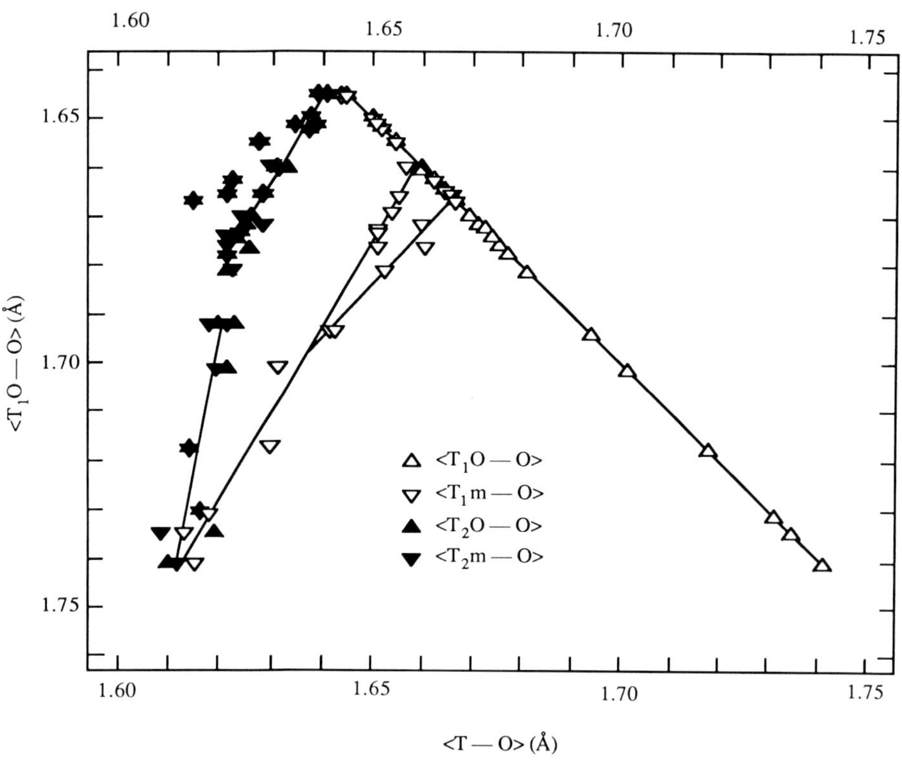

Figure 2-19. Two-step ordering path for K-rich feldspars. No attempt is made to calculate Al distribution in this type of plot, but it is implicitly assumed that longer mean *T*-O bond lengths indicate greater Al concentration. After [8], with data from [30] added.

sites, with monoclinic symmetry being maintained. In the second step, the feldspar behaves more or less like an albite that is ordering, with aluminum migrating from T_1m, T_2O, and T_2m to T_1O. The bond length (or, alternatively, the Al content) at which the aluminum occupancy of T_1O and T_1m begin to diverge and the mineral becomes triclinic is evidently not constrained to a single value, but Figure 2-19 indicates only two values that have been observed so far—~1.659 and ~1.667 Å, corresponding approximately to 0.77 and 0.90 Al in T_1 sites. It is interesting to note that $2t_1 = 0.75$ is nearly equal to the structural state at which O_A1 and O_A2 are equally overbonded (p. 46), while $2t_1 = 0.89$ results in a charge-balanced O_A1 atom. Whether these observations are significant is not clear from work that has been done on feldspar ordering to date.

It is also worth noting that the experimental disordering of microcline to sanidine has been observed to be a one-step process [7]. The ordering and disordering processes for natural potassic feldspars are thus not mirror images of one another, and it is possible that the well-documented two-step ordering path reflects alternative metastable behavior (p. 20). In other words, it may be that the equilibrium behavior is a one-step process, but that it is prevented by kinetic barriers from occurring.

LATTICE-PARAMETER VARIATIONS IN ALKALI FELDSPARS

Changes in the lattices of minerals in the alkali feldspar series result either from changes in composition, or in (Al,Si) distribution, or both. It is evident, therefore, that knowledge of the symmetry and lattice parameters can provide information about both of these crystal-chemical variables, and a matter that must be addressed early in the investigation of any alkali feldspar is its crystal system. The most convenient method for determining whether a feldspar is monoclinic or triclinic is x-ray powder diffraction. (See Chapter 10 for a general discussion of x-ray powder diffractometry.) For monoclinic crystals, the hkl and $h\bar{k}l$ x-ray reflections are symmetrically identical by virtue of the mirror plane in the monoclinic Laue group, and so they superimpose to yield a single sharp peak on a powder-diffraction pattern. Triclinic crystals lack the mirror plane, so that hkl and $h\bar{k}l$ peaks have different 2θ values, in general.

Observation of any hkl reflection (or, for that matter, any $hk0$, $0kl$, or $0k0$ reflection) to determine if a single or double peak were present would, in principle, suffice to discriminate between the two crystal systems for feldspars. In actuality, though, many peaks are not consistently observed because of inherently low intensities, and only two—the 131 and $\bar{2}41$ peaks—are commonly used. For monoclinic K-rich alkali feldspars diffracting Cu$K\alpha$ x-rays, the 131 peak occurs at ~29.9° 2θ, and in maximum microcline it is split into two peaks, the 131 and $1\bar{3}1$, at ~29.5° 2θ and ~30.2° 2θ, respectively. For the $\bar{2}41$ peak and its split counterparts, the 2θ values are respectively ~34.9°, and ~34.2° and ~35.5°. For feldspars at the onset of ordering to triclinic symmetry (i.e., the initial loss of the diad and mirror from the monoclinic crystal class), the 131 and $\bar{2}41$ peaks present a broadened appearance because the doublets are not sufficiently separated to be resolved, and increasing divergence of the crystallographic angles α and γ from 90° results in wider separation of the doublets. The separation between the 131 and $1\bar{3}1$ peaks, in degrees of 2θ as measured with Cu$K\alpha$ radiation, is often called *obliquity;* it is generally symbolized $\Delta 2\theta_{131}$, or simply $\Delta 131$, and we shall encounter it again as a very useful parameter for plagioclase feldspars. (The term *triclinicity* has sometimes been used instead of obliquity, but it should be avoided because a mineral is obviously either triclinic to x-ray diffraction or it is not, and one mineral cannot be "more triclinic" than another.)

Figure 2-20 is a projection of the feldspar structure, in part, onto the (100) plane. If we begin with the T_2m tetrahedron at the center left edge and move to the right side of the unit cell by following a continuous chain of corner-sharing tetrahedra, then we encounter the sequence of tetrahedra T_2m–T_1O–T_2m–T_2O–T_1m–T_2O. If the (Al,Si) distribution is random, there will be 6(0.25Al + 0.75Si) = 1.5Al + 4.5Si in this sequence of sites, which essentially constitutes the b cell edge. But if the crystal is completely ordered, there will be 1.0Al + 5.0Si along b. Because the ionic radius of Al is larger than that of Si, the b cell dimension will decrease with ordering. Following a continuous chain of tetrahedra along the c direction, we encounter T_1O–T_2m–T_1m. A disordered alkali feldspar will then have 3(0.25Al + 0.75Si) = 0.75Al + 2.25Si along c, whereas an ordered polymorph will have 1.0Al + 2.0Si. Thus the c cell edge will increase in length with ordering. Each structural state on

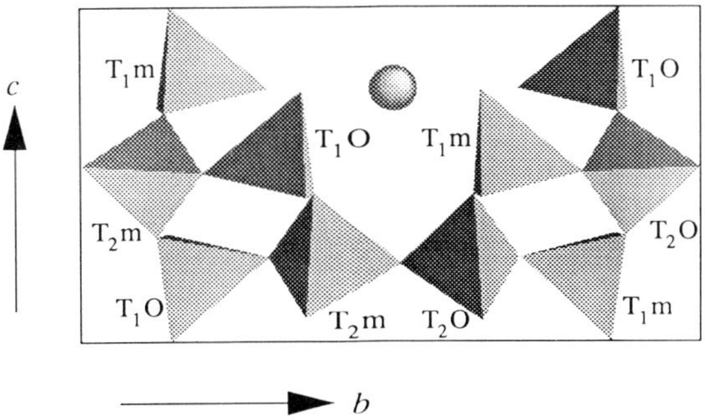

Figure 2-20. Part of the feldspar framework projected onto (100).

the continuum from $t_1o = t_1m = t_2o = t_2m = 0.25$ to $t_1o = 1.0$, $t_1m = t_2o = t_2m = 0.0$ will result in a unique *b* and *c* unit cell dimension for constant composition. Moreover, each different composition on the continuum from $KAlSi_3O_8$ to $NaAlSi_3O_8$ will have a unique set of *b*,*c* values over the range from ordered to disordered.

This leads us to the *b–c* plot as a means of determining the structural states of alkali feldspars of any compositions. The *b–c* plot is shown in Figure 2-21, with the values for *b* horizontal and those for *c* vertical, and it is contoured for values of the *a* cell edge. K-feldspars plot to the right and Na-feldspars to the left. (It is, of course, not necessary to know composition in advance to use the plot. Only *b* and *c* are necessary.) The disordered polymorphs (high structural states) plot between sanidine and analbite, and completely ordered polymorphs fall between maximum microcline and low albite. Alkali feldspars with different compositions or structural states, or both, will plot at different positions on the diagram. The contours for *a*

Figure 2-21. The *b–c* plot. Solid contours yield $t_1o + t_1m$, total Al in T_1 sites. Dotted contours estimate *a*. (After [74].)

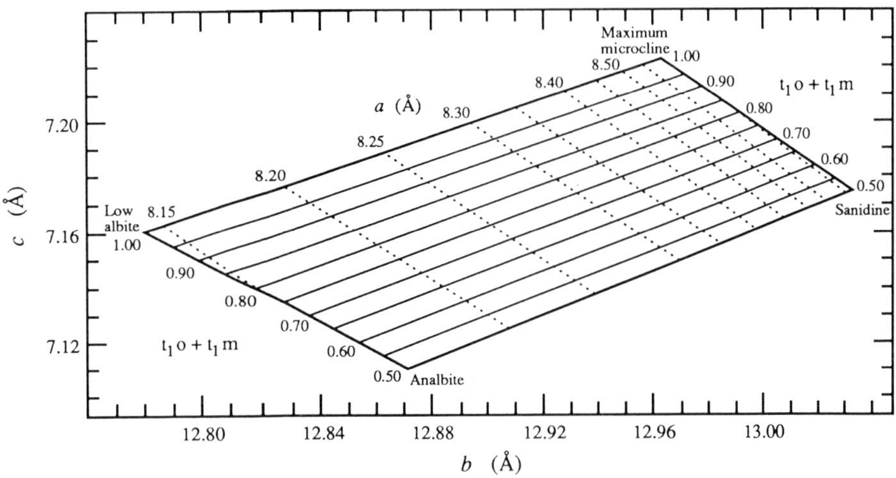

(straight lines trending "NW–SE") are accurate to perhaps 0.02 Å. If the observed value of the *a* cell edge differs more than 0.05 Å from the one predicted by the contours, the feldspar is said to be "strained"; we shall address the significance of strain later.

The contours that trend "NE–SW" yield total Al in T_1 sites ($2t_1$ or t_1o+t_1m) [73]. If the feldspar is monoclinic, then $t_1o=t_1m$, and the *b-c* plot is sufficient to specify structural state. (For all alkali feldspars so far investigated, $t_2o=t_2m$.) For a triclinic alkali feldspar, $t_1o \neq t_1m$, and, since the *b-c* plot yields only the sum of the two, more information is needed to specify either. Reexamination of Figure 2-20 will confirm that a difference in size between T_1O and T_1m should skew the unit cell so as to make K-feldspar metrically triclinic, and likewise increase the obliquity of Na-feldspar. The magnitudes of the α and γ crystallographic angles are a measure of the deviation from monoclinicity, and therefore of the difference between t_1o and t_1m [73]. There are advantages for the x-ray crystallographer if the reciprocal lattice angles α^* and γ^* are used instead of the direct lattice angles. [α^* is the angle between b^* and c^*, which are perpendicular to (010) and (001), respectively. γ^* is the angle between a^* and b^*, which are perpendicular to (100) and (010), respectively.] The α^*–γ^* plot is shown in Figure 2-22, and is similar in form to the *b–c* plot except that (1) all monoclinic alkali feldspars plot at one corner, and (2) the contours are for t_1o-t_1m rather than for t_1o+t_1m. To use the *b–c* and α^*–γ^* plots, one usually obtains both the direct- and reciprocal-lattice parameters from x-ray powder-diffraction data (although any other experimental method yielding those data would do as well). The *b* and *c* cell edges are then plotted on the *b–c* plot, and a value of t_1o+t_1m, say X, is read from the contours; similarly, a value for t_1o-t_1m, say Y, is read from the α^*–γ^* plot. It is then a simple matter to solve the two simultaneous equations

$$
\begin{aligned}
t_1o + t_1m &= X \\
t_1o - t_1m &= Y,
\end{aligned}
$$

and obtain both t_1o and t_1m for a triclinic alkali feldspar; t_2o and t_2m are then obtained by difference, under the constraints that $t_1o+t_1m+t_2o+t_2m = 1.0$, and $t_2o=t_2m$.

Although one could estimate composition from position on the *b–c* and α^*–γ^* plots, there is a better way. Figure 2-23 shows how mean (K,Na)–O distance varies with composition for the disordered alkali feldspar series. The obvious implication is that substitution of K for Na distends the tetrahedral framework, and this should be apparent from lattice parameters. The *a–b* projection in Figure 2-6 shows that the sequence of tetrahedral sites along *a* is T_1m–T_2O–T_1O–T_2m. Since there is one of each type of site, the *a* cell edge is insensitive to structural state, but it should change with composition owing to the effect of alkali atoms of different sizes on the framework. Because the same number of Al and Si atoms are present in the unit cell regardless of structural state, unit-cell volume should also be a function of composition but not of structural state, at least to a very good first approximation. Figure 2-24 shows that either *a* cell edge or unit-cell volume can be used to obtain good estimates of composition. For reasons to be discussed later in connection with

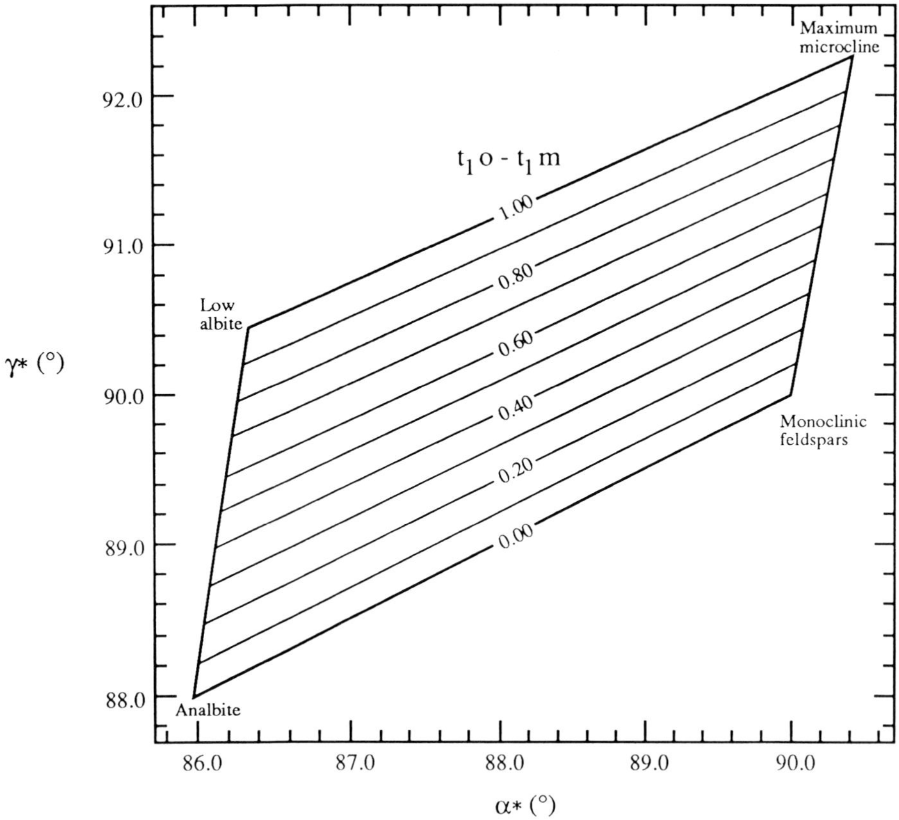

Figure 2-22. The α^*–γ^* plot. Contours yield $t_1o - t_1m$, the difference between in Al in T_1O and that in T_1m. (After [74].)

Figure 2-23. Variation of mean (K,Na)–O distance for alkali feldspars as a function of composition.

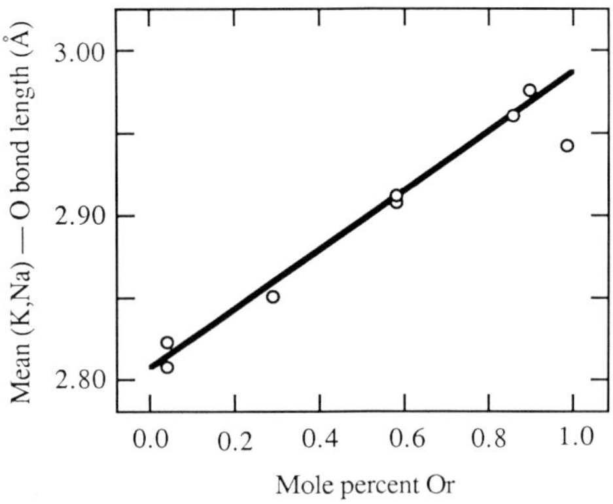

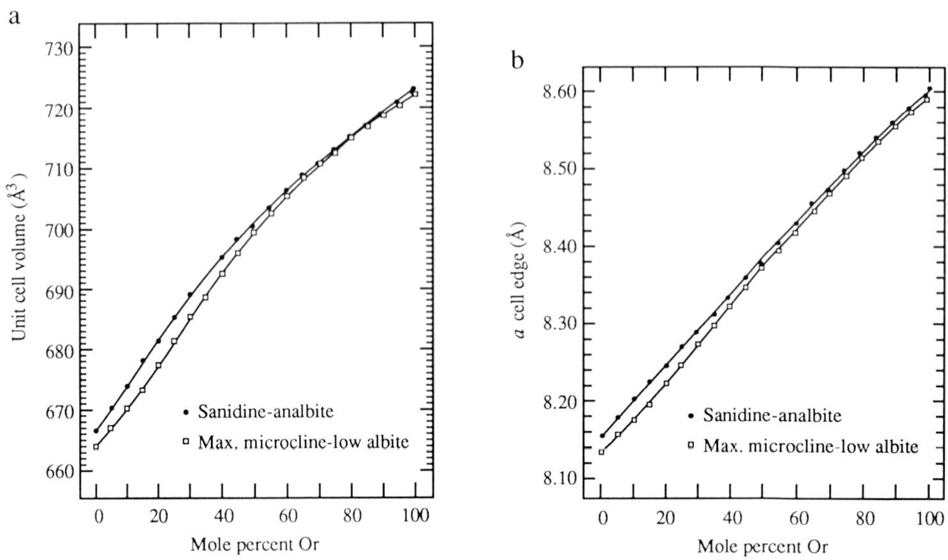

Figure 2-24. Unit cell volume and *a* cell edge as functions of composition for alkali feldspars. Note that structural state has a small effect, and that the prediction of composition from either plot is more accurate for potassic compositions, because it is less sensitive to structural state. (After [47].)

strained feldspars, cell volume is a more reliable indicator of composition, and the relation [47]

$$n_{Or}(\text{mol }\%) = -872.33 + 3.82791V - 5.61695 \times 10^{-3}V^2 + 2.75683 \times 10^{-6}V^3 \quad (2.1)$$

is useful for calculating composition as a function of cell volume (V), although the graph of Figure 2-24 is perfectly adequate. A somewhat more convenient but slightly less accurate equation is [74]

$$n_{Or}(\text{mol }\%) = [0.2962 - \sqrt{(0.953131 - 0.0013V)}]/0.0018062.$$

Of course if unit-cell axial dimensions convey information about (Al,Si) order in alkali feldspars, then it is likely that some other unit-cell translations do also. A continuous chain of tetrahedra along the unit cell translation [110] [i.e., parallel to a vector from the origin to a point at ($1a$, $1b$, $0c$) from the origin; refer to Figure 2-6] consists of the sequence T_2m-T_1m-T_2O-T_2m-T_1m-T_2O-T_2m, while a similar chain parallel to [1$\bar{1}$0] consists of the sequence T_2m-T_1O-T_2O-T_2m-T_1O-T_2O-T_2m [41]. Note that each translation comprises four T_2 sites (considering the chain to include only one-half of each of the identical sites at each end) and two T_1 sites. The translations are identical and ~7.81 Å long for a disordered monoclinic K-feldspar (sanidine). At the Na end of the series the translations are not equal because even the disordered Na-feldspar (analbite) is triclinic owing to the collapse of the tetrahedral framework about the small Na atom; the [110] and [1$\bar{1}$0] translations are, respectively, ~7.605 Å and ~7.635 Å. As aluminum begins to order into T_1 sites, maintaining monoclinic symmetry, the [110] and [1$\bar{1}$0] translations both decrease because each chain possesses only half as many T_1 sites as T_2 sites. As ordering yields triclinic topochemistry, the [1$\bar{1}$0] translation continues to decrease in length, but the [110] translation increases until it reaches a maximum for maximum microcline (K end) or low albite (Na end).

If accurate unit-cell dimensions are available, for example from routine x-ray powder-diffraction work (see Chapter 10), the necessary translations can be calculated easily as follows:

$$\mathrm{tr}[110] = \tfrac{1}{2}(a^2 + b^2 + 2ab\cos\gamma)^{1/2}$$
$$\mathrm{tr}[1\bar{1}0] = \tfrac{1}{2}(a^2 + b^2 - 2ab\cos\gamma)^{1/2}.$$

These formulas follow directly from the application of the law of cosines to the *a*–*b* projection of the unit cell (Figure 2-6). The factor of $\frac{1}{2}$ in each expression arises because each translation consists of two identical three-tetrahedron repeats due to the *C*-centering of the unit cell.

Figures 2-25 and 2-26 allow one to use these two translations, along with composition [which may be found from equation (2.1)] to determine t_1o (Figure 2-25) and t_1m (Figure 2-26). Again, $t_2o \approx t_2m \approx \frac{1}{2}(1.0\text{-}t_1o\text{-}t_1m)$—that is, aluminum not in the T_1 sites is evenly divided between the two T_2 sites. This method has the advantages that no use of reciprocal unit cell dimensions is involved, and that t_1o and t_1m can be read directly without solving any equations. Of course, composition must be found, but one would likely obtain this from the unit-cell volume anyway in the course of investigating a feldspar. On the other hand, it is not as easy to recognize strained feldspars using this method as it is using the *b*–*c* plot, as we shall see below.

Other methods for estimating structural state have been devised, and in some ways are superior to those described here. Many also have limitations, however; some can be used only for K-rich feldspars [27], some only with monoclinic feldspars [75], and some only if a crystal structure refinement has already yielded *T*–O bond lengths [45]. Given the convenience and generality of the methods just described, we shall concern ourselves only with one method that offers an improvement over the *b*–*c* plot.

The *b*–*c* plot provides only a good approximation to $t_1o + t_1m$, because the paths followed on it by series of alkali-exchanged feldspars are not linear, but slightly curved. This shortcoming has largely been remedied by the construction of *b*–*c** plots (or the corresponding regression equations) on which the paths followed by alkali-exchanged series are essentially linear [46]. The equation for monoclinic alkali feldspars is

$$\Sigma t_1 \equiv 2t_1 = (b - 24.8095 + 74.9054c^*)/(-3.3261 + 19.5012c^*), \quad (2.2)$$

and that for triclinic alkali feldspars is

$$\Sigma t_1 = t_1o + t_1m = (b - 22.5398 + 53.8405c^*)/(2.1567 + 15.8583c^*). \quad (2.3)$$

The α^*–γ^* plot and the unit-cell volume are still used for Δt_1 and composition, respectively, by this method, although the equation

$$\Delta t_1 \equiv t_1o - t_1m = (\gamma^* - 44.778 - 0.50246\alpha^*)/(6.646 - 0.05061\alpha^*) \quad (2.4)$$

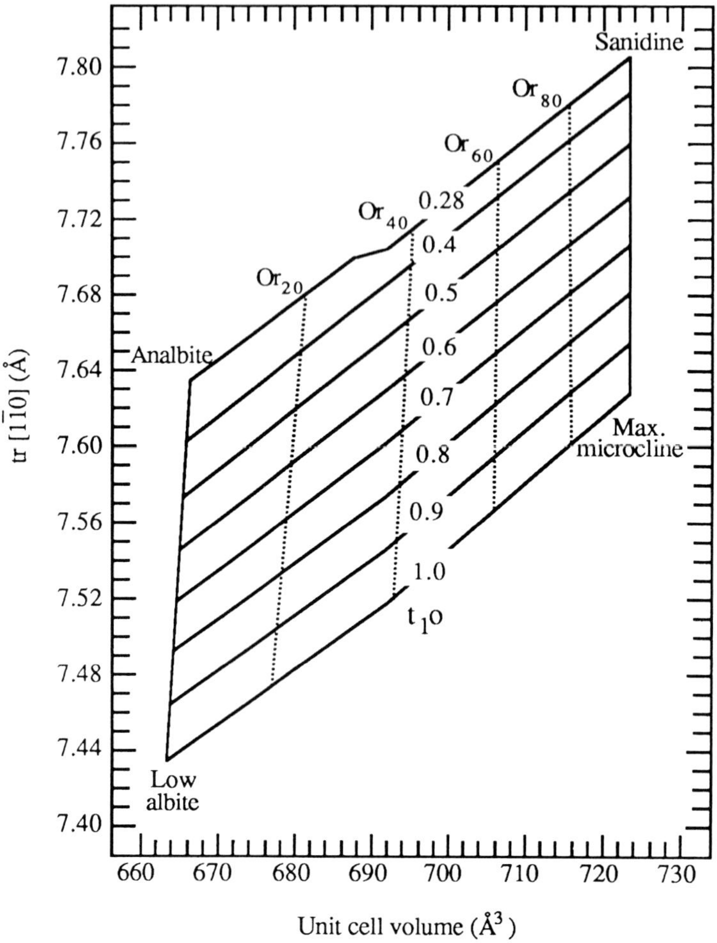

Figure 2-25. Using and [110] translation, obtained from unit cell parameters as described in the text, and unit cell volume, t_1o is read from the solid contours. Composition may be estimated from the dotted contours. (After [45].)

may be substituted for the α^*–γ^* plot if desired. An accuracy of ± 0.02 in estimating structural state is claimed. Equations are also presented in reference [46] for estimating Σt_1 using only the positions of two x-ray powder-diffraction peaks (assuming the use of Cu$K\alpha_1$ radiation). They are, for monoclinic and triclinic alkali feldspars, respectively

$$\Sigma t_1 \equiv 2t_1 = \{2\theta(060) + 12.1814 - 1.04093[2\theta(\bar{2}04)]\}/\{0.6112 + 0.01592[2\theta(\bar{2}04)]\},$$

$$\Sigma t_1 \equiv t_1o + t_1m = \{2\theta(060) + 8.3063 - 0.96459[2\theta(\bar{2}04)]\} /\{-6.5616 + 0.15724[2\theta(\bar{2}04)]\}.$$

Reference has been made to "strained" feldspars. These are alkali feldspars for which the observed a differs from the value of a predicted from the b–c plot by 0.05

Å or more. The difference is called Δa, and thus strained feldspars have $\Delta a \geq 0.05$ Å. Because the [110] and [1$\bar{1}$0] translations are calculated from a and b, Δa values obtained from Figures 2-25 or 2-26 are smaller than those obtained using Figure 2-21, and a criterion for recognizing strained feldspars from the former is not easily arrived at. Hence strain is commonly referred to in terms of Δa obtained from the b–c plot. Strained feldspars are not at all uncommon, and they present a problem in interpretation of the b–c (or b–c^*) and α^*–γ^* plots. The first question is, "What causes feldspars to be strained?" The second is, "If the b–c plot has predicted the a cell dimension incorrectly for a strained feldspar, has it also yielded an incorrect structural state?" We shall now consider these questions.

EXSOLUTION IN ALKALI FELDSPARS

At high temperatures both Ab and Or are monoclinic, and so are feldspars whose compositions lie between these. The reason, as explained earlier, is that the less

Figure 2-26. Using the difference between the [110] and [1$\bar{1}$0] translations, calculated as described in the text from unit-cell parameters, and unit cell volume, ($t_1o - t_1m$) may be read from the solid contours. This and t_1o from Figure 2-25 enable one to determine t_1o, t_1m, and t_2o (= t_2m). (After [45].)

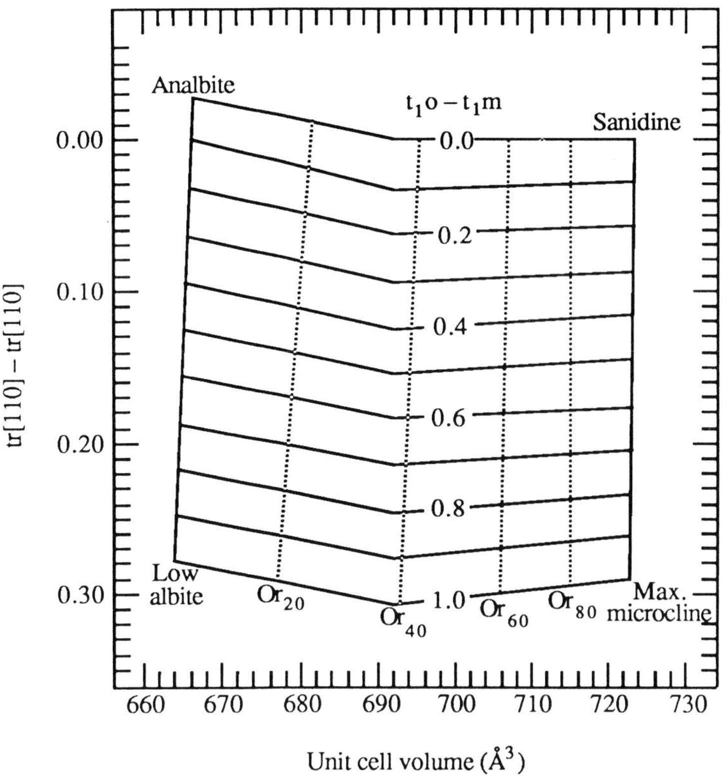

massive sodium atom experiences more thermal motion than potassium, thus increasing its effective size. Consequently, with crystal structures, radii of substituent atoms, and oxidation numbers of substituent atoms identical or nearly so, a complete solid solution between Ab and Or exists at high temperature. As temperature decreases, the structure of a compositionally intermediate alkali feldspar has to adjust to increasingly divergent sizes of *M* cations in adjacent *M* sites. Eventually the crystallographic strain becomes excessive and a compositional adjustment is made. Sodium atoms begin to migrate in the solid state to form regions that are Na-rich and consequently K-rich regions develop. This results in exsolution lamellae of Na-rich feldspar in a host of K-rich feldspar (called perthite, mesoperthite, microperthite, or cryptoperthite, depending on the scale), or K-rich lamellae in an Na-rich host (antiperthite, and so forth).

Exsolution can occur by either nucleation and growth of one phase in another, or by the development of initially small but increasingly divergent compositional variations in a process called *spinodal decomposition.* If the crystal structures of the host and exsolving phases are dissimilar, then nucleation is the only mechanism by which exsolution can occur, but nucleation may also occur even if the two phases have similar structures. Nucleation on specific crystal defects is called *heterogeneous nucleation,* whereas *homogeneous nucleation* develops in regions of perfect crystal structure. Spinodal decomposition involves initially sinusoidal compositional variations and takes place simultaneously over the entire crystal. It is thus a continuous transformation (see p. 16), although it is not a polymorphic transformation because composition is not constant.

The alkali feldspar phase diagram (Figure 2-2) shows a solvus between Or and Ab. If the bulk composition of a cooling feldspar lies underneath the solvus at a particular temperature, then the equilibrium compositions of the stable exsolution products (i.e., the two phases with lowest free energy) are given by the intersection of a horizontal line with the solvus at the temperature of interest. Equilibrium exsolution involves nucleation of a new phase in a host mineral, and is a discontinuous process, generally with a high activation energy. Spinodal decomposition, involving the continuous development of small compositional modulations, involves little or no activation energy, and is a form of alternative metastable behavior (p. 20)—it lowers free energy without requiring the energy necessary to lower it to a minimum. Figure 2-27 shows schematically an equilibrium, or chemical, solvus (curve a) and the chemical spinodal (curve b). The chemical spinodal, derived from free energy versus temperature relations, is the curve outside of which heterogeneous nucleation must occur and spinodal decomposition cannot. Inside curve b, spinodal decomposition may occur. Curves c and d are the coherent solvus and coherent spinodal, respectively, and are discussed below.

If the compositions of the two exsolving phases are not too dissimilar, the lattice parameters may be sufficiently alike to allow the two lattices to be continuous across the interfaces between them, although distortion of one or both lattices is obviously required. Such lattices are said to be *coherent,* and the strain inherent in fitting the two lattices together is called *coherency strain.* Figure 2-28 displays the relationships defining coherency, noncoherency, and semicoherency. Curve c in Figure 2-27 is the coherent solvus. The coherent solvus shows the compositional limits, as a function of temperature, of phases whose lattice parameters allow them to remain

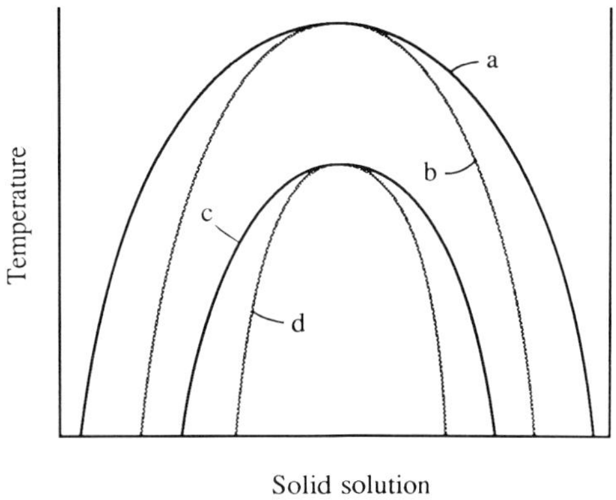

Figure 2-27. A schematic illustration of an equilibrium (or chemical) solvus (curve a), a chemical spinodal (curve b), a coherent solvus (curve c), and a coherent spinodal (curve d). Within the equilibrium solvus, exsolution is the stable behavior. Between the equilibrium solvus and the chemical spinodal, exsolution can occur only by heterogeneous nucleation and growth. Within the chemical spinodal, spinodal decomposition may occur. The coherent solvus shows the compositional limits, as a function of temperature, for which two phases may remain structurally coherent on homogeneous nucleation and growth, and the coherent spinodal is analogous for spinodal decomposition. (After [61].)

Figure 2-28. (a) Two lattices with different lattice parameters, meeting at a noncoherent boundary. The lattices are discontinuous across the interface between them. (b) Two lattices that meet across a zone of partial coherency. (c) Two lattices that exhibit coherency across the interface separating them. Note that lattice parameters have adjusted in both phases to achieve coherency. After [85].

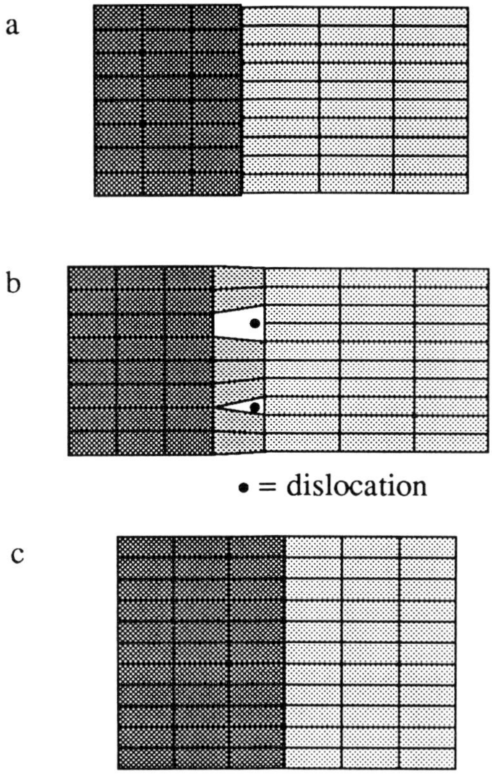

coherent upon homogeneous nucleation. Curve d in that figure is analogous, but refers to spinodal decomposition.

In decomposing spinodally, the compositional modulations will spontaneously adjust their orientations so that the K-rich and Na-rich phases experience the least lattice strain. The initial modulations have wavelengths of 50 to 500 Å, and low amplitudes (that is, small compositional differences). As wavelengths become larger and amplitudes greater, a point is reached at which sharp interfaces separate the phases, whose compositions are by this point quite different. The resulting two-phase mixture is cryptoperthite. The plane of coherency is a nonrational plane

Figure 2-29. Transmission electron micrograph of a cryptoperthite. The sodium-rich lamellae are finely twinned on the albite law. [Photo from G. W. Lorimer and P. E. Champness (1973). "The Origin of the Phase Distribution in Two Perthitic Alkali Feldspars." *Philosophical Magazine* 28, 1391–1403.]

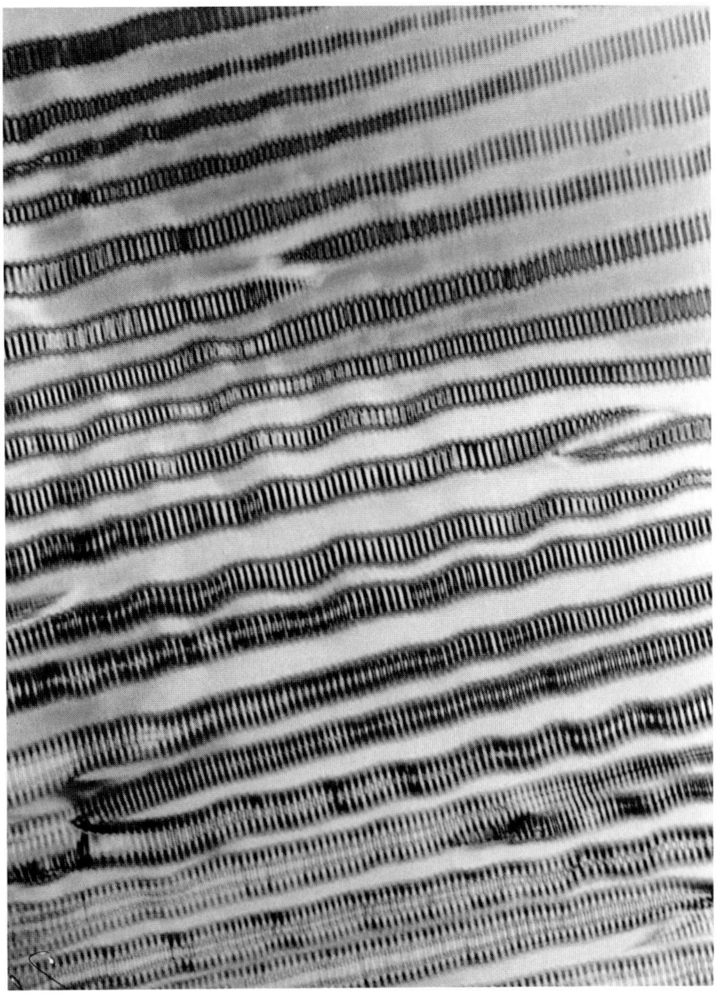

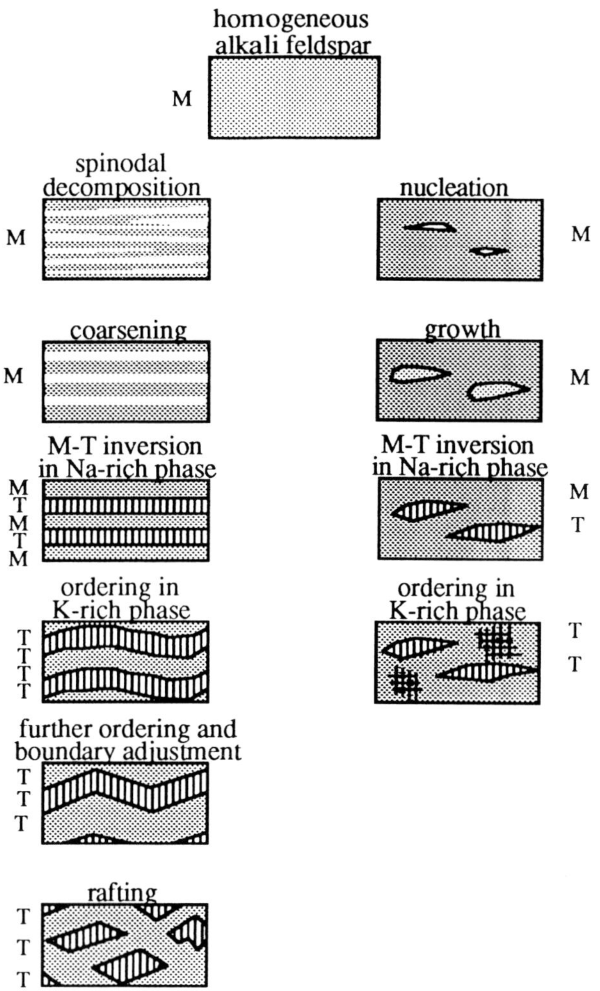

Figure 2-30. A schematic comparison of textures arising from spinodal decomposition (left) and nucleation and growth (right) in alkali feldspar. (After [61].)

approximately parallel to $(60\bar{1})$, which is near the (100) plane. Thus it is the *b* and *c* dimensions that must match most closely across the lamellar interface. The electron micrograph shown in Figure 2-29 shows a cryptoperthite in which the Na-rich component has twinned on the albite law in order to maintain pseudomonoclinic symmetry, and therefore better coherency, with the K-rich lamellae; the latter have subsequently become triclinic by (Al,Si) ordering, causing adjustment to the wavy interfaces. The sequence from homogeneous alkali feldspar to exsolved microstructure is shown schematically in Figure 2-30 for both spinodal decomposition and nucleation. Note that the microstructures are usually diagnostic of the exsolution mechanism that produced them. Composition and cooling rate determine the path taken and the stage attained. Figure 2-31 shows some electron micrographs and

a
b
1 µm

Table 2-1. Lattice Parameters (Å, °) of Unstrained Alkali Feldspar [30] and Strained Alkali Feldspar [62]

	Unstrained: $Or_{88}Ab_{11}Cn_1$	Strained: $Or_{91}Ab_6Or_3$
a	8.552(2)[a]	8.643(3)
b	12.975(3)	12.929(4)
c	7.205(1)	7.190(3)
α	90.09(2)	90.13(3)
β	115.92(1)	116.24(3)
γ	89.51(4)	89.60(3)

[a]Numbers in parentheses are one estimated standard deviation in the last decimal place cited.

their interpretations in light of Figure 2-30. Exsolution textures are sometimes very complex, reflecting complicated exsolution mechanisms (e.g., [12]).

It is apparent from the *b–c* plot (Figure 2-21) that, if Na-rich and K-rich lamellae are coherently intergrown, then the *b* and *c* dimensions of the K-rich phase must be compressed and those of the Na-rich phase must be expanded. Elasticity theory [66;76;86] then demands that the *a* cell edge of the K-rich phase expand to compensate for compression of *b* and *c*, and that it contract in the Na-rich phase for analogous reasons. This gives rise to strained feldspars (p. 68), those for which $\Delta a \geq 0.05$ Å. For potassic phases Δa is positive, and for sodic phases it is negative. A comparison of the crystal structure of a *s*trained *a*lkali *f*eldspar (*saf*) [62] and an *u*nstrained *a*lkali *f*eldspar (*uaf*) [30] of essentially the same compositions and structural states has revealed the ways in which the unit-cell distortions are accomplished by the crystal structure. Table 2-1 compares the unit-cell geometries of the two, and confirms the general predictions of elasticity theory.

Figure 2-32 shows a repeat along the *b* direction. Numbers alongside tetrahedra are differences in tetrahedral edge lengths, *saf* minus *uaf*. Numbers parallel to *b* are projections of those distances onto the *b* axis. The latter obviously add to the difference in *b* cell edges given in Table 2-1. The former add to zero, showing that the compression of *b* with strain cannot be attributed to changes in sizes of tetrahedra. The arrows indicate the sense of tetrahedral rotation in *saf* relative to *uaf*, and it is these tetrahedral rotations that account for shortening of *b* due to elastic strain.

Along the *c* cell edge (Figure 2-33) the situation is somewhat different. The only dimension parallel to *c* that is shorter in *saf* than in *uaf* is the T_1O tetrahedral dimension (−0.019 Å). When added to the other two distances parallel to *c* [that is, (0.010+0.002)/2 across the "square" and 0.001 Å at the T_1m site], this accounts

←

Figure 2-31. Transmission electron micrographs of textures similar to those shown in Figure 3-30. (a) A texture typical of spinodal decomposition. [Photo from G. W. Lorimer and P. E. Champness (1973). "The Origin of the Phase Distribution in Two Perthitic Alkali Feldspars." *Philosophical Magazine* 28, 1391–1403.] (b) A texture typical of nucleation and growth. [Photo from C. Willaime, W. L. Brown, and M. Gandais (1976). "Structural Aspects of Exsolution in Natural Alkali Feldspars." In H.-R. Wenk (ed.) *Electron Microscopy in Mineralogy*, 248–57. Springer-Verlag, Berlin.]

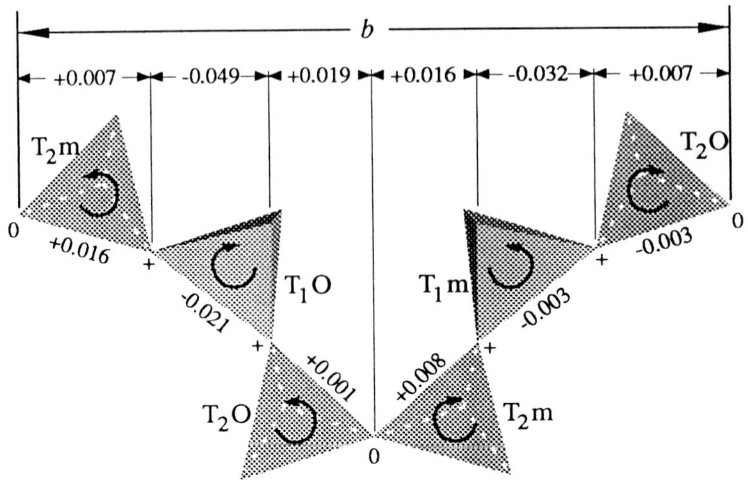

Figure 2-32. The *b*-axis repeat in alkali feldspar, showing the crystallographic effect of strain on that axis. Curved arrows show the direction of rotation of tetrahedra in a strained feldspar. Numerical labels show differences between corresponding distances and angles in strained and unstrained intermediate microcline. (After [30].)

for 80% of the observed contraction in *c*. The other 20% is explained by out-of-plane torsion of the "square." Thus the T_1O site undergoes a strain-generated compression. The *a* dimension in *saf* is expanded due to elastic strain required by coherency. Two factors accomplish this from a crystallographic point of view (Figure 2-34). Note that the differences in O···O···O intertetrahedral angles are all positive,

Figure 2-33. The *c* axis repeat in alkali feldspar. Numerical labels show differences between corresponding distances and angles in strained and unstrained intermediate microcline. (After [30].)

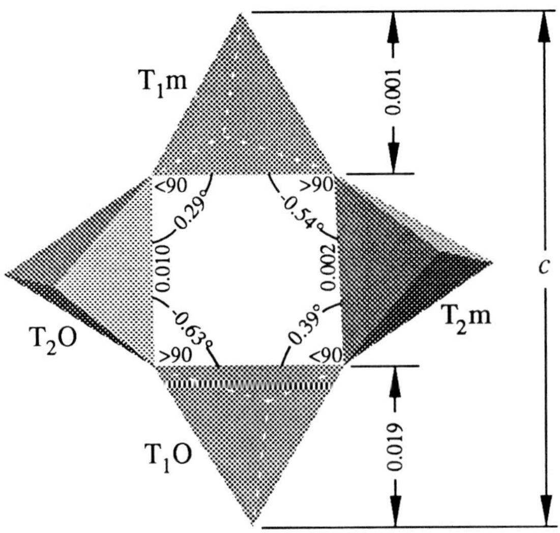

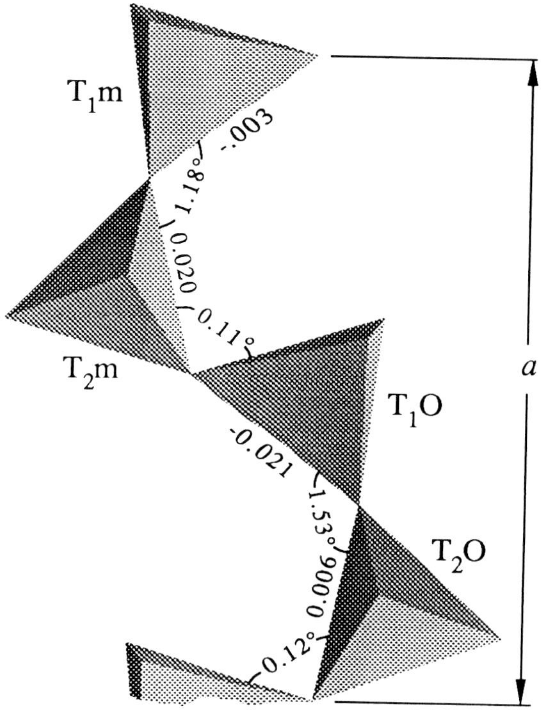

Figure 2-34. The *a* axis repeat in alkali feldspar. As in the previous two figures, numerical labels show differences between corresponding distances and angles in strained and unstrained intermediate microcline. (After [30].)

meaning that they are greater in *saf*. This amounts to a stretching of the double-crankshaft chain. Additionally, the two tetrahedral edges that most nearly parallel *a* are longer in *saf*, while those that are shorter are less nearly parallel to *a*, and so affect it less.

Evidence available at present suggests that the b–c and α^*–γ^* plots yield reasonable estimates of structural states for strained feldspars, and are as good as any other methods currently in use. There are undoubtedly inaccuracies, but they appear to be within or near the limits of precision of the method. Composition of strained feldspars should obviously not be estimated by the *a* cell parameter because the *a* cell edge is strained. Cell volume, on the other hand, is much less sensitive to strain because the changes in *b* and *c* are counteracted to a large degree by an opposite change in *a*. Equations (2.2), (2.3), and (2.4) also lead to a reasonable estimate of structural state for strained feldspars.

THE SPACE GROUPS OF FELDSPARS

The alkali feldspars have an [Al]:[Si] ratio of 1:3 and unit-cell dimensions of $a \sim 8.5$ Å, $b \sim 13$ Å, and $c \sim 7$ Å. If both geometry and topochemistry are monoclinic, the space group is $C2/m$. Loss of either monoclinic geometry or monoclinic topochemistry (or both) results in reduction of symmetry to space group $C\bar{1}$, but unit-cell dimensions remain nearly unchanged. The diagrams in Figure 2-35 show space-group symmetry elements and schematic representations of the double-crankshaft chains for $C2/m$ and $C\bar{1}$. These are the only space groups in which feldspars crystallize if the [Al]:[Si] ratio is 1:3.

In anorthite ($CaAl_2Si_2O_8$) and celsian ($BaAl_2Si_2O_8$) the [Al]:[Si] ratio is 1:1, and this limits the possible ordering schemes for tetrahedral sites. An empirical rule called the "aluminum avoidance principle" states that in a crystal structure containing both Al and Si in tetrahedral coordination aluminum-containing tetrahedra will tend not to share corners with one another [48]. One rationale behind this is that the presence of Al–O–Al linkages will create an unfavorable distribution of electrostatic charge. In a framework silicate with [Al]:[Si] ratio 1:1, the only way in which Al–O–Al linkages can be avoided is to have a perfect alternation of Al- and Si-containing tetrahedra; there can be no disordered (Al,Si) distribution. Figure 2-36 schematically shows aluminum-containing tetrahedral sites as light spheres, with oxygen atoms omitted for clarity. The repeat in albite is 7 Å; this is also true for sanidine, orthoclase, and microcline since an identical site, geometrically as well as chemically, is repeated every 7 Å. In anorthite and celsian, however, the alternation of Al and Si results in a topochemical repeat only every 14 Å. Figure 2-37 displays space-group symmetry elements and the double-crankshaft chain in a manner similar to that in Figure 2-28, but for the celsian and anorthite structures. The barium atom is not very different in size from potassium, so it prevents the displacive collapse we have seen in albite, and celsian is monoclinic. Calcium is about the same size as sodium; so anorthite is "collapsed" and this results in differ-

Figure 2-35. Space-group symmetry elements and the double-crankshaft chain for $C2/m$ and $C\bar{1}$ feldspars (with only tetrahedral sites shown). Dotted lines are unit cell boundaries, not glide planes. Atoms connected by heavy lines are at $y \approx 0.35$, those connected by light lines are at $y \approx 0.15$. Equivalent sites in each space group are indicated by intensity of shading. For space group $C\bar{1}$, Al is in the lightest site (T_1O). (After [63].)

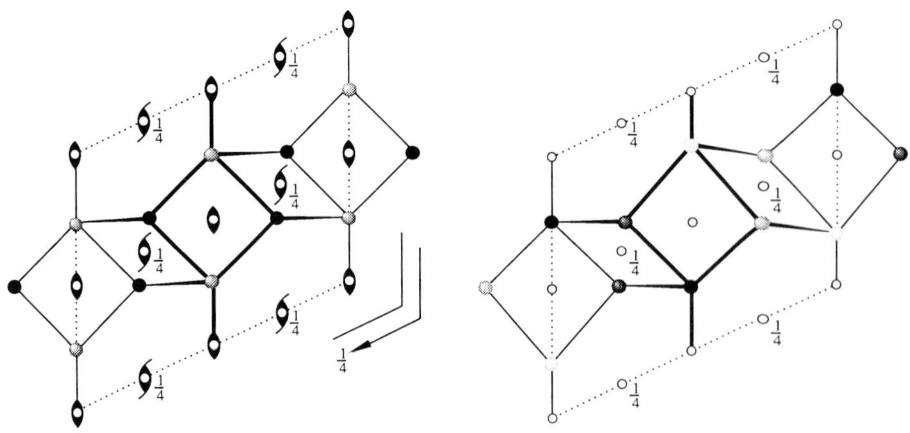

$C2/m$ $C\bar{1}$

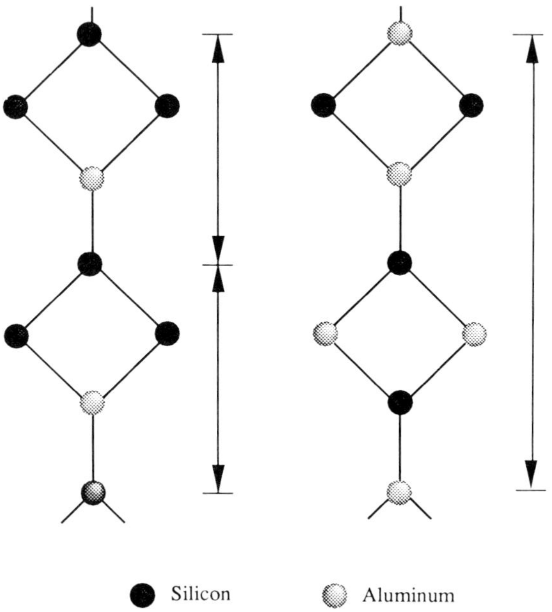

Figure 2-36. The difference in the 7 Å c repeat (left) in feldspars with 3:1 Si/Al ratios, and the 14 Å c repeat (right) in those with 1:1 Si/Al ratios.

ent geometries for the 4-membered rings of tetrahedra. Thus the body-centering lattice point is lost, and the space group is $P\bar{1}$. Rings A and C, and B and D, are topochemically similar and geometrically different, whereas the opposite is true of pairs A and D, and B and C.

A schematic diagram showing how lattice points are related from one space group to another is shown in Figure 2-38. No mention has thus far been made of space group $I\bar{1}$, depicted in this figure. It is the space group in which certain plagioclases of intermediate to calcic compositions crystallize, and more will be said about it in the following. For now, imagine translating a duplicate of the $P\bar{1}$ lattice from the origin to $(a+b+c)/2$, then superimposing it on the original lattice. The average of the two would be an $I\bar{1}$ cell. Just how this averaging comes about will be considered on p. 85.

Comparison of the schematic representations of $C2/m$ feldspars (Figure 2-35) and $P\bar{1}$ feldspars (Figure 2-37) shows that the numbers of symmetrically unique tetrahedral sites proliferate from 2 in $C2/m$ to 16 in $P\bar{1}$, though the unit cell volume only doubles. The reason is, of course, that symmetry elements disappear, so that tetrahedra that are equivalent in one structure are not in another. The total number of tetrahedra per given volume of feldspar does not differ from one structure to another, but the numbers of symmetrically different tetrahedra change. This leads to the potentially confusing site nomenclature shown in Figure 2-39. The O stands for original, and the other suffixes indicate the ways in which sites are related to original sites: m for mirror pseudosymmetry, z for a $c/2$ translational pseudosymmetry, and i for a body-centering translational pseudosymmetry.

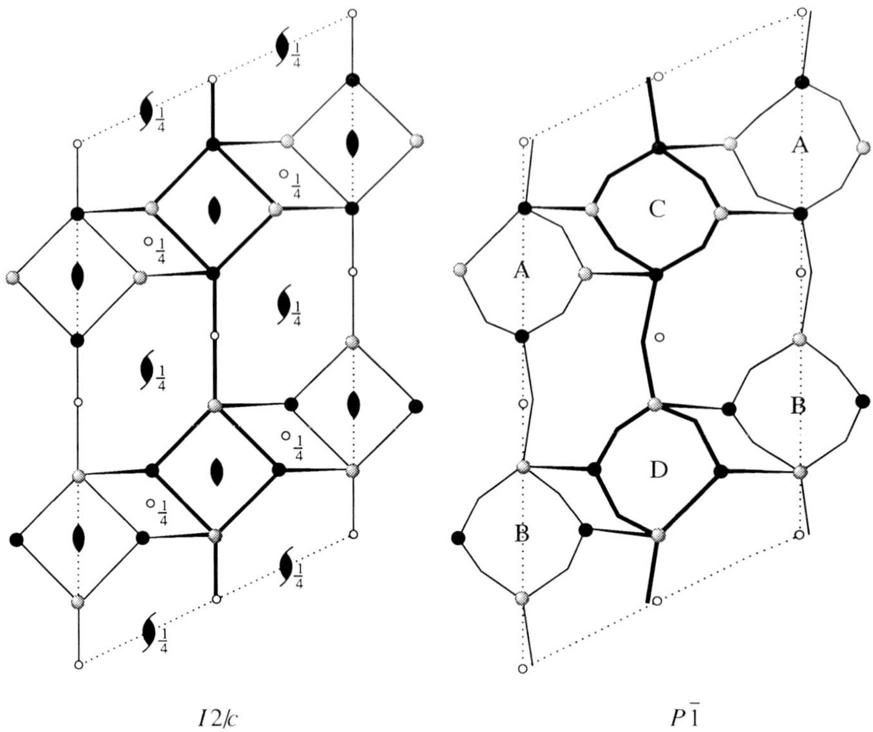

Figure 2-37. Space-group symmetry elements and the double-crankshaft chain for *I*2/*c* and $P\bar{1}$ feldspars. Dotted lines are unit-cell boundaries, not glide planes. Atoms connected by heavy lines are at $y \approx 0.35$, those connected by light lines are at $y \approx 0.15$. Darker spheres are Si and lighter ones are Al, which alternate following the Al-avoidance rule. Unlike Figure 2-35, shading here indicates only tetrahedral occupancy, not equivalence of sites. (After [63].)

Figure 2-38. Correspondence of lattice points in the feldspar space groups. Black circles represent lattice points, and gray circles represent pseudolattice points. (After [63].)

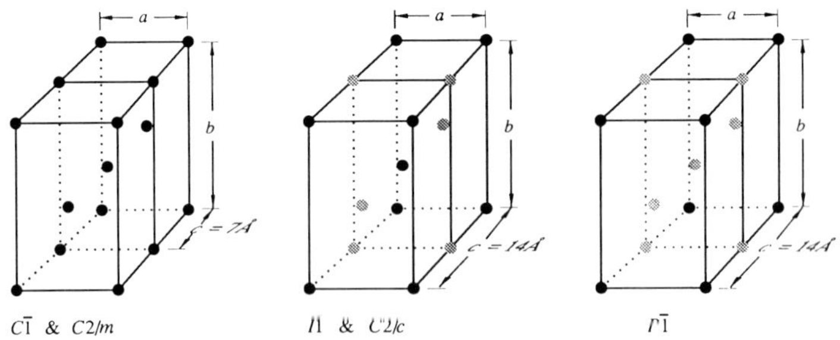

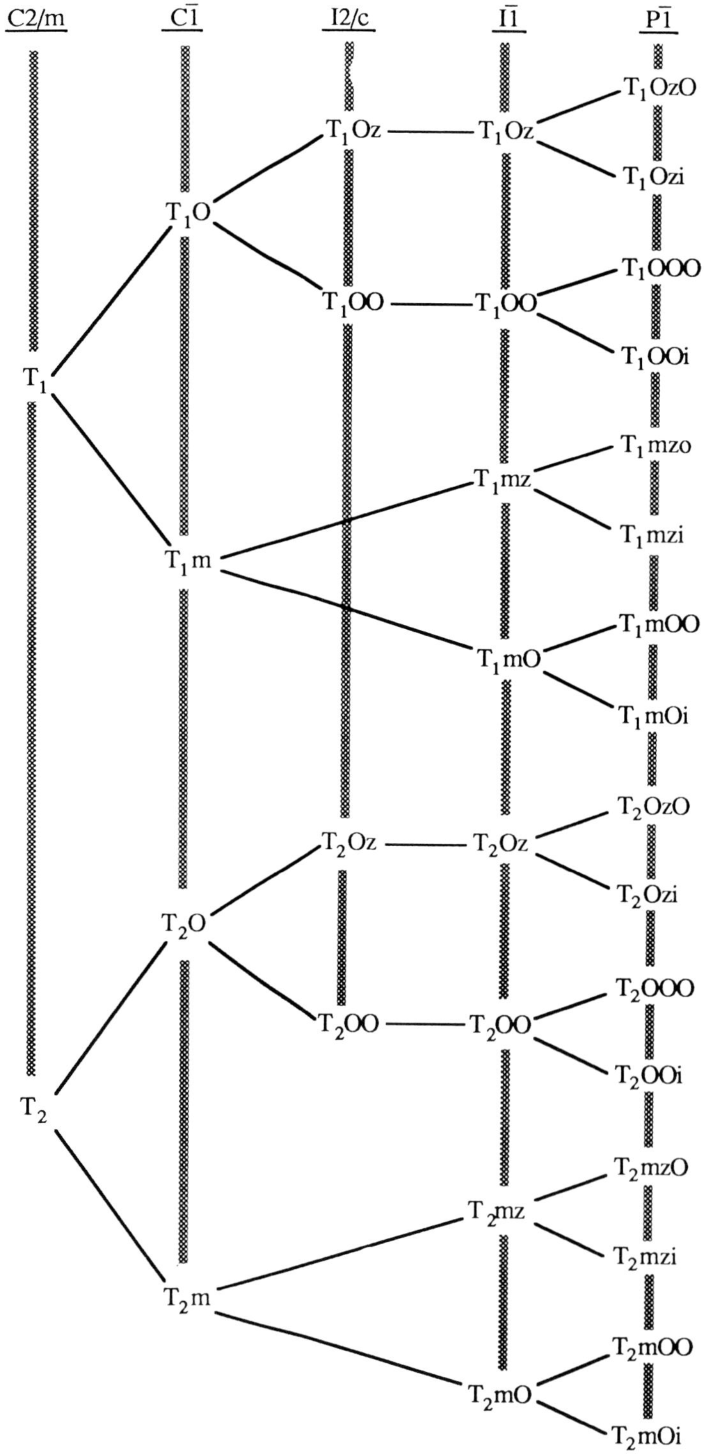

Figure 2-39. Tetrahedral site nomenclature in feldspars. O stands for original, and the lower case letters stand for various types of pseudosymmetrical relationships of sites to the "original" site: m stands for mirror pseudosymmetry, z for *C*-centering pseudosymmetry, and i for body-centering pseudosymmetry.

STRUCTURES OF END-MEMBER PLAGIOCLASES

We have already considered the crystal structure and polymorphism of albite as a member of the alkali feldspar series (pp. 53–56), so there is no need to go into detail here. Briefly, albite is monoclinic and disordered above ~950–1100°C (monalbite), triclinic and partially ordered just below that (high albite), and fully ordered (low albite) at temperatures perhaps as high as 650°C but certainly by 400°C. High albite is somewhat arbitrarily considered to have Δ131 values (p. 62) in the range 1.70–2.00, whereas low albites have Δ131 < 1.15. "High albites" with Δ131 between 2.00 and 2.03 invert directly to monalbite on heating, and thus are completely disordered and are called *analbites*. Albites of intermediate order have not been observed in nature, but have been synthesized and are apparently stable. (Interestingly, even though low albite departs more from monoclinic symmetry than does high albite, in the sense of having α and γ angles further from 90°, the effect on the difference in d spacings of the 131 and $1\bar{3}1$ planes is just the opposite: Albites closer to monoclinic geometry have greater Δ131 values. The apparent incongruity is caused largely by the greater b cell edge in high albite and the cumulative effects of other differences in unit-cell geometry.)

The other end member is anorthite, and at low temperatures and pressures its space group is $P\bar{1}$. The schematic topology and topochemistry are shown in Figure 2-37. Aluminum is essentially fully ordered into half of the 16 tetrahedral sites. As the temperature is raised to ~245°C (±10°C?), anorthite undergoes a reversible, apparently displacive, phase transition from $P\bar{1}$ to $I\bar{1}$ as thermal motion causes rings A and C to become geometrically identical, as well as rings B and D. The strict ordering of Al and Si is maintained. Conflicting interpretations of x-ray-diffraction and neutron-diffraction evidence have raised a question as to whether $I\bar{1}$ anorthite is truly body centered or actually primitive. X-ray-diffraction data (e.g., [24]) suggest that $I\bar{1}$ anorthite is the superposition, or average, of domains whose tetrahedral framework is essentially identical, but in which the Ca atoms occupy one of two slightly different positions in different domains. Since the positions are $P\bar{1}$ positions, $I\bar{1}$ anorthite would actually be the average of two kinds of $P\bar{1}$ domains if this interpretation were correct. Neutron-diffraction results [3;26] indicate that all of the atoms, including Ca, are at body-centered positions, so that $I\bar{1}$ anorthite is not made up of $P\bar{1}$ domains. The matter needs further investigation, but since neutron diffraction yields nuclear positions rather than electron-density maxima, it is likely that the strict $I\bar{1}$ interpretation is the more reliable one.

The existence of certain types of antiphase domains (see discussion of calcic bytownites to follow) suggests that $I\bar{1}$ anorthite inverts to $C\bar{1}$ with a 7 Å c cell edge at very high temperatures [37]; it is not likely a stable phase for the anorthite composition, inasmuch as it would require the (Al,Si) distribution to be disordered and the aluminum avoidance principle to be violated [54]. Nonetheless, a refinement of an anorthite quenched from 1530°C has revealed a slight preference of Al for "T_1O" (where "T_1O" here is the average of the four pseudosymmetrically related T_1O-type sites in anorthite) [17]; it appears that the partially ordered (Al,Si) distribution is a compromise between the aluminum-avoidance principle and charge-balance requirements imposed by a range of Ca–O distances.

Anorthite has been found to also undergo a pressure-induced phase transformation at 26.1 ± 0.6 kbar. It appears that this involves a transformation to a true body-centered cell in space group $I\bar{1}$ (as opposed to the average $I\bar{1}$ cell previously discussed) [4]. The mechanism for the transition is largely the tilting of tetrahedra that remain essentially rigid—that is, the *T*–O–*T* angles of the framework are modified while *T*–O bond lengths and O–*T*–O angles remain relatively fixed.

PLAGIOCLASES OF INTERMEDIATE COMPOSITION

We have seen (p. 78) that the ordering schemes of albite and anorthite are quite different, and we might expect one or the other of two possibilities for intermediate plagioclases: Either (1) each intermediate composition would have a unique ordering scheme that was some linear combination of the Ab and An arrangements, and that obeyed the aluminum avoidance principle, or (2) plagioclases of intermediate bulk compositions would consist of macroscopic exsolution intergrowths of albite and anorthite. Neither of these scenarios actually occurs. The first cannot, because the albite and anorthite ordering schemes are so different that it is impossible to generate linear combinations of the two that obey the Al avoidance principle at arbitrary bulk compositions. The second fails because of the very large kinetic barriers to the movement of Al and Si at subsolvus temperatures. (This problem is not encountered with alkali feldspars because Ab and Or have the same [Al]:[Si] ratio, so Al and Si do not need to exchange sites for exsolution to occur, and migration of alkali atoms presents little kinetic difficulty.)

What does occur appears to be alternative metastable behavior, and may be a kind of spinodal decomposition. A cooling plagioclase of intermediate composition "decomposes" on the 20–80 Å scale into regions of albitelike structure and regions of anorthitelike structure. The more calcic the bulk composition, the relatively wider are the anorthitelike domains. Figure 2-40 shows the general idea, although it is uncertain whether the domains have the form of wavelike modulations [72],

Figure 2-40. A schematic representation of An-rich and Ab-rich domains in plagioclases of various bulk compositions. (After [72].)

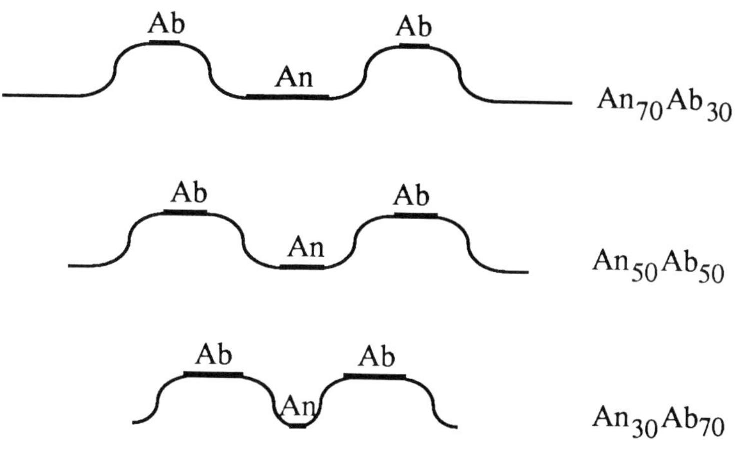

as shown, or whether they are more like alternating slabs [32;40]. A little consideration of the albite and anorthite ordering schemes will confirm that it is easier to replace some Al with Si in the anorthite scheme and not violate the aluminum avoidance principle than it is to replace any Si by Al in the albite scheme. It is therefore likely that deviation from the ideal end-member albite composition is much more restricted in the albitelike domains than is deviation from the ideal anorthite composition in the anorthitelike domains.

Calcic bytownites (An_{80-90}) are sufficiently calcic that their ordering patterns are based on that of anorthite (An_{90-100}). As the mineral orders from high temperature, either of two equivalent patterns of (Al,Si) distribution is possible and presumably equally likely, as depicted in Figure 2-41. As the two kinds of ordering microdomains meet, they are out of step, or out of phase, with each other. These are there-

Figure 2-41. Antiphase boundaries result when domains that have ordered out of phase with one another meet and there is a mismatch in ordering sequences in the tetrahedral sites.

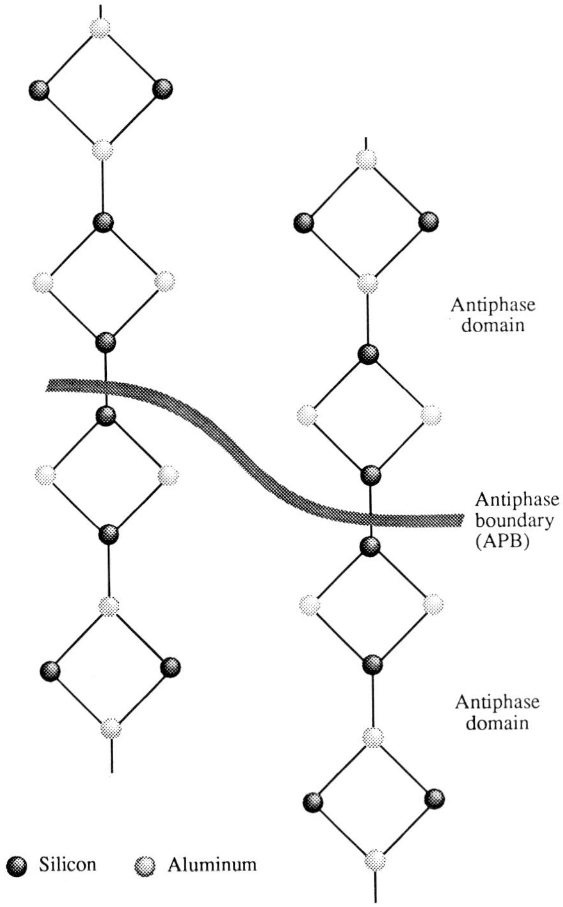

fore called *antiphase domains*, and the boundaries are called *antiphase boundaries* (APBs). (The α_1 and α_2 domains of quartz (p. 7) are really antiphase domains, although we did not call them that in Chapter 1.) At low temperature, each domain has the space group $P\bar{1}$, but averaging of the two domains superimposes the A and D, and B and C, rings (Figure 2-37). The result is truly primitive, but it appears body centered. Note that this apparent body-centering phenomenon is observed at room temperature, and should not be confused with the $I\bar{1}$ transformation in end-member anorthite, which occurs around 245°C, is reversible, and is probably a genuinely primitive–to–body-centered transition (p. 82).

Plagioclases with compositions between $An_{\sim 76}$ and $An_{\sim 25}$ are called *e plagioclases,* because of the so-called type *e* reflections (diffraction maxima) that appear on single-crystal x-ray and electron-diffraction photos (Figure 2-42). The *e* reflections are non-Bragg reflections produced not by planes of atoms in the unit cell, but by the slabs or modulations suggested schematically in Figure 2-40. Figure 2-43(a) is an electron diffraction pattern of an *e* plagioclase, and part (b) of the same figure is an electron micrograph that shows the periodicity causing the *e* reflections. Note the regularity of the spacing of the lamellae within a region of constant bulk composition.

The phase diagram for the plagioclase feldspars, shown in Figure 2-44, is schematic, but is probably reasonably close to correct. There are three kinds of intergrowths, indicated by Pe (*peristerite*), Bø (*Bøggild*), and Hu (*Huttenlocher*).

Peristerites

Many plagioclases whose bulk compositions are between An_2 and $An_{\sim 17}$ unmix in some way, probably spinodally, to $An_{1\pm 1}$ and $An_{24\pm 6}$. The calcic phase is an *e* plagioclase, and is thus itself an intergrowth. The initial stage of spinodal decomposition is on a very fine scale, and results in a "tweed" texture due to modulations parallel to (041) and (100). (The planes are indexed on the $c \approx 14$ Å unit cell.) The (100) modulation is weaker, and as the intergrowth coarsens, the (100) modulations disappear in favor of the (041) intergrowth. In peristerites of bulk composition $An_{8\pm 4}$, the lamellae may become coarse enough (500–1000 Å thick) to produce interference among light waves, resulting in iridescence. In the range An_{17-24} where the phase diagram suggests that exsolution is thermodynamically favored, it is not observed. This is presumed to be because of some sort of kinetic barriers. Transmission electron micrographs of peristerites (e.g., Figure 2-45) show the lamellae to be tabular with boundaries sharp and nearly parallel.

Huttenlocher intergrowths

At the other end of the plagioclase composition range are the Huttenlocher intergrowths, and these have much in common with the peristerites. One end member is a fully ordered plagioclase (in this case, body-centered anorthite solid solution), and the other is an *e* plagioclase with composition $An_{\sim 65}$. Again, the kinetics control the exsolution, so that the full range of thermodynamically possible "end mem-

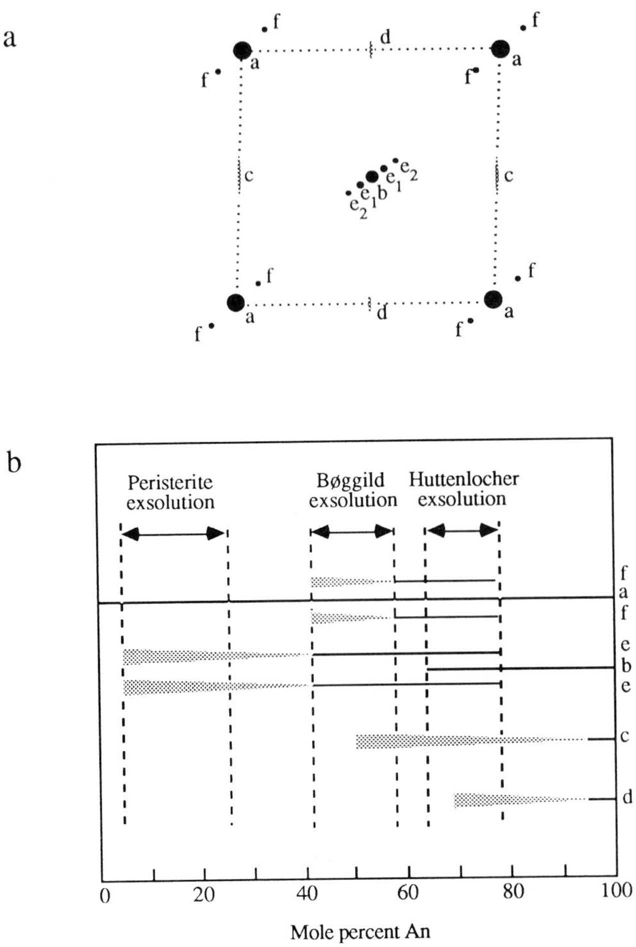

Figure 2-42. (a) A diagrammatic representation of all of the diffraction maxima observed in plagioclases. A given plagioclase crystal would not display all of these reflections. (b) Existence and diffuseness of the various reflections shown above as functions of composition of plagioclase. Solid lines indicate the presence of reflections shown at the right edge of the diagram, and shaded areas represent the presence of diffuse reflections, with narrower shaded regions indicating sharper reflections. (After [38].)

Figure 2-43. (a) An electron diffraction pattern showing various types of reflections diagrammed in Figure 2-42(a). (b) Transmission electron micrograph showing the modulated structure of an "e" plagioclase. The center of the photograph is slightly more sodic than the top and bottom. [Photos from T. L. Grove (1976) "Exsolution in Metamorphic Bytownite." In H.-R. Wenk (ed.) *Electron Microscopy in Mineralogy*, 266–70. Springer-Verlag, Berlin.]

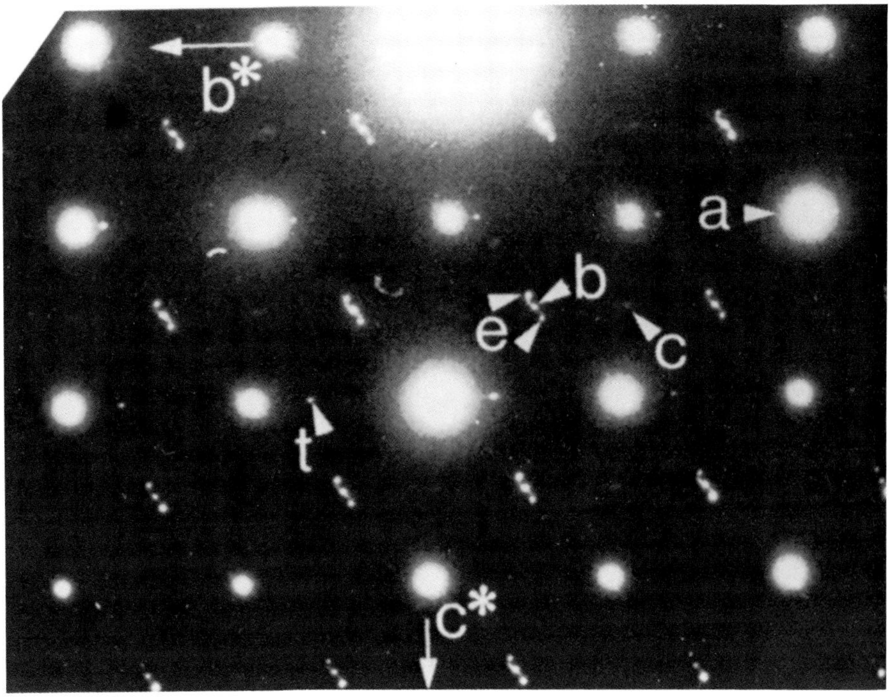

a

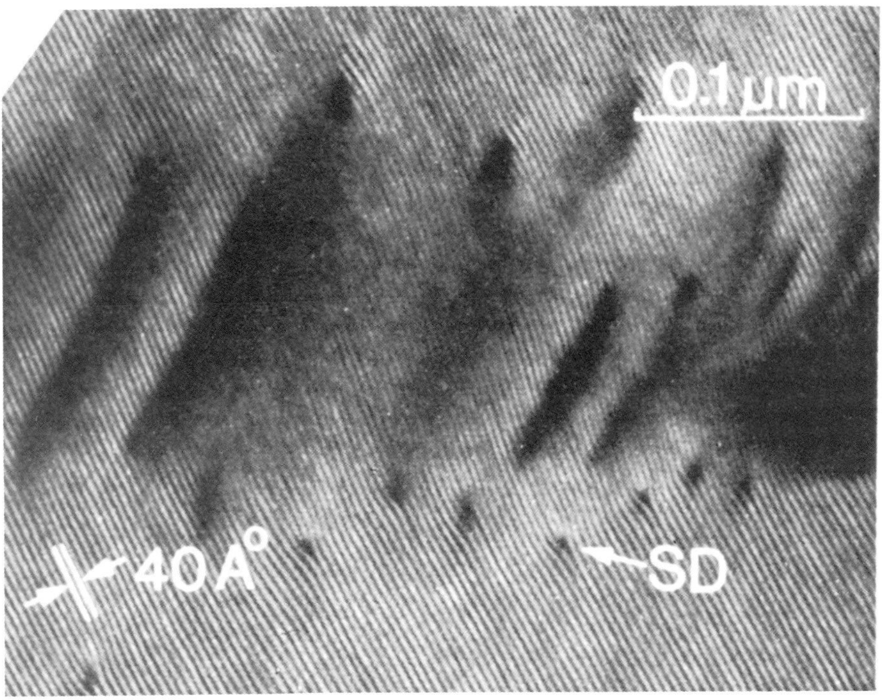

b

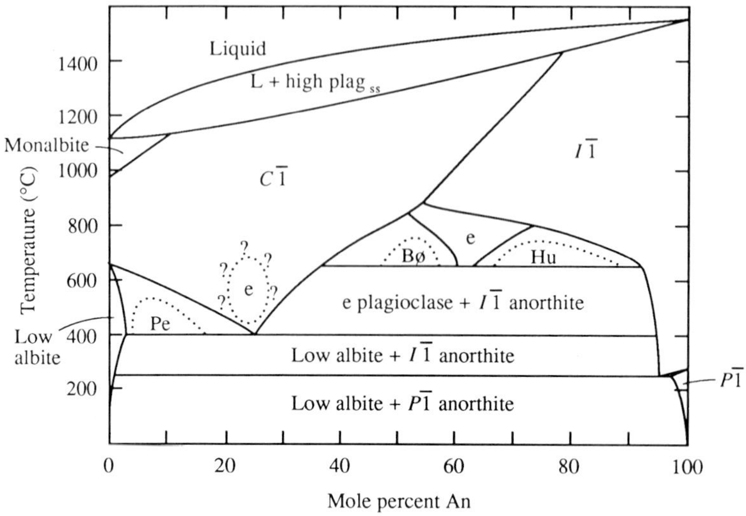

Figure 2-44. The phase diagram for the plagioclase system. The diagram is largely schematic, but may be close to reality. (Compiled from [13;14;15;34;70].)

Figure 2-45. Transmission electron micrograph of peristerite lamellae. [Photo from R. A. Yund (1986). "Interdiffusion of NaSi–CaAl in Peristerite." *Physics and Chemistry of Minerals* 13, 11–16.]

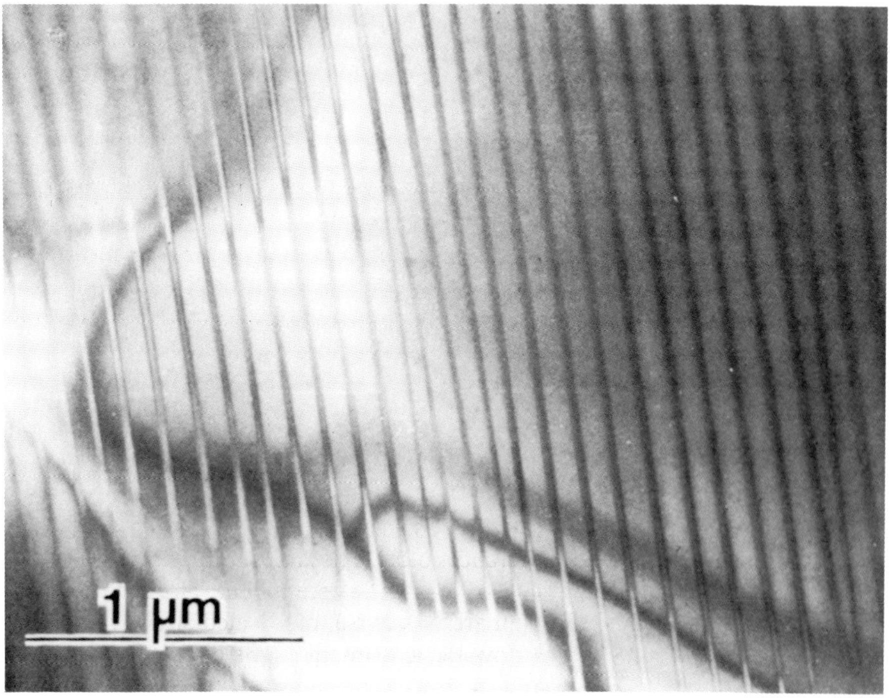

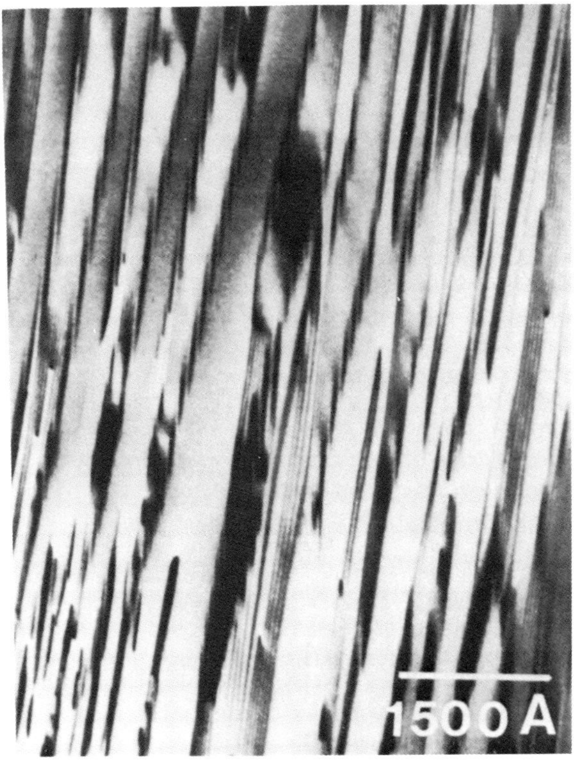

Figure 2-46. Transmission electron micrograph of a Huttenlocher intergrowth. [Photo from T. L. Grove, J. M. Ferry, and F. S. Spear (1983). "Phase Transitions and Decomposition Relations in Calcic Plagioclase." *American Mineralogist* 68, 41–59.]

bers" is not realized. As in peristerites, it appears that very fine-scale spinodal decomposition initiates the texture as two different modulations, but that only one set ultimately coarsens at the expense of the other. Huttenlocher intergrowths rarely display iridescent colors because the slabs of $An_{\sim 65}$ and $An_{\sim 90}$ are too thick. Transmission electron micrographs often show Huttenlocher intergrowths as less geometrically regular than peristerite intergrowths (Figure 2-46).

Bøggild intergrowths

In the range $An_{48 \pm 3}$ to $An_{58 \pm 3}$, exsolution to two *e* plagioclases takes place, and this produces the common iridescence ("labradorescence") of labradorite. The thicknesses of lamellae governs the interference colors. The lamellae appear to be coherent and to differ in composition by 10–12 mol % An component. The *e* plagioclases that constitute a Bøggild intergrowth are, of course, modulated structures themselves. Little is understood about how Bøggild intergrowths come about, but it would be surprising if it did not involve spinodal decomposition.

COMPOSITION AND STRUCTURAL STATES OF PLAGIOCLASES

The determination of structural state [i.e., (Al,Si) order] and composition from x-ray-diffraction data is not nearly as simple and straightforward for plagioclases as it is for alkali feldspars. We have seen that all low plagioclases of intermediate composition apparently consist of some complex intergrowth of ordered albitelike domains alternating with ordered domains of anorthitelike structure. If this is true, then all low plagioclases of any composition is fully ordered in two ways—part on the albite scheme and part on the anorthite scheme. If the two intergrown feldspars that constitute an intermediate plagioclase occur on such a fine scale that they are not resolvable, however, then the two structures and ordering schemes are superimposed, and the plagioclase does not *appear* to be ordered. No current analytical techniques permit the unambiguous determination of order in the two phases, and the methods generally available for routine examinations do not resolve the two physically distinct components at all. Thus whether a compositionally intermediate plagioclase may be ordered in its separate compositional domains becomes rather an academic question. For plagioclase feldspars then, the concept of a structural state will of necessity be applied to the somewhat artificial, but quite observable (and therefore practical) *average* structure, which will be described.

All feldspars may be considered as based on a $c \approx 7$ Å average unit cell, although ordering may double the true c cell dimension as in anorthite. The "average structure" of a plagioclase feldspar is the true unit cell (or unit cells, in an intergrowth) projected onto a ~7 Å albite-type unit cell. As an example, consider $P\bar{1}$ anorthite. We have seen that there are 16 tetrahedral sites in primitive anorthite, 4 each of the T_1O, T_1m, T_2O, and T_2m types (Figure 2-39). In each of these four sets of four sites, two sites contain Al and two contain Si. When the 14 Å anorthite cell is projected, or averaged, onto the 7 Å albite-type cell, the 16 sites degenerate to 4 distinct sites that each appear to contain 0.5 Si + 0.5 Al. Thus the *average* structure of $P\bar{1}$ anorthite is disordered, even though the true structure has perfect tetrahedral order. The average and true structures of low albite are, of course, the same, because albite really does have a ~7 Å unit cell; it thus has an ordered average structure. Between Ab and An, then, plagioclase average structures will show an amount of apparent disorder as required by the chemical composition. Moreover, the low structural state of every compositionally intermediate plagioclase will, in terms of the average structure, be different from that of every other plagioclase. Unlike with alkali feldspar, therefore, the observable structural state is not independent of composition, and the determination of composition is critical for the determination of structural state. [Note also that the disorder of analbite is properly reflected in its ~7 Å unit cell, so that the average structures of *all high plagioclases of any composition are disordered.*]

Determination of Composition

Because K and Na are very different in size and the a cell dimension is independent of structural state in alkali feldspars, the a cell edge or the unit cell volume can be used to determine composition in that series. For plagioclase, no function of lattice parameters or *d* spacings is useful for estimation of composition because (1) Ca and Na are nearly identical in size, and (2) composition and structural state are intimately related, as explained above.

Short of some direct chemical analytical technique like electron microprobe analysis, the most accurate (and, happily, the most convenient) method for assessing compositions of plagioclases is the measurement of either the α refractive index or α', the index of refraction closest to α measured on (001) cleavage fragments. Figure 2-47 provides a nomogram relating wavelength of light and measured α' index to composition. With a refractive-index measurement accurate to ± 0.001 using monochromatic light, compositions are accurate to $\pm 2\%$ An component. Refractive indices accurate to ± 0.001 are not difficult to measure, and the α' index

Figure 2-47. Plagioclase composition as a function of the α' index of refraction [measured on (001) cleavage fragments] and wavelength of light. Wavelengths for Na_D and "white" light are shown. (After [53].)

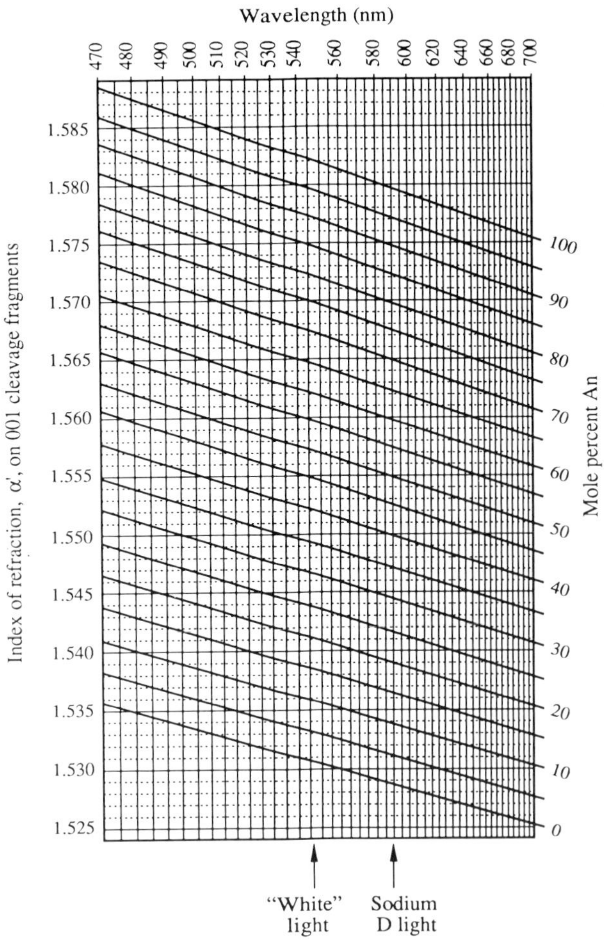

has been shown to be essentially independent of both structural state and minor Or content in the plagioclase.

Determination of Structural State

It will be recalled that $t_1o - t_1m$ for alkali feldspars could be determined from the α^*–γ^* plot (p. 64). The utility of that plot derived from the fact that α^* is primarily a function of composition, whereas γ^* is dependent mostly on structural state. Because of the similarity in sizes of Ca and Na, α^* loses its compositional dependence in plagioclases, but the γ^* angle is still a good indicator of (Al,Si) distribution. In fact, one would expect that, if γ^* is a good indicator of structural state, then γ, the direct-lattice angle between *a* and *b*, would be also. It turns out that γ is actually a more reliable indicator for plagioclases, because it is independent of Or content (substitution of K for Ca–Na), whereas γ^* is not. Figure 2-48 shows γ as a function of composition, and is contoured for $t_1o - \langle t_1m \rangle$, where $\langle t_1m \rangle = (t_1m + t_2o + t_2m)/3$. (In all plagioclases for which structure refinements have been done so far, the average structure has $t_1m \approx t_2o \approx t_2m$.) Since both γ and the α' index of refraction are independent of minor Or content, these may be used to determine composition and $t_1o - \langle t_1m \rangle$. We shall see in the following how to use $t_1o - \langle t_1m \rangle$ to get structural state, but first, a word about γ^*.

The foregoing would seem to suggest that γ^* need not be used as an indicator of structural state, but there is good reason not to abandon it altogether. There are

Figure 2-48. The γ crystallographic angle as a function of composition. The diagram is used to obtain $t_1o - \langle t_1m \rangle$, where $\langle t_1m \rangle = (t_1m + t_2o + t_2m)/3$. (After [44].)

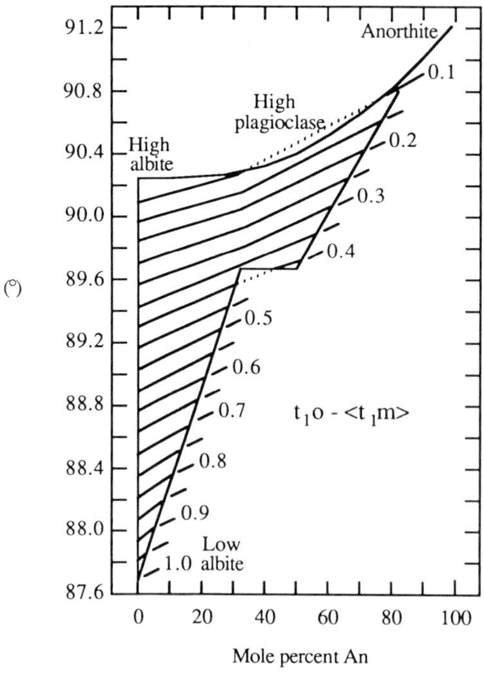

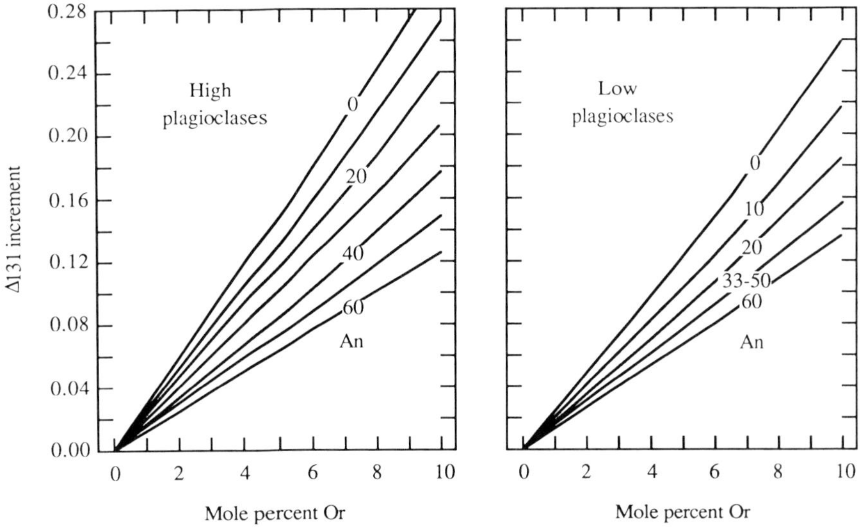

Figure 2-49. Diagrams for correcting Δ131 for Or content in plagioclases. Given Or content and bulk composition, the increment read on the ordinate is added to the Δ131 value obtained from an x-ray powder-diffraction experiment, and then used in Figure 2-50. (After [44].)

several easily measured parameters that are directly related to γ^*, the most convenient to obtain of which is $\Delta 131 = 2\theta(131) - 2\theta(1\bar{3}1)$. Measuring Δ131 with an x-ray powder diffractometer takes only a few minutes, and is thus much more easily done than obtaining lattice parameters. If Or content is known from an electron microprobe analysis or some other analytical technique, then Figures 2-49(a) or (b) can be used to find an approximate correction to Δ131 for the presence of potassium, and Figure 2-50 can be used to find $t_1o - \langle t_1m \rangle$. In terms of accuracy, the γ method is superior to the Δ131 method, and is recommended. For estimates where accuracy is less critical than convenience and speed, the Δ131 method may suffice. (For using the figures given here, it is important that Δ131 be measured with Cu$K\alpha$ radiation.)

Once $t_1o - \langle t_1m \rangle$ is determined by either of the above methods, structural state in the average 7 Å cell can be found. Total aluminum is 1.00 plus the mole fraction of An. (For example, $1.00+0.00=1.00$ for An_0, $1.00+1.00=2.00$ for An_{100}, and $1.00+0.66=1.66$ for An_{66}.) Also, since

$$\langle t_1m \rangle = (t_1m + t_2o + t_2m)/3 \approx t_1m \approx t_2o \approx t_2m,$$
$$t_1o - t_1m = \Delta Al,$$
$$\text{and } t_1o + 3t_1m = \Sigma Al$$

where ΔAl is read from contours on one of the figures (Figure 2-48 or 2-50), and $\Sigma Al = 1.00 + $ mole fraction An. Solving for t_1o and t_1m now becomes a simple matter of solving two simultaneous equations in two unknowns, and

$$t_1o = 0.25(\Sigma Al) + 0.75(\Delta Al),$$
$$t_1m \approx t_2o \approx t_2m = 0.25(\Sigma Al - \Delta Al).$$

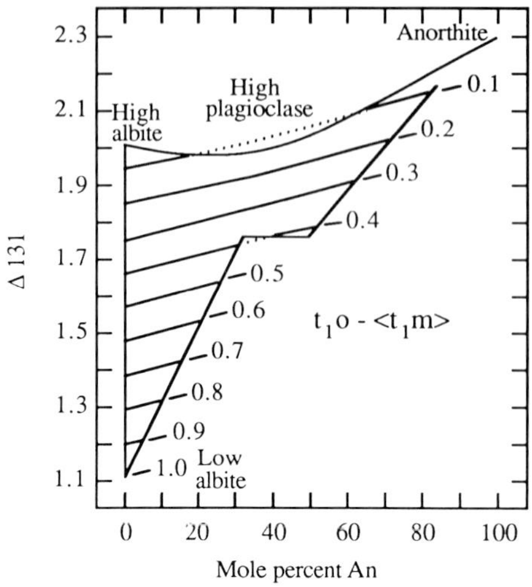

Figure 2-50. A diagram identical in purpose to Figure 2-49, but using Δ131 instead of the γ angle. (After [44].)

The data from which the γ versus An and Δ131 versus An plots were constructed are crystal-structure refinements of natural and synthetic high and low plagioclases. A somewhat surprising result of some of the recent work on high plagioclases is that they do not appear as disordered as they could be. This is supported by nuclear magnetic resonance (NMR) studies [84]. Note that, except for anorthite, the $t_1o - \langle t_1m \rangle$ contours yield values greater than 0.00. The interpretation for anorthite seems paradoxical at first glance, but makes sense if carefully thought through as follows: To obey the aluminum avoidance rule, anorthite must be fully ordered on a ~14 Å unit cell, but it will *appear* completely disordered when projected onto the ~7 Å unit cell. If high anorthite were actually partially *disordered*, that would result in differences in the Al occupancy of different average sites in the ~7 Å cell, $t_1o - \langle t_1m \rangle$ would not be zero, and the mineral would appear as if it were partially *ordered* on Figures 2-48 and 2-50. This is evidently the case for plagioclases in the composition range $An_{\sim 80-100}$, and pure anorthite is thus partly disordered between the crystallization temperature (1547°C) and ~1350°C [5].

Available evidence suggests that, like high albite, high *e* plagioclases are topochemically as well as geometrically triclinic on the ~7 Å cell. Thus, as indicated by Figures 2-48 and 2-50, they are not fully disordered.

BARIUM FELDSPARS

Because of their relative scarcity, much less work has been done with barium feldspars than with alkali feldspars or plagioclases, and there are consequently many gaps in the current understanding of these minerals. The barian end member is called *celsian* (abbreviated Cn), and members intermediate in composition between celsian and about $Cn_{15}Or_{85}$ are called *hyalophanes*. Feldspars with compositions more potassic than that are called *barian orthoclase* or *barian microcline*, depending on structural state. Natural hyalophanes tend to contain significant amounts of Na substituting for K, and of Fe^{3+} substituting for Al.

Celsian is, of course, similar to anorthite in composition, and therefore has a perfectly alternating arrangement of Al- and Si-containing tetrahedra, and a ~14 Å *c* cell edge. Because Ba is a large cation and prevents collapse of the tetrahedral framework, celsian (Cn_{100} to around $Cn_{85}Or_{15}$) is not triclinic, like anorthite, but monoclinic with space group *I*2/*c* (p. 78). Barian orthoclases (or barian microclines) have [Al]:[Si] ratios of essentially 1:3, and so they are monoclinic with space group *C*2/*m* and a ~7 Å *c* cell edge. The composition at which the *I*2/*c*-*C*2/*m* transition occurs is not known. Substitution of Na for K sometimes results in triclinic geometry [79].

Early studies of barium feldspars (e.g., [28]) indicated the possibility of a miscibility gap in the mid-composition range, but the paucity of specimens suggested that this might just be an artifact of sparse sampling. Recent observations [25;77;79] have confirmed, however, that barium-rich and barium-poor feldspars sometimes coexist either in exsolution relationships, or in mutual paragenetic contact. At low temperature, the edges of the solvus appear to be around $Cn_{85}Or_{15}$ and $Cn_{45}Or_{55}$, although some of the Or component is often replaced by Ab. The behavior at higher temperature is not known, although it is safe to assume that the solvus narrows. Additionally, observed exsolution [55] provides evidence for the existence of a narrow solvus at the barium-poor end of the compositional range; the limits are not yet well-defined, but may be something like $Or_{98}Cn_2$ for the barian microcline and $Or_{90}Cn_{10}$ for the hyalophane.

Determinative curves relating composition and structural parameters are not as well developed for barium feldspars as they are for other feldspars, largely because of the ternary (Ba,K,Na) nature of most natural hyalophanes. The unit-cell volume increases with Cn content, but the volume is of course also dependent on Ab concentration. The β angle decreases with increasing Cn concentration, and appears to be nearly independent of both the Ab component and structural state [28;79]; it may thus turn out to be a useful indicator of Cn content, but will not discriminate between Ab and Or in the remainder of the composition.

Like the plagioclases, in which the range of [Al]:[Si] ratios is the same as it is in the Or–Cn series, variation in structural state can be defined in terms of (Al,Si) order projected onto a ~7 Å *c* dimension (an "average" cell). Unlike the plagioclases, the barium feldspars have not been shown to consist of intergrowths of essentially end-member phases; a striking difference in orientations of the optical indicatrices of $Cn_{\sim 38}$ and $Cn_{\sim 90}$ suggests some very significant structural differences, however [65]. Celsian, with space group *I*2/*c*, is presumably fully ordered (see, however, the discussion in the following paragraph), and hyalophanes and barian orthoclase–barian microcline, with space group *C*2/*m*, can show a range of structural states from completely disordered (high) to ordered to the extent allowed by composition (low); with increasing Cn content of hyalophanes, even the most ordered specimens are evidently increasingly disordered [80]. Preliminary indications are that the *c* dimension holds the greatest promise as an indicator of structural state, but this has not been quantified adequately [78].

There is a change in optic axial angle with structural state [28], but the structural change causing the variation is not apparent; crystal-structure refinements of a $Cn_{39}Or_{43}Ab_{18}$ crystal before (low) and after heating (high) show very little difference in ordering [20]. This enigmatic result prompts some fundamental questions about the nature of both hyalophanes and celsians containing some Or (or Ab) component. Space group $C2/m$ implies that $t_1o = t_1m$, and that the aluminum is disordered. With >1 Al per formula unit, however, this makes the probability of Al–O–Al linkages high. Yet, as stated above, there is at present no evidence that a plagioclaselike intergrowth exists to alleviate violations of the aluminum avoidance principle. Structure refinements of ($I2/c$) $Cn_{85}Or_{15}$ [57] and $Cn_{95}Or_5$ [31] show apparent disorder in the (Al,Si) distribution beyond that required by the composition. In both, even after the Si in excess of 2.00 atoms is assigned to the Al-rich sites, there is an additional ~0.35 atom of Si in the Al sites and, correspondingly, ~0.35 atom of Al in the Si sites. Again, the aluminum avoidance principle seems violated. An alternative possibility is that there are antiphase domains that are averaged by structure refinement to yield *apparent* disorder that does not really exist. Whether such microstructures occur in barium feldspars remains to be determined.

REFERENCES

[1] Akizuki, M. (1981). "Crystal Growth and Phase Transition of Intermediate Microcline." *Neues Jahrbuch für Mineralogie, Monatshefte,* 181–89.

[2] Akizuki, M., and I. Sunagawa (1978). "Study of the Sector Structure in Adularia by Means of Optical Microscopy, Infra-Red Absorption, and Electron Microscopy." *Mineralogical Magazine* 42, 453–62.

[3] Aldhart, W., F. Frey, and H. Jagodzinski (1980). "X-ray and Neutron Investigations of the $P\bar{1}$–$I\bar{1}$ Transition in Pure Anorthite." *Acta Crystallographica* A36, 450–60.

[4] Angel, R. J. (1988). "High-Pressure Structure of Anorthite." *American Mineralogist* 73, 1114–19.

[5] Benna, P., G. Zanini, and E. Bruno (1985). "Cell Parameters of Thermally Treated Anorthite. Al,Si Configurations in the Average Structures of the High Temperature Calcic Plagioclases." *Contributions to Mineralogy and Petrology* 90, 381–85.

[6] Bernotat, W. H., and G. Morteani (1982). "The Microcline/Sanidine Transformation Isograd in Metamorphic Regions: Western Tauern Window and Merano-Mules-Anterselva Complex (Eastern Alps)." *American Mineralogist* 67, 43–53.

[7] Blasi, A., A. Brajkovic, and C. de Pol Blasi (1984). "Dry-Heating Conversion of Low Microcline to High Sanidine via a One-Step Disordering Process." *Bulletin de Minéralogie* 107, 423–35.

[8] Blasi, A., C. De Pol Blasi, and P. F. Zanazzi (1981). "Structural Study of a Complex Microperthite from Anatexites at Mt. Caval, Argentera Massif, Maritime Alps." *Neues Jahrbuch für Mineralogie, Abhandlungen* 142, 71–90.

[9] Brown, G. E., and P. M. Fenn (1979). "Structure Energies of the Alkali Feldspars." *Physics and Chemistry of Minerals* 4, 83–100.

[10] Brown, I. D. (1981). "The Bond-Valence Method: An Empirical Approach to Chemical Structure and Bonding." In M. O'Keefe and A. Navrotsky (eds.), *Structure and Bonding in Crystals*, Vol. 2. Academic Press, New York.

[11] Brown, W. L., R. E. Openshaw, P. F. McMillan, and C. M. B. Henderson (1984). "A Review of the Expansion Behavior of Alkali Feldspars: Coupled Variations in Cell Parameters and Possible Phase Transitions." *American Mineralogist* 69, 1058–71.

[12] Brown, W. L., and I. Parsons (1983). "Nucleation on Perthite-Perthite Boundaries and Exsolution Mechanisms in Alkali Feldspars." *Physics and Chemistry of Minerals* 10, 55–61.

[13] Carpenter, M. A. (1981). "A 'Conditional Spinodal' within the Peristerite Miscibility Gap of Plagioclase Feldspars." *American Mineralogist* 66, 553–60.

[14] Carpenter, M. A. (1986). "Experimental Delineation of the 'e'-$I\bar{1}$ and 'e'-$C\bar{1}$ Transformation in Intermediate Plagioclase Feldspars." *Physics and Chemistry of Minerals* 13, 119–39.

[15] Carpenter, M. A., and J. D. C. McConnell (1984). "Experimental Delineation of the $C\bar{1}$-$I\bar{1}$ Transformation in Intermediate Plagioclase Feldspars." *American Mineralogist* 69, 112–21.

[16] Cherry, M. E., and L. T. Trembath (1978). "Structural State and Composition of Alkali Feldspars in Granites of the St. George Pluton, South-Western New Brunswick." *Mineralogical Magazine* 42, 391–99.

[17] Chiari, G., A. Facchinelli, and E. Bruno (1976). "Anorthite Quenched from 1530°C. II. Discussion." *Acta Crystallographica* B34, 1757–1764.

[18] Collerson, K. D. (1976). "Composition and Structural State of Alkali Feldspars from High-Grade Metamorphic Rocks, Central Australia." *American Mineralogist* 61, 200–11.

[19] De Pieri, R., and S. Quareni (1973). "The Crystal Structure of an Anorthoclase: An Intermediate Alkali Feldspar." *Acta Crystallographica* B29, 1483–87.

[20] De Pieri, R., S. Quareni, and K. M. Hall (1977). "Refinement of the Structures of Low and High Hyalophanes." *Acta Crystallographica* B33, 3073–76.

[21] Donnay, G., and Allmann, R. (1970). "How to Recognize O^{2-}, OH^{-}, and H_2O in Crystal Structures Determined by X-rays." *American Mineralogist* 55, 1003–15.

[22] Eggleton, R. A., and P. R. Buseck (1980). "The Orthoclase-Microcline Inversion: A High-Resolution Transmission Electron Microscope Study and Strain Analysis." *Contributions to Mineralogy and Petrology* 74, 123–33.

[23] Fenn, P. M., and G. E. Brown (1977). "Crystal Structure of a Synthetic, Compositionally Intermediate, Hypersolvus Alkali Feldspar: Evidence for Na,K Site Ordering." *Zeitschrift für Kristallographie* 145, 124–45.

[24] Foit, F. F., and D. R. Peacor (1973). "The Anorthite Crystal Structure at 410 and 830°C." *American Mineralogist* 58, 665–75.

[25] Fortey, N. J. and B. Beddoe-Stephens (1982). "Barium Silicates in Stratabound Ba-Zn Mineralization in the Scottish Dalradian." *Mineralogical Magazine* 46, 63–72.

[26] Frey, F., H. Jagodzinski, and W. Prandl (1977). "Dynamic Character of the Primitive to Body-Centered Phase Transition in Anorthite. *Physics and Chemistry of Minerals*, 1, 227–31.

[27] Ferguson, R. B. (1980). "From Unit-Cell Parameters to Si/Al Distribution in K-Feldspars." *Canadian Mineralogist* 18, 443–58.

[28] Gay, P., and N. N. Roy (1968). "The Mineralogy of the Potassium–Barium Feldspar Series III: Subsolidus Relationships." *Mineralogical Magazine* 36, 914–32.

[29] Goldsmith, J. R., and D. M. Jenkins (1985). "The High–Low Albite Relations Revealed by Reversal of Degree of Order at High Pressures." *American Mineralogist* 70, 911–23.

[30] Griffen, D. T., and B. T. Johnson (1984). "Strain in Triclinic Alkali Feldspars: A Crystal Structure Study." *American Mineralogist* 69, 1072–77.

[31] Griffen, D. T., and P. H. Ribbe (1976). "Refinement of the Crystal Structure of Celsian." *American Mineralogist* 61, 414–18.

[32] Grove, T. L. (1977). "A Periodic Antiphase Structure Model for the Intermediate Plagioclases (An_{33} to An_{75})." *American Mineralogist* 62, 932–41.

[33] Grove, T. L., J. M. Ferry, and F. S. Spear (1983). "Phase Transitions and Decomposition Relations in Calcic Plagioclase." *American Mineralogist* 68, 41–59.

[34] Grove, T. L., J. M. Ferry, and F. S. Spear (1983). "Phase Transitions in Calcic Plagioclase: A Correction and Further Discussion." *American Mineralogist* 71, 1049–50.

[35] Harlow, G. E. (1982). "The Anorthoclase Structures: The Effects of Temperature and Composition." *American Mineralogist* 67, 975–96.

[36] Hazen, R. M. (1976). "Sanidine: Predicted and Observed Monoclinic–to–Triclinic Reversible Transformations at High Pressure." *Science* 194, 105–07.

[37] Heuer, A. H., and G. L. Nord, Jr. (1976). "Polymorphic Phase Transitions in Minerals." In H.-R. Wenk et al. (eds.), *Applications of Electron Microscopy in Mineralogy*. Springer-Verlag, Berlin.

[38] Horst, W. T. Tagai, and M. Korekawa (1981). "Modulated Structure of a Plagioclase An_{52}: Theory and Structure Determination." *Zeitschrift für Kristallographie* 157, 233–50.

[39] Hovis, G. L. (1980). "Angular relations of alkali feldspar series and the triclinic–monoclinic displacive transformation." *American Mineralogist* 65, 770–78.

[40] Kitamura, M. and N. Morimoto (1977). "The Superstructure of Plagioclase Feldspars. A Modulated Coherent Structure of the *e*-Plagioclase." *Physics and Chemistry of Minerals* 1, 199–212.

[41] Kroll, H. (1980). "Estimation of the Al,Si Distribution of Alkali Feldspars from Lattice Translations tr[110] and tr[$\bar{1}$10]. Revised diagrams." *Neues Jahrbuch für Mineralogie, Monatshefte*, 31–36.

[42] Kroll, H., and H.-U. Bambauer (1981). "Diffusive and Displacive Transformation in Plagioclase and Ternary Feldspar Series." *American Mineralogist* 66, 763–69.

[43] Kroll, H., H.-U. Bambauer, and U. Schirmer (1980). "The High Albite–Monalbite and Analbite–Monalbite Transitions." *American Mineralogist* 65, 1192–1211.

[44] Kroll, H., and P. H. Ribbe (1980). "Determinative Diagrams for Al,Si Order in Plagioclases." *American Mineralogist* 65, 449–57.

[45] Kroll, H., and P. H. Ribbe (1983). "Lattice Parameters, Composition and Al,Si Order in Alkali Feldspars." In P. H. Ribbe (ed.), *Feldspar Mineralogy*, 2nd ed. *Reviews in Mineralogy, Volume* 2. Mineralogical Society of America, Washington, D.C.

[46] Kroll, H., and P. H. Ribbe (1987). "Determining (Al,Si) Distribution and Strain in Alkali Feldspars Using Lattice Parameters and Diffraction-Peak Positions: A Review." *American Mineralogist* 72, 491–506.

[47] Kroll, H., I. Schmeimann, and G. von Cölln (1986). "Feldspar Solid Solutions." *American Mineralogist* 71, 1–16.

[48] Loewenstein, W. (1954). "The Distribution of Aluminum in the Tetrahedra of Silicates and Aluminates." *American Mineralogist* 39, 92–96.

[49] Martin, R. F. (1974). Controls of "Ordering and Subsolidus Phase Relations in the Alkali Feldspars." In W. S. MacKenzie and J. Zussman, (eds.), *The Feldspars.* Manchester University Press, Manchester.

[50] McConnell, J. D. C. (1971). "Electron Optical Study of Phase Transformations." *Mineralogical Magazine* 38, 1–20.

[51] Megaw, H. D. (1974). "The Architecture of the Feldspars." In W. S. MacKenzie and J. Zussman, (eds.), *The Feldspars.* Manchester University Press, Manchester.

[52] Merkel, G. A., and J. G. Blencoe (1982). "Thermodynamic Procedures for Treating the Monoclinic/Triclinic Inversion as a High-Order Phase Transition in Equations of State for Binary Analbite-Sanidine Feldspars." In S. K. Saxena (ed.), *Advances In Physical Geochemistry*, Volume 2. Springer-Verlag, New York.

[53] Morse, S. A. (1968). "Revised Dispersion Method for Low Plagioclase." *American Mineralogist* 53, 105–15.

[54] Müller, W. F., R. J. John, and H. Kroll (1984). "On the Origin and Growth of Antiphase Domains in Anorthite." *Bulletin de Minéralogie* 107, 489–94.

[55] Nakano, S. (1979). "Intergrowth of Barium Microcline, Hyalophane and Albite in the Barium-Containing Alkali Feldspar from the Noda–Tamagawa mine, Iwate Prefecture, Japan." *Mineralogical Journal* 9, 409–16.

[56] Nemec, D. (1978). "Obliquity of Potassium Feldspars in the Differentiation Series of Minettes." *Tshermaks Mineralogische und Petrographische Mitteilungen* 25, 63–69.

[57] Newnham, R. E., and H. D. Megaw (1960). "The Crystal Structure of Celsian (Barium Feldspar)." *Acta Crystallographica* 13, 303–12.

[58] Post, J. E., and C. W. Burnham (1987). "Structure–Energy Calculations on Low and High Albite." *American Mineralogist* 72, 507–14.

[59] Prewitt, C. T., S. Sueno, and J. J. Papike (1977). "The Crystal Structures of High Albite and Monalbite at High Temperatures." *American Mineralogist* 61, 1213–25.

[60] Prince, E., G. Donnay, and R. F. Martin (1973). "Neutron Diffraction Refinement of an Ordered Orthoclase Structure." *American Mineralogist* 58, 500–07.

[61] Putnis, A., and J. D. C. McConnell (1980). *Principles of Mineral Behaviour.* Elsevier, New York.

[62] Ribbe, P. H. (1979). "The Structure of a Strained Intermediate Microcline in Cryptoperthitic Association with Twinned Plagioclase." *American Mineralogist* 64, 402–08.

[63] Ribbe, P. H. (1983). "The Chemistry, Structure and Nomenclature of Feldspars." In P. H. Ribbe (ed.), *Feldspar Mineralogy*, 2nd ed. *Reviews in Mineralogy*, Volume 2. Mineralogical Society of America, Washington, D.C.

[64] Ribbe, P. H. (1983). "Aluminum–Silicon Order in Feldspars: Domain Textures and Diffraction Patterns." In P. H. Ribbe, (ed.), *Feldspar Mineralogy*, 2nd ed. *Reviews in Mineralogy*, Volume 2. Mineralogical Society of America, Washington, D.C.

[65] Roberts, M. J. T. (1975). "Indicatrix Orientation for Barium Feldspars Referred to Crystallographic Axes by Computer Using X-ray and Extinction Data." *American Mineralogist* 60, 1113–17.

[66] Robin, P.-Y. (1974). "Stress and Strain in Cryptoperthite Lamellae and the Coherent Solvus of Alkali Feldspars." *American Mineralogist* 59, 1299–1318.

[67] Salje, E. (1986). "Raman Spectroscopic Investigation of the Order Parameter Behaviour in Hypersolvus Alkali Feldspar: Displacive Phase Transition and Evidence for Na–K site Ordering." *Physics and Chemistry of Minerals* 13, 340–46.

[68] Salje, E., B. Kuscholke, B. Wruck, and H. Kroll (1985). "Thermodynamics of Sodium Feldspar II: Experimental Results and Numerical calculations." *Physicsand Chemistry of Minerals* 12, 99–107.

[69] Seck, H. A. (1971). "Koexistierend Alkali-Feldspäte und Plagioklase im System $NaAlSi_3O_8$–$KAlSi_3O_8$–$CaAl_2Si_2O_8$–H_2O bei Temperaturen von 650°C bis 900°C." *Neues Jahrbuch für Mineralogie Abhandlungen* 115, 315.

[70] Smith, J. V. (1974). *Feldspar Minerals*, Volume I. Springer-Verlag, Berlin.

[71] Smith, J. V., G. Artioli, and Å. Kvick (1986). "Low Albite, $NaAlSi_3O_8$: Neutron Diffraction Study of Crystal Structure at 13K." *American Mineralogist* 71, 727–33.

[72] Smith, J. V., and P. H. Ribbe (1969). "Atomic Movements in Plagioclase Feldspars: Kinetic Interpretation." *Contributions to Mineralogy and Petrology* 21, 157–202.

[73] Stewart, D. B., and P. H. Ribbe (1969). "Structural Explanation for Variations in Cell Parameters of Alkali Feldspar with Al/Si Ordering." *American Journal of Sciences* 267-A, 444–62.

[74] Stewart, D. B., and T. L. Wright (1974). "Al/Si Order and Symmetry of Natural Alkali Feldspars, and the Relationship of Strained Cell Parameters to Bulk Composition." *Bulletin de la Societe françqise Minéralogie et Cristallographie* 97, 356–77.

[75] Su, S. C., F. D. Bloss, P. H. Ribbe, and D. B. Stewart (1984). "Optic Axial Angle, a Precise Measure of Al,Si Ordering in T_1 Tetrahedral Sites of K-Rich Alkali Feldspars." *American Mineralogist* 69, 440–48.

[76] Tullis, J. (1975). "Elastic Strain Effects in Coherent Perthitic Feldspars." *Contributions to Mineralogy and Petrology* 49, 83–91.

[77] Viswanathan, K. (1978). "Intergrowth of Ba-Rich and Ba-Poor Phases in Barium Feldspars from Otjosondu, South-West Africa." *Mineralogical Magazine* 42, 294.

[78] Viswanathan, K. (1984). "The Behaviour of the Lattice Edges of Plagioclases and (Ba,K)-Feldspars." *Bulletin de Minéralogie* 107, 447–54.

[79] Viswanathan, K., and H.-M. Kielhorn (1983). "Variations in the Chemical Compositions and Lattice Dimensions of (Ba,K,Na)-Feldspars from Otjosondu, Namibia and their Significance." *American Mineralogist* 68, 112–21.

[80] Viswanathan, K., and H.-M. Kielhorn (1983). "Al,Si Distribution in a Ternary (Ba,K,Na)-Feldspar as Determined by Crystal Structure Refinement." *American Mineralogist* 68, 122–24.

[81] Weitz, G. (1972). "Die Struktur des Sanidins bei Vershiedenen Ordnungsgraden." *Zeitschrift für Kristallographie* 136, 418–26.

[82] Winter, J. K., S. Ghose, and F. P. Okamura (1977). "A High-Temperature Study of the Thermal Expansion and the Anisotropy of the Sodium Atom in Low Albite." *American Mineralogist* 62, 921–31.

[83] Winter, J. K., F. P. Okamura, and S. Ghose (1979). "A High-Temperature Structural Study of High-Albite, Monalbite, and the Analbite–Monalbite Phase Transition." *American Mineralogist* 64, 409–23.

[84] Yang, W.-H., R. J. Kirkpatrick, and D. M. Henderson (1986). "High-resolution ^{29}Si, ^{27}Al, and ^{23}Na NMR Spectroscopic Study of Al-Si Disordering in Annealed Albite and Oligoclase." *American Mineralogist* 71, 712–26.

[85] Yund, R. A., and J. Tullis (1983). "Subsolidus Phase Relations in the Alkali Feldspars with Emphasis on Coherent Phases." In P. H. Ribbe (ed.), *Feldspar Mineralogy*, 2nd ed. *Reviews in Mineralogy*, Volume 2. Mineralogical Society of America, Washington, D.C.

[86] Yund, R. A., and J. Tullis (1983). "Strained Cell Parameters for Coherent Lamellae in Alkali Feldspars and Iron-Free Pyroxenes." *Neues Jahrbuch für Mineralogie Monatshefte*, 22–34.

3
Micas

Micas are the most important rock-forming members of a structurally and chemically diverse group of minerals classified as sheet silicates (also called layer silicates or phyllosilicates) because their crystal structures contain "two-dimensional" parallel sheets that control most of their physical properties. While all sheet silicates do not possess topologically identical kinds of sheets, those in micas are all of essentially the same topology. They are much like tridymite sheets (Figure 1-7), except that all tetrahedra of a given sheet point in the same direction. Figure 3-1 shows a portion of the mica tetrahedral sheet. The oxygen atoms that are shared between tetrahedra are called bridging oxygens (O_{br}, because they form "bridges" between tetrahedra) or basal oxygens (because they form the bases of the tetrahedra that are parallel to the sheets), whereas the others are called nonbridging oxygens (O_{nbr}) or apical oxygens. At the centers of the hexagonal rings, and in the plane of the apical oxygens, are hydroxyls, OH. These are not bonded to components of the tetrahedral sheets in any way, but we shall nevertheless consider them part of the apical plane of anions.

A mica layer consists of two tetrahedral sheets stacked so that the apices of the tetrahedra point toward each other, but they are not vertically superposed. Rather, the "top" sheet is shifted laterally (i.e., parallel to the sheets) so that its apical oxygens and hydroxyls lie over triangular spaces between two apical oxygens and a hydroxyl of the "bottom" sheet. Figure 3-1 shows that the anions of the apical plane form a topologically close-packed layer (although they are not close enough to really be close packed), so that when two tetrahedral sheets are stacked, the same kinds of octahedral and tetrahedral voids are found between them as are found between close-packed layers. In the sheet silicates the tetrahedral voids of this intermediate layer are empty, but there are cations in some or all of the octahedral voids; this layer, formed by the juxtaposition of apical oxygens, is thus called the *octahedral sheet.* It is shown in Figure 3-2, with **m** indicating a mirror plane in the anion array. The mica layer, then, consists of a tetrahedral–octahedral–tetrahedral sequence, sometimes called a *T–O–T* layer, or even a *T–O–T* "sandwich." It is important to recognize that both the tetrahedral and octahedral sheets are produced by stacking only four, not six, planes of anions and filling appropriate voids with cations; the apical oxygens of the tetrahedral sheets are the very same oxygens that,

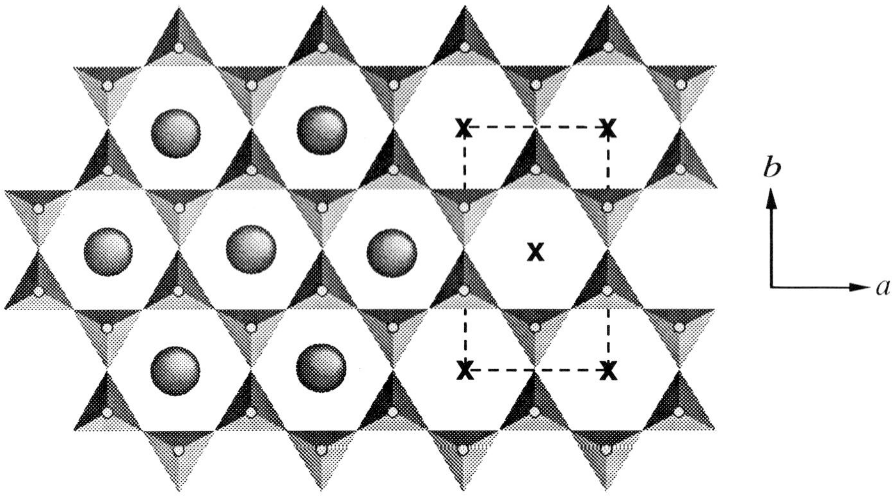

Figure 3-1. The tetrahedral sheet of micas, viewed along **c***. Shaded circles represent hydroxyl groups (OH) in the plane of the apical oxygens. (Apical oxygen atoms are not drawn to scale.) The outline of the unit cell is indicated on the right side of the figure, with lattice points represented by Xs.

with the hydroxyls, bound the octahedral sheets. Therefore, as we shall see later, any change in the tetrahedral sheets results in a change in the octahedral sheets, and vice versa. Because the mica layer contains two tetrahedral sheets and one octahedral sheet, micas are called 2:1 layer silicates; this distinguishes them from some clays, chlorites, and other sheet silicates that have different T:O sheet ratios.

Figure 3-2. The octahedral sheet of micas, viewed along **c***. The sheet comprises two parallel apical oxygen planes, including hydroxyl groups. Three of the octahedra so formed are shown at the right. The ideal sheet has mirror symmetry.

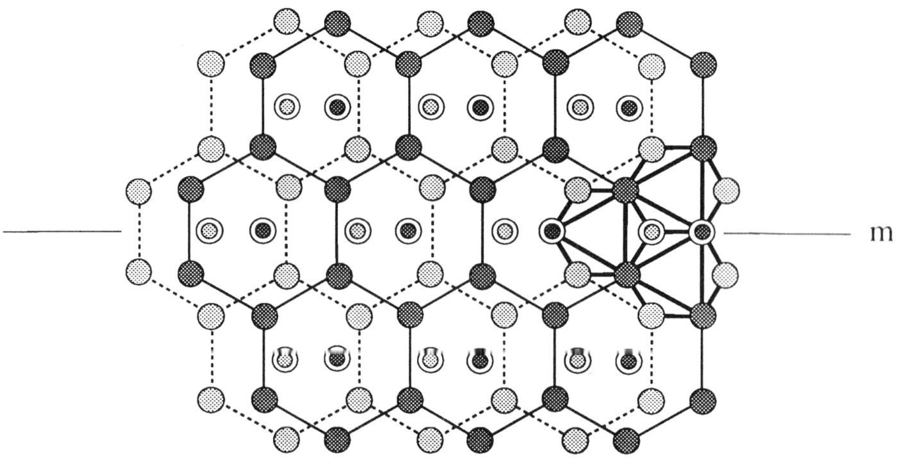

Note the precise terminology of the preceding paragraphs. A monolayer of atoms is referred to as a *plane.* A layer of linked polyhedra, made by stacking two anion planes and filling some of the voids with cations, is called a *sheet.* A *layer,* in this terminological scheme, is made up of linked sheets. A layer plus the interlayer atoms that bind two layers together is collectively named a *unit structure.* This has been internationally approved terminology since 1972, and should be used consistently to avoid confusion. Fortunately, context usually makes meaning clear when dealing with micas, even if the conventions are ignored. In particular, because *unit structure* is a somewhat cumbersome phrase, the term *layer* is often used to refer either to a layer, *sensu stricto,* or to a unit structure.

Because the O_{nbr} and OH atoms do not stack directly atop their counterparts in the sheet below, the hexagonal symmetry of the ideal tetrahedral sheets is not the actual symmetry of the mica layer. The offset of the sheets produces monoclinic symmetry, with a and b axes as shown in Figure 3-1. Note that a lattice point at the center of each projected ring makes the lattice C centered, so that the "space group" of the ideal mica layer is $C2/m$. Ordering or geometrical distortion could lower the symmetry of either a single layer or a stack of layers, and stacking of unit structures in special ways could result in higher symmetry than monoclinic.

There are two kinds of octahedral sites in the mica layer (Figure 3-3). One lies astride the mirror plane, and the other lies adjacent to it. The one on the mirror plane has hydroxyls on opposite corners, and is called a *"trans site"* (a term borrowed from organic chemistry). The other has hydroxyls on adjacent corners, and is called a *"cis site"* (thanks, again, to the organic chemists). Because *cis* sites in a $C2/m$ mica layer must occur in pairs, owing to the mirror plane, there are three octahedral sites available for occupancy. (On this basis alone, we might expect some difference in cation preference for the sites, and we shall investigate that possibility later.) Conventionally the *trans* sites are designated $M1$ and the *cis* sites $M2$, M standing for "metal atom." Unnecessary (and fortunately infrequent) confusion exists in the mineralogical literature because a few authors have reversed these designations, apparently in an attempt to make mica site nomenclature somehow more consistent with that of amphiboles. Such "consistency" is only partial because their mica "$M2$" then corresponds not to amphibole $M2$, but to amphibole $M3$.

Figure 3-3. The *trans* and *cis* sites of the octahedral layer. The former are called $M1$ sites, and the latter are called $M2$ sites. Hydroxyls are shaded, and oxygens are unshaded.

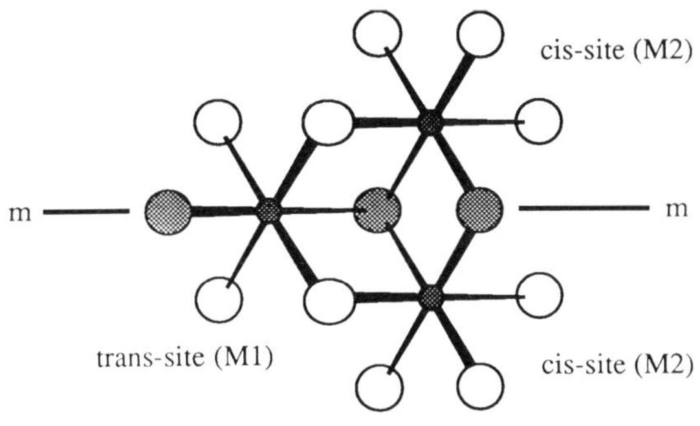

Such an exercise in the face of a well-established precedent for site names in micas seems pointless and confusing. We shall use the conventional designations cited above.

To provide some idea of mica unit-cell geometry, it is easy to calculate "ideal" lattice parameters based on regular six-membered tetrahedral rings. Referring again to Figure 3-1, it can be seen that the a and b axes are easily identified (b being necessarily perpendicular to the mirror plane in the second setting of the monoclinic system, and Figures 3-1 and 3-2 being oriented identically). Assuming that the oxygen anions "touch," the a cell dimension is four times the radius of an oxygen; the latter is about 1.32 Å, so $a \approx 5.28$ Å. In the b direction the cell edge can either be considered as the tangent of 60° times the a cell edge, or four times the height of the base of a tetrahedron. Either way, $b \approx 9.14$ Å. The c cell dimension can be calculated only if we know how the mica layers stack. Figure 3-4 shows that large cations of low oxidation number—the interlayer cations—hold one mica layer to another by fitting into the hexagonal (or at least *ideally* hexagonal) depressions in the backs of the layers. The bottom tetrahedral sheet of one mica layer, then, stacks above the top tetrahedral sheet of the mica layer below, with interlayer cations between. Recall, however, that the tetrahedral sheets of a single mica T–O–T layer are offset. From Figure 3-4, the offset can be seen to be $a/3$, and this offset is characterized by a vector called the *stacking vector*. The thickness of the mica layer can be calculated

Figure 3-4. The mica structure viewed down the **b** direction.

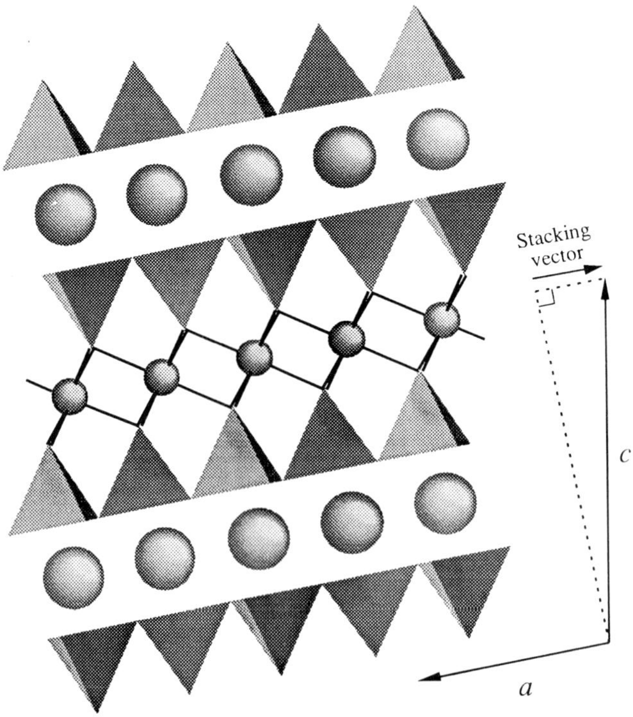

assuming hard-sphere ions and a standard set of ionic radii (e.g., from [54]; including the radii of the interlayer cations on either side, it is ~9.8 Å, so that $c \approx [(9.8)^2 + (a/3)^2]^{1/2} \approx 9.96$ Å. The β angle is $\{90° + \sin^{-1}[(a/3)/c]\} \approx 100°$. Simplistic as this "hard-sphere" model of mica unit-cell geometry is, it gives results that agree within about 2% with observed values.

MICA CHEMISTRY

The general formula for the micas is $W_1Y_{2-3}Z_4O_{10}(OH,F,Cl)_2$, where

$$W = \text{K, Na, Ca, Ba, Rb, } H_3O^+\text{, Li}$$
$$Y = \text{Al, Mg, } Fe^{2+}\text{, Li, } Mn^{2+}\text{, } Fe^{3+}\text{, Ti, } Mn^{3+}\text{, Ni, Co}$$
$$Z = \text{Si, Al, } Fe^{3+}\text{, B, Be, Ge, Ga, Mg, } Fe^{2+}.$$

This long list of elements contains not only the most common found, but some that occur only in synthetic micas, or only rarely or in trace amounts in natural ones. W represents the interlayer cation, and potassium is by far the most common. Micas in which the interlayer cation is either K or Na are called *common micas;* those in which it is Ca (the only other reasonably common interlayer cation) are called *brittle micas,* because the increased interlayer charge renders the mica sheets brittle rather than elastic. Barium micas are rare, but not unknown; neither rubidium nor hydronium (H_3O^+) has ever been found as a major interlayer cation in a natural mica. Lithium may substitute for potassium to a substantial extent in the interlayer regions of micas—particularly those with a high tetrahedral aluminum content—but it does not do so in the actual interlayer (K) site; rather, it is thought [47] to occupy a pseudo-octahedral site located between the bases of tetrahedra in opposing layers.

Y stands for the octahedrally coordinated cations. If there are two of them in the formula, then two of the three available octahedral sites are filled and the mica is said to belong to the *dioctahedral group.* The octahedral sheets of dioctahedral micas are, except for the substitution of oxygen for some of the hydroxyls, topologically like the sheets that constitute the mineral gibbsite, $Al(OH)_3$, so they are often called *gibbsite sheets.* If there are three Y cations in the formula, then all of the octahedral sites are filled and the mineral is a member of the *trioctahedral group.* Because the octahedral sheets of trioctahedral micas resemble, in topology, the mineral brucite, $Mg(OH)_2$, they are often called *brucite sheets.* Nearly all natural micas have formulas close to either the dioctahedral or trioctahedral ideals, and an extensive simple solid solution between the two does not appear to exist for any compositions. The only notable exceptions are the lithium micas (meaning, in this case, those with Li in the *octahedral* sheet), in which values of total octahedral occupancy between 2.5 and 3.0 are common. The most common Y cations in natural micas are the first five in the list above; minor Fe^{3+} often substitutes for octahedral Al, and Ti is likewise also a fairly common minor substituent.

Z sites contain tetrahedrally coordinated atoms, generally only silicon or aluminum, with [Si] $\geq$ [Al] for most micas. Fe^{3+} as a trace or minor tetrahedral substituent is not uncommon, and B and Be have also been found as major tetrahedrally coordinated elements in natural specimens.

In addition to cationic substitutions, many micas exhibit substitutions for OH in the hydroxyl site of the apical oxygen plane. In some cases the substituent anions are F^- or Cl^-, but for many micas the sum of OH, F, and Cl is considerably less than 2.0, on the basis of 12 anions. For these micas, O^{2-} has replaced some OH. This may occur by oxidation of some Fe^{2+} to Fe^{3+}, requiring loss of H^+ to balance charges, or it may be a response to excess positive charge in the tetrahedral or octahedral sheets, caused by substitutions of Si for Al, or (Al,Fe^{3+}) for Mg. Figure 3-5 shows how the (OH,F,Cl) content of biotites and phlogopites varies with respect to total cation charge, which is ideally 21.0 in either of these micas. It is apparent from the figure that some biotites and phlogopites have excess (OH,F,Cl), although this is not as common as a deficiency.

Substitution of fluorine for hydroxyl is more extensive in trioctahedral than in dioctahedral micas, and may be essentially complete in fluorine-rich phlogopite (fluorphlogopite). In muscovites, the ratio of fluorine to (OH,F,Cl) reaches only about 0.2. For chlorine substitution, the analogous ratio for biotites is rarely over 0.1, and is an order of magnitude less for muscovites. The extent of halogen substitution for OH is obviously dependent upon the concentrations of halogens available during the crystallization of a given mica, but there are several crystal chemical factors that are also of great importance. They fall into two categories, structural and chemical, and these categories are interrelated. We shall see in our discussion of distortions of the ideal mica sheet that the tetrahedral sheet is "crimped" to form rings with ditrigonal rather than hexagonal shape. While it is premature to go into detail on this matter, it is sufficient at this point to state that the room available for the OH or its substituents decreases as the ditrigonal distortion increases. The amount of such distortion is related to the chemistry of the mica in that the distortion decreases with increasing size of the octahedral cations and with decreasing size of the tetrahedral cations. Because chlorine is substantially larger than either hydroxyl or fluorine, its substitution into the distorted structure is qualitatively gov-

Figure 3-5. Anionic substitutions in biotite and phlogopite as a function of layer charge. (After [13].)

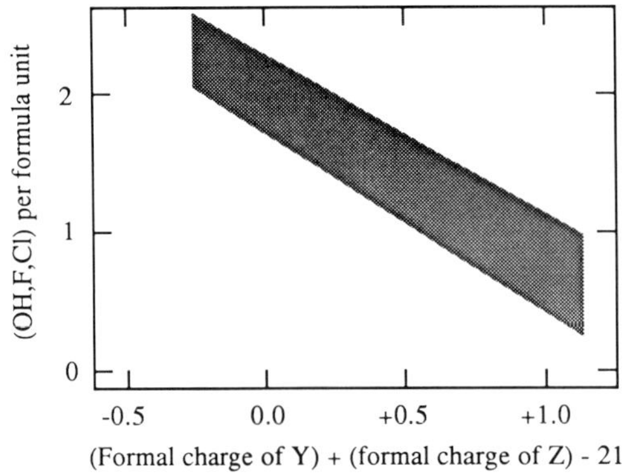

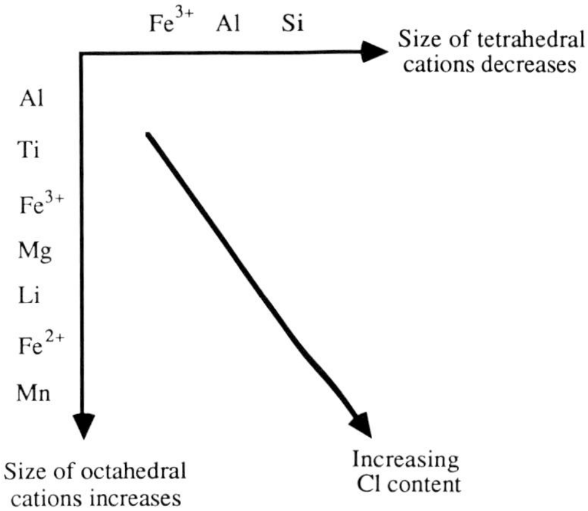

Figure 3-6. The effect of changing the sizes of tetrahedral or octahedral cations on chlorine content of micas. (After [64].)

erned as shown in Figure 3-6; the more the distortion (later to be referred to as *tetrahedral rotation*), the less chlorine can be incorporated.

Figure 3-7 shows that fluorine concentration increases and chlorine concentration decreases with Mg content for some biotites of a porphyry copper deposit, and this is true of biotites in general. Because Fe is the other principal cation in the octahedral sheet of biotite, this figure implies opposite trends relating halogen concentration to Fe content. These observations are sometimes referred to as the "Fe–F

Figure 3-7. Variation in the amount of fluorine and chlorine as a function of cation chemistry for a suite of biotites. (After [39].)

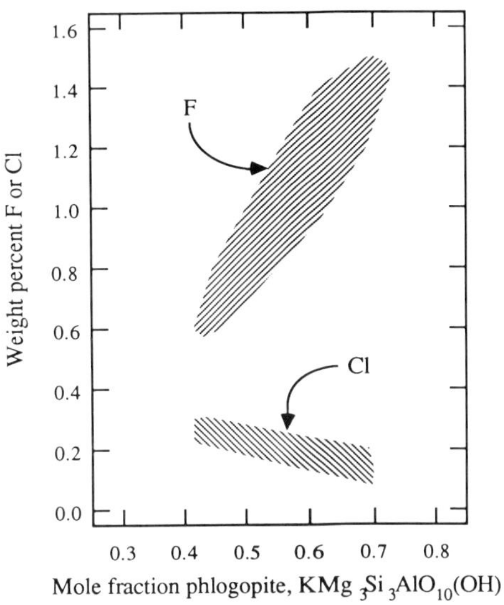

avoidance" and the "Mg–Cl avoidance" rules, and they seem to have broad application to F- and Cl-bearing silicates. Calculations based on thermodynamic data for the model reactions

$$FeF_2 + Mg(OH)_2 = Fe(OH)_2 + MgF_2, \qquad \text{and}$$
$$FeCl_2 + Mg(OH)_2 = Fe(OH)_2 + MgCl_2$$

yield a positive equilibrium constant for the first reaction and a negative one for the second [39]. This indicates that Mg–F bonds are stronger than, and thermodynamically favored over, Fe–F bonds, and that Fe–Cl bonds are stronger than, and favored over, Mg–Cl bonds. A secondary effect may be related to details of the crystal structure. Increasing the iron content of the octahedral sheet in biotites enlarges it, and we shall later learn that the larger the octahedral sheet, the less the ditrigonal distortion of the tetrahedral layer; inasmuch as less ditrigonal distortion corresponds to a larger hydroxyl site, it follows that greater Fe substitution would make room for more Cl. (This latter argument can only be applied to silicates in which, like the micas, addition of Fe alters the geometry of the site in which Cl substitutes. The thermodynamic calculations do not suffer from that limitation.)

It has also been suggested [48] that the Fe–F avoidance rule can be rationalized on the basis of crystal-field theory (see Chapter 11), because the crystal-field splitting, and thus the crystal-field stabilization energy, of Fe^{2+} in an octahedron of OH anions is considerably greater than that of Fe^{2+} in an octahedron of F^- anions; thus it is energetically advantageous for Fe^{2+} to occupy $MO_4(OH)_2$ octahedra rather than either $MO_4(OH)F$ or MO_4F_2 octahedra. On the other hand, Cl^- is closer to the low end of the spectrochemical series than is F^-, and so the crystal-field splitting of an $FeCl_6$ octahedron would be even less than that of an FeF_6 octahedron. On this basis, Fe would be expected to avoid occupancy of Cl-containing octahedra even more than it would F-containing octahedra, but this is inconsistent with Figure 3-7. Evidently crystal-field theory does not explain Fe–F and Mg–Cl avoidance well.

It is to be noted that the trends shown in Figure 3-7 are usually evident only when data are plotted for genetically related biotites; plots of micas randomly chosen from many different areas generally show no such correlations at all. The reason is that the halogen content of biotite depends primarily on the activity of halogens in the systems from which the micas crystallize—essentially, that is, on the amount of halogen available. The absolute concentrations of fluorine and chlorine in biotite are evidently not governed by the [Mg]:[Fe] ratio, but, given a constant total amount of F and Cl available in a given geologic setting, the relative concentrations in biotite do depend on that ratio.

Micas that are either fairly abundant or of some particular interest to us later in this text are shown in Table 3-1. Lithium micas are, ideally, trioctahedral, but for our purposes it is convenient to place them in a group by themselves. Solid solution between and among trioctahedral end-member micas is quite extensive, as it is between muscovite and lithium micas; Figure 3-8 shows some of the solid solution ranges observed in natural micas, as well as in some synthetic micas with compositions that shed light on the common natural minerals. Natural common micas that fall compositionally between the dioctahedral and trioctahedral groups are

Table 3-1. Chemical Formulas of Some Members of the Mica Family[a]

Common micas	
Dioctahedral: $(K,Na)Y_2Z_4O_{10}(OH)_2$	
Muscovite	$KAl_2Si_3AlO_{10}(OH)_2$
Paragonite	$NaAl_2Si_3AlO_{10}(OH)_2$
Phengite	$KAl_{1.5}(Mg,Fe)_{0.5}Si_{3.5}Al_{0.5}O_{10}(OH)_2$
Celadonite	$K(Mg,Fe^{2+})(Fe^{3+},Al)Si_4O_{10}(OH)_2$
Trioctahedral: $(K,Na)Y_3Z_4O_{10}(OH)_2$	
Phlogopite	$KMg_3Si_3AlO_{10}(OH)_2$
Annite	$KFe_3^{2+}Si_3AlO_{10}(OH)_2$
Biotite	$K(Mg,Fe^{2+})_3Si_3AlO_{10}(OH)_2$
Ferriannite	$KFe_3^{2+}Si_3Fe^{3+}O_{10}(OH)_2$
Siderophyllite	$KFe_2^{2+}AlSi_2Al_2O_{10}(OH)_2$
Polylithionite	$KLi_2AlSi_4O_{10}(OH)_2$
Trilylithionite	$KLi_{1.5}Al_{1.5}Si_3AlO_{10}(OH)_2$
Taeniolite	$KMg_2LiSi_4O_{10}(OH)_2$
Zinnwaldite	$KFe^{2+}Li(Al,Fe^{3+})Si_3AlO_{10}(OH)_2$
Ephesite	$NaLiAl_2Si_2Al_2O_{10}(OH)_2$
Brittle micas	
Dioctahedral: $(Ca,Ba)Y_2Z_4O_{10}(OH)_2$	
Margarite	$CaAl_2Si_2Al_2O_{10}(OH)_2$
Trioctahedral: $(Ca,Ba)Y_3Z_4O_{10}(OH)_2$	
Clintonite	$CaMg_2AlSi_2Al_2O_{10}(OH)_2$
Bityite	$CaAl_2LiSi_2AlBeO_{10}(OH)_2$
Anandite	$BaFe_3^{2+}Si_3Fe^{3+}O_{10}(OH)S$
Kinoshitalite	$Ba(Mg,Mn,Al)_3Si_2Al_2O_{10}(OH)_2$

[a]Notes: (1) Micas containing octahedral Li are often grouped as "lithium micas" separate from other common micas. (2) Formulas given are idealized. Natural specimens exhibit compositional ranges as discussed in the text. (3) In most micas F (and minor Cl) may substitute for OH.

rare, except for lithium micas, as shown by Figure 3-8(a). The same is true of the brittle micas, for which a chemically intermediate example [34] is $(Ca_{0.95}Na_{0.02})(Al_{2.04}Li_{0.55})(Si_{2.02}Be_{0.64}Al_{1.34})O_{10}(OH)_2$; its composition lies between those of margarite and bityite (Table 3-1).

Solid solution from muscovite toward lithium-free micas is either by substitution of a biotite component, which increases octahedral sheet occupancy by the reaction $2Al = 3(Mg,Fe^{2+})$ while leaving constant the tetrahedral sheet chemistry, or by substitution of a celadonite component by the reaction $^{[6]}Al + {}^{[4]}Al = {}^{[6]}(Mg,Fe^{2+}) + {}^{[4]}Si$, which preserves the dioctahedral character of the octahedral sheet but substitutes Si for Al in the tetrahedral sheet. These are called *biotitic substitutions* and *phengitic substitutions,* respectively. Similar substitutions may, of course, be envisioned from the biotite end member toward muscovite and celadonite. In practice, these substitutions are complicated by the additional substitutions of Fe^{3+}, Ti^{4+}, and interlayer cations, as well as the variability in (OH,F,Cl). Figure 3-8(b), which is based in part on experiments in the synthetic systems noted in the figure caption, reveals that solid solution in any of these ideal directions is quite restricted. It turns out to be dependent on temperature of formation, and the range of solid solution decreases with increasing temperature for both muscovite and biotite. The biotitic

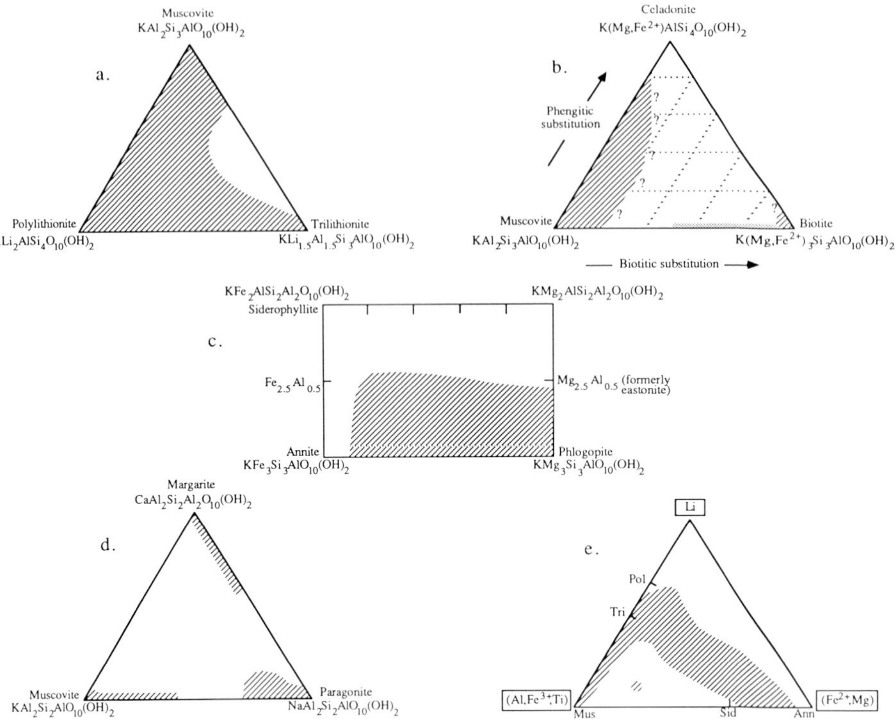

Figure 3-8. Solid solution ranges for various mica series. (a) Solid solution among muscovite and lithium micas. (b) Phengitic and biotitic substitutions in muscovite. Hatchured area refers to natural subsitutions, shaded area to synthetic aluminuos biotites. (c) The natural biotite composition range. (d) Muscovite–paragonite–margarite solid solution ranges. (e) Composition ranges in the octahedral layers of natural micas. (Data from [10;16;21;36;46].)

substitution in muscovite is more important at higher temperatures (above ~600°C) than is the phengitic substitution; the lower limit of octahedral occupancy for synthetic micas between phlogopite and muscovite is about 2.5 atoms [$KMg_{1.5}Al_{1.0}Si_3AlO_{10}(OH)_2$], and in the synthetic minerals the range of octahedral occupancy from 2.5 to 3.0 appears complete, though this greatly exceeds the range that has been observed in natural aluminum-rich biotites. In contrast, solid solution within the trioctahedral compositional field [Figure 3-8(c)] is extensive.

The solid solution behavior of muscovite and biotite can be viewed as unusual, and it has an interesting consequence. Recall that for alkali feldspars, the range of solid solution decreased as the temperature was lowered because the waning thermal vibrations of the substituent atoms (K and Na, in this case) caused their effective sizes to diverge; the same is true for many of the silicates to be discussed later. A consequence of this behavior of alkali feldspars was that exsolution occurred on cooling as the limits of solid solution narrowed. For micas, exsolution is virtually unknown because the solid solution ranges become *greater* as temperature is lowered.

DISTORTIONS IN MICAS

Thus far, with a few allusions to the contrary, we have discussed mica architecture in rather simplistic terms: hexagonal tetrahedral rings with O_{nbr}-OH planes that fit together to make octahedral sheets. In fact, it turns out that the *a* and *b* unit-cell dimensions as measured in a tetrahedral sheet made of ideal hexagonal rings differ substantially from those measured in an independent, undistorted octahedral sheet of appropriate mica chemistry. Obviously, however, *a* and *b* in the two sheets must be identical in a real mica in order for the layer to fit together properly; so there must be distortions of some sort in one or both of the sheets that render the repeat distances identical.

The ideal *b* cell edge, for example, of the sheet in gibbsite (dioctahedral occupancy by aluminum) is 8.64 Å, and that of the sheet in brucite (trioctahedral occupancy by magnesium) is 9.36 Å (Figure 3-9). The difference is due to two factors: First, Mg is inherently larger than Al, and so the oxygens that surround it are farther apart, on the average, than those that surround Al; $^{[6]}$Al–O bonds are typically ~1.9 Å in length, while $^{[6]}$Mg–O bonds are ~2.1 Å in length. Second, dioctahedral and trioctahedral micas have octahedra that share different numbers of edges with adjacent occupied octahedra. In dioctahedral micas only the *cis* sites (*M*2 sites) are filled. Because the *trans* sites (*M*1) are empty, each *M*2 site of a dioctahedral mica shares three edges, the ones shown in heavy lines, with other filled octahedra. In trioctahedral micas, both *M*1 and *M*2 sites are occupied, with the consequence that six edges of each octahedron are shared with adjacent filled octahedra. It is conventionally understood from Pauling's rules that repulsion between cations in adjacent edge-sharing polyhedra generally causes shortening of those edges. The rationale is that cation–anion bond lengths are much more difficult to change than anion–cation–anion angles; as the cations repel each other, they move apart by closing the angles rather than by increasing bond lengths (Figure 3-10). Because the edges in question are not parallel to the basal triangular faces, shortening of the them flattens the octahedra in a direction perpendicular to the plane of the sheets, which is ~(001). The elastic properties of the octahedra require that flattening, or compression, perpendicular to (001) result in expansion of the octahedra in the (001) plane.

Figure 3-9. Shared edges are shown as heavy lines in dioctahedral (gibbsite) and trioctahedral (brucite) layers (left and right, respectively).

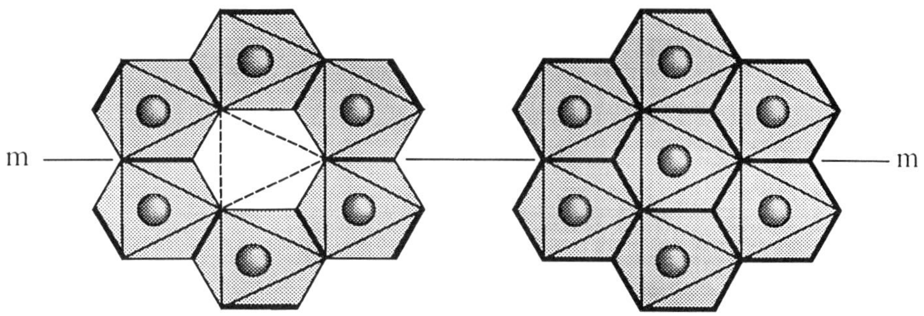

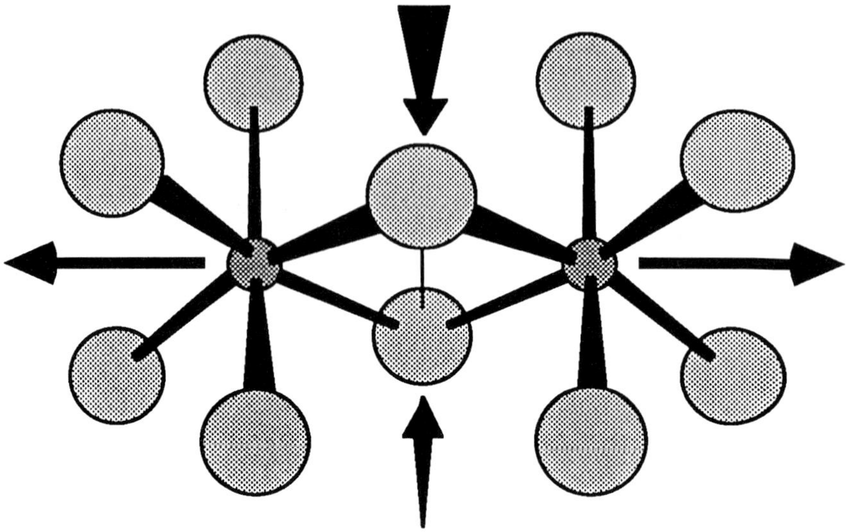

Figure 3-10. The shortening of shared edges by mutual repulsion of cations.

The trioctahedral sheet, with twice as many shared edges as the dioctahedral sheet, is the more flattened of the two, and therefore also proportionately the more laterally expanded. Thus the brucite sheet is larger than the gibbsite sheet (in the *a* and *b* directions) both because of the relative sizes of the cations and because of different amounts of octahedral flattening.

For a tetrahedral sheet containing only Si and Al, the *b* cell dimension of the ideal sheet consisting of hexagonal rings (Figure 3-1) is approximately

$$b = 9.15 + 0.74x,$$

where the cation occupancy of the sheet, based on a formula containing 12 anions, is $4(Si_{1-x}Al_x)$ [6]. For example if $x = 0.25$, then the tetrahedral sheet contains Si_3Al, corresponding to muscovite $[KAl_2Si_3AlO_{10}(OH)_2]$, biotite $[K(Mg,Fe)_3Si_3AlO_{10}(OH)_2]$, and many other micas. Table 3-2 shows the *b* cell edges for tetrahedral sheets of different occupancies, as well as the *b* cell edges for brucite and gibbsite octahedral sheets.

Table 3-2. *b* Cell Dimensions of Unconstrained Sheets

Octahedral Sheet	Tetrahedral Sheet	
Brucite $[Al(OH)_3]$	Si_4	9.15 Å
9.36 Å	Si_3Al	9.34 Å
	Si_2Al_2	9.52 Å
Gibbsite $[Mg(OH)_2]$	$SiAl_3$	9.71 Å
8.64 Å	Al_4	9.89 Å

Note that an undistorted tetrahedral sheet of any composition is too large for the gibbsite sheet; therefore, dioctahedral micas can only exist if the gibbsite sheet is expanded or the tetrahedral sheet is contracted, or both, so that they match. For trioctahedral micas, the match between a brucite sheet and a Si_3Al sheet is very good, and corresponds compositionally to the mica phlogopite. Introduction of some octahedral Fe^{2+}, corresponding compositionally to the biotites, produces only minor additional misfit. (Biotites have greater thermal stability than muscovite, and perhaps this minimal tetrahedral–octahedral misfit is part of the reason.) Some trioctahedral micas with less tetrahedral Si than Si_3Al also have some Al in the octahedral sheet; this facilitates charge balance, but aggravates the misfit. In the brittle mica xanthophyllite $[Ca(Mg_2Al)SiAl_3O_{10}(OH)_2]$, for example, the unconstrained tetrahedral *b* dimension would be 9.71 Å, and the octahedral *b* dimension for a sheet of ideal octahedra would be about 9.12 Å, assuming a linear variation from brucite to gibbsite; the actual *b* dimension of a natural specimen with a somewhat more complicated formula than the ideal one turns out to be 9.01 Å (Joswig et al., 1986).

Clearly, in both dioctahedral and trioctahedral micas of virtually any fixed composition, there must be some way of altering sheet dimensions to achieve the required fit between tetrahedral and octahedral sheets. The data just cited for xanthophyllite suggest that the octahedral sheet is the more rigid of the two, and that most of the adjustment must occur in the tetrahedral sheet; nevertheless, the possibility of adjustment in the octahedral sheet should not be overlooked.

Five ways to counteract misfit may be suggested:

1. Lengthening or shortening of average cation–anion bond lengths.
2. Elongation of tetrahedra perpendicular to the sheets, with concomitant shortening of tetrahedral basal edges as an elastic response.
3. Tilting of tetrahedra.
4. Flattening or thickening of the octahedra, with resulting lateral expansion or contraction.
5. Rotation of the tetrahedral rings away from hexagonal symmetry.

Force constants for cation–anion bonds, anion–cation–anion angles, and anion–anion–anion angles have relative orders of magnitude of 10^0, 10^{-1}, and 10^{-2}, respectively. Therefore, possibility (1) does not seem likely if other possibilities are available, and it is not observed to occur to any important extent.

With regard to possibility (2), a survey of over fifty well-refined mica structures determined to date confirms that virtually all tetrahedra are, in fact, elongated perpendicular to the plane of the sheets, meaning that for a given tetrahedron the average lengths of the apical edges exceeds the average lengths of the basal edges. Moreover, for tetrahedra of the same compositions, there seems to be no significant difference in the amount of elongation from one mica to another. As one would expect, this means that O–*T*–O bond angles involving the apical bond (often called τ_{tet} by those who specialize in mica crystallography) are wider than 109.47°, the ideal tetrahedral bond angle, while those involving only basal bonds are narrower. The effect does not compensate markedly for misfit, but does reduce the magnitudes of the other adjustments that must occur. A related distortion that occurs within the tetrahedra has to do with the T–O bond lengths. (Here T represents the tetrahedrally coordinated atom.) For trioctahedral common micas the difference $[(T\text{–}O_{apical}) - \langle T\text{–}O_{basal}\rangle]$, where $\langle T\text{–}O_{basal}\rangle$ is the average of the three bond lengths to basal oxygens, is positive, whereas it is negative for dioctahedral common micas and most lithium micas. Consistent with this observation, the mean τ_{tet} angles show a general increase in the order trioctahedral common micas < dioctahedral common micas < lithium micas. It would seem that this must be related to the nature of the octahedral sheet, but no thorough investigation of these phenomena has been done.

Tilting of tetrahedra is observed, and results in corrugation of the tetrahedral sheet (Figure 3-11). Such corrugation provides only minor changes in cell parameters, and may well be less a response to tetrahedral–octahedral misfit than to the type of octahedral sheet involved: In dioctahedral sheets one of every three sites is empty, and the empty one ($M1$) is located in a different relative direction from each of the six tetrahedra (three above and three below) that shares an oxygen atom with

Figure 3-11. Corrugation of the tetrahedral sheet. Spheres represent octahedral cations. (a) A trioctahedral mica, fluorphlogopite, viewed along b and showing virtually no out-of-plane tilting of the tetrahedral bases. (b) A dioctahedral mica, paragonite, viewed along a, with obvious out-of-plane tilting of the tetrahedral bases.

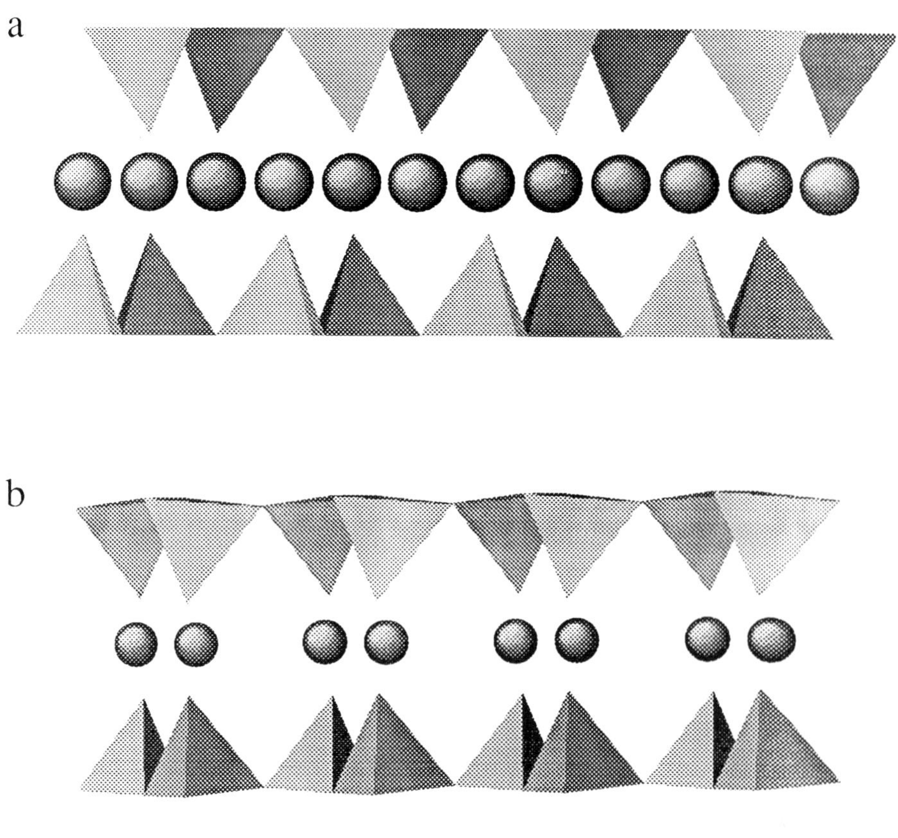

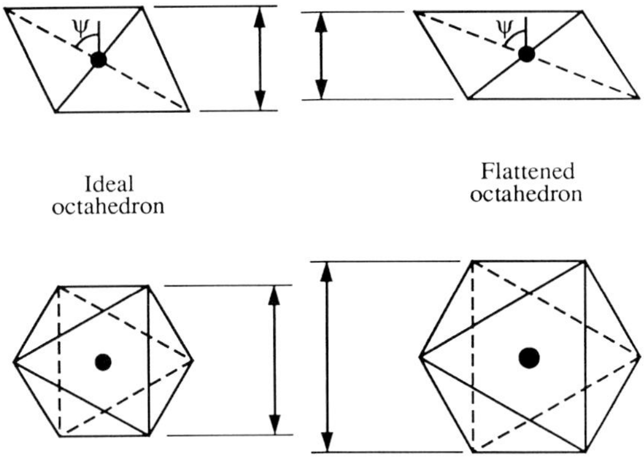

Figure 3-12. The effect of flattening an octahedron. The octahedral compression angle ψ is 54° 44′ for an undistorted octahedron.

it. These oxygen atoms are pulled toward the filled octahedra and away from the empty ones, resulting in a distortion of the octahedral sheet and rigid-body tilting of the attached tetrahedra. (Compare with discussion on octahedral counter-rotations, p. 127.) In trioctahedral micas, tilting is less pronounced because each tetrahedron shares its apical oxygen atom with three nearly identical octahedra, and so the apical oxygens are not "pulled" one way or the other. In the synthetic trioctahedral mica fluor-polylithionite, $KAlLi_2Si_4O_{10}F_2$, however, the octahedral sites are not identically occupied because lithium preferentially orders into $M1$; the tetrahedral sheet in fluor-polylithionite is thus slightly more corrugated than in trioctahedral micas with nearly identical M sites (see discussion on homo-octahedral micas, p. 130) because, though not empty as in dioctahedral micas, $M1$ is occupied by a different kind of atom than $M2$. The quantitative measure of tetrahedral tilt is Δz_o, the maximum difference in the z fractional coordinates of the basal oxygen atoms of a tetrahedron multiplied by $c \sin \beta$.

Flattening of the octahedral sheet is an important factor in the mutual adjustment of tetrahedral and octahedral sheets, but it is more likely a consequence of octahedral expansion than a cause of it. The amount of compression is measured by the angle ψ (Figure 3-12), which is 54°44′ for an undistorted octahedron, and greater than 54°44′ in a flattened one. Observed values range from ~56.5° to ~62°, indicating that all micas contain flattened octahedral sheets. Given that most of the flattening results from shortening of shared octahedral edges (p. 111), any additional flattening would seem to be better described as an elastic *response* to lateral expansion as the octahedral sheet adjusts to the tetrahedral sheet geometry, than as the *cause* of the octahedral expansion. In fact, there is a tendency for octahedra containing larger cations to be more flattened than those with smaller ones.

The most important deviation from the "ideal" structure that facilitates dimensional matching between the tetrahedral and octahedral sheets is rotation in the tetrahedral sheet (Figure 3-13). Rotation is measured by the angle α. For ideally hexagonal rings, $\alpha = 0°$, while for a fully rotated tetrahedral sheet, $\alpha = 30°$. "Fully rotated" means that the tetrahedra have been rotated as far as is physically possible

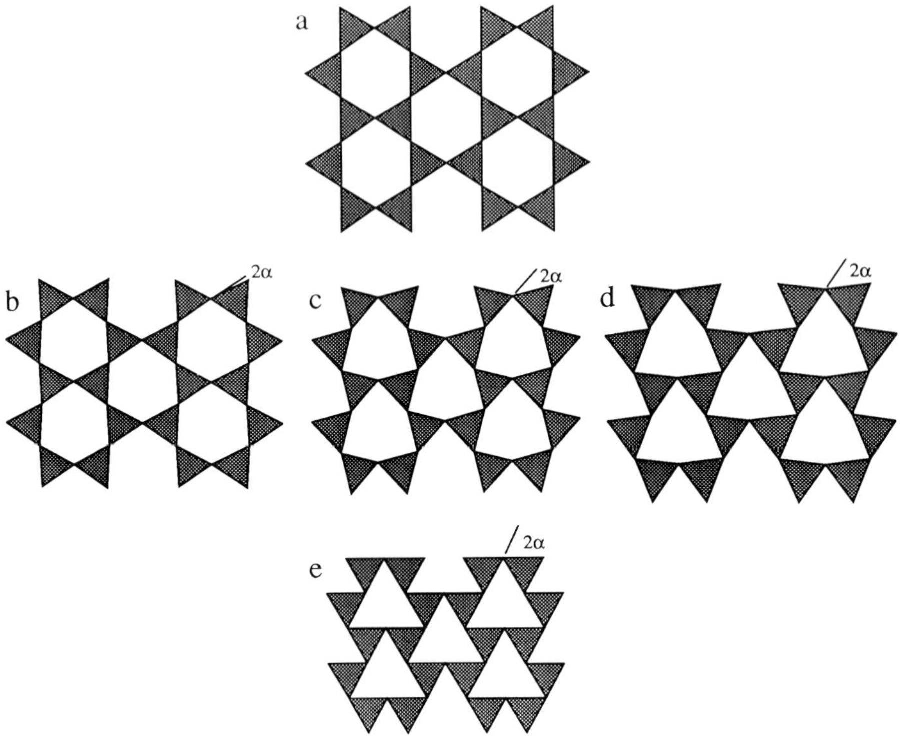

Figure 3-13. (a) The ideal unrotated tetrahedral sheet, $\alpha = 0°$. (b) The tetrahedral sheet of fluor-polylithionite, $\alpha = 3°$. (c) The tetrahedral sheet of paragonite, $\alpha = 16°$. (d) The tetrahedral sheet in clintonite, $\alpha = 23°$. (e) An ideal fully rotated tetrahedral sheet, $\alpha = 30°$.

for "hard-sphere" anions, resulting in a close-packed arrangement of oxygens in the tetrahedral sheet. The maximum rotation angles so far found in micas range from 21° to 23°, and occur in the brittle micas. The effects of tetrahedral rotation on the dimensions of the sheet are seen in Figure 3-14. With rotation, both *a* and *b* contract, but *c* is unaffected. Thus, tetrahedral rotation facilitates adjusting a larger "ideally hexagonal" tetrahedral sheet to an inherently smaller octahedral sheet. For trioctahedral potassium micas α tends to be small (1°–7°), and it is 10°–13° for dioctahedral potassium micas. For sodium micas it is ~16°, and increases to around 22° for calcium (brittle) micas. Pyrophyllite, a dioctahedral 2:1 mica-like layer silicate that has no interlayer cation, has a tetrahedral rotation angle of 10.2°; in talc, the trioctahedral analog of pyrophyllite, $\alpha = 3.4°$.

The rotation may be made in either of two ways, as shown in Figure 3-15. The small circles represent octahedral cations beneath the tetrahedra whose bases are shown. The letters O and S mean "opposite" and "same." In O rotations the triangles forming the tetrahedral bases and the ~(001) octahedral triangular face point in opposite directions; in S rotations, they point in the same direction. An O rotation of $\alpha = 30°$ results in cubic close packing, while an S rotation of the same amount yields hexagonal close packing. A few chlorites (which, of course, are also

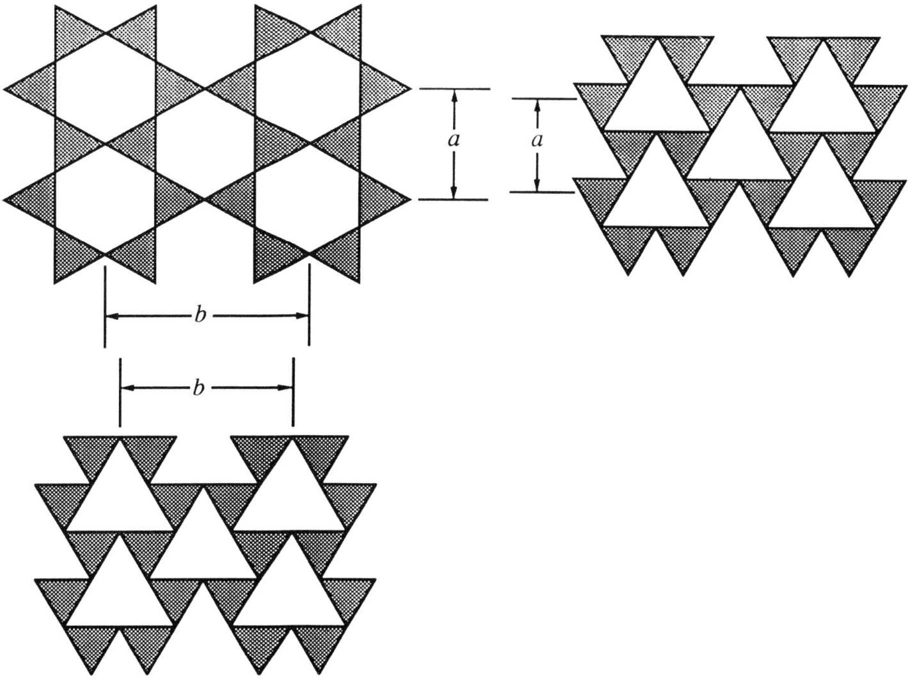

Figure 3-14. The reduction in the *a* and *b* cell edges of the tetrahedral sheet due to rotation. (Modified from [35].)

sheet silicates but are not micas) have S rotations, but all dioctahedral micas and most trioctahedral micas have O rotations. In Figure 3-16 we see that O rotations cause oxygens of the tetrahedral bases to move toward, rather than away from, octahedral cations of the same layer. An explanation that has been proposed for this preference is that these oxygen atoms are pulled toward the cations by ionic forces,

Figure 3-15. Relative orientations of tetrahedra and octahedra for an O rotation and an S rotation.

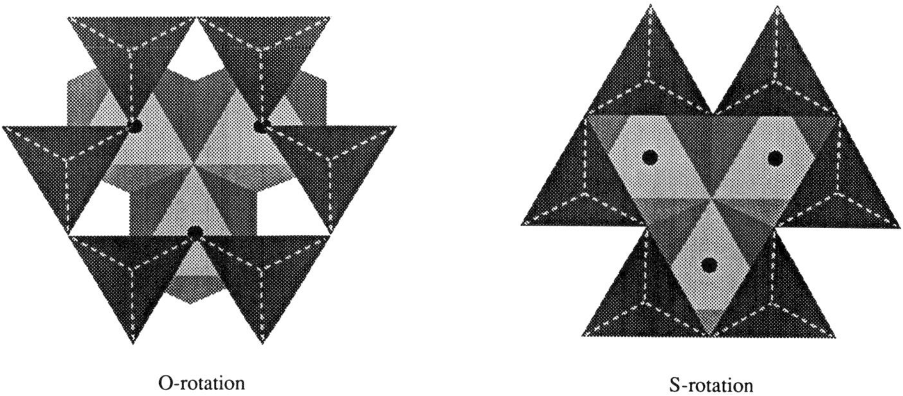

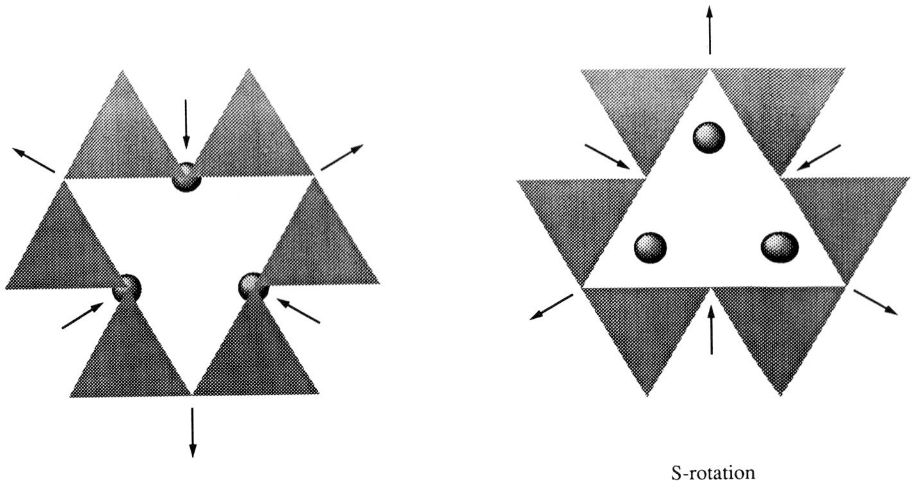

Figure 3-16. Positions of the basal oxygen atoms of the tetrahedral sheet relative to the octahedral cations for O and S rotations. Note that oxygens of the tetrahedral sheet shield the octahedral cations from the overlying interlayer cation only for O rotations.

but mitigating against this explanation are the facts that (1) the distance between octahedral cations and basal oxygens exceeds 3 Å, far longer than any expected bond length, (2) the tetrahedral rotation angle is larger in dioctahedral micas, which have empty *M*1 sites, than in trioctahedral micas, and (3) molecular-orbital calculations carried out on a "molecular fragment" designed to mimic an octahedral sheet connected to a tetrahedral sheet (but with no interlayer cation) yielded the

Figure 3-17. Variation in Δ, the difference between the longer and shorter bonds between the interlayer cations and surrounding oxygen atoms, with tetrahedral rotation angle. (Modified from [35].)

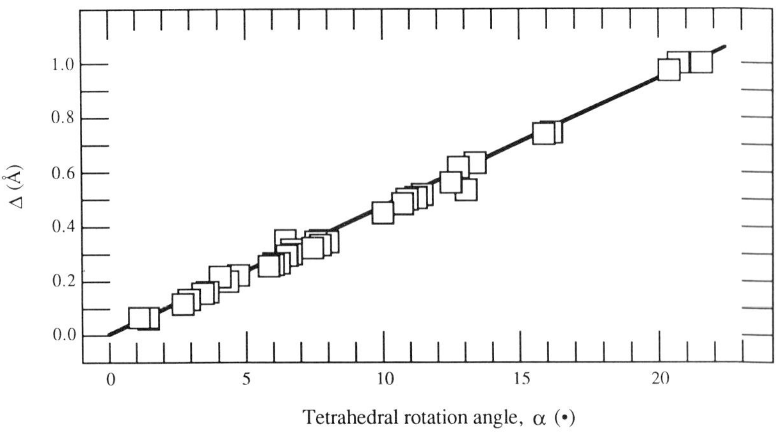

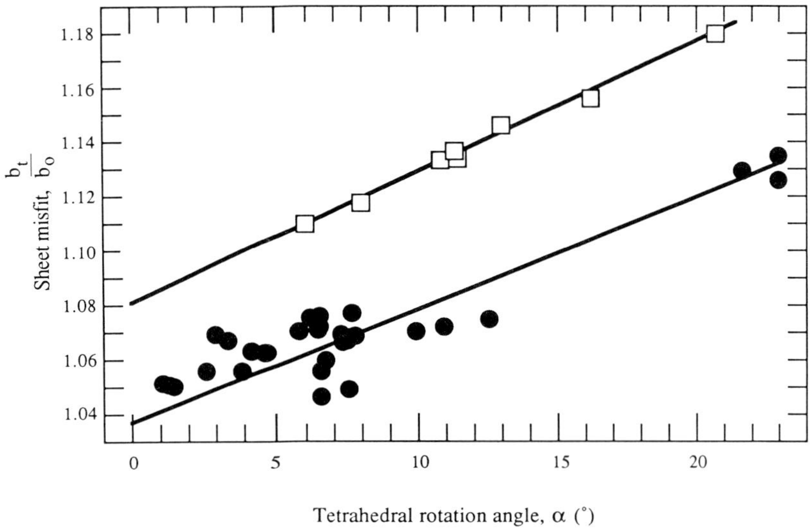

Figure 3-18. Relationship between the tetrahedral–octahedral misfit and the tetrahedral rotation angle. Instead of the formulas given in the text, the mean bond lengths from refined crystal structures were used to determine b_T/b_O; so scatter in the plot indicates that tetrahedral rotation is not the only way in which the crystal structure adjusts for misfit, although it is clearly an important way. Open squares are for dioctahedral micas, and filled circles are for trioctahedral ones.

lowest energy for an S rotation, not an O rotation [42]. A more likely interpretation may be that the positions of the bridging oxygens in O-rotated sheets allows them partially to shield the interlayer cation, which is directly above the ditrigonal ring, from the octahedral cations. Chlorites, which do not have interlayer cations, would not require such shielding.

Clearly, there are at least two possible factors that may influence α. The rotation angle changes with the size of the interlayer cation, and also with the kind of octahedral sheet (dioctahedral or trioctahedral). The first effect is because the interlayer cation fits into hexagonal holes (ditrigonal holes if $\alpha > 0°$) in the backs of the tetrahedral sheets, and the greater the rotation, the smaller the holes. The second effect is a function of sheet misfit. The next two diagrams (Figures 3-17 and 3-18) show these correlations graphically. In the first, Δ is the difference in the means of the six short and six long bonds between the interlayer cation and surrounding basal oxygens. If $\alpha = 0°$, the cation is coordinated by 12 oxygens at equal distances, and $\Delta = 0$. As α goes from 0° to 30°, six of the oxygens converge on the interlayer cation while six move away from it. At $\alpha = 30°$, the cation is strictly six coordinated. In the second figure, b_T/b_O is the ratio between the "ideal" b dimensions of the tetrahedral and octahedral sheets, assuming unconstrained geometry (i.e., no requirement that they fit together) and observed chemistry. The equations

$$b_T = 4\sqrt{2}\langle T\text{–O}\rangle$$

$$b_O = 3\sqrt{2}\langle M\text{–O}\rangle,$$

where $\langle T\text{–O}\rangle$ and $\langle M\text{–O}\rangle$, the average bond lengths of chemically identical and geometrically regular tetrahedra and octahedra, respectively, are often used to determine the ratio b_T/b_O, although observed bond lengths were used in Figure 3-18. The correlations for both variables with α are obviously very strong, and two questions arise:

1. Which variable, if either, *causes* the tetrahedral rotation?
2. Which variable, if either, *limits* the tetrahedral rotation?

Several lines of investigation, pursued by various investigators, have shed some light on these questions:

1. Statistical analyses [35] of the data used in Figures 3-17 and 3-18 indicate that sheet misfit, b_T/b_O, is more responsible for variation in α than is Δ. In addition, high resolution transmission electron microscopy of an unusual muscovite–biotite interface in a crystal intergrowth [28] revealed that the octahedral sheets and interlayer planes were continuous across the interface, whereas about 60% of the tetrahedral sheets were discontinuous; this suggests that the tetrahedral sheets were unable to maintain geometrical integrity because of the different tetrahedral rotation angles of muscovite and biotite. With the same interlayer cations but different amounts of misfit to accommodate, misfit appears to be the only explanation for the different rotation angles.
2. Electrostatic energy calculations [4] suggest that the interlayer cation experiences a more favorable potential as α decreases, and therefore the interlayer cation should not cause rotation. This is supported by the observation that talc and pyrophyllite, which have no interlayer cations, nonetheless possess rotated tetrahedral sheets. However, the electrostatic energy calculations, along with comparison of known mica structures, indicate that the size of the interlayer cation may limit the amount of tetrahedral rotation possible.
3. Molecular-orbital calculations carried out on fragments of the mica structure (three edge-sharing octahedra and a six-membered tetrahedral ring) [42] showed that flattened octahedra and a ditrigonally distorted ring of either O or S rotation yield lower calculated energies than an undistorted ring. As mentioned above, the S-rotated ring turned out to have somewhat lower energy than the O-rotated ring, but this might be because no attempt was made to model the effects of the interlayer cation.

Taken together, these results suggest the following conclusions:

1. The primary driving force behind tetrahedral rotation is the elimination of tetrahedral–octahedral sheet misfit. One must be careful, however, not to envision this as fitting the tetrahedral sheet *to* the octahedral sheet, because apical oxygens of the tetrahedral sheet *constitute* most of the anions of the octahedral sheet. There is simply a compromise in which two tetrahedral sheets assume the geometrical configuration that minimizes the sum of their own energies and that of the enclosed octahedral cations (plus any secondary effects caused by interlayer cations or other factors), and that configuration is almost always a rotated one. Some adjustment also occurs in the octahedral sheet, but the greatest amount

occurs in the tetrahedral sheet—not a surprising conclusion given the relative magnitudes of force constants associated with bond lengths, bond angles, and nonbonded O–O–O angles (p. 113).

2. The size of the interlayer cation limits the amount of tetrahedral rotation possible in the order (most to least rotation allowed) Ca>Na>K≈Ba.
3. It appears that short range bonding forces make ditrigonal rings and flattened octahedra energetically more favorable than undistorted ones, but it is not obvious that the calculated low-energy configurations would eliminate misfit for a range of octahedral cation sizes. Note, however, that two synthetic compounds, α-$BaAl_2Si_2O_8$ and hexagonal $CaAl_2Si_2O_8$ (hexacelsian), are 2:0 layer silicates—that is, the tetrahedral sheets share apical oxygens and there is no octahedral layer in between at all; in neither of these is there a need for rotation to adjust for misfit, but the tetrahedra are rotated 9.9° and 24.3°, respectively, anyway. In these cases, of course, adjustment *to* an octahedral layer is unnecessary, so if fully rotated tetrahedral rings *were* more stable than unrotated ones, only the sizes of the interlayer cations would stop the tetrahedral rotation for these two compounds.

MICA POLYTYPES

The illustrations given thus far indicate that both the backs of the tetrahedral sheets and the interlayer cations are disposed in a hexagonal or ditrigonal arrangement. This should make it possible to stack one layer on top of another in more than one way; that is, the stacking vectors would not always have to be parallel from layer to layer, but could point in any of six different directions within the (001) plane. The stacking of structural modules that are essentially identical except for their rotational orientations with respect to one another is a special kind of polymorphism called *polytypism*, and it is common among all sheet silicates and many minerals that are not sheet silicates. [Tridymite and cristobalite are polytypes, as are many minerals (e.g., the enstatite polymorphs to be considered in Chapter 4) in which the stacked modules are not layers, but blocks.] The least constrained definitions of polytypism do not require successive modules to be chemically or even structurally identical, in which case polytypic structures would not necessarily be polymorphic. In this introductory discussion of polytypism in micas, however, it will be most useful to assume that successive layers of a given polytype have the same composition, though slight structural differences among layers may be accommodated. We shall delay any substantial discussion of factors that influence polytypism until the end of the chapter, and introduce now only the concepts and notation necessary for the sections that follow.

Before considering the ways in which mica layers may be stacked to yield polytypes, we shall describe a shorthand notation that has been developed to describe polytypes of the micas. While not the only such notation to have been devised, it is a simple and convenient one that will suffice for most purposes. It is based on a string of symbols of the general form *n*_[ ], where *n* is the number of mica layers (or unit structures) in a unit cell of the polytype; _ is a symbol representing the crystal system of the polytype; and [ ] encloses n numbers that give the angles (in multiples of 60°) between stacking vectors of successive layers.

The symbols used for crystal system are H for hexagonal, T for trigonal, Or (sometimes O, but that is too easily confused with zero) for orthorhombic, M for monoclinic, and Tc for triclinic. Thus the notation for a group of single mica layers all stacked parallel to the layer above and the layer below would be $1M[0]$, indicating that there is one layer in the unit cell, the crystal system is monoclinic, and the angle between successive stacking vectors, when projected onto a (001) plane, is 0° (0 × 60°). The symbol $3T[222]$ (or, equivalently, $3T[2]^3$) describes a three-layer unit cell with trigonal symmetry; the angles between stacking vectors of successive layers are all +120° (that is, 2 × 60°).

Although it would seem that the number of possible polytypes must be very large, perhaps even infinite, there are only six of the so-called simple polytypes, and they are found to account for the majority of polytypes observed in micas. The six are $1M[0]$, $2M_1[2\bar{2}]$, $2M_2[1\bar{1}]$, $2Or[33]$, $3T[222]$, and $6H[111111]$. The bar over some of the numerals indicates a negative rotation (that is, one performed in a negative sense or direction with respect to those without the bar). The subscripts 1 and 2 serve only to differentiate between the two kinds of two-layer monoclinic polytypes. The illustrations in Figure 3-19 show how the layers are stacked to yield these simple polytypes. Because there are only six simple polytypes, and their angular sequences are straightforward, the angular portions in brackets are usually omitted, and the symbols are shortened to $1M$, $2M_1$, $2M_2$, $2Or$, $3T$, and $6H$. If one assumes that each polytype consists of identical ideal $1M$ layers stacked differently, the ideal space groups are then $C2/m$, $C2/c$, $C2/c$, $Ccmm$, $P3_121$ (or $P3_112$) and $P6_122$, respectively.

Sometimes the stacking sequence is random, or disordered, in which case the notation is as follows: $1Mrn(120)$ is a nonperiodic structure in which successive layers are related by rotations of $n \times 120°$, where n takes on the values 0, 1, and 2 randomly; $1Mrn(60)$ is similar, but the successive layers are related by random rotations of $n \times 60°$, where n is any integer from 0 to 5. The less explicit notation $1Md$ is also used for any disordered stacking sequence. We shall say no more about random stacking sequences; while they certainly occur in nature, our purpose here is to gain some understanding of periodically repeating polytypes.

The "ideal" geometries of the unit cells of the simple polytypes, and examples of micas that are represented by them, are shown in Table 3-3. Several micas appear under more than one polytype, and Table 3-4 gives the primary and secondary polytypes for the micas we will discuss.

Three facts stand out immediately. First, there are no known $6H$ polytypes among either the natural or synthetic micas; second, there is a preponderance of micas that crystallize with the $1M$, $2M_1$, and $3T$ polytypes, whereas $2Or$ and $2M_2$ are quite uncommon; and third, dioctahedral micas tend to favor the $2M_1$ polytype, while trioctahedral varieties tend to favor $1M$. Figure 3-20 shows two (O_{nbr} + OH) planes that form the octahedral sheet, and indicates the six possible ways of stacking one plane on another. The heavy arrows are the stacking vectors. Thus we could, instead of considering polytypes to be made up of rotated layers, consider them to be made up of layers stacked in one or more of these ways. Note, however, that

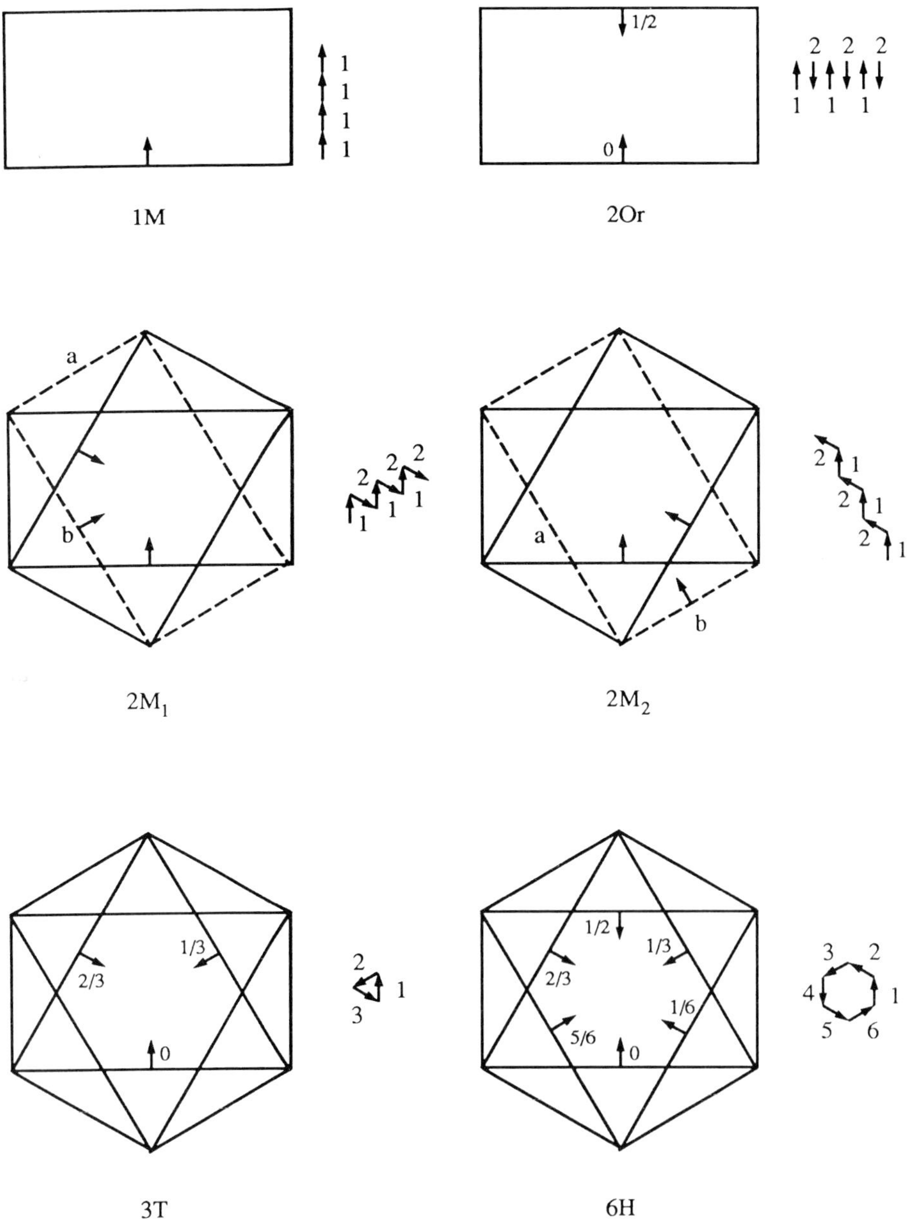

Figure 3-19. Stacking sequences for the six simple mica polytypes. Arrows represent the stacking vectors (see Figure 3-4). Solid lines indicate the (001) outline of the basic unit (a single mica layer). For the $2M_1$ and $2M_2$ polytypes, the dotted line shows the unit cell of the polytype; fractions indicate heights of stacking vectors about screw axes.

Table 3-3. Ideal Cell Dimensions of the Six Simple Polytypes and Examples of Micas that Occur in Them

Polytype	Cell dimensions	β	Examples
$1M$	$a = 5.28$ Å $b = 9.14$ Å $c = 9.96$ Å	$\beta = 100°$	Muscovite, biotite, phlogopite, paragonite, clintonite, lepidolite, phengite
$2M_1$	$a = 5.28$ Å $b = 9.14$ Å $c \approx 20$ Å	$\beta = 95°$	Muscovite, biotite, phlogopite, paragonite, phengite, lepidolite, margarite
$2M_2$	$a = 9.14$ Å $b = 5.28$ Å $c \approx 20$ Å	$\beta = 98°$	Lepidolite, phengite
$2Or$	$a = 5.28$ Å $b = 9.14$ Å $c \approx 20$ Å		Anandite
$3T$	$a = 5.28$ Å $c \approx 30$ Å		Muscovite, paragonite, lepidolite, phlogopite, biotite, zinnwaldite, clintonite
$6H$	$a = 5.28$ Å $c \approx 60$ Å		(None known)

Table 3-4. Primary and Secondary Ordered Simple Polytypes of Some Micas[a]

	Primary	Secondary
Common micas		
Dioctahedral		
Muscovite	$2M_1$	$1M$, $3T$
Paragonite	$2M_1$	$3T$
Phengite	$1M$, $3T$	$2M_1$, $2M_2$
Celadonite	$1M$	
Triocatahedral		
Phlogopite	$1M$	$2M_1$, $3T$
Annite	$1M$	
Biotite	$1M$	$2M_1$, $3T$
Ferriannite	$1M$	
Siderophyllite	$1M$	
Lepidolite	$1M$, $2M_2$	$2M_1$, $3T$
Taeniolite	$1M$, $3T$	$2M_1$
Zinnwaldite	$1M$	$2M_1$, $3T$
Ephesite	$2M_1$	$1M$
Brittle micas		
Dioctahedral		
Margarite	$2M_1$	
Trioctahedral		
Clintonite	$1M$	$2M_1$, $3T$
Bityite	$2M_1$	
Anandite	$2Or$	$2M_1$, $1M$
Kinoshitalite	$1M$, $2M_1$	

[a]Data from [6].

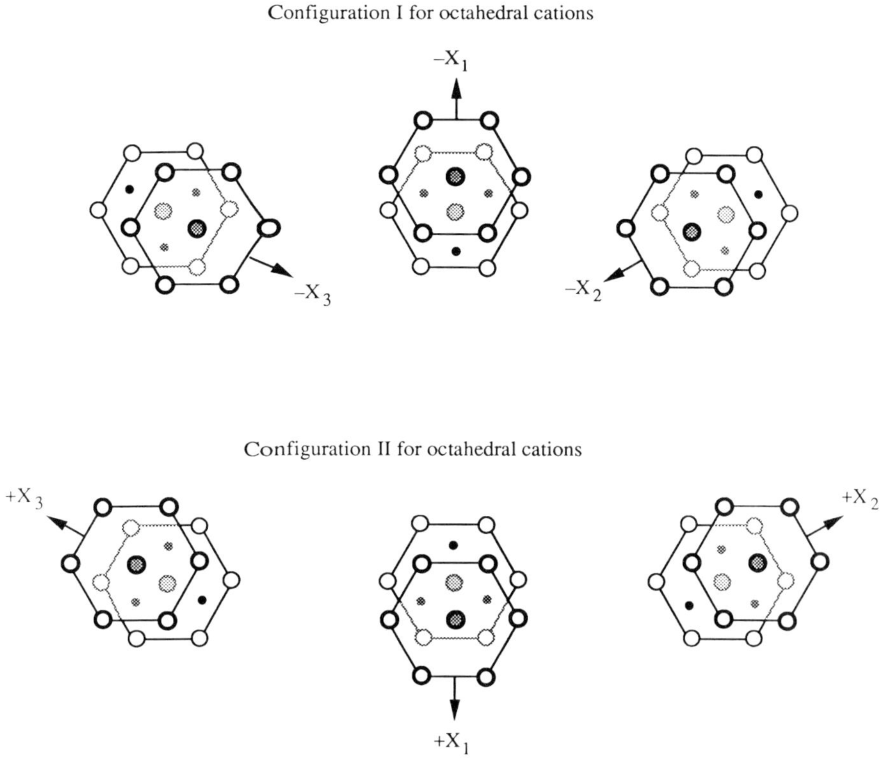

Figure 3-20. An alternative view of polytypic stacking. Open circles represent apical oxygens, filled circles represent hydroxyls, and dots represent octahedrally coordinated cations. Arrows denote the stacking vectors (see Figure 3-4). Note that, within each row of figures, the stacking differs only in orientation of the top layer, but the two rows differ in the relative placement of octahedrally coordinated cations (either configuration I or II).

octahedral cations can be either in configuration I or configuration II in any given layer, and that either configuration permits only three (not six) ways of stacking (O_{nbr} + OH) planes.

The polytypes $1M$, $2M_1$, and $3T$ all generate stacking sequences I–I–I– . . . , whereas $2M_2$, $2Or$, and $6H$ result in I–II–I–II–. . . . We have seen that in real micas the open hexagonal rings of Figure 3-1 are crimped to form ditrigonal rings, resulting in the effective coordination number of the interlayer cations being 6 instead of 12. If the stacking sequence is I–I–I . . . , then the six oxygens about an interlayer cation are disposed in an octahedron; if it is I–II–I–II . . . , then they form the corners of a trigonal prism. Certainly, simple qualitative ideas about electrostatic (ionic) repulsions between oxygen ions appear to favor octahedral coordination, in which the oxygens are farther apart if the polyhedral thicknesses are the same. Unpublished molecular-orbital calculations carried out for an ideal $(Na \cdot 6H_2O)^+$ octahedron and a trigonal prism of the same composition, in which the Na–O bond lengths of the two were constrained to be equal but the trigonal prism was allowed to assume a height of minimum energy, resulted in the octahedron having a lower total energy than the trigonal prism by 4 kcal/mol. This amount of energy is not

Table 3-5. The 36 Polytypes with 1 to 4 Mica Layers per Repeat

$1M[0]$	$C2/m$	$4M_5[1122]$	$C2$
$2Or[33]$	$Ccmm$	$4M_6[1\bar{1}\bar{1}1]$	$C2/c$
$2M_1[2\bar{2}]$	$C2/c$	$4M_7[01\bar{2}1]$	$C2$
$2M_2[1\bar{1}]$	$C2/c$	$4M_8[010\bar{1}]$	$C2/c$
$3T[222]$	$P3_112$	$4M_9[1131]$	$C2$
$3M_1[033]$	$C2/m$	$4M_{10}[12\bar{1}\bar{2}]$	$C2/c$
$3M_2[11\bar{2}]$	$C2$	$4M_{11}[1\bar{2}3\bar{2}]$	$C2$
$3Tc_1[02\bar{2}]$	$C\bar{1}$	$4Tc_1[2\bar{2}33]$	$C\bar{1}$
$3Tc_2[0\bar{1}\bar{1}]$	$C1$	$4Tc_2[1\bar{1}\bar{2}2]$	$C1$
$3Tc_3[02\bar{2}]$	$C1$	$4Tc_3[13\bar{2}\bar{2}]$	$C1$
$4Or_1[0303]$	$Ccmm$	$4Tc_4[0213]$	$C1$
$4Or_2[13\bar{1}3]$	$C2cm$	$4Tc_4[0132]$	$C1$
$4Or_3[23\bar{2}3]$	$Cc2m$	$4Tc_6[1\bar{1}33]$	$C\bar{1}$
$4Or_4[1212]$	$C222_1$	$4Tc_7[0123]$	$C1$
$4M_1[020\bar{2}]$	$C2/c$	$4Tc_8[002\bar{2}]$	$C\bar{1}$
$4M_2[2220]$	$C2$	$4Tc_9[\bar{1}1\bar{2}2]$	$C\bar{1}$
$4M_3[2\bar{2}\bar{2}2]$	$C2/c$	$4Tc_{10}[001\bar{1}]$	$C\bar{1}$
$4M_4[0033]$	$C2/m$	$4Tc_{11}[011\bar{2}]$	$C1$

large, but it is on the order of the energies involved in some chemical reactions. Moreover, the nonbonding overlap between electron orbitals of oxygen atoms in the trigonal prism was found to be more than five times that in the octahedron. Based on this evidence and the relative abundances of the two groups of polytypes, it is reasonable to suppose that the octahedral arrangement yields a lower energy than the trigonal prism, and that this accounts for the observed relative abundances. In support of this idea is the observation that for micas crystallizing in the $2Or$ or $2M_2$ polytypes, the rings are more nearly hexagonal (i.e., there is less ditrigonal distortion) than in $1M$, $2M_1$, or $3T$ micas, and the coordination number of the interlayer cation is more like 12 than 6.

Polytypes with more complicated stacking sequences than those so far considered are possible, of course, and many have been observed. They are generally called *complex polytypes,* and the number of possibilities becomes very large as the number of layers in one unit cell increases. Table 3-5 identifies and gives ideal space groups for the 36 possible polytypes with one to four layers in a repeat unit [49]. In a study of 42 oxybiotites (biotites with OH replaced by oxygen) from a silicic lava flow, about one-third were $1M$, one-third were $2M_1$, and the other third comprised $4M_2$, $8M_8$, $8Tc_{12}$, and other complex polytypes; these relative proportions of common and complex polytypes seem typical.

ORDERING IN MICAS

Octahedral Ordering

We have noted that the ideal octahedral sheet of micas possesses two crystallographically different kinds of octahedral sites, $M1$ and $M2$. "Crystallographically different" implies the strong probability that the sites are energetically different, and they are decidedly geometrically different, inasmuch as one is a *trans* site and the other is a *cis* site (Figure 3-3). This appears to set the stage for ordering of cations in the octahedral sheet.

In dioctahedral micas, the ordering scheme is straightforward. The two octahedral cations occupy the *M*2 sites (of which there are two for every one *M*1 site), and the *M*1 sites remain empty. Because the *M*1 site is unoccupied, its oxygen atoms are pulled away from it and toward the *M*2 cations to which they are bonded, and this results in the *M*1 site being larger than the *M*2 site by some 12% or 13%. A consequence of this is seen in Figure 3-21(a), which shows how the empty *M*1 is distorted, first because anions are attracted to *M*2 cations, and second, in response to the shortening of edges shared between adjacent *M*2 sites. Note that the upper and lower triangular faces are rotated in opposite directions—a phenomenon known as *counter-rotation,* measured in degrees and often symbolized as either ω or δ. Figure 3-21(b) shows how the counter-rotation angle is determined. Because the oxygens of the triangular faces are also apical oxygens of tetrahedra, counter-

Figure 3-21. (a) The enlargement of *M*1 in dioctahedral micas owing to attraction of oxygen atoms to the *M*2 cations, and shortening of shared *M*2–*M*2 edges. (b) The method of determining octahedral counter-rotation, ω (or δ). The angles labeled ε are measured in the plane of the projection and are 60° for an octahedron with no counter-rotation.

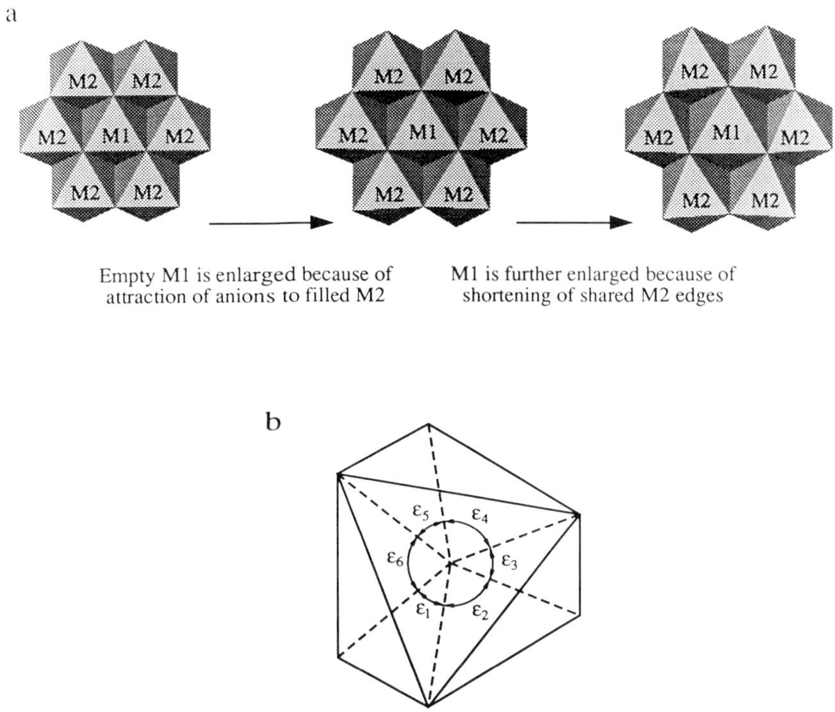

$$\omega\ (\equiv \delta) = \left| \frac{\varepsilon_1 + \varepsilon_3 + \varepsilon_5}{3} - 60 \right| = \left| \frac{\varepsilon_2 + \varepsilon_4 + \varepsilon_6}{3} - 60 \right|$$

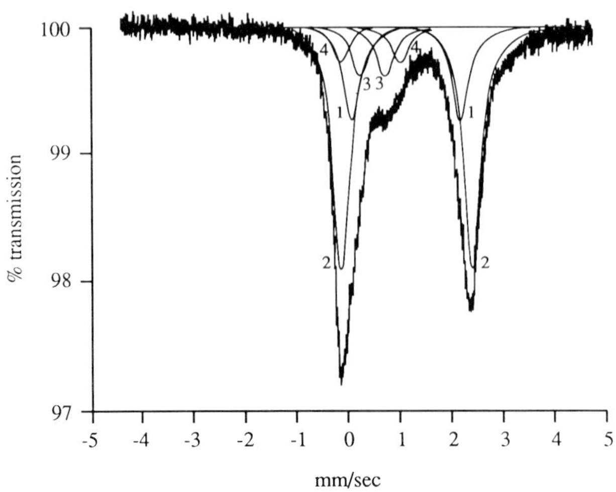

Figure 3-22. A Mössbauer spectrum of a biotite from an igneous rock. The spectrum has been fitted with four doublets with assignments as follows: 1-1, Fe^{2+} in *M*2; 2-2, Fe^{2+} in *M*1; 3-3, Fe^{3+} in *M*2; 4-4, Fe^{3+} in *M*1. The area of peaks 2-2 is greater than twice the area of peaks 1-1, indicating that the ratio of Fe^{2+} in *M*2 to Fe^{2+} in *M*1 is greater than 2:1, the value it would have if Fe^{2+} were completely disordered, given that there are twice as many *M*2 sites as *M*1 sites. (After [11].)

rotation tilts the tetrahedra, perhaps thus facilitating the tetrahedral–octahedral fit (p. 114). It is again important not to confuse causes and effects: The *M*1 site is large because it is vacant, and the *M*2 edges are short because they are shared by occupied octahedra. The second factor simply reinforces the first so that the vacant *M*1 site is even larger than it would have been without the sharing of edges between occupied *M*2 octahedra. Counter-rotation is a response to these interactions within the octahedral sheet, not a response to sheet misfit; there is, in fact, evidence that the amount of counter-rotation in a mica is a function of the differences in the sizes of neighboring octahedral sites.

An illustrative comparison of the influences of ionic size and crystal-field stabilization energy is provided by a consideration of the distribution of iron in the octahedral sheets of biotite and muscovite. Fe^{3+} is a smaller ion than either Fe^{2+} or Mg, and so it is not surprising that Fe^{3+} prefers the slightly smaller *M*2 site over *M*1 in biotites and ferrian muscovites (see, for example, [15], but note that the octahedral site nomenclature in that paper is reversed). Fe^{2+} is larger than Mg, however, so on the basis of size alone Fe^{2+} would not be expected to prefer *M*2; yet Mössbauer spectroscopic studies indicate that it does (Figure 3-22). An important contributing reason is that the distortion of the *M*2 site is greater than that of the *M*1 site, and the crystal-field stabilization energy (Chapter 11) of Fe^{2+} in a more distorted octahedral site exceeds that of Fe^{2+} in a less distorted one. Thus the ordering of Fe^{2+} preferentially into *M*2 of biotites instead of into *M*1 results in lower total energy for the crystal. In muscovites that contain Fe^{2+}, there are more than two octahedral cations in the formula (barring the concomitant substitution of a quadrivalent ion such as titanium by the exchange reaction $Ti^{4+} + R^{2+} = 2Al$ observed in some muscovites [37]). Here the smaller ionic size of Al^{3+}, as well as an electrostatic advantage in surrounding a vacant site entirely with trivalent cations [32], favor ordering of aluminum ions into *M*2, but the favorable lowering of the crystal-field stabilization energy also causes Fe^{2+} to prefer *M*2. In this case, size turns out to be the more important factor; all of the Al^{3+} enters *M*2, and Fe^{2+} preferentially occupies *M*2 as long as there is room for it, entering *M*1 only as the number of octahedral cations exceeds two [12,15]. In neither the dioctahedral nor trioctahedral cases is the advantage in crystal-field stabilization energy sufficient to completely negate size effects: In biotite, in which the distortion of *M*2 is comparatively slight, the $[Fe_{M2}]$:$[Fe_{M1}]$ ratio is generally between about 7:3 and 4:1 (although higher values have been reported); in muscovite, where *M*2 is very distorted and crystal-field effects are thus stronger, the ratio is generally close to 4:1. (Here $[Fe_{M2}]$:$[Fe_{M1}]$ is the ratio of *divalent* Fe in *both* of the two *M*2 sites to divalent Fe in the *M*1 site, so that a ratio of 2:1 indicates no site preference.) In aluminous biotites, which are essentially trioctahedral but contain some octahedral vacancies, NMR studies [51] show that the vacancies occur in *M*1, consistent with the Mössbauer evidence cited previously.

In trioctahedral micas that contain three identical octahedral cations—for instance, $KMg_3Si_3AlO_{10}(OH)_2$ (phlogopite) or $KFe_3Si_3AlO_{10}(OH)_2$ (annite)—*M*1 and *M*2 are obviously occupied by the same species of atom and there is nothing to order. Even so there is a slight, though somewhat inconsistent, tendency for *M*1 to be larger than *M*2. If the octahedral cations are chemically different but of essentially the same size—such as in Fe–Mg biotites or the synthetic mica $KLiMg_2Si_4O_{10}F_2$ [63]—then the cations are found to be more or less randomly distributed over *M*1 and *M*2.

In lepidolites, which are trioctahedral micas in which the octahedral cations are of quite different sizes, ordering proceeds so as to make *M*1 the larger site and, because Li is not only larger but also of lower oxidation number than Al, *M*1 also has a lower charge than *M*2. Site occupancies for six lepidolites that possess only two types of *M* sites are shown in Table 3-6. In the case of fluor-polylithionite, a synthetic mica with the formula $KLi_2AlSi_4O_{10}F_2$, it might seem intuitively logical to order the two lithium atoms into the two *M*2 octahedra and the aluminum into *M*1. That would make *M*1 the smaller site with *M*1–*M*2 shared edges shorter than *M*2–*M*2 shared edges. Since Li–Li repulsions are "softer" than Li–Al repulsions, this

Table 3-6. Octahedral Site Occupancies in Lithium Micas

Mica	*M*1 Occupancy	*M*2 Occupancy	Reference
Fluor-polylithionite-1*M*	$Li_{0.89}Al_{0.11}$	$2\times(Li_{0.55}Al_{0.45})$	[51]
Lepidolite-1*M*	$Li_{0.95}Al_{0.05}$	$2\times(Li_{0.37}Al_{0.63})$	[52]
Lepidolite-2M_2	$Li_{0.95}Al_{0.05}$	$2\times(Li_{0.37}Al_{0.63})$	[47]
Lepidolite-2M_2	$Li_{0.93}(Fe,Mg)_{0.07}$	$2\times(Li_{0.35}Al_{0.58}\quad{}_{0.07})$	[50]
Lepidolite-1*M*	$Li_{0.91}(Mn,Mg)_{0.09}$	$2\times(Li_{0.28}Al_{0.65}\quad{}_{0.07})$	[15]
Lepidolite-2M_2	$Li_{1.00}$	$2\times[Li_{0.24}Al_{0.65}\quad{}_{0.07}$ $(R^{2+},R^{3+})_{0.05}]$	[15]

would appear to be an acceptable arrangement. Because nature chooses a different ordering scheme, we conclude that some combination of the following factors makes the observed cation distribution better than any alternative:

1. $M1$ no smaller than $M2$.
2. Lower charge in $M1$ than in $M2$, resulting in $M2$–$M2$ shared edges being shorter than $M1$–$M2$ shared edges.
3. More nearly equal average charges in $M1$ and $M2$.

The first two factors complement each other, inasmuch as larger ions tend to be the ones with lower oxidation numbers; so the choice is evidently between favorable geometry (factors 1 and 2) and favorable charge distribution (factor 3). Given that more nearly equal average charges for $M1$ and $M2$ could certainly be obtained by other than the observed ordering pattern in, say, dioctahedral micas, we conclude that favorable geometry (probably involving several of the distortions from "ideal" geometry) is the governing consideration. It appears to be nearly inviolable, because thus far only one mica structure [63] for which a high-quality refinement is available shows the reverse octahedral ordering pattern that we might have expected for fluor-polylithionite. (An early refinement of clintonite did show the reverse pattern [61], but more recent data do not support that interpretation [e.g., 29].)

We have so far considered octahedral ordering as if all micas had two kinds of octahedral sites—the $M1$ site that lies on a mirror plane, and the two $M2$ sites related by that mirror. But in the $3T$ polytype, for example, the ideal space group is $P3_112$ (or its enantiomorph, $P3_212$), and the mirror is gone, resulting in three different M sites—$M1$, $M2$, and $M3$. In the dioctahedral mica muscovite-$3T$, $M1$ is again large and vacant. $M2$ and $M3$ are no longer equivalent as in the $1M$, $2M_1$, and $2M_2$ polytypes, and it is found that larger substituent cations (Mg, Fe, Ti, Mn) prefer $M3$ over $M2$. In trioctahedral lepidolite-$3T$, $M1$ and $M3$ are lithium-rich and large, while $M2$ is aluminum rich and small. It appears that wherever octahedral ordering can occur in micas, it nearly always does.

(Parenthetically, the discussion of the past several paragraphs suggests another way of classifying micas. If, on average, three identical cations populate the octahedral sites, the mica is said to be *homo-octahedral*. If two of the three sites are filled identically and the third is different, the mica is said to be *meso-octahedral*. If there are three symmetrically different octahedra populated differently, then the mica is said to be *hetero-octahedral*. Note that the terms *dioctahedral* and *trioctahedral* refer to numbers of octahedrally coordinated atoms in the chemical formula, not to site occupancies of different M sites. Thus a trioctahedral mica could be homo-octahedral, or it could be meso-octahedral or hetero-octahedral because of ordering of different chemical species.)

Tetrahedral Ordering

Table 3-7 shows the restrictions on cation ordering that are required by the *ideal* space groups of simple polytypes. The ideal space group of any polytype permits octahedral ordering of some sort and prohibits any ordering in the interlayer sites, but the ideal space groups for $1M$, $2Or$, and $6H$ do not allow tetrahedral ordering

Table 3-7. The Permissibility ("Yes") and Nonpermissibility ("No") of Ordering in Simple Polytypes that Crystallize with Ideal Symmetry

Polytype	Ideal Space Group	Tetrahedral Sheet	Octahedral Sheet	Interlayer
$1M$	$C2/m$	no	yes	no
$2M_1$	$C2/c$	yes	yes	no
$2M_2$	$C2/c$	yes	yes	no
$2Or$	$Ccmm$	no	yes	no
$3T$	$P3_112$	yes	yes	no
$6H$	$P6_122$	no	yes	no

because all tetrahedra must be identical in these space groups. NMR spectroscopy [25,26] verifies the disordered Si–Al arrangement for many micas for which x-ray-diffraction experiments indicates the space group to be $C2/m$, and confirms that the aluminum avoidance principle is obeyed in these cases. The ideal space groups for $2M_1$, $2M_2$, and $3T$ each possess two symmetrically different tetrahedral sites. Table 3-8 shows that, judging from mean bond lengths, tetrahedral ordering appears to occur in a large fraction of the micas that crystallize in a space group that permits it.

The preceding paragraphs bring us to an argument of considerable significance. The assumption of a space group during the refinement of the crystal structure of a mineral imposes all the symmetry restrictions of that space group on the diffraction data. Thus, if we have a $2M_1$ mica and we assume the ideal space group $C2/c$, then the tetrahedral sites will be averaged and constrained to appear disordered, *even if they are not really disordered.* Refinement in space group Cc, however, will allow the tetrahedral sites to refine independently and may show that there are real differences in bond lengths. Recall that we encountered an analogous situation in the average structures of calcic plagioclases (p. 85). Inasmuch as Table 3-8 suggests that many micas display tetrahedral ordering of some sort when refined in a space

Table 3-8. Tetrahedral Bond Lengths in Micas for Which the Space Group Permits Tetrahedral Ordering

Mica	Space Group	Bond Lengths (Å)	Reference
Muscovite-$3T$	$P3_121$	1.670, 1.603	[20]
Lepidolite-$2M_2$	$C2/c$	1.630, 1.627	[47]
Lepidolite-$2M_2$	$C2/c$	1.620, 1.632	[52]
Biotite-$2M_1$	$C2/c$	1.656, 1.663	[54]
Muscovite-$2M_1$	$C2/c$	1.646, 1.639	[45]
Lepidolite-$2M_1$	$C2/c$	1.628, 1.631	[50]
Paragonite-$2M_1$	$C2/c$	1.652, 1.651	[9]
Phengite-$2M_1$	$C2/c$	1.637, 1.636	[50]
Muscovite-$2M_1$	$C2/c$	1.645, 1.645	[9]
Paragonite-$2M_1$	$C2/c$	1.653, 1.652	[33]
Lepidolite-$2M_2$	$C2/c$	1.629, 1.629	[15]

group that allows order, one must ask whether it is wise to accept any mica structure refinement that indicates disorder when carried out in an ideal space group, if the chemical composition of the mica suggests that ordering in the tetrahedral sheet is possible under the constraints of a subgroup of the ideal space group and refinement in that space group was not attempted.

Consider, for example, anandite-2*Or* [14], with the composition as shown in Table 3-1. The ideal space group is *Ccmm*, which has only one unique tetrahedral site. A crystal-structure refinement in the true space group, *Pnmn* (which is a subgroup of *Ccmm* and results from distortion of the ideal structure), shows that the average bond lengths in the tetrahedral sites (two different ones in *Pnmn*) are 1.675 Å and 1.766 Å. We conclude from these bond lengths that iron is clearly ordered into the larger of the two sites. Similarly margarite-2M_1 [18,19], which has an Si_2Al_2 tetrahedral sheet composition, appears disordered if refined in the ideal space group *C*2/*c*. If refined in the subgroup *Cc*, however, it is shown to have a perfect alternation of Si and Al (Figure 3-23), as would be predicted from the aluminum avoidance principle (p. 78). This is confirmed by NMR spectroscopy [25], which provides information on the numbers of different kinds of tetrahedra surrounding the silicate tetrahedra (Figure 3-24).

Refinement in subgroup symmetry does not always reveal an ordered tetrahedral sheet. For instance, the refinement of lepidolite-2M_1 in space group *Cc* (a subgroup of *C*2/*c*) failed, but refinement in *C*2/*c* was successful and gave tetrahedral bond lengths of 1.629Å in both tetrahedra [17]. An even more interesting example is

Figure 3-23. Ordering in the tetrahedral sheets of margarite, as refined in space group *Cc*, a subgroup of *C*2/*c*. The view is down **c***. The lighter shaded tetrahedra contain silicon, and the darker shaded ones contain aluminum. Topochemically similar tetrahedra in the two sheets are not crystallographically identical, so that each sheet shown contains one unique Si site and one unique Al site, a total of four different tetrahedral sites altogether.

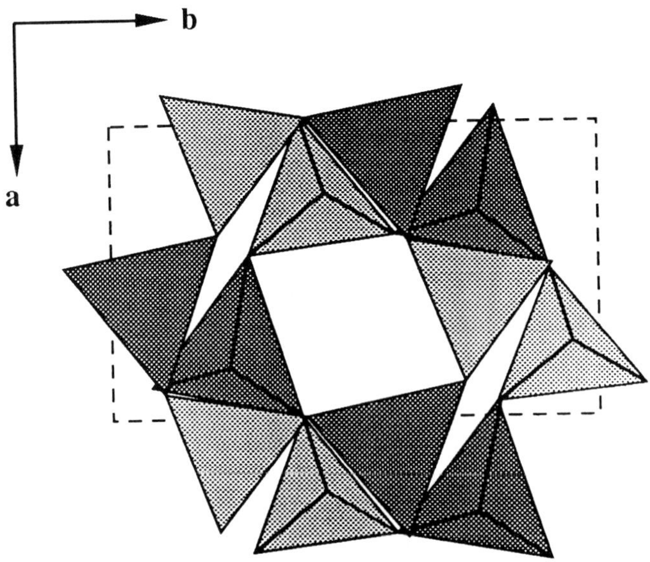

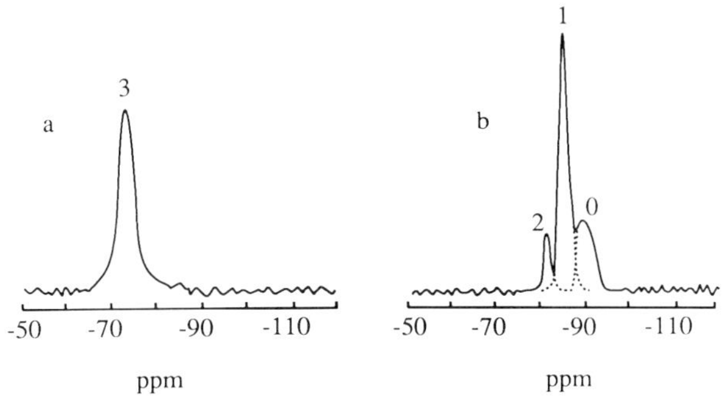

Figure 3-24. ^{29}Si MAS–NMR spectra of (a) margarite and (b) muscovite, included for comparison. Numbers above peaks indicate the number of next-nearest-neighbor Al atoms around Si. Spectrum (a) shows that in margarite each SiO_4 tetrahedron is linked to three AlO_4 tetrahedra; because the Si:Al ratio is 1:1, each AlO_4 tetrahedron is thus surrounded by three SiO4 tetrahedra and margarite is completely ordered. Spectrum (b) indicates a disordered Si–Al distribution for muscovite, with most SiO_4 tetrahedra linked to one AlO_4 tetrahedron, and the others linked to either two or no AlO_4 tetrahedra. (After [25].)

kinoshitalite-1*M*, a Ba-containing brittle mica with a 1:1 ratio of tetrahedral silicon and aluminum [20,30]. All other brittle micas with similar [Si]:[Al] ratios show ordering in the tetrahedral sheet, and there is no *a priori* reason to suspect that this one would not. Nevertheless, in a refinement of the crystal structure of kinoshitalite-1*M* in space group $C\bar{1}$, a subgroup of $C2/m$ that permits tetrahedral ordering in which the aluminum avoidance principle is obeyed, the atomic positions refined to those consistent with space group $C2/m$; it is not possible in a disordered 1:1 [Si]:[Al] tetrahedral sheet to construct an arrangement that does not violate the aluminum avoidance principle, and it is not at all obvious why kinoshitalite-1*M* should not be ordered. It is conceivable that we could be dealing with the average structure of two ordered domains (compare the "$\alpha_1 + \alpha_2$" model of β-quartz, p. 8, or the structure of $I1$ bytownite, p. 85), although there is no indication of it so far in this case. Evidence for antiphase domains in micas is at present not strong, but one possible indication is the presence of very weak electron-diffraction maxima that violate the extinction conditions for space group $C2/m$ in a particular biotite-1*M* [1]. It can be argued on the basis of the aluminum avoidance principle and the favorability of maximizing distances between aluminum-containing tetrahedra that the most likely space group for a trioctahedral 1*M* mica with an ordered 3:1 [Si]:[Al] ratio is $P2_1$, and it has been suggested that the particular biotite examined may consist of $P2_1$ (or perhaps $P2$ or $P\bar{1}$) domains.

Mössbauer spectra of some annites and ferriannites have confirmed the presence of tetrahedrally-coordinated Fe^{3+} [11]. In the case of ferriannite, this is not surprising, because the sum of the Si and Al atoms is insufficient to fill the tetrahedral sites. The annites, however, have more than enough Si and Al to fill the tetrahedral sites; so this means that the Fe^{3+} atoms compete successfully with Al for tetrahedral coordination, and this is not entirely expected—Fe^{3+} is somewhat large for the tetrahedral site, and its half-filled *d* subshell experiences no crystal-field stabilization that would favor tetrahedral coordination. One possible explanation may be gleaned from Table 3-2. Note that the *b* cell edge for a brucite octahedral layer (all Mg) is slightly larger than that of a Si_3Al tetrahedral layer. The presence of a large amount of Fe^{2+} in the octahedral sheets of annites would worsen the misfit, and the accommodation of Fe^{3+} in the tetrahedral sheet would alleviate that somewhat.

COMPOSITIONAL VARIATIONS AND PHYSICAL PROPERTIES

The geometrical constraints imposed by the requirement that tetrahedral and octahedral sheets fit together leads to strong relationships between the chemistry and the geometrical details of micas. Given the chemistry, many geometrical details should be predictable. Conversely, measurement of crystallographic parameters should make it possible to estimate chemical composition in at least some cases. The investigation of such relationships has proved particularly fruitful for trioctahedral micas [23;24;26].

Figure 3-25 shows how various unit-cell dimensions vary with the radii of cations substituted at octahedral, tetrahedral, or interlayer sites. We have already observed that the tetrahedral rotation angle α is related to both octahedral–tetrahedral sheet misfit (which is dependent on both octahedral and tetrahedral chemistry) and to the size of the interlayer cation. Given the chemistry of a trioctahedral mica, then, it is possible to predict at least some of the unit-cell parameters.

Of more practical significance is the other side of the same problem: Given some unit-cell dimensions or other measurable structural parameters, is it possible to determine the composition of a mica? For Fe–Mg–Al biotites, *a*, *b*, and *V* (cell volume) are strongly related to composition. An even better structural parameter, however, is $d_{\bar{3}31}$ the interplanar spacing of the $\bar{3}31$ planes. Unfortunately, the $\bar{3}31$ peak on an x-ray powder-diffraction pattern overlaps the 060 peak, making the measurement of $2\theta_{\bar{3}31}$ difficult and uncertain. If a least squares refinement of the unit cell parameters is carried out, however, the calculated value of $d_{\bar{3}31}$ is more reliable as a compositional indicator than any single unit-cell edge. Another property, more easily obtained than the refined unit-cell dimensions and the calculated $d_{\bar{3}31}$, is the refractive index γ, which is measured on cleavage flakes. Figure 3-26 is a plot for determining the compositions of Fe–Mg–Al biotites given γ and the calculated value of $d_{\bar{3}31}$. The letters *x*, *y*, and *z* stand for subscripts in the formula $K(Mg_xFe_yAl_z)Al_{1+z}Si_{3-z}O_{10}(OH)_2$. Values of *z* and of $x/(x+y)$ are read directly from the plot. In the biotite formula, $x+y+z = 3$. If *z* and $x/(x+y)$, as read from the graph, are N_1 and N_2, respectively, then

$$x + y = 3 - N_1, \quad \text{and}$$
$$(N_2 - 1)x + N_2 y = 0.$$

When these equations are solved simultaneously, all three subscripts are found. Minor constituents, such as Mn or Ti, may have substantial effects, particularly on the refractive index γ.

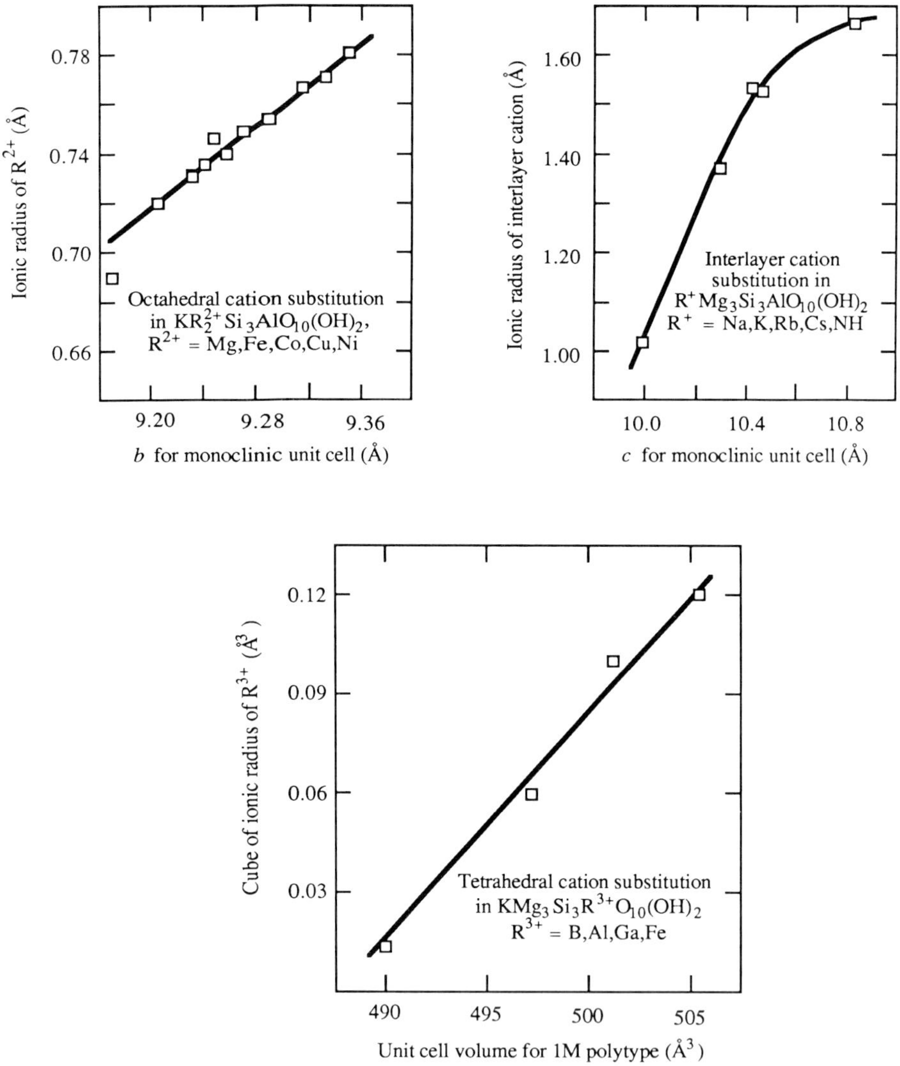

Figure 3-25. Some representative relationships among ionic sizes and unit cell dimensions (related to the 1*M* unit cell) for trioctahedral micas. (After [23].)

It appears that there is an upper limit for aluminum content in these Fe–Mg–Al biotites, corresponding to the formula $K(Fe,Mg)_{2.25}Al_{0.75}Si_{2.25}Al_{1.75}O_{10}(OH)_2$. The likely explanation lies in the decreasing size of the octahedral sheet as the aluminum content is increased. This requires the tetrahedral rotation to increase, making the interlayer site progressively smaller. At $z = 0.75$Al atom, corresponding to $\alpha \approx$ 13°–14°, the interlayer site is as small as it can be and still accommodate K. If the interlayer cation is made smaller, as in $NaMg_2AlSi_2Al_2O_{10}(OH)_2$, then a value of z as high as 1.0 Al atom is possible.

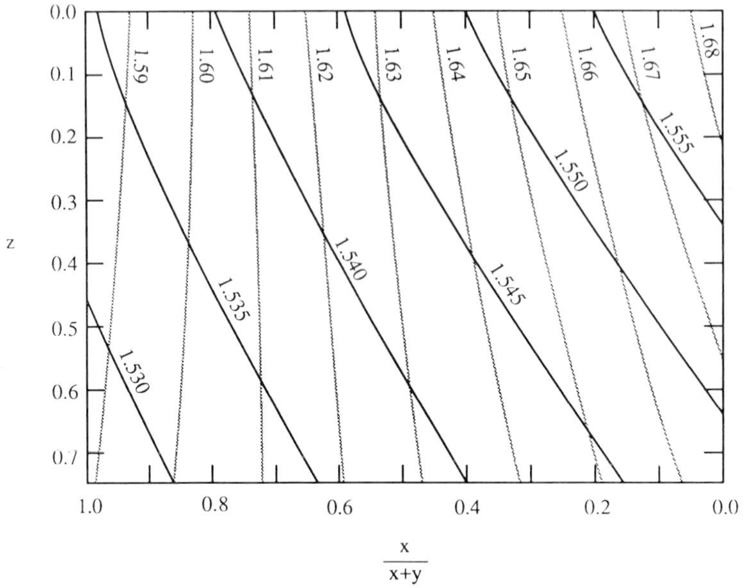

Figure 3-26. This plot is contoured for the index of refraction γ (light lines) and the $\bar{3}31$ d spacing (dark lines) of aluminous biotites with the general formula $K(Mg_xFe_yAl_z)(Si_{3-z}Al_{1+z})O_{10}(OH)_2$; $z = 3 - x - y$. Given values of the contoured variables, values of z and $x/(x + y)$ are read from the ordinate and abscissa, respectively. See the text for an explanation of how to transform these values into a chemical formula. (After [27].)

From the preceding paragraph, and much of the discussion in the section on distortions in micas, it is evident that there must be a relationship among the octahedral cation radius, the tetrahedral cation radius, and the tetrahedral sheet rotation angle that determines the geometrical stabilities of micas. For trioctahedral potassic micas, such a relationship is known, and is shown in Figure 3-27. If the octahedral radius [54] and tetrahedral bond length are specified, then the cosine of α can be read. If the cosine of α exceeds 1.0, the mica is predicted to be unstable. In the same way, either of the other pairs of variables can be specified to predict values of the third that produce a geometrically stable mica. As an example, the radius of Fe^{2+} is 0.77 Å and the mean bond length of the T–O bond in a Si_3Al sheet is 1.643 Å. The cos α predicted for this combination is greater than 1.0, and so annite, $KFe_3^{2+}Si_3AlO_{10}(OH)_2$, is predicted to be unstable. In fact, pure annite has never been found in nature, nor has it been synthesized (although compositions close to it with smaller mean octahedral cation radii, such as $K(Fe_{2.7}^{2+}Fe_{0.3}^{3+})Si_3AlO_{10}(OH)_2$, have been).

THE PROPAGATION OF POLYTYPES

We have already discussed mica polytypism from a descriptive point of view (pp. 121–126). We have found that there are six simple polytypes, only five of which have so far been found in nature or synthesized, and that large numbers of complex polytypes can conceivably exist. Although we considered a crystallochemmically reasonable explanation for the predominance of the $1M$, $2M_1$, and $3T$ varieties in micas, relative to $2M_2$, $2Or$, and $6H$, no explanation was given for the relative abundance of simple polytypes compared to complex ones. Moreover, just how the precise order of stacking vector orientations is maintained over the hundreds of thousands of layers in a macroscopic mica crystal was not discussed.

A significant observation is that, among complex polytypes, certain series of stacking vector repeats appear more commonly than others. Notably, those with layer repeats of $N=3n+1$, $N=3n+2$, and $N=n+2$, and stacking sequences of $[(222)_n0]$, $[(222)_n2\bar{2}]$, and $[(0)_n2\bar{2}]$, respectively, are common (where n is an integer). Figure 3-28 illustrates stacking vector diagrams for three of these. This suggests regular, periodic mistakes, or stacking faults, of 120°, and prompts one to wonder how the crystal "knows" that it should begin the same sequence of layers it started, say, 11 layers (~110 Å!) below. How do sheet silicates "keep track" of stacking sequences?

A related but somewhat different question is, "What controls polytypism in the first place? Given that polytypes can be propagated in some way, what determines the kind of polytype that will form, and why should the simple ones be most abun-

Figure 3-27. A nomogram relating the mean radius of the octahedral cation, the cosine of the tetrahedral rotation angle, and the mean tetrahedral bond length for micas. Because the cosine cannot exceed unity, the nomogram can be used to predict combinations of octahedral and tetrahedral sheet chemistry that are structurally unstable because the predicted cosine of the rotation angle is greater than one. (After [23].)

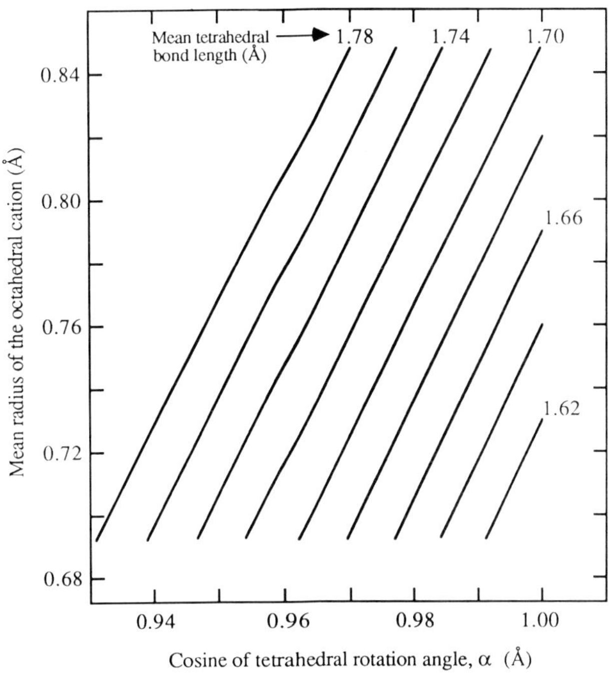

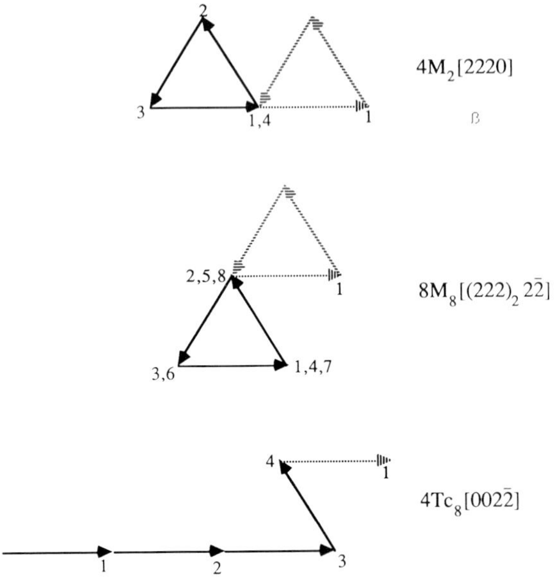

Figure 3-28. Stacking vector diagrams for three representative complex polytypes.

dant?" The tendency for micas of any given composition to occur mostly in one or two simple polytypes suggests that polytypism may be an equilibrium phenomenon that can be treated in terms of classical thermodynamics. If this is true, however, it must be immediately admitted that the energy differences among observed polytypes would be very small because the crystal structures of individual layers are so similar, and thus the frequent occurrence and virtually interminable persistence of metastable polytypes would make it very difficult to assess attainment of thermodynamic equilibrium. While it is possible that temperature and pressure of formation, or the chemical environment in which the mica grew, are important, these arguments are seriously weakened by the coexistence of two or more polytypes in the same rock; surely temperature, pressure, and chemical gradients between such coexistent minerals must have been small to zero. Control of polytypism has been attributed to ditrigonal distortion of the tetrahedral sheet, but a high-temperature structure refinement of a phlogopite [59] indicates that ditrigonal distortion decreases with increasing temperature; inasmuch as micas form at high temperature where ditrigonal distortion is thus reduced, such distortions may exert less control on polytypism than their room-temperature magnitudes would suggest. It is possible, of course, that polytypism, particularly where complex polytypes are concerned, is not an equilibrium phenomenon at all. Perhaps there is no single explanation that applies to all micas, but it would seem that any explanation that is proposed must address the following observations:

1. The most common polytype for dioctahedral micas is $2M_1$, while the most common one for trioctahedral micas is $1M$.
2. The octahedral sheets, and therefore the tetrahedral sheets attached to them, are

distorted differently for dioctahedral micas than for trioctahedral micas, due to the differences in cation ordering.

3. There appears to be some connection between the particular polytype and the symmetry of individual mica layers. For example, a $1M$ muscovite necessarily has individual layers of symmetry $C2/m$, but for two particular muscovites of other polytypes (a $2M_1$ and a $3T$), the symmetries of individual mica layers were found to be $C\bar{1}$ and $C2$, respectively [22]. In a coexisting $1M$–$2M_1$ biotite pair [60], layer symmetries were found to be $C2/m$ and $C\bar{1}$, respectively. Thus it appears that *intra*layer effects are sometimes important.
4. In contrast to (3), layer structures were found to be identical in coexisting $1M$–$2M_1$ oxybiotites [40] even though the polytypes were different. Oxybiotite contains no OH, so perhaps the interactions between the OH and interlayer cations are important. The tendency for dioctahedral micas to crystallize in the $2M_1$ polytype and for trioctahedral micas to prefer $1M$ may be related to the orientation of the O–H vector in the hydroxyl group in the apical oxygen plane. In trioctahedral micas the hydrogen atom is repelled equally by the three octahedral cations to which the OH is bonded, so that the O–H vector is perpendicular to (001) and thus directed toward the interlayer cation and away from the plane of the octahedral cations. This orientation would tend to weaken the interlayer bond. In dioctahedral micas, the $M1$ site is empty, and so the repulsion due to the two $M2$ cations and the interlayer cation causes the O–H vector to be inclined toward the empty site. While this description is admittedly oversimplified because it ignores complications arising from ordering of different chemical species between or among otherwise similar octahedral sites, it at least points out a difference that may be significant in the context of the present question. Exactly how hydroxyl orientation might influence polytype preference is not understood.

The theories that have been proposed to account for polytypism fall generally into one of two quite different categories. One category assumes that the generation of polytypes is a nonequilibrium crystal growth process—one that depends upon "accidental," but enumerable and thus somewhat predictable, malformations in the initial stages of crystallite formation. The other considers growth of polytypes to result from equilibrium competition between interactions among structural units such as sheets or layers. Both approaches have strengths and weaknesses, and we shall briefly outline four approaches to the problem, the first of which is a nonequilibrium treatment, and the other three of which assume that polytypes result from equilibrium phenomena.

Let us begin by imagining a mica crystal nucleus that is growing in a layer-upon-layer mode in one of the six simple polytypic modifications. Let us then assume that one or more "mistakes"—so-called *stacking faults*—are made in the stacking order, breaking the polytypic sequence; and further, that this is followed by the development of a screw dislocation, or spiral growth step, as depicted in Figure 3-29. (Stacking faults are not an integral feature of all spiral growth models (e.g., [7]), but are required in the one that currently enjoys the most predictive success,

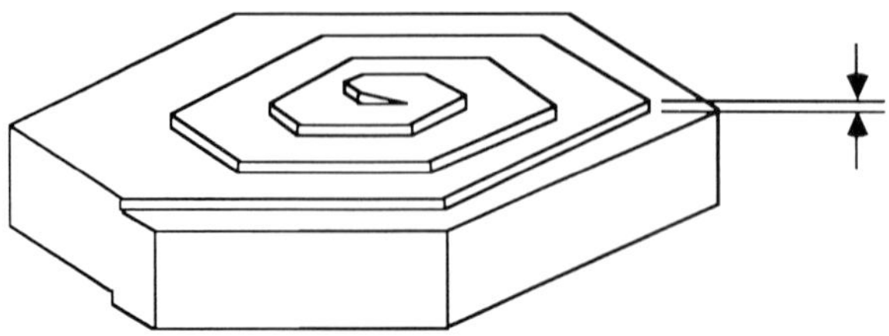

Figure 3-29. A schematic illustration of a spiral growth step on a mica crystallite. The height of the step is the repeat distance of the polytypic layer sequence.

which will be described below.) The ledge on the side of the screw dislocation will then expose the near-surface mica layers, including the stacking fault or faults. As atoms are added to the growth ledge and the spiral dislocation sweeps around the surface of the mica crystal in ramplike fashion, the precise stacking order exposed by it will be propagated, resulting in a complex polytype with a period dependent on the height of the growth step. This could obviously yield polytypes of very long periods.

Because relatively few of the large number of possible complex polytypes have been observed, acceptance of the spiral growth model of polytype generation should depend on its degree of success at predicting the predominance of observed polytypes. An evaluation of the success of one version of the model has indeed been carried out for the cases in which the initial (i.e., prefaulted) mica platelet crystallizes in either the $1M$, $2M_1$, or $3T$ modifications [41]. The $2M_2$, $2Or$, and $6H$ polytypes were not considered because they are relatively uncommon, and because none of the complex polytypes so far known is based on any of them. The scheme used in this study for predicting the occurrence of complex polytypes was: first, determine all geometrically possible stacking faults that can occur in each of the three simple polytypes considered; second, determine the most probable stacking fault configurations by estimating stacking fault energies; third, deduce the possible complex polytypes that can result from a screw dislocation containing these most likely stacking fault configurations at different positions within the exposed ledge of the step; and finally, predict the most probable complex polytypes by estimating the relative stacking fault energies of all polytypes derived in the third step that are generated by screw dislocations of equal strength in crystals having the same initial simple polytype.

The results of this study are somewhat encouraging. About 60% of observed complex polytypes are found in the list of "most probable" polytypes. Another 10% or so are included in the list of "second most probable" polytypes. The other 30% result from higher-energy stacking fault configurations and should occur less frequently than the others. It is possible that some of these could be low-energy polytypes resulting from multiple exposed stacking faults on the spiral growth step, but in this study only single exposed faults were considered.

Two problems are evident. First, not only are some apparently high-energy polytypes observed, but some that are predicted in the low-energy "most probable" list have not been found. This may be merely a happenstance of limited sampling, and these others may yet appear. The second problem is more serious for a general theory: While a reasonably good job is done at predicting which complex polytypes should occur, the theory is absolutely silent about the simple ones, which constitute most of the polytypes found in nature. Direct observation leaves no question that spiral growth steps occur in sheet silicates, but it is quite unlikely that a spiral growth mechanism can be called on to explain the simple polytypes.

As a second example, we consider an approach based on the ionic bonding model (Chapter 11), in which simple polytypes are considered as products of equilibrium crystal growth [2]. Ionic forces include Coulombic, short-range repulsive, and van der Waals contributions. In this particular study sets of model structures were constructed so as to minimize differences between the latter two contributions within any given set, and only Coulombic forces were considered. Model sets were constructed for pyrophyllite, talc, muscovite, biotite, and anandite, and energies were calculated in two or more of the simple polytypes $1M$, $2M_1$, $2M_2$, and $2Or$. The results are consistent with the observation that $1M$ and $2M_1$ polytypes are much more abundant than $2M_2$ and $2Or$ polytypes. Moreover, it appears that the most influential structural variable influencing the very strong tendency of dioctahedral micas to occur in the $2M_1$ polytype and trioctahedral micas to be $1M$ is the corrugation of the tetrahedral layer, measured as Δz_0 (p. 115). Most micas in which Δz_0 exceeds 0.10 Å occur as $2M_1$ polytypes regardless of other structural parameters such as octahedral flattening or tetrahedral rotation. For micas in which Δz_0 is less than 0.10 Å the usual polytype is $1M$.

Some models of polytypic behavior that emphasize the short-range interactions necessary in order to consider polytypism as an equilibrium phenomenon have been developed for and applied in detail only to structures that are unrelated to micas. We shall review two of these that appear to have the potential for elucidating the formation of some complex polytypes as well as simple ones, although they are not currently applicable to micas per se.

We begin consideration of the "interaction energy model" [43] by assuming a polytypic structure consisting of identical layers that can be stacked only either "up" or "down." Because up and down are strictly relative terms, the sequence "all up" is identical in every way to the sequence "all down." The notation is straightforward: A "band" is an unbroken sequence of layers that are all stacked in the same way, and one or more numbers enclosed in brackets represent the numbers of layers in each band of the repetitive sequence. For example, $\cdots \uparrow\uparrow\downarrow\downarrow\downarrow \cdots$ is written $\langle 23 \rangle$, indicating that a band of two "up" (or "down") is followed by a band of three "down" (or "up") layers. $\langle 2 \rangle$ and $\langle 1 \rangle$ represent $\cdots \uparrow\uparrow\downarrow\downarrow \cdots$ and $\cdots \uparrow\downarrow\uparrow\downarrow \cdots$, respectively. A polytype in which all layers are stacked identically, as in the $1M$ micas, is represented as $\langle \infty \rangle$.

The stacking of layers together must involve some sort of interaction energy W_1, $W_2, \ldots, W_i$ between any module and the ith adjacent layer. Moreover, the interaction will be different for the case of parallel stacking ($\uparrow\uparrow$) than it will for antiparallel stacking ($\uparrow\downarrow$). If we let N represent number of layers (the total number of layers, if not subscripted) and W the interaction energy, and the subscripts p and a stand respectively for parallel and antiparallel stacking arrangements, then the total interaction energy over i layers is

$$\begin{aligned} E &= -\Sigma_i(N_{p,i}W_{p,i} + N_{a,i}W_{a,i}) \\ &= -\Sigma_i(NW_{p,i} - N_{a,i}W_{p,i} + N_{a,i}W_{a,i}) \\ &= -\Sigma_i[NW_{p,i} - N_{a,i}(W_{p,i} - W_{a,i})] \\ &= -N\Sigma_i[W_{p,i} - (N_{a,i}/N)(W_{p,i} - W_{a,i})] \\ &= N\Sigma_i(Z_iW_i - W_{p,i}), \end{aligned}$$

where Z_i is the proportion of ith neighbor layers with antiparallel stacking and W_i is the difference between parallel and antiparallel interaction energies for the ith neighbors.

The interaction energy per layer is then just $E' = \Sigma_i(Z_iW_i - W_{p,i})$. If we assume that the values of $W_{p,i}$ and $W_{a,i}$ are independent of the overall arrangement of layers, then $E' = W_1\Sigma_i(Z_iW_i/W_1) - \Sigma_iW_{p,i}$, and the last term is a constant for any given number of layers, i. If we include first-, second-, and third-nearest neighbors the interaction energy per layer is

$$E' = W_1[Z_1 + Z_2(W_2/W_1) + Z_3(W_3/W_1)] - C,$$

and the stable polytype will be the one with the lowest interaction energy. If $W_1 < 0$, then energy minimization will occur when the bracketed quantity is maximized, whereas if $W_1 > 0$, a minimum in the bracketed quantity will yield minimum energy.

Although W_1, W_2, and W_3 are not known for any polytypic structures, Z_i ($i =$ 1,2,3) are easily determined for any polytype by inspection of the stacking arrangement. By postulating ranges of values of W_2/W_1 and W_3/W_1, it is then a simple (though tedious) matter to find which of any given set of possible polytypes has the lowest energy for each pair of ratios, and to plot phase diagrams in "W_2/W_1–W_3/W_1 space" for $W_1 < 0$ and $W_1 > 0$. In a study directed at spinelloids, but applicable to any polytypes in which each layer can be stacked in one of only two opposite ways, 22 polytypes were evaluated in this way, for ratio values of +20 to −20, and the resulting phase diagrams are shown in Figure 3-30. The top phase diagram is for the case in which $W_1 < 0$, and the other is for $W_1 > 0$. Note that the symmetry of the problem imposes symmetry relationships between the geometries of the two phase diagrams, but not necessarily among the polytypes occurring in corresponding fields.

Observe that only eight or nine polytypes have stability fields at all, while three (a) or four (b) of those are very narrow, actually constituting degeneracies along phase boundaries. In Figure 3-30(a), four of the five major fields do indeed represent stable, observed spinelloid structures, and $\langle 1 \rangle$, which has the largest field of stability in the phase diagram, is the most commonly observed one. Thus it is assumed that $W_1 < 0$ for spinelloids. The polytype $\langle \infty \rangle$, predicted by the phase diagram, has never been observed in spinelloids, however. Among the phases represented on the figures, only $\langle \infty \rangle$ and $\langle 1 \rangle$ correspond to simple mica polytypes—1*M* and 2*Or*, respectively—if layers must be stacked in one of two opposite ways. Because the

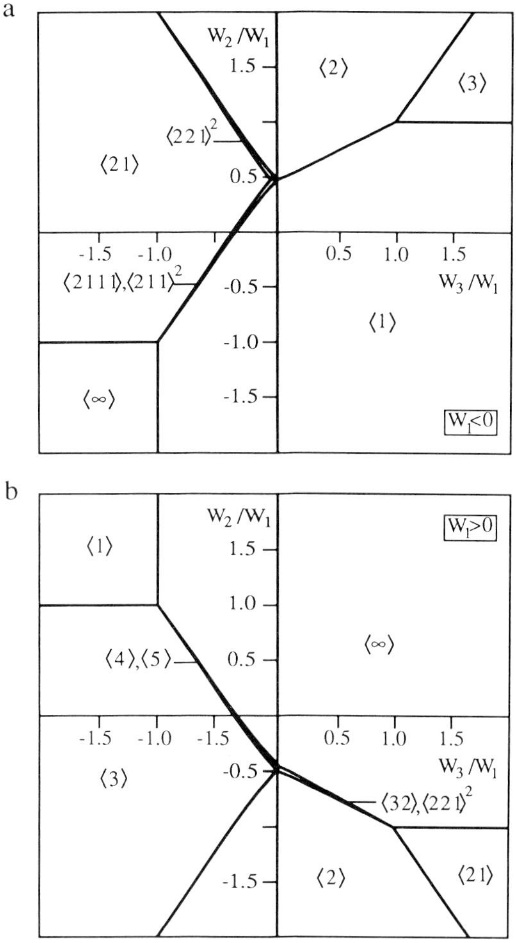

Figure 3-30. A phase diagram in W_3/W_1–W_2/W_1 space depicting regions of stability for polytypes that can be stack in only two ways ("up" or "down"), if $W_1 > 0$. See text for definitions of symbols. (After [43].)

$1M$ polytype is vastly more abundant than the $2Or$ polytype, we would assume that Figure 3-30(b) is the appropriate diagram, and that W_1 is therefore larger than zero for micas. The other polytypes shown are complex polytypes that have not been observed to be abundant.

This equilibrium model of polytype generation has some strong points with regard to micas, although it cannot be applied to them in its present form because mica layers can stack in six ways, not two. It does predict that relatively few polytypes will be stable despite very large numbers of possible sequences, and that is substantiated by observation. On the other hand, it does not in its present form predict very-long-period complex polytypes, which though less common than simple polytypes, do exist.

Another model of equilibrium growth is called the axial next-nearest-neighbor Ising (ANNNI) model [44; 66], and it may hold the potential for predicting both simple and complex polytypes. Both the name and the model are taken from a statistical-mechanical treatment of magnetic phase transitions, for which the model was originally proposed. Like the interaction energy model, it has been applied to spinelloids because each layer is stacked in one of two possible ways and, correspondingly, the Ising magnetic spin variable can take on one of two values, $+1$ and -1. All spinelloids consist of discrete structural modules stacked in three mutually orthogonal directions, but for any given spinelloid the stacking arrangement in two of the three is constant, yielding sheets that may be stacked differently in the remaining direction. In the ANNNI model, each module is reduced to a lattice point, and a spin variable is associated with each lattice point of the resulting cubic lattice. The interactions between modules can then be treated as if they were interactions between spin variables. A Hamiltonian that describes the system is written as follows:

$$H = H_0 - J_1 \Sigma_{NN} s_i s_j - J_2 \Sigma_{NNN} s_i s_j.$$

The H_0 term describes interactions within layers, but is independent of the stacking arrangement of layers. J_1 and J_2 are nearest-neighbor (NN) and next-nearest-neighbor (NNN) spin interaction parameters along the axial (stacking) direction, and the spin variables, s_i and s_j, are $+1$ or -1 depending upon whether modules are stacked "up" or "down."

The phase diagram in J_1–J_2 space has been determined, and is shown in Figure 3-31(a) for T=0 K (absolute zero of temperature). Only three stable polytypes appear, but as in the phase diagram of the interaction energy model, some of the boundaries (namely, between $\langle 1 \rangle$ and $\langle 2 \rangle$, and between $\langle \infty \rangle$ and $\langle 2 \rangle$) turn out to be highly degenerate; that is, several phases of equal energy occur exactly on them. At high temperature [(Figure 3-31(b)], these boundaries become bands of finite width that contain stability fields for several longer-period polytypes. It can be shown that, in the region of these bands, other polytypes ($\langle 4 \rangle$ and $\langle 5 \rangle$, for example) have free energies only very slightly higher than those shown to be stable, suggesting the metastable persistence of many long period polytypes produced at temperatures at which spinelloids—or micas—grow.

The ANNNI model is similar in many respects to the interaction energy model, and, while exhibiting similar strengths, suffers from most of the same shortcomings. It predicts the observed preponderance of simple polytypes from equilibrium growth, but also provides for the existence of complex (long-period) polytypes. It does no better at predicting specific spinelloid polytypes, and, because of the restriction to only two layer (or module) orientations, is, in its current state of development, equally inapplicable to micas. On the other hand, if it could be adapted to six possible layer orientations—or even three—, and proper account taken of crystal chemical variables such as layer chemistry and cationic substitution, the ANNNI model might be the key to understanding both simple and complex polytypism in micas.

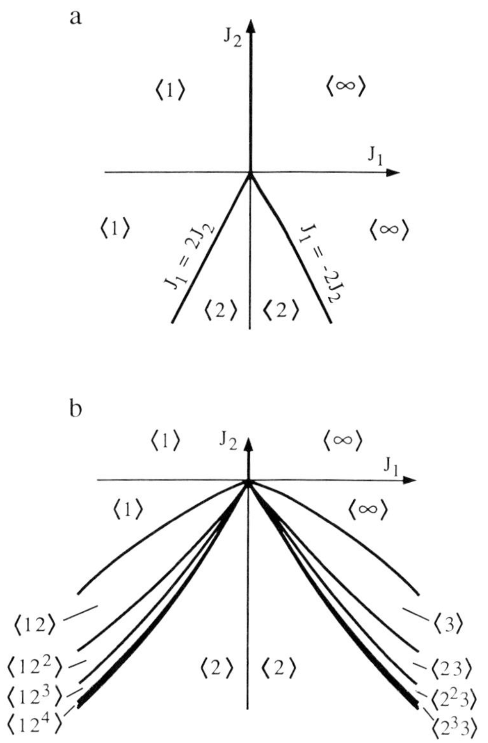

Figure 3-31. (a) A phase diagram in J_1–J_2 space at a temperature of 0 K. Only three polytypes are stable. (b) A similar diagram, but for elevated temperature, which allows a greater range of polytypes. See text for definitions of symbols. (After [44].)

That crystal chemical variables cannot be ignored in understanding polytype generation is clear, as evinced not only by the data of Tables 3-3 and 3-4, but by studies of coexisting micas. Of the latter, the following two examples are typical:

1. Phengite is a silica-rich, alumina-poor dioctahedral mica that is common in metamorphic rocks. In Cazadero, California (Franciscan Formation), it has been observed to coexist with the relatively uncommon sodium mica, paragonite, where the former occurs either as discrete crystals or intergrown in the latter [3]. The chemical formula for this phengite, obtained by electron microprobe, is $(K_{0.83}Na_{0.15})(Al_{1.49}Fe^{3+}_{0.25}Mg_{0.23}Ti_{0.22}Mn_{0.01})(Si_{3.21}Al_{0.79})O_{9.97}(OH)_{2.03}$, and that of the paragonite is $(Na_{0.91}K_{0.07}Ca_{0.01})(Al_{1.88}Fe^{3+}_{0.08}Mg_{0.02}Ti_{0.01}Fe^{2+}_{0.01})(Si_{3.02}Al_{0.98})O_{10}(OH)_2$. The phengite crystals form exclusively in the $2M_1$ polytype, whereas the paragonite (which is mostly $2M_1$, also) displays $1Mrn(120)$, $3T(?)$, and complex polytypes. Phengite intergrown in paragonite shows two-, three-, and four-layer periodicities, and random stacking. Clearly, chemical and structural differences must play some significant role here where external variables were presumably identical for the two phases. The paragonite has a smaller

average interlayer cation and octahedral sheet, and a larger tetrahedral sheet, suggesting that it likely experiences greater tetrahedral rotation. While structure refinements of these particular crystals were not carried out, the structures of noncoexisting phengite and paragonite of somewhat different compositions from these showed that the amounts of octahedral flattening were nearly identical, but yielded tetrahedral rotation angles of 7.9° and 16.2°, respectively.

2. In the Western Alps both the $3T$ and $2M_1$ polytypes of a phengitic mica occur in gneisses of high pressure paragenesis [55]. In some locations the two polytypes coexist with virtually no differences in composition. Careful microprobe analyses reveal that only the $2M_1$ polytype crystallizes if Mg/(Al+Fe+Mg) is less than 0.015. If that ratio exceeds 0.025, then only the $3T$ polytype forms. For the case in which $0.015 <$ Mg/(Al+Fe+Mg) < 0.025, $2M_1$ and $3T$ forms coexist, but almost always as discrete rather than interstratified crystals. Again, compositional control, perhaps because of its influence on structural parameters, is obvious, but the underlying reasons are not.

In quite a different sort of study, the crystal energies of dioctahedral $1M$ and $2M_1$ polytypes of fluor-mica were calculated using an electrostatic model, and the results were applied to natural micas [4]. It was found, first, that smaller octahedral cations produced more counter-rotation than larger ones, and second, that the energy of the $2M_1$ polytype was less than that of the $1M$ polytype at counter-rotation angles over 4° (although the energy difference was calculated to be <0.2 kcal/mol up to 6°). This was used to rationalize why muscovites and phengites, which typically have counter-rotations of at least 6°, occur most commonly in the $2M_1$ polytype, whereas celadonite, which has a larger average octahedral cation due to the presence of Mg, and thus presumably less counter-rotation, occurs as the $1M$ polytype. While this approach to polytypism does not answer all questions either, the implication is clear that crystal chemical variables have a substantial effect on polytype stability.

In summary, polytype configurations can be determined and fully described, but the theory of generation of polytypes largely remains an area with many more questions than answers. At this stage of development, it would seem prudent to accept some nonequilibrium phenomenon like spiral growth around one or more faulted screw dislocations as responsible for the propagation of long-period complex polytypes. It seems likely that the growth of simple polytypes in micas is an equilibrium phenomenon. While models such as the ANNNI and interaction energy models provide phase diagrams that possess some necessary properties of the correct phase diagrams, they are not at present applicable to micas and, even if they were, it is by no means clear how to quantitatively evaluate interaction energies or magnetic spin parameters for micas. Moreover, these models take no account of compositional variations at either the layer or the individual cation-site levels, although such variations are obviously very important. It is quite possible—perhaps even likely—that current research efforts are on the right track, and they are unquestionably among the most exciting in mineralogy, but a complete understanding of polytype generation in micas seems a distant goal indeed.

REFERENCES

[1] Abbott, R. N., Jr. (1984). "Al–Si Prdering in 1*M* Trioctahedral Micas." *Canadian Mineralogist* 22, 659–67.
[2] Abbott, R. N., Jr., and C. W. Burnham (1988). "Polytypism in Micas: A Polyhedral Approach to Energy Calculations." *American Mineralogist* 73, 105–18.
[3] Ahn, J. H., D. R. Peacor, and E. J. Essene (1985). "Coexisting Paragonite–Phengite in Blueschist Eclogite: A TEM Study." *American Mineralogist* 70, 1193–1204.
[4] Appelo, C. A. J. (1979). "Layer Deformation and Crystal Energy of Micas and Related Minerals. II. Deformation of the Coordination Units." *American Mineralogist* 64, 424–31.
[5] Bailey, S. W. (1975). "Cation Ordering and Pseudosymmetry in Layer Silicates." *American Mineralogist* 60, 175–87.
[6] Bailey, S. W. (1984). "Crystal Chemistry of the True Micas." In S. W. Bailey (ed.), *Micas. Reviews in Mineralogy*, Volume 13. Mineralogical Society of America, Washington, D. C.
[7] Baronnet, A. (1975). "Growth Spirals and Complex Polytypism in Micas. I. Polytypic Structure Generation." *Acta Crystallographica* A31, 345–55.
[8] Brown, B. E. (1978). "The Crystal Structure of a 3*T* Lepidolite." *American Mineralogist* 63, 332–36.
[9] Burnham, C. W., and E. W. Radoslovich (1964). "Crystal Structures of Coexisting Muscovite and Paragonite." *Carnegie Institution of Washington Year Book* 63, 232–36.
[10] Cerny, P., and D. M. Burt (1984). "Paragenesis, Crystallochemical Characteristics, and Geochemical Evolution of Micas in Granitic Pegmatites." In S. W. Bailey (ed.), *Micas. Reviews in Mineralogy*, Volume 13. Mineralogical Society of America, Washington, D. C.
[11] Dyar, M. D., and R. G. Burns (1986). "Mössbauer Spectral Study of Ferruginous One-Layer Trioctahedral Micas." *American Mineralogist 71, 955–65.*
[12] Finch, J., A. R. Gainsford, and W. C. Tennant (1982). "Polarized Optical Absorption and ^{57}Fe Mössbauer Study of Pegmatitic Muscovite." *American Mineralogist* 67, 59–68.
[13] Forbes, W. C. (1972). "An Interpretation of the Hydroxyl Contents of Biotites and Muscovites." *Mineralogical Magazine* 38, 712–720.
[14] Giuseppetti, G., and C. Tadini (1972). "The Crystal Structure of 2*O* Brittle Mica: Anandite." *Tschermaks Mineralogische und Petrographische Mitteilungen* 18, 169–84.
[15] Goodman, B. A. (1976). "The Mössbauer Spectrum of a Ferrian Muscovite and its Implications in the Assignment of Sites in Dioctahedral Micas." *Mineralogical Magazine* 40, 513–17.
[16] Green, T. H. (1981). "Synthetic High-Pressure Micas Compositionally Intermediate between the Dioctahedral and Trioctahedral Mica Series." *Contributions to Mineralogy and Petrology* 78, 452–58.
[17] Guggenheim, S. (1981). "Cation Prdering in Lepidolite." *American Mineralogist* 66, 1221–32.
[18] Guggenheim, S., and S. W. Bailey (1975). "Refinement of the Margarite Structure in Subgroup Symmetry." *American Mineralogist* 60, 1023–29.
[19] Guggenheim, S., and S. W. Bailey (1978). Refinement of the Margarite Structure in Subgroup Symmetry: Correction, Further Refinement, and Comments. *American Mineralogist* 63, 186–87.
[20] Guggenheim, S., and T. Kato (1984). "Kinoshitalite and Mn Phlogopites: Trial Refinements in Subgroup Symmetry and Further Refinement in Ideal Symmetry." *Mineralogical Journal* 12, 1–5.
[21] Guidotti, C. V. (1984). "Micas in Metamorphic Rocks." In S. W. Bailey (ed.), *Micas. Reviews in Mineralogy*. Volume 13, Mineralogical Society of America, Washington, D. C.
[22] Güven, N., and C. W. Burnham (1967). "The Crystal Structure of 3*T* Muscovite." *Zeitschrift für Kristallographie* 125, 163–83.
[23] Hazen, R. M., and D. R. Wones (1972). "The Effect of Cation Substitutions on the Physical Properties of Trioctahedral Micas." *American Mineralogist* 57, 103–29.
[24] Hazen, R. M., and D. R. Wones (1978). "Predicted and Observed Compositional Limits of Trioctahedral Micas." *American Mineralogist* 63, 885–92.
[25] Herrero, C. P., J. Sanz, and J. M. Serratosa (1985a). "Si,Al Distribution in Micas: Analysis by High-Resolution ^{29}Si NMR Spectroscopy." *Journal of Physics C: Solid State Physics* 18, 13–22.
[26] Herrero, C. P., J. Sanz, and J. M. Serratosa (1985b). "Tetrahedral Cation Ordering in Layer Silicates by ^{29}Si NMR Spectroscopy." *Solid State Communications* 53, 151–54.

[27] Hewitt, D. A., and D. R. Wones (1975). "Physical Properties of Some Synthetic Fe–Mg–Al Trioctahedral Biotites." *American Mineralogist* 60, 854–62.

[28] Iijima, S. and J. Zhu (1982). "Electron Microscopy of a Muscovite–Biotite Interface." *American Mineralogist* 67, 1195–1205.

[29] Joswig, W., G. Amthauer, and Y. Takéuchi (1986). "Neutron-Diffraction and Müssbauer Spectroscopic Study of Clintonite (Xanthophyllite)." *American Mineralogist* 71, 1194–97.

[30] Kato, T., Y. Miúra, M. Yoshii, and K. Maeda (1979). "The Crystal Structures of a 1M-Kinoshitalite, a New Barium Brittle Mica and 1M-Manganese Trioctahedral Micas." *Mineralogical Journal* 9, 392–408.

[31] Kinsey, R. A., R. J. Kirkpatrick, J. Hower, K. A. Smith, and E. Oldfield (1985). "High Resolution Aluminum-27 and Silicon-29 Nuclear Magnetic Resonance Spectroscopic Study of Layer Silicates, Including Clay Minerals." *American Mineralogist* 70, 537–548.

[32] Krzanowski, W. J., and A. C. D. Newman (1972). "Computer Simulation of Cation Distribution in the Octahedral Layers of Micas." *Mineralogical Magazine* 38, 926–35.

[33] Lin, C.-Y., and S. W. Bailey (1984). "The Crystal Structure of Paragonite-$2M_1$." *American Mineralogist* 69, 122–27.

[34] Lin, J.-C, and S. Guggenheim (1983). "The Crystal Structure of a Li,Be-Rich Brittle Mica: A Dioctahedral-Trioctahedral Intermediate." *American Mineralogist* 68, 130–42.

[35] McCauley, J. W., and R. E. Newnham (1971). "Origin and Prediction of Ditrigonal Distortions in Micas." *American Mineralogist* 56, 1626–38.

[36] Monier, G., and J.-L. Robert (1986). "Muscovite Solid Solutions in the System K_2O–MgO–FeO–Al_2O_3–SiO_2–H_2O: An Experimental Study at 2 kbar P_{H_2O} and Comparison with Natural Li-Free White Micas." *Mineralogical Magazine* 50, 257–66.

[37] Monier, G., and J.-L. Robert (1986). "Titanium in Muscovites from Two Mica Granites: Substitutional Mechanism and Partition with Coexisting Biotites." *Neues Jahrbuch für Mineralogie Abhandlungen* 153, 147–61.

[38] Munoz, J. L. (1968). "Physical Properties of Synthetic Lepidolites." *American Mineralogist* 53, 1490–1512.

[39] Munoz, J. L. (1984). "F-OH and Cl-OH Exchange in Micas with Applications to Hydrothermal Ore Deposits." In S. W. Bailey (ed.), *Micas. Reviews in Mineralogy*, Volume 13, Mineralogical Society of America, Washington, D. C.

[40] Ohta, T., H. Takeda, and Y. Takéuchi (1982). "Mica Polytypism: Similarities in the Crystal Structures of Coexisting $1M$ and $2M_1$ Oxybiotite." *American Mineralogist* 67, 298–310.

[41] Pandey, D., A. Baronnet, and P. Krishna (1982). "Influence of Stacking Faults on the Spiral Growth of Polytype Structures in Mica." *Physics and Chemistry of Minerals* 8, 268–78.

[42] Peterson, R. C., R. J. Hill, and G. V. Gibbs (1979). "A Molecular-Orbital Study of Distortions in the Layer Structures Brucite, Gibbsite and Serpentine." *Canadian Mineralogist* 17, 703–11.

[43] Price, G. D. (1983). "Polytypism and the Factors Determining the Stability of Spinelloid Structures." *Physics and Chemistry of Minerals* 10, 77–83.

[44] Price, G. D., and J. Yeomans (1984). "The Application of the ANNNI Model to Polytypic Behaviour." *Acta Crystallographica* B40, 448–54.

[45] Richardson, S. M., and J. W. Richardson, Jr. (1982). "Crystal Structure of a Pink Muscovite from Archer's Post, Kenya: Implications for Reverse Pleochroism in Dioctahedral Micas." *American Mineralogist* 67, 69–75.

[46] Robert, J.-L. (1976). "Phlogopite Solid Solutions in the System K_2O–MgO–Al_2O_3–SiO_2–H_2O." *Chemical Geology* 17, 195–212.

[47] Robert, J.-L., M. Volvinger, J.-N. Barrandon, and M. Basutçu (1983). "Lithium in the Interlayer Space of Synthetic Trioctahedral Micas." *Chemical Geology* 40, 337–51.

[48] Rosenberg, P. E., and F. F. Foit (1977). "Fe^{2+}–F Avoidance in Silicates." *Geochimica et Cosmochimica Acta* 41, 345–46.

[49] Ross, M., H. Takeda, and D. R. Wones (1966). "Mica Polytypes: Systematic Description and Identification." *Science* 151, 191–93.

[50] Rule, A. C., and S. W. Bailey (1985). "Refinement of the Crystal Structure of Phengite-$2M_1$." *Clays and Clay Minerals* 33, 403–09.

[51] Sanz, J., C. de la Calle, and W. E. E. Stone (1984). "NMR Applied to Minerals. V. The Localization of Vacancies in the Octahedral Sheet of Aluminous Biotites." *Physics and Chemistry of Minerals* 11, 235–40.

[52] Sartori, F. (1976). "The Crystal Structure of a $1M$ Lepidolite." *Tschermaks Mineralogische und Petrographische Mitteilungen* 23, 65–75.

[53] Sartori, F., M. Franzini, and S. Merlino (1973). "Crystal Structure of a $2M_2$ Lepidolite." *Acta Crystallographica* B29, 573–78.
[54] Shannon, R. D. (1976). "Revised Effective IUonic Radii and Systematic Studies of Interatomic Distances in Halides and Chalcogenides." *Acta Crystallographica* A32, 751–67.
[55] Stöckert, B. (1985). "Compositional Control on the Polymorphism ($2M_1$–$3T$) of Phengitic White Mica from High Pressure Parageneses of the Sesia Zone (Lower Aosta Valley, Western Alps; Italy)." *Contributions to Mineralogy and Petrology* 89, 52–58.
[56] Swanson, T. H., and S. W. Bailey (1981). "Redetermination of the Lepidolite-$2M_1$ Structure." *Clays and Clay Minerals* 29, 81–90.
[57] Takeda, H., and C. W. Burnham (1969). "Fluor-polylithionite: A Lithium Mica with Nearly Hexagonal $(Si_2O_5)^{2-}$ Ring." *Mineralogical Journal* 6, 102–09.
[58] Takeda, H., N. Haga, and R. Sadanaga (1971). "Structural Investigation of Polymorphic Transition Between $2M_2$-, $1M$-Lepidolite, and $2M_1$-Luscovite." *Mineralogical Journal* 6, 203–15.
[59] Takeda, H., and B. Morosin (1975). "Comparison of Observed and Predicted Structural Parameters of Mica at High Temperature." *Acta Crystallographica* B31, 2444–52.
[60] Takeda, H., and M. Ross (1975). "Mica Polytypism: Dissimilarities in the Crystal Structures of Coexisting $1M$ and $2M_1$ Biotite." *American Mineralogist* 60, 1030–40.
[61] Takéuchi, Y., and R. Sadanaga (1966). "Structural Studies of Brittle Micas (I) The Structure of Xanthophyllite Refined." *Mineralogical Journal* 4, 424–37.
[62] Toraya, H. (1981). "Distortions of Octahedra and Octahedral Sheets in $1M$ Micas and the Relation to their Stability." *Zeitschrift für Kristallographie* 157, 173–90.
[63] Toraya, H., S. Iwai, F. Marumo, and M. Hirao (1977). "The Crystal Structure of Taeniolite, $KLiMg_2Si_4O_{10}F_2$." *Zeitschrift für Kristallographie* 146, 73–83.
[64] Volfinger, M., J.-L. Robert, D. Vielzeuf, and A. M. R. Neiva (1985). "Structural Control of the Chlorine Content of OH-Bearing Silicates (Micas and Amphiboles)." *Geochimica et Cosmochimica Acta* 49, 37–48.
[65] Weiss, Z., M. Rieder, M. Chmielova, and J. Krajicek (1985). "Geometry of the Octahedral Coordination in Micas: A Review of Refined Structures." *American Mineralogist* 70, 747–57.
[66] Yeomans, J., M., and G. D. Price (1985). "An Equilibrium Theory of Polytypism." Department of Theoretical Physics, University of Oxford, Publication 26/85.

4
Pyroxenes

GENERAL CHEMISTRY OF PYROXENES

The general chemical formula for the chemically diverse pyroxene group is XYZ_2O_6, where X is the cation that occupies a 6- to 8-coordinated polyhedron conventionally labeled $M(2)$; Y is the cation that occupies the $M(1)$ site, which is always a somewhat distorted octahedron; and Z is the tetrahedrally coordinated cation of the T site. In natural pyroxenes

$$X = \text{Mg, Fe}^{2+}\text{, Ca, Na, Li, Mn}$$

$$Y = \text{Mg, Fe}^{2+}\text{, Fe}^{3+}\text{, Mn, Al, Cr}^{3+}\text{, Zn, Sc, Ti}^{4+}\text{, Ti}^{3+}\text{, V}^{3+}\text{, Zr}$$

$$Z = \text{Si, with substitution by Al and Fe}^{3+}.$$

Individual elements listed as X and Y cations may occur exclusively in their respective polyhedra (except for Ti, V, and Zr, which occur only as major or minor substituents), in which cases the formulas represent end-member pyroxenes; in addition, they may, and commonly do, occur together in several extensive solid solution series produced by mutual substitution of the atoms listed in a given position. The Z cation is almost always mostly Si, although Al substitution of up to 0.8 atoms per 2.0 tetrahedral cations has been observed [21] and there may be minor substitution of Fe^{3+} (for example, [27]). As in most rock-forming silicates, many other elements occur in trace amounts in natural pyroxenes, but it is not our purpose to detail the trace-element chemistry. Pyroxenes with up to 1.00 atom of tetrahedrally coordinated aluminum (e.g., $CaAlSiAlO_6$) have been synthesized, as have pyroxenes of unusual X and Y chemistry. Some nonmineralogical inorganic compounds like cobalt germanate ($CoGeO_3$) and sodium vanadate (α-$NaVO_3$) are known to have the pyroxene structure. (Note that if the X and Y cations are of the same chemical species, it is not uncommon to see the pyroxene formula written as YZO_3; for example, enstatite, $Mg_2Si_2O_6$, is often written simply $MgSiO_3$. This is not incorrect, but because it obscures the nature of solid solutions with pyroxenes for which the expanded formula must be used, we shall use the expanded formulas throughout.)

It is convenient to group the pyroxenes into a few broad chemical families. The vast majority of natural pyroxenes fall within or near the "pyroxene quadrilateral" shown in Figure 4-1. The abbreviations at the corners are En for enstatite ($Mg_2Si_2O_6$), Fs for ferrosilite ($Fe_2Si_2O_6$), Di for diopside ($CaMgSi_2O_6$), and Hd for hedenbergite ($CaFeSi_2O_6$). The En and Fs compositions occur in different polymorphic modifications, so En and Fs are used here strictly as compositional abbreviations without implication as to crystal structure—much as Or was used in the discussion of the feldspar compositions (p. 37).

Substantial solid solution (to at least $En_{20}Fs_{80}$) exists between En and Fs, although pure $Fe_2Si_2O_6$ is a synthetic high-pressure phase that has not been observed in the pyroxene polymorphic modification in nature. Natural pyroxenes essentially of the composition $En_{10}Fs_{90}$, but containing very small amounts of Ca and Mn, have been documented (e.g., [1;32]). Complete solid solution occurs between Di and Hd. Solid solution between the high-calcium (Di–Hd) and the low calcium (En–Fs) compositional joins is quite limited, however; its extent is indicated by the miscibility gap in Figure 4-1. The high-calcium intermediate members of this restricted solid solution series are the *augites*, and the low-calcium intermediate members are the *pigeonites* if monoclinic; if orthorhombic, they are called by the end-member names, either *enstatite* or *ferrosilite* (whichever is closer in composition to the pyroxene in question, although the compositional modifiers "ferroan" and "magnesian" can be used with enstatite and ferrosilite, respectively, to indicate intermediate compositions). Those that crystallize in orthorhombic space groups are collectively called the *orthopyroxenes* (often abbreviated as opx), while the high-calcium pyroxenes are the *clinopyroxenes* (cpx). To avoid confusion the term *cli-*

Figure 4-1. The pyroxene quadrilateral. For natural pyroxenes with compositions represented on the quadrilateral, the shaded area is a miscibility gap as determined from chemical analyses given in [24]. Compositions of rapidly quenched metastable subcalcic augites fall in the shaded area between augite and pigeonite. Pyroxenes of virtually any composition within the miscibility gap can be synthesized, but they do not appear to exist stably in nature.

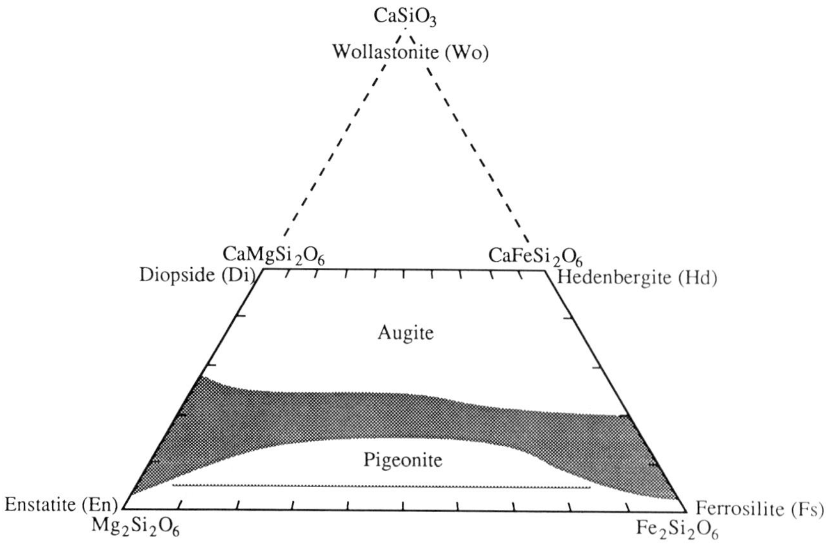

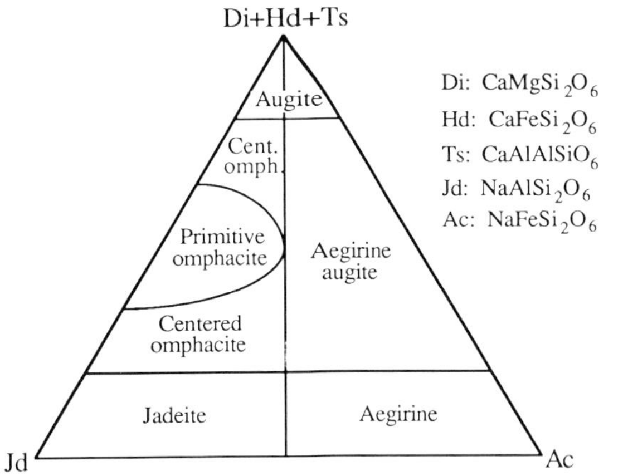

Figure 4-2. The jadeite–aegirine–augite compositional triangle. The compositional limits of omphacite and aegirine augite are Jd_{20}–Jd_{80} and Ac_{20}–Ac_{80}, respectively. The intersections of the primitive omphacite solvus with the triangle occur at about Jd_{35} and Jd_{65}.

nopyroxene is sometimes restricted to high-Ca quadrilateral pyroxenes, and is avoided for pigeonites or for pyroxenes with compositions that do not fall on or near the pyroxene quadrilateral, even though these are indeed monoclinic. Where this is not done, the context clarifies the type of monoclinic pyroxenes under discussion.

If the En–Di and Fs–Hd edges of the quadrilateral were extended, they would meet at a corner representing the composition $CaSiO_3$. While the formula is pyroxenelike, the structure is slightly different from that of a pyroxene, and $CaSiO_3$ (wollastonite, abbreviated Wo) belongs to the pyroxenoid group. (We shall shortly see that pyroxenes have a tetrahedral chain with a two-tetrahedron repeat; pyroxenoids have topologically similar chains, but the repeat is greater than two tetrahedra.) Compositions within the quadrilateral are often given in mole percentages of En, Fs, and Wo; thus diopside is $En_{50}Wo_{50}$, pigeonites do not usually exceed 15% Wo, and augites fall in the range 25–50% Wo. Pyroxenes with compositions that fall within the miscibility gap are metastable at surface conditions, and are sometimes called *subcalcic augites* or *subcalcic diopsides*. Usually pyroxenes that have bulk compositions within this range are found to consist of two exsolved phases—either augite plus pigeonite or augite plus orthopyroxene. Where this is not so, the subcalcic augites have been very rapidly quenched so that exsolution was prevented. Pyroxenes with compositions that fall within the miscibility gap have been synthesized, but they are metastable [69].

Although the bulk of common pyroxenes fall near the pyroxene quadrilateral, some that are of considerable interest in mineralogy and petrology do not. The sodic pyroxenes are *jadeite* ($NaAlSi_2O_6$) and *aegirine* ($NaFe^{3+}Si_2O_6$), and there is complete solid solution between them. (Aegirine is also called acmite, although aegirine is currently the preferred name.) The compositional triangle in Figure 4-2 shows that incorporation of $CaMgSi_2O_6$ (Di), $CaFeSi_2O_6$ (Hd), and $CaAl(SiAl)O_6$

compositions into the aegirine-rich sodic pyroxenes yields *aegirine–augite*, whereas on the jadeitic side of the diagram it yields *omphacites*. ($CaAlSiAlO_6$ is *calcium Tschermak's pyroxene*—CaTs—which has been synthesized [52] but has not been found in nature). All of these minerals are monoclinic, and omphacite is represented by polymorphs that crystallize in two different monoclinic space groups, as indicated in the figure. Some ophacites that apparently originate in the mantle exhibit a deficiency in total $X+Y$ cations, and thus have nonstoichiometric formulas [39]. Two other relatively common end-member compositions represented among pyroxenes are *johannsenite*, $CaMnSi_2O_6$, and *spodumene*, $LiAlSi_2O_6$, both of which are monoclinic. There are a few other end members that are less common (see Table 4-1).

Table 4-1 summarizes the compositions of the thirteen natural pyroxene end members and the seven recognized intermediate pyroxenes, as well as their space groups, as currently defined by the International Mineralogical Association's Subcommittee on Pyroxenes of the Commission on New Minerals and Mineral Names [41]. For pyroxenes that contain concentrations of other elements in excess of limits defined in reference [41], adjectives such as manganoan, ferroan, sodian, and so forth, may be used. (The work of this subcommittee, incidentally, made over 100

Table 4-1. Accepted Names for Pyroxene End Members and Solid Solutions

End Member	Solid Solution	Composition	Space Group
Mg–Fe^{2+}–Mn pyroxenes			
Enstatite		$Mg_2Si_2O_6$	*Pbca*
Clinoenstatite		$Mg_2Si_2O_6$	$P2_1/c$
Ferrosilite		$Fe_2^{2+}Si_2O_6$	*Pbca*
Clinoferrosilite		$Fe_2^{2+}Si_2O_6$	$P2_1/c$
	Pigeonite	$(Mg,Fe^{2+},Ca)_2Si_2O_6$ $(Ca<0.2)$	$P2_1/c$
Kanoite		$MnMgSi_2O_6$	$P2_1/c$
	Donpeacorite	$(Mn,Mg)MgSi_2O_6$	*Pbca*
Ca pyroxenes			
Diopside		$CaMgSi_2O_6$	$C2/c$
Hedenbergite		$CaFe^{2+}Si_2O_6$	$C2/c$
	Augite	$(Ca,Mg,Fe^{2+})_2Si_2O_6$ $(Ca\geq0.2)$	$C2/c$
Johannesenite		$CaMnSi_2O_6$	$C2/c$
Petedunnite		$CaZnSi_2O_6$	$C2/c$
Esseneite		$CaFe^{3+}(AlSi)O_6$	$C2/c$
Ca–Na pyroxenes			
	Omphacite	$(Ca,Na)(R^{2+},Al)Si_2O_6$	$C2/c$ $P2/n$
	Aegirine–augite	$(Ca,Na)(R^{2+},Fe^{3+})Si_2O_6$	$C2/c$
Na pyroxenes			
Jadeite		$NaAlSi_2O_6$	$C2/c$
Aegirine		$NaFe^{3+}Si_2O_6$	$C2/c$
Kosmochlor		$NaCr^{3+}Si_2O_6$	$C2/c$
Jervisite		$NaSc^{3+}Si_2O_6$	$C2/c$
Li pyroxene			
Spodumene		$LiAlSi_2O_6$	$C2$

formerly used names obsolete. Those names are found throughout literature older than 1988, of course, and may persist somewhat beyond that.) Most natural pyroxenes fall somewhere in between the end member compositions given in the table, and even those for which the pyroxene quadrilateral can be used appropriately usually contain some nonquadrilateral chemical components.

To divide a pyroxene formula between quadrilateral and nonquadrilateral components, the structural formula must first be constructed. This consists of three steps [41]: (1) Sum the tetrahedrally coordinated atoms to 2.000 using Si, Al, and then Fe^{3+}, in that order. (2) Sum the cations in $M(1)$ to 1.000 using all remaining Al and Fe^{3+}, followed by Ti^{4+}, Cr^{3+}, V^{3+}, Ti^{3+}, Zr^{4+}, Sc^{3+}, Zn^{2+}, Mg, Fe^{2+}, and Mn^{2+} (in that order) until the sum of 1.000 is reached. (3) Finally sum the cations in $M(2)$ to 1.000 if possible, using all remaining Mg, Fe^{2+}, and Mn^{2+}, then add all Li, Ca, and Na. If the sum in this site is not close to 1.000, the chemical analysis should be viewed with suspicion. With the formula thus constructed, the division into quadrilateral and nonquadrilateral components is accomplished in four further steps (modified from [7]): (1) Add the number of atoms of Ca, Fe^{2+}, and Mg in $M(2)$ and multiply by 100. (2) Divide the number of atoms of Si (per six oxygens) by 2 and multiply by 100. (3) Add Mg and Fe^{2+} in $M(1)$ and multiply by 100. (4) The smallest of the three numbers obtained in steps (1), (2), and (3) is the percent quadrilateral component in the pyroxene. Naturally, the nonquadrilateral component is 100 minus the quadrilateral component.

CRYSTAL STRUCTURE—AN OVERVIEW

The principal structural units of all pyroxenes are single chains of tetrahedra that share polyhedral corners with, and are parallel to, bands or strips of edge-sharing octahedra that run along the **c** direction. As in micas, the apical oxygens of the tetrahedra constitute the oxygens of the octahedra, but here there are no hydroxyls to supplement them as there are in the micas. The tetrahedral chain, in a somewhat idealized form that lacks the distortions found in the chains of real pyroxenes, is shown in Figure 4-3. The repeat distance along the **c** direction can be seen to be two tetrahedra, or about 5.2 Å.

Figure 4-4 is a view down the **c** direction, that is, with the chains of tetrahedra seen end-on. Along the **b** direction the tetrahedral chains are stacked alternately "up" and "down" adjacent to one another (apices pointing alternately along $+\mathbf{a}^*$ and $-\mathbf{a}^*$). Along the **a** direction the same sort of alternation is seen, so that tetrahedral chains abut bases to bases or apices to apices. This juxtaposition creates two kinds of octahedral sites: The smaller is $M(1)$, and is found between the apices of two abutting tetrahedral chains; the larger (which may become a 7- or 8-coordinated site with distortion of the tetrahedral chains) is $M(2)$, which is located at the edges of abutting tetrahedral bases. Figure 4-4 reveals that the pyroxene structure is a type of layered structure, in that tetrahedra and octahedra occupy different layers, which are parallel to (100). In terms of structural integrity and cation occupancy, however, the layers are discontinuous, and so they do not exert the same degree of control over physical properties that the layers in sheet silicates do.

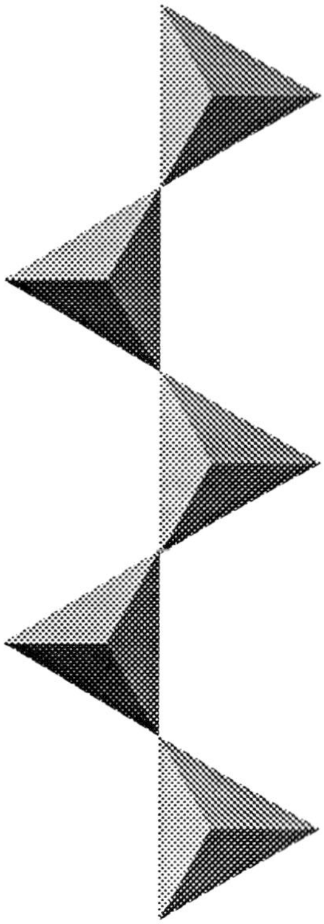

Figure 4-3. The pyroxene tetrahedral chain. Apical oxygens are designated O(1), the unshared basal oxygens are O(2), and bridging oxygens are O(3).

Space Groups

The space groups in which pyroxenes crystallize are shown in Figure 4-5, along with schematic representations of the crystal structures. In each case, the *b* axis is horizontal in the plane of the page, and the vertical axis in that plane is $a \sin \beta$; for the two orthorhombic space groups, $a \sin \beta$ is, of course, just *a*, because β is 90°. Accordingly, *c* comes directly out of the page for each of the structure drawings, and the view in each is down the tetrahedral chain (represented as the triangular cross-sections of tetrahedra). The patterns with which the tetrahedra, the *M*(1) sites (smaller circles), and the *M*(2) sites (larger circles) are filled indicate crystallographic equivalence or nonequivalence among those sites in the various space groups. Two space groups show more than one type of tetrahedron per tetrahedral chain, and two have two distinct types of tetrahedral chain, referred to as the *A* and *B* chains. Likewise, some space groups contain only one type of *M*(1) site and one type of *M*(2) site, whereas others have two of each type.

Nonequivalence suggests, of course, geometrical and perhaps chemical difference, and therefore the possibility of ordering. We shall postpone a discussion of that topic in detail (see p. 172), but here make a cautionary observation concerning a point that is sometimes confused in the professional literature because of careless choices of words. Because the space group can be determined—or at least narrowed to a few choices—by single-crystal x-ray techniques (see Chapter 10) without knowing the crystal structure of a mineral, it is sometimes taken to be a cause rather than an effect. It is not correct to say that a particular space group, say $P2_1/c$, *allows* the ordering of cations in the tetrahedral sites. Rather, it is the ordering of cations in the tetrahedral sites that *causes* the space group of the pyroxene to be $P2_1/c$. The ordering pattern determines the space-group symmetry. In some cases it is not ordering, but simply geometrical differences in coordination polyhedra that result in one particular space group rather than another of higher symmetry. Figure 4-5 thus shows the various distributions of nonequivalent sites observed in pyroxenes, and the space groups in which those distributions result.

In passing, we note two space groups that have been suggested, but probably do not occur among pyroxenes. The orthorhombic space group $P2_1ca$ has been

Figure 4-4. The general features of the pyroxene atomic arrangement are shown in this pyroxene structure, projected down **c**, with **a** vertical and **b** horizontal (the dashed box outlines the clinopyroxene unit cell). The small spheres represent the Y cations, which occupy the $M(1)$ octahedra, and the larger ones represent the X cations, which occupy $M(2)$. The pyroxene chains are seen end-on. For orthopyroxenes, the repeat along **a** is twice what it is in monoclinic pyroxenes, but the general features of the structure are the same.

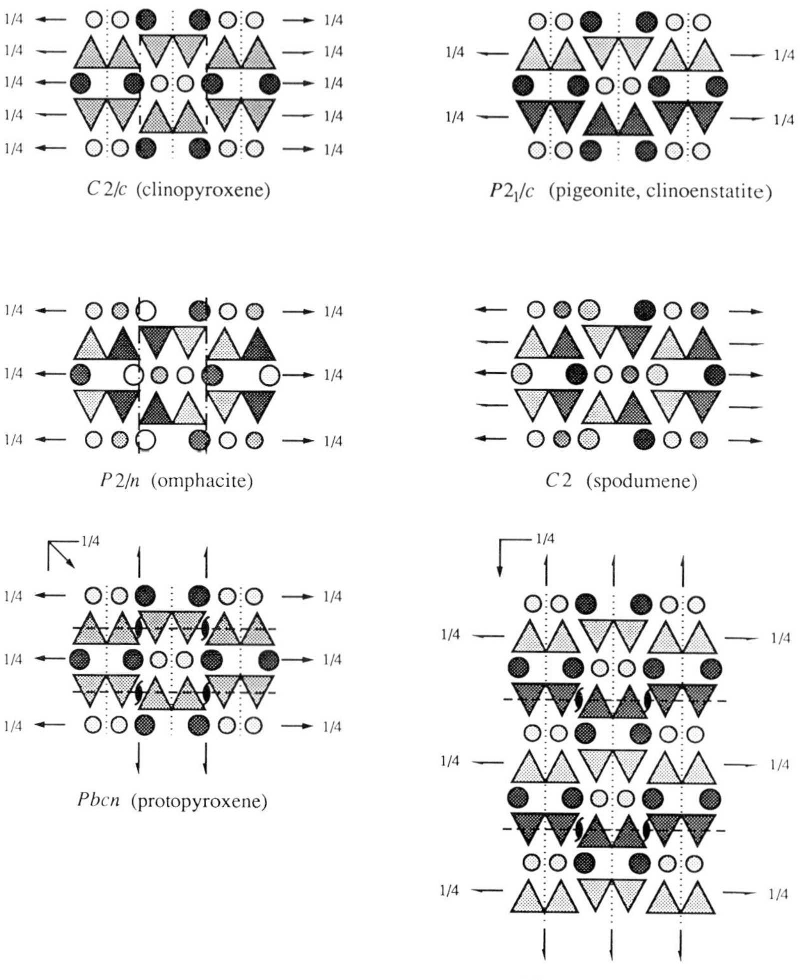

Figure 4-5. The space groups of pyroxenes, in schematic views down the **c** direction. Triangles represent tetrahedra, small circles the $M(1)$ cations, and large circles the $M(2)$ cations. The b axis and $a \sin \beta$ are horizontal and vertical in the plane of the page, respectively. Symmetry elements are indicated for each space group. Different patterns or shadings indicate symmetrical nonequivalence of geometrically similar sites.

reported several times for certain orthopyroxenes, owing to the presence in some single-crystal x-ray diffraction patterns of reflections forbidden in the space group *Pbca*; this has been encouraged by the prediction of pyroxenes in this space group, based on close-packed topological models (e.g., [68]). These "extra" reflections have been evaluated in two studies [57;58], and virtually all of them have been ascribed to non-space-group diffraction effects. In the absence of evidence to the contrary, we shall consider $P2_1ca$ as a spurious and unlikely space group for pyroxenes.

Early reports of omphacite crystal structures gave *P*2 as the space group. Later work revealed that some omphacites crystallize in *P*2/*n*, and reinvestigation of the crystals assigned to *P*2 showed that they, too, were actually *P*2/*n* pyroxenes. It is now considered unlikely that any omphacites crystallize in *P*2, and so our discussions of omphacites will assume that all primitive ones crystallize in *P*2/*n*.

Polyhedral Interrelationships

Figure 4-6 depicts parts of the crystal structures of a *C*2/*c* pyroxene (jadeite) and a *Pbca* pyroxene (enstatite), viewed down the **a*** direction [that is, perpendicular to the (100) planes]. Careful study of these illustrations will reveal the three-dimensional relationships between the tetrahedral chains and the octahedral bands. Note that in the clinopyroxene, the *M*(2) site is 8-coordinated, rather than octahedral as in the orthopyroxene. In jadeite, the *X* cation, Na, is large; in enstatite, *X* is Mg and *M*(2) is thus a smaller, although very distorted, octahedral site. The *M*(1) sites are also of different sizes, being occupied by Al and Mg, respectively. Because the *M* sites are constructed from oxygens of the tetrahedral chain, the chain must adjust to the required sizes of the *M*(1) and *M*(2) sites by (1) altering the size of the tetrahedra themselves, or (2) changing the angular geometry of the chain. The latter is energetically very much easier to effect, and it is the O(br)–O(br)–O(br) angle that changes. (Compare tetrahedral–octahedral adjustment in the micas, previously discussed.) Here again, O(br) stands for bridging oxygens—those shared by adjacent

Figure 4-6. The crystal structures of jadeite and enstatite, projected down **a***, with **c** vertical and **b** horizontal. Tetrahedral chains are lightly shaded, *M*(1) octahedra are shown in medium shading, and *M*(2) has the darkest shading. Note that in jadeite (as well as in aegirine and in the diopside–hedenbergite series), where *M*(2) is occupied by a large Na (or Ca) atom, the site is eight-coordinated; in enstatite (as well as in all members of the enstatite–ferrosilite series and low-Ca pigeonites), the *M*(2) site is occupied by smaller Mg or Fe atoms, and is six-coordinated.

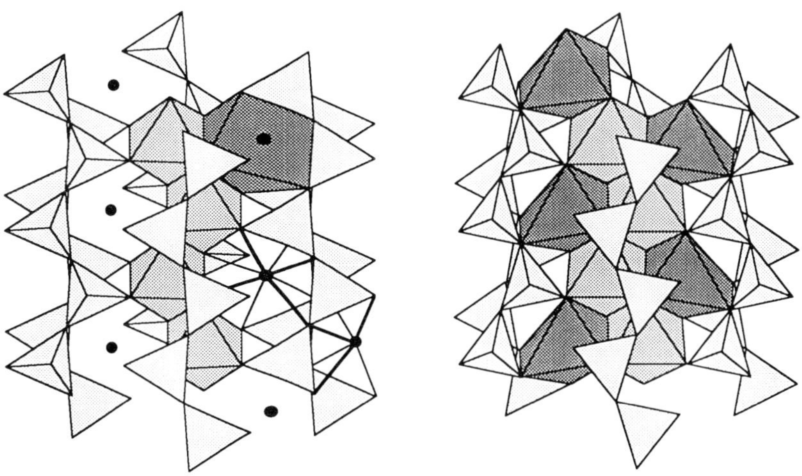

Jadeite Orthohypersthene

tetrahedra—and O(nbr) for nonbridging oxygens, which are shared between tetrahedra and $M(1)$ or $M(2)$ sites. Inasmuch as O(br) has been designated as O(3) in pyroxenes [with the basal O(nbr) being O(2) and the apical oxygen O(1)], it is common to call this angle either the O(3)–O(3)–O(3) angle, or simply the chain angle. The tetrahedral chain is nearly fully extended (i.e., it has a chain angle close to 180°) in the monoclinic structure of Figure 4-6, because of its particular combination of X and Y cations, and the resulting sizes of their coordination polyhedra; the chains in the orthopyroxene are kinked considerably because both $M(1)$ and $M(2)$ are small. Given sufficient flexibility in X and Y chemistry, the tetrahedra could conceivably be rotated until close packing were achieved, the chain angle were 120°, and $M(1)$ and $M(2)$ octahedra were of identical size. Although no tetrahedral chains in natural pyroxenes have a chain angle of 120°, none is fully extended either. Synthetic $LiFeSi_2O_6$ has been found to have a fully extended chain [17], and synthetic $CoGeO_3$ ($\equiv Co_2Ge_2O_6$) has a chain angle of 130°, so that it is very nearly close packed. Figure 4-7 illustrates the difference between a fully extended chain and a fully rotated one.

The rotations may occur in either of two directions, resulting in O rotations and S rotations, as previously defined for micas and illustrated in Figure 4-8. An O rotation of 30° (chain angle = 120°) results in cubic close packing, whereas a fully S-rotated structure is hexagonal close packed. O rotations are much more common

Figure 4-7. (a) A fully rotated and (b) a fully extended pyroxene chain. The O(3)–O(3)–O(3), or tetrahedral chain, angles are labeled.

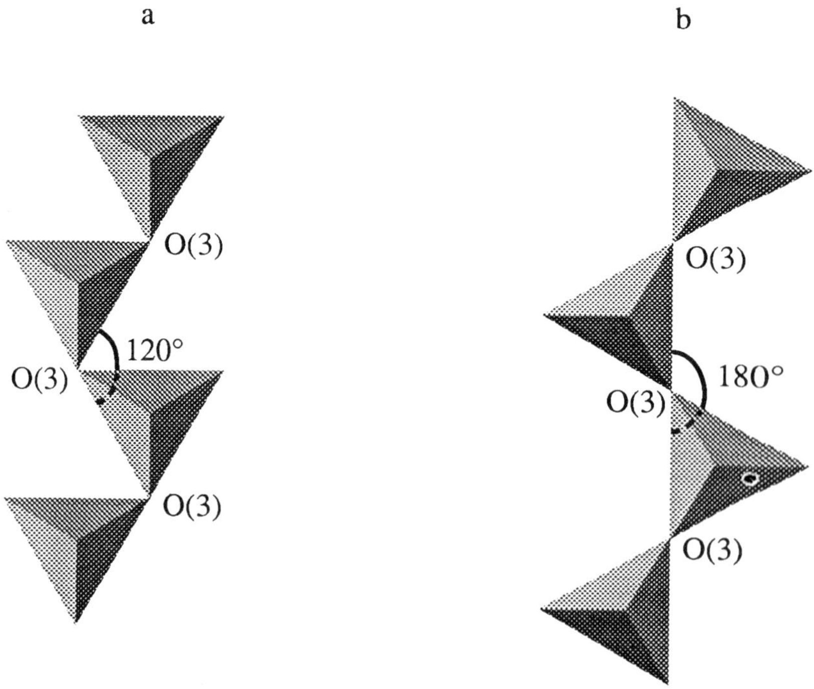

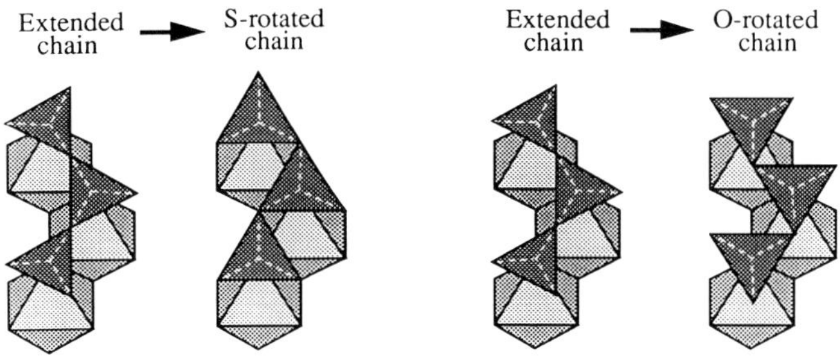

Figure 4-8. Tetrahedral chains in the extended, S-rotated, and O-rotated geometries. Note that the designation S or O depends upon the relative orientations of the tetrahedral and octahedral triangular faces linked through O(1).

than S rotations in pyroxenes, but S rotations do occur. Note that if the sizes of the octahedra were to remain constant through a rotation, then the tetrahedra would have to become larger as the rotation increased. We have seen already that the chain extension depends on the sizes of the octahedral sites. Thus the sizes of the octahedral and tetrahedral sites, and the magnitude of the O(3)–O(3)–O(3) angle, must all be interrelated, and the structural details that result from that interrelationship will prove interesting, as we shall investigate in the following.

DETAILS OF THE CRYSTAL STRUCTURE

The previous summary of pyroxene crystallography and chemistry has shown that the two are interrelated, and the breadth of chemical variation makes the pyroxenes good candidates for the elucidation of many crystal chemical principles involving phenomena such as ordering and polymorphism. Before treating these subjects, there are some crystallographic details that need to be understood, particularly as they relate to variations in chemistry.

The Tetrahedral Chain

Not only is the tetrahedral chain parallel to, and linked to, the octahedral strip, but two of four oxygens of each tetrahedron also belong to octahedra. As mentioned above, then, any change in the tetrahedral chain must affect the octahedral strip, and vice versa. Most pyroxenes contain, exclusively or largely, Si in the tetrahedral chain, and the major chemical variation occurs in the octahedral sites. The tetrahedral chain may adjust to the geometrical constraints imposed by the octahedral strip either by distortion of the tetrahedra themselves, or by some adjustment in chain configuration, such as alteration of the chain angle, O(3)–O(3)–O(3).

We have already pointed out above that most pyroxenes have O-rotated tetrahedral chains. Recall that fully O-rotated arrangements are cubic close packed, and that fully S-rotated ones are hexagonal close packed. In the HCP structure, tetrahedra share edges with $M(2)$ octahedra, whereas in the CCP structure the two share only corners. Thus, if all other factors were equal, one might expect O rotations to be energetically advantageous, although in some real pyroxenes even partially O-rotated chains share edges with $M(2)$. S rotations appear to occur only in chains that must rotate in that way in order to attain proper coordination for the $M(2)$ cation, and that is necessary only when that cation is small.

A final symbolic representation of tetrahedral rotation with respect to adjacent octahedra depends on the relationship between $M(2)$ and the tetrahedra adjacent to it. Observe, in Figure 4-9, that the tetrahedral chain is linked to $M(1)$ through O(2). If we look at the same (100) triangular faces of $M(1)$ again, the (100) tetrahedral bases either "point" in the same direction (P) or the opposite direction (N) as those faces. [Recall that it was a different octahedral strip—the one connected to the tetrahedra through O(1)—that we used for identifying O and S rotations. And, as with O and S rotations, the triangular bases being compared will point strictly in the same or opposite directions only if the chains are fully rotated, so it is often helpful to mentally turn the tetrahedra toward full rotation to determine the sense of rotation in a crystal structure drawing.] The figure reveals that for each octahedron there are two tetrahedra to consider—one on the "top" and one on the "bottom" of $M(1)$ and its adjacent $M(2)$—and that these may yield the rotation sequences PP, NP, PN, or NN going from "top" to "bottom" through $M(2)$. Even though it was $M(1)$ that we used to determine the P or N rotations of the tetrahedra, it is $M(2)$ about which the determination provides the most significant information. For N configurations, a tetrahedron shares an edge with $M(2)$, and for P configurations it does not. While PP, NP, and PN sequences occur in common pyroxenes,

Figure 4-9. P and N rotations. Note that the designation P or N depends upon the relative orientations of the tetrahedral and octahedral triangular faces linked through O(2), and the the consequences are important not for $M(1)$, but for the coordination of $M(2)$.

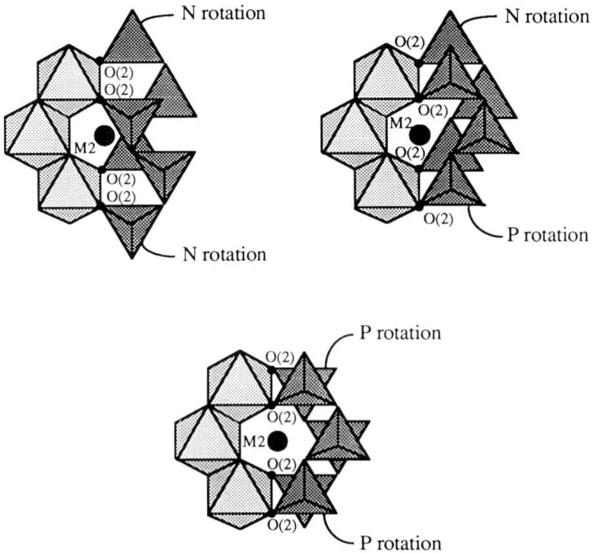

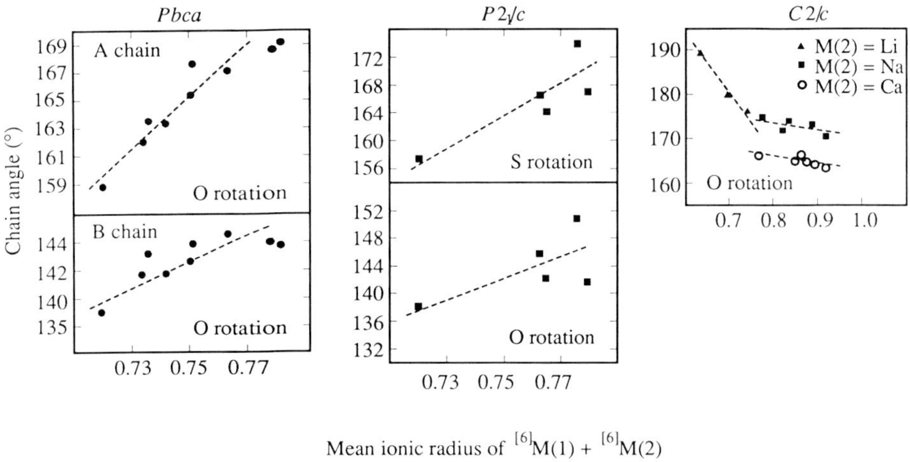

Figure 4-10. Tetrahedral chain angles as functions of the mean ionic radius of the *M* cations (considered as six-coordinated) for selected *Pbca*, $P2_1/c$, and $C2/c$ pyroxenes. The *Pbca* and $P2_1/c$ pyroxenes are Mg–Fe varieties, some with minor Ca or other elements. For $C2/c$ pyroxenes, the chemical occupancy of *M*(2) is indicated by the symbol. *M*(1) sites are occupied by (Li pyroxenes, lowest to highest mean radius) Al, Fe^{3+}, and Sc; (Na pyroxenes) Al, Cr, Fe^{3+}, Sc, and In; and (Ca pyroxenes) Al, Ni, Mg, Co, Fe^{2+}, and Mn. (After [7].)

NN is found only in protoenstatite (one of the enstatite polymorphs, p. 180), and it produces a small and very distorted octahedral *M*(2) site because two edges are shared with tetrahedra.

The plots of Figure 4-10 show, as expected, that the chain angle increases as the increase in size of *M*(1) and *M*(2) cations forces a lengthening of the *c* cell dimension. The fact that O(3)–O(3)–O(3) increases with mean octahedral cation radius in *Pbca* and $P2_1/c$ pyroxenes but decreases in $C2/c$ pyroxenes seems incongruous, and we shall return to it after discussing changes in the geometry of individual tetrahedra.

Figure 4-11 shows how the mean of the two bridging Si–O bonds is related to the Si–O–Si angle between adjacent linked tetrahedra. Once the meanings of all the symbols and numbers on the plot have been ascertained (see the figure caption), the following observations can be made. (1) There is a general trend for longer mean Si–O(br) distances to be associated with narrower Si–O(3)–Si angles. This can be rationalized as an adjustment necessary to maintain an acceptably large Si–Si separation. (2) The Si–Si distances (unparenthesized numbers) actually tend to increase as mean Si-O(br) decreases. (3) For *Pbca* and $P2_1/c$ pyroxenes, there seems to be no strong correlation between either mean Si-O(br) or Si-O(3)-Si and the mean octahedral cation radius (parenthesized numbers). (4) For $C2/c$ pyroxenes, separate trends for different *M*(2) cations are evident, but for all three there is a strong correlation between mean Si–O(br) and the radius of the *M*(1) cation. The Ca and Na pyroxenes exhibit little variation in Si–O(3)–Si angles, but it appears that for these, and certainly for the Li pyroxenes, Si-O(3)-Si increases with the radius of the *M*(1) cation. (5) As in *Pbca* and $P2_1/c$ pyroxenes, the Si-Si distance in

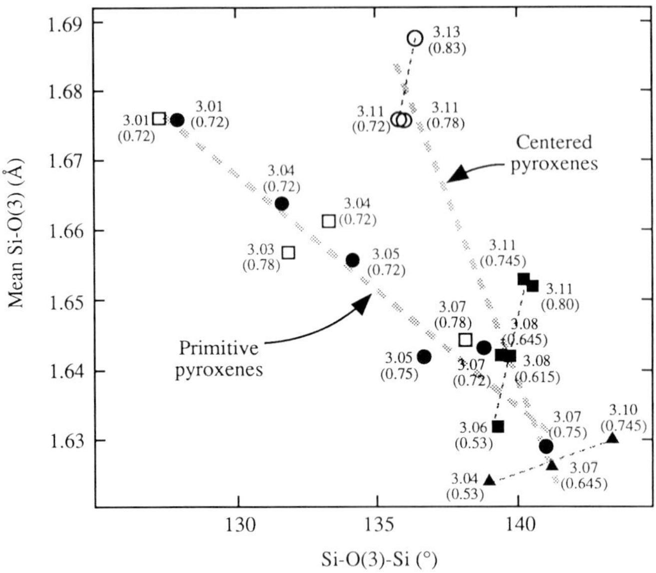

Figure 4-11. Mean Si–O(3) (that is, mean bridging bond lengths) versus the angle Si–O(3)–Si. Symbols are as in Figure 4-10. Heavy dashed lines indicate trends for primitive (*Pbca* and $P2_1/c$) and centered ($C2/c$) pyroxenes. Light dashed lines indicate trends for $C2/c$ pyroxenes with different $M(2)$ cations. Unparenthesized numbers are Si-Si distances within the tetrahedral chains. Parenthesized numbers are the mean ionic radii of both M cations for primitive pyroxenes, and mean radii of $M(1)$ cations for centered ones. (After [7].)

$C2/c$ pyroxenes increases as the Si–O(3)–Si angle increases. But unlike the pyroxenes with primitive space groups, the $C2/c$ pyroxenes show an increase in Si–Si distance as mean Si–O(br) increases, rather than as it decreases.

Figure 4-12 displays how Si–O bond lengths—bridging, nonbridging, and mean—vary with the mean octahedral cation size. Again the trends for primitive and $C2/c$ pyroxenes seem to be opposite in general.

By synthesizing the trends shown in Figures 4-10 through 4-12, we can arrive at a reasonably cogent picture of how the tetrahedral chains adjust to the size requirements imposed by the octahedral strip. In $C2/c$ pyroxenes, $M(2)$ is large and 8-coordinated (except in spodumene—which is really $C2$—where it is 6-coordinated), whereas in *Pbca* and $P2_1/c$ pyroxenes both $M(1)$ and $M(2)$ are about the same size and 6-coordinated. [$M(2)$ in some pigeonites may be 7-coordinated, but that does not alter the present discussion.] It is perhaps not surprising, then, that $C2/c$ pyroxenes require the largest range of tetrahedral adjustment. In order to adjust to octahedral compositions from LiAl to CaMn, the tetrahedra must not only enlarge, but the chain angle must increase. In the primitive pyroxenes, the range of octahedral cation sizes is much more restricted, and a straightening of the chain angle is sufficient. In fact, as the chain is straightened, the tetrahedra may even become smaller in order to stabilize the Si–Si distances and optimize other steric details. (Note that in $P2_1/c$ pyroxenes, the A tetrahedral chain is S-rotated, so that straightening the chain consists in *decreasing* the O(3)–O(3)–O(3) angle from something greater than 180°.) Because rotation and dilation of the tetrahedra both occur in response to requirements imposed by octahedral chemistry, it is hard to say which is the more important factor in achieving the geometrical fit, although there is certainly a greater range in flexibility for bending the chain than for stretching bond lengths (p. 113).

Figure 4-12. Mean lengths of bridging, nonbridging, and all Si–O bonds as functions of the mean radii of the M cations. As in Figure 4-10, the 6-coordinated radii are used, even though $M(2)$ might not be 6-coordinated. In the plots for $C2/c$ structures triangles represent Li pyroxenes, squares Na pyroxenes, and circles Ca pyroxenes. All ⟨Si–O⟩ bond lengths are in Å. (After [7].)

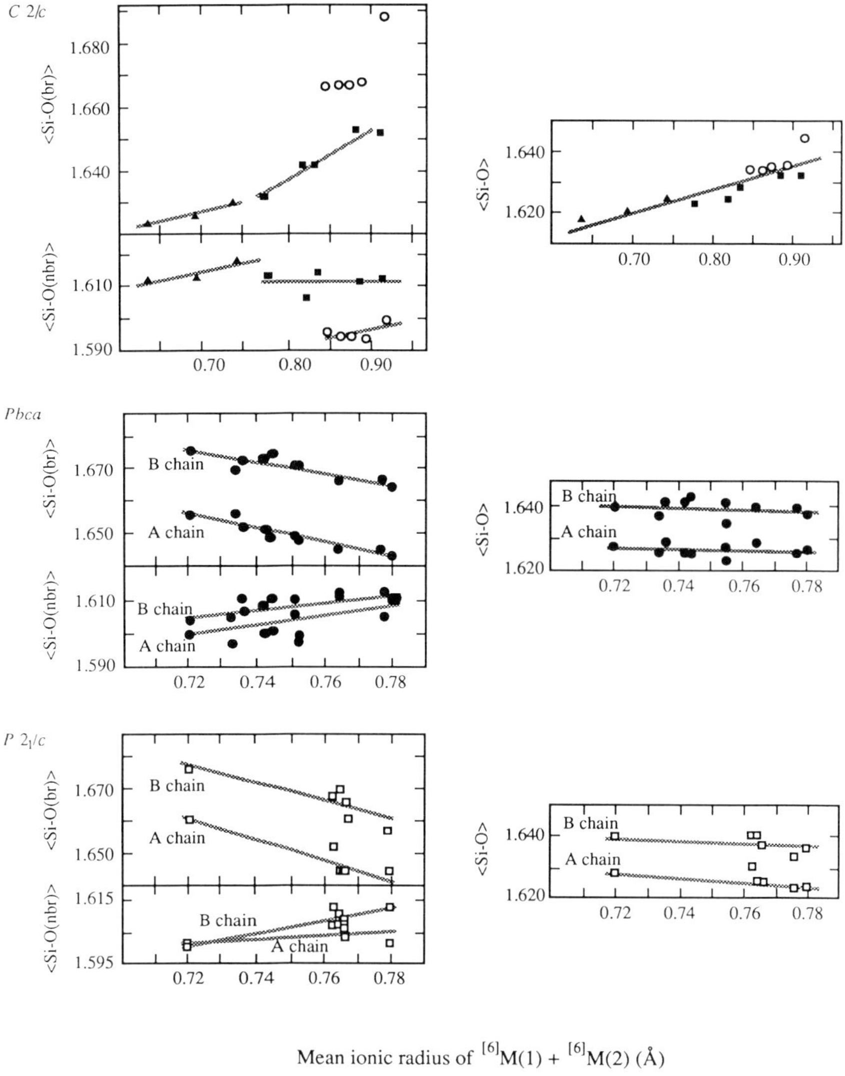

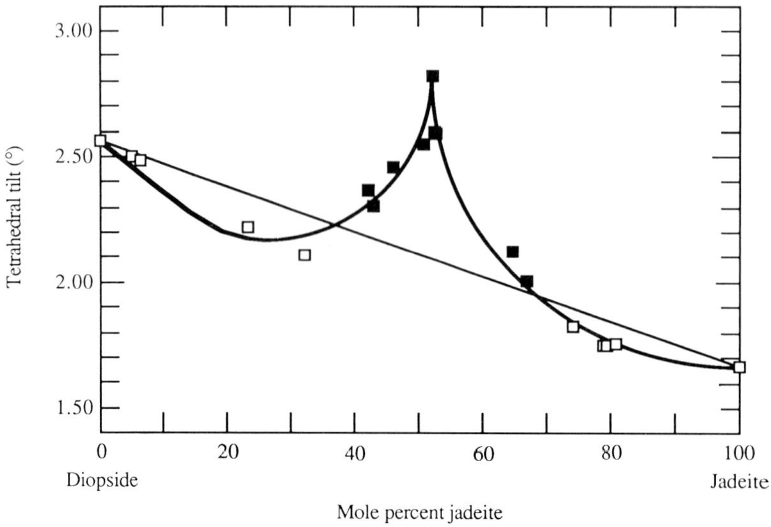

Figure 4-13. Tetrahedral tilt [the dihedral angle between (100) and the tetrahedral base] as a function of composition for the diopside–jadeite series. Open squares represent disordered (*C*2/*c*) omphacites, and filled squares represent ordered (*P*2/*n*) omphacites. Note that more tilting is required of the tetrahedra in order to fit the more varied octahedral geometry of the ordered omphacites. (After [55].)

In addition to tetrahedral rotation, which changes the chain angle, the tetrahedral chain can also adjust to octahedral cation size by tilting of tetrahedra, just as is observed in micas. The amount of tilt is measured as the dihedral angle between the tetrahedral bases and the (100) crystallographic planes. Its magnitude seems to be directly proportional to the extension of the chain angle, but is probably a geometrically less important variable than chain angle. Figure 4-13 shows tetrahedral tilt plotted as a function of chemistry for 16 sodic diopsides and omphacites that fall nearly on the Di–Jd compositional join [55]. Without exception, all of the points that lie below the line joining diopside and jadeite are for *C*2/*c* (disordered) pyroxenes, and those that lie above it are for *P*2/*n* (ordered) pyroxenes. We have not yet discussed ordering and space groups in pyroxenes (that discussion follows very shortly), but it suffices here to state that the disordered pyroxenes have octahedral sites of more uniform sizes than the ordered ones. Figure 4-13 thus illustrates the greater amount of tetrahedral tilt required when it is necessary to fit a greater variety of octahedral sizes in a single octahedral strip.

The Octahedral Strip

The *M*(1) and *M*(2) sites are symmetrically distinct, and if the anion arrangement is not strictly close packed (which, as we have seen, it is not in natural pyroxenes), then they are also geometrically different. The *M*(1) site is always octahedrally coordinated, is always the smaller of the two sites, and is more geometrically regular than *M*(2). The coordination number of the latter is six, seven, or eight, depending at least on the size of the cation. There is not universal agreement that seven is an appropriate coordination number for $M(2)$ in *any* pyroxenes, however, and a short diversion into that question is instructive.

When the $M(2)$ site is filled by magnesium, which is the smallest of all the major cations that occupy $M(2)$ in natural pyroxenes, then that site is octahedrally coordinated; this situation obtains in the enstatite polymorphs. When the cation in $M(2)$ is large—for example, calcium in diopside or sodium in jadeite—then the site is clearly eight-coordinated and is a distorted cubic antiprism. Within the En–Di–Hd–Fs pyroxene quadrilateral, the $M(2)$ cation becomes, on the average, slightly larger from enstatite to ferrosilite, and considerably larger from the En–Fs join to the Di–Hd join. One of two situations must thus exist with respect to the coordination of the $M(2)$ site when both Ca and Mg–Fe are required to occupy it: Either (1) two types of $M(2)$ polyhedra exist simultaneously, so that the Mg–Fe atoms that occupy $M(2)$ are in sixfold coordination while the Ca atoms are in eightfold coordination, or (2) there is but one type of $M(2)$ polyhedron, and it experiences a continuous change from six- to eightfold coordination, from the bottom to the top of the quadrilateral, respectively. Because x-ray- and neutron-diffraction experiments yield information on only the average geometry of an $M(2)$ site in the crystal being investigated, this technique is unable to distinguish clearly between the two alternatives. Crystal structure refinements by x-ray diffraction do suggest a seven-coordinated $M(2)$ site in pigeonite, but the results are somewhat enigmatic. First, in crystal structure refinements of pigeonite, a seventh oxygen is found to be close enough to $M(2)$ to be considered bonded, while the eighth is not; but second, this seventh oxygen atom is further from $M(2)$ than one of the Si atoms of the tetrahedral chain, and the conventional wisdom is that no cations should be within the primary anion coordination sphere of another cation.

Mössbauer spectroscopy is a more diagnostic approach than x-ray or neutron diffraction for this problem (see Chapter 10) and that method has been used on a series of synthetic pyroxenes ranging in composition from hedenbergite to ferrosilite [26]. The spectrum for $Hd_{60}Fs_{40}$ is shown in Figure 4-14, and we see that four doublets are required to fit the experimental data. Each doublet is presumed to represent Fe^{2+} in some particular type of coordination, but of course the only two sites available are $M(1)$ and $M(2)$. Because $M(1)$ is always occupied by Fe^{2+} in this series, three of the doublets are interpreted as representing Fe^{2+} in $M(1)$ with three different next-nearest-neighbor [that is, $M(2)$] configurations: two calciums, one calcium and one iron, and two irons. The remaining doublet is then attributed to Fe^{2+} in $M(2)$, and is the doublet of interest here. The quadrupole splitting for that doublet as a function of composition is given in Figure 4-15. The quadrupole splitting is sensitive to changes in the coordination geometry around iron atoms (but *not* about Ca atoms), and it is apparent that it has changed continuously and significantly, although not monotonically. Our first alternative above [that there are two types of $M(2)$ sites, an Fe^{2+} $M(2)$ and a Ca $M(2)$, that are simply averaged by x-ray-diffraction methods] would require that the geometry, and therefore the quadrupole splitting, of the $M(2)$ sites containing Fe^{2+} remain constant, and that clearly is not the case. If this is accepted as evidence that the coordination geometry changed smoothly, then it must have changed from six- through seven- to eightfold from

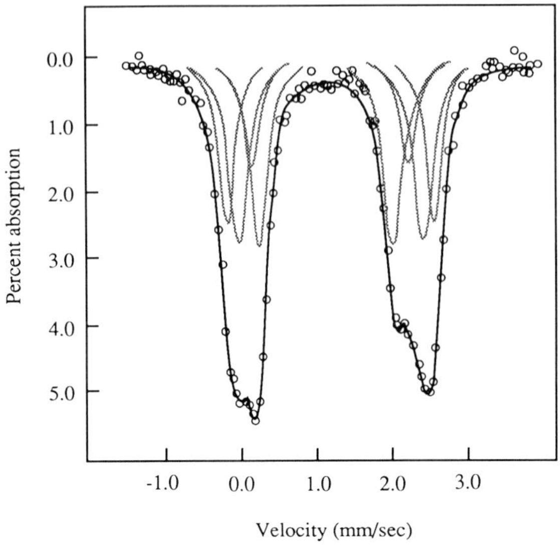

Figure 4-14. The Mössbauer spectrum of synthetic $Hd_{80}Fs_{20}$. The circles are experimental data points, and the gray curves are the doublets used to fit the data. The black curve is the sum of the gray curves. The innermost doublet is the one assigned to *M*(2). (After [26].)

ferrosilite to hedenbergite, and we can conclude that the coordination of any particular *M*(2) site is influenced not only by the chemical species of the atom occupying it, but also by the chemistry of the sites around it. Seven appears, on the basis of this evidence, to be a valid coordination number for *M*(2) in pyroxenes of the Hd–Fs series. Reenforcing this interpretation is the fact that high-temperature crystal structure refinements of ferrosilite [64] reveal that the coordination of *M*(2) goes

Figure 4-15. Quadrupole splitting at the *M*(2) site for synthetic compositions form $Hd_{80}Fs_{20}$ through Hd_0Fs_{100}. (After [26].)

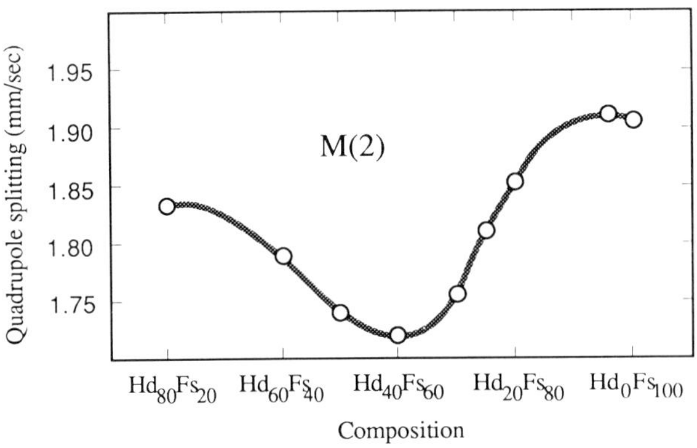

from six to seven, and back to six as temperature increases and atomic positions shift. If one assumes that increased temperature yields effects similar to substitution of larger cations (not necessarily a safe assumption), then seven-coordination again seems vindicated.

Figure 4-4 suggests that vacancies separate adjacent $M(2)$ polyhedra, but this is true only if $M(2)$ is 6-coordinated. Figure 4-16 shows the octahedral strips of a $P2_1/c$ pyroxene (pigeonite) and a $C2/c$ pyroxene (diopside), projected onto (100). Note the continuous rift between adjacent strips of M sites in the primitive mineral; above and below the plane of this strip, the rift is bridged by tetrahedral chains that run parallel to it and have their bases toward it. In the C-centered structure, $M(2)$ polyhedra share edges across this interstrip zone, eliminating the continuous rift, and strengthening the polyhedral linkage in the **b** direction. Octahedral strips are thus sometimes described as *separate* or *linked.*

Another geometrical descriptor used to characterize the configuration of the octahedral strip is *skew*. Recall that we looked at the triangular faces of the octahedra that were approximately parallel to (100) to determine whether the connected tetrahedral chain was O or S rotated. Each octahedron has two such faces, one on the "top" (+**a** side) and one on the "bottom" (−**a** side). All those on the top point in the same direction, opposite of course to those on the bottom. If, with respect to the usual right-handed coordinate system of crystallographic axes, all the top faces point along +**c** , the skew is said to be *positive* (+). Otherwise, it is *negative* (−). Figure 4-17 schematically illustrates the idea and sense of skew.

Figure 4-16. The crystal structures of diopside ($C2/c$) and low pigeonite ($P2_1/c$). In pyroxenes with eight-coordinated $M(2)$ sites, those sites share edges between base-to-base tetrahedral chains, linking adjacent octahedral bands. In pyroxenes with six-coordinated $M(2)$ sites, there is a rift between adjacent octahedral bands.

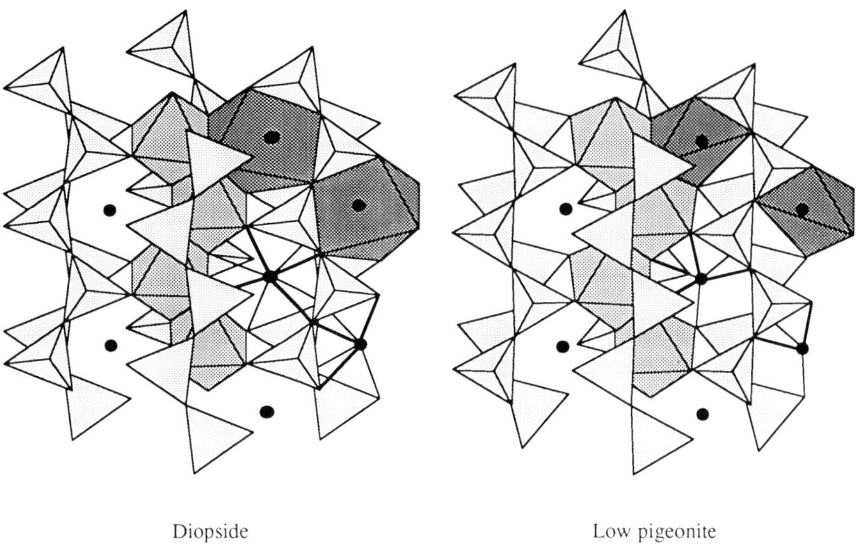

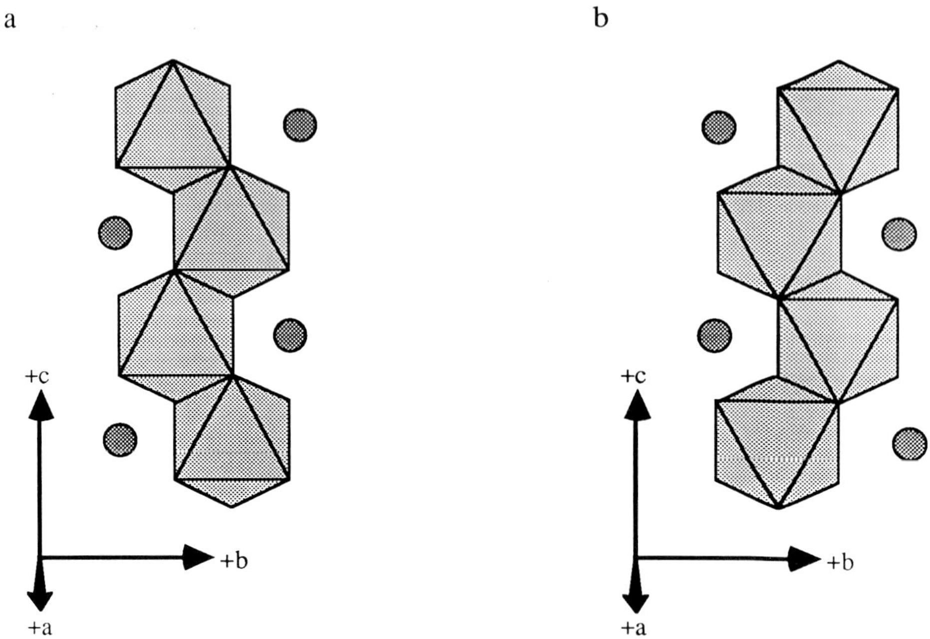

Figure 4-17. (a) Positive and (b) negative skews in the octahedral band. If the top triangular faces (those furthest along $+\mathbf{a}$) "point" in the $+\mathbf{c}$ direction, then the skew is said to be positive. If those faces "point" in the $-\mathbf{c}$ direction, the skew is negative. Note that the usual right-hand orientation of crystallographic axes is used in both cases.

STRUCTURAL SUMMARY

The structural details of the tetrahedral chains and octahedral strips that have been discussed in the preceding pages are illustrated in Figure 4-18 in schematic views that have come to be called I-beam drawings of the pyroxene structures. They are not intended to represent the structures very accurately, but they provide a simple way of displaying the relationships among the space groups.

Changes in the geometrical details of pyroxenes can be effected in the following ways:

1. Chemical substitutions that result in altered unit-cell dimensions without alteration of the basic structure. The diopside-hedenbergite solid solution series is an example; the substitution of Fe^{2+} for Mg requires only very minor adjustment of the structure.
2. Changes in the relative sizes of $M(1)$ and $M(2)$ cations that result in significant variations in both unit cell dimensions and the geometrical details of the silicate chain, but that leave the space group unchanged. $C2/c$ pyroxenes with compositions as different as $NaAlSi_2O_6$ and $CaMnSi_2O_6$ are examples.

3. Changes in space group that accompany substitution in $M(2)$ of cations less than ~ 1.0 Å in radius. This is typified by the change in space group that accompanies the substitution of Mg (in enstatite) for Ca (in diopside).
4. Ordering of cations.
5. Polymorphic changes arising from a sort of polytypism.

The first two have already been covered adequately, and the third partially so. We shall discuss the last two of these in the following, with some additional discussion of the third.

Figure 4-18. "I-beam" representations of the six pyroxene space groups. Trapezoids represent tetrahedral chains, with O or S indicating the sense of tetrahedral rotation with respect to octahedral faces joined to tetrahedra through O(1). Where two types of tetrahedral chains exist, they are labeled *A* and *B*. Octahedral *M*(1) sites, schematically compressed to fit entirely between tetrahedral chains, show skew as either positive (+) or negative (−). *M*(2) sites are shown as dots, with P or N indicating the sense of rotation of the tetrahedra with respect to octahedral faces joined to them through O(2) atoms.

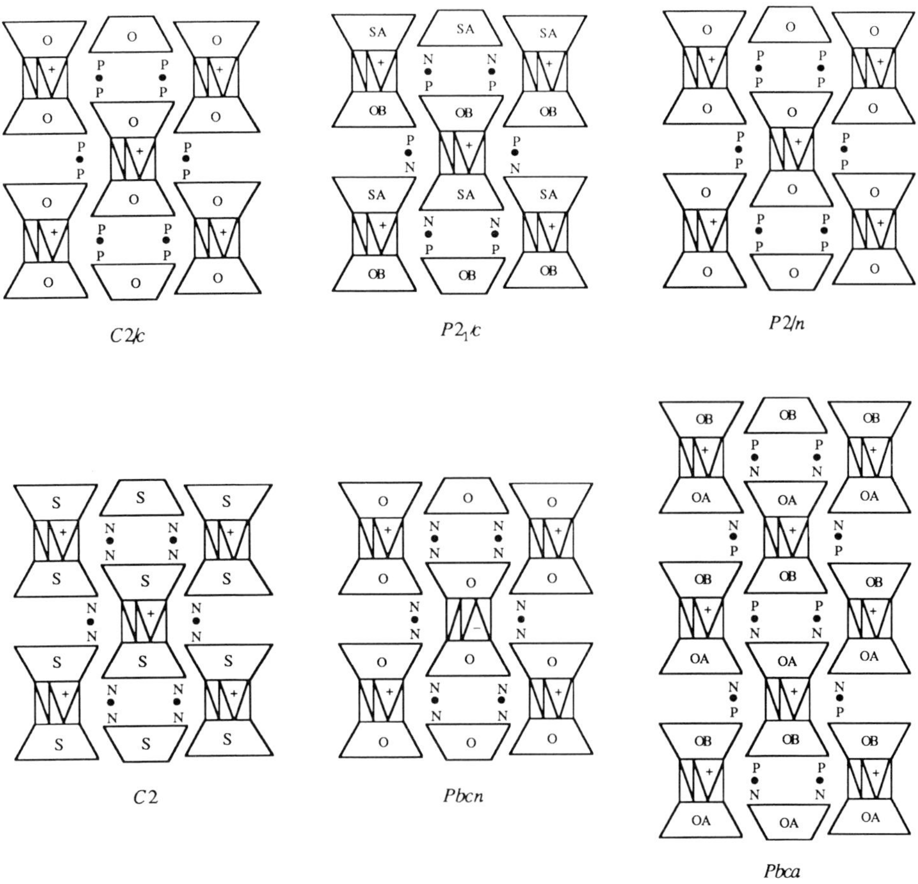

ORDERING IN PYROXENES

Cation Order and Space Group

The space group of a mineral is determined by its atomic arrangement, but it may be ascertained (or at least narrowed down to two or three possibilities) without any knowledge of that arrangement. If the basic topology of a structure is known, which it is for pyroxenes, and if the space group can be determined, which it can be from single-crystal x-ray-diffraction data, then certain predictions about the detailed atomic arrangement can be made with confidence. Consider, for example, the pyroxene chain shown in Figure 4-3. (The fact that a fully extended chain is shown is not relevant to the present discussion—a fully rotated chain would yield the same results.) In space group $C2/c$, the presence of the c-glide is conclusive evidence that all tetrahedra in a chain are symmetrically equivalent. Thus any aluminum that may be present in the tetrahedral chains of natural augites, for instance, cannot be ordered but must be randomly distributed (as it is in the alkali feldspar sanidine, for example). If, however, the single-crystal x-ray-diffraction data indicate the absence of a c glide (as in space groups $C2$ or $P2/n$), then the successive tetrahedra along the chains are not symmetrically equivalent and need not be chemically identical on the average. While such non-equivalence can be purely geometrical and does not *require* ordering (as, for example, in $C2$ spodumene, where there is only Si in the chain, so there is nothing to order), it would certainly allow it in pyroxenes with order-prone chemistry.

Figure 4-19 shows some of the symmetry elements present in space group $C2/c$. In addition to the c glides and diads, there are twofold screw axes and n glides. The $M(1)$ and $M(2)$ cations lie exactly on diad axes at $x=0$, and the screw axes at $x=\frac{1}{4}$ pass through the tetrahedral chains in a direction perpendicular to them. Recall that the structure consists of alternating layers parallel to (100), one layer containing only tetrahedral chains, and the next only octahedral bands (p. 155). The diads map the octahedral bands into themselves, but each tetrahedral layer into the tetrahedral layer on the opposite side of the octahedral layer. Similarly, the presence of the screw diads reveals that successive (100) octahedral layers are alike, and that all tetrahedral chains in a given layer are identical. Consequently, in $C2/c$ pyroxenes there is only one kind of octahedral band, which has only one kind of $M(1)$ site and one kind of $M(2)$ site in it, and there is only one kind of tetrahedral chain, which has one kind of tetrahedron in it. If the diad axes are absent (as in pyroxenes that crystallize in space group $P2_1/c$), then there is one kind of octahedral band, but there are two kinds of tetrahedral layers and hence two kinds of tetrahedral chains. If the diads are present but the screw axes are not (as in pyroxenes that crystallize in space group $C2$), then it is because all (100) tetrahedral layers are equivalent, but there are two kinds of tetrahedral chains in each layer; in addition, there are two kinds of octahedral layers.

In a rather schematic way, Figure 4-5 (p. 158) summarizes the symmetrically distinct sites in the six pyroxene space groups. In $C2/c$ pyroxenes there is one kind of tetrahedron per chain and only one kind of tetrahedral chain; also, there is one kind of octahedral strip containing one kind of $M(1)$ site and one kind of $M(2)$ site. In $P2_1/c$ the diads are absent, indicating that there are two types of silicate chains, but

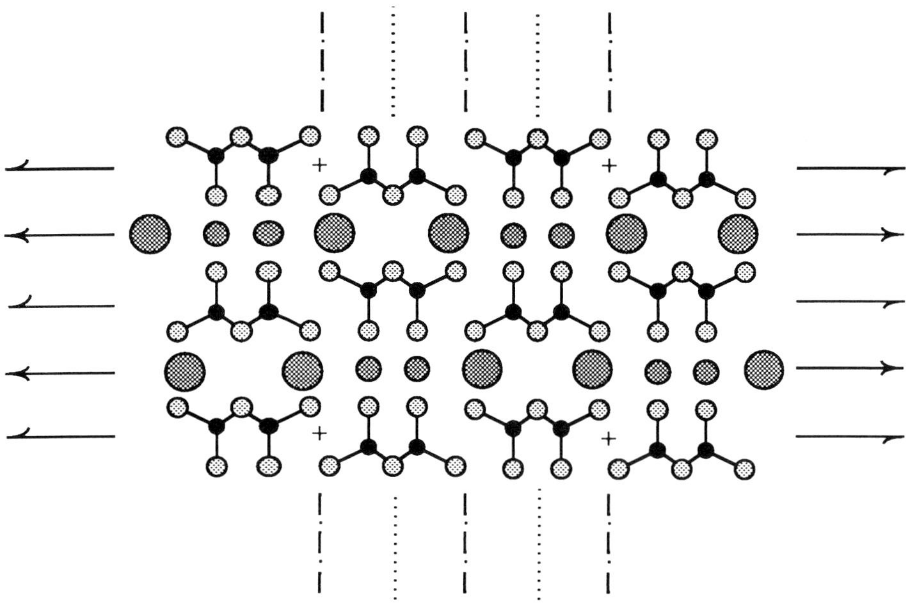

Figure 4-19. A *C*2/*c* pyroxene, depicted schematically in a projection down the *c* axis, with space group symmetry elements shown. Dots are tetrahedrally coordinated atoms, small shaded circles are oxygens, intermediate shaded circles are *M*(1) cations, and large shaded circles are *M*(2) cations. Plus signs indicate corners of the unit cell.

still one kind of tetrahedron per chain. The *C*2 pyroxenes lack the *c*-glide because there are two kinds of tetrahedra per chain (but only one kind of chain) and two kinds of *M*(1) and *M*(2) sites per strip (but only one kind of strip). The *P*2/*n* pyroxenes possess exactly the same numbers of each type of site as *C*2 pyroxenes, but the arrangement of those sites is different. For orthorhombic pyroxenes, the same numbers and kinds of sites are present in *Pbca* as in $P2_1/c$, and in *Pbcn* as in *C*2/*c*.

Order in the Octahedral Strip

For *C*2/*c* end-member pyroxenes, the ordering in the *M* sites is very simple: The large cations (Ca or Na) occupy the large and distorted *M*(2) polyhedra and the smaller cations (Fe^{2+}, Fe^{3+}, Mg, Mn, Cr^{3+}, or Al) occupy *M*(1). Because there is but one type of *M*(1) site and one type of *M*(2) site, the presence of more than one species of atom in either site requires disorder at that site. For solid solutions among end-member *C*2/*c* pyroxenes, then, the large cations are disordered on *M*(2) and the small ones are disordered on *M*(1).

If, instead of being large like Ca and Na, the *M*(2) cation is of intermediate size and univalent (namely, if it is lithium), then the *c*-glide is absent and the space group is *C*2. Spodumene is the only mineral so far found in this group, although lithium pyroxenes of other compositions have been synthesized in space group *C*2. The reason for the loss of the *c* glide is not definitely known, because the extra x-ray-diffraction maxima that are present due to the loss of symmetry are few and weak; they have thus not been used in the structure refinements of spodumene, and so only the "average" $C2/c$ structure (the one that averages sites that are actually distinct, so as to simulate the presence of a *c* glide) has been determined [17]. Even though space group $C2$ requires two symmetrically distinct kinds of $M(1)$ and two distinct kinds of $M(2)$ sites, it seems likely that the only octahedral ordering that occurs is the segregation of Li into both of the $M(2)$ sites and Al into both of the $M(1)$ octahedra.

As chemistry renders $M(1)$ and $M(2)$ closer in size (i.e., as we move from anywhere on the Di–Hd boundary of the pyroxene quadrilateral toward the En–Fs boundary), $M(2)$ becomes smaller, changing from eight-, to seven(?)-, to six-coordination (p. 167), because there is insufficient calcium to occupy $M(2)$ fully. As this happens, the tetrahedral chains become more kinked in order to reduce the coordination number of $M(2)$, eliminating the diad axis; the space group becomes $P2_1/c$, that of low pigeonite. Along the Hd–Fs join, the symmetry change takes place in synthetic crystals at a composition around $Hd_{40}Fs_{60}$ [49]. The miscibility gap on the Mg-rich side of the quadrilateral (Figure 4-1) prevents observation of such a symmetry change, but the space group of even the most magnesian pigeonites is $P2_1/c$ at low temperatures. The lower symmetry yields two types of tetrahedral chains (called the Si*A* and Si*B* chains, with N and P rotations, respectively), but there is still only one kind of $M(1)$ site and one kind of $M(2)$ site; so calcium again preferentially orders into $M(2)$ on account of its size. [The $M(2)$ site, while larger than $M(1)$, is smaller than it is in $C2/c$ pyroxenes both because it is occupied by a smaller average cation, and because the N-rotated Si*A* tetrahedron shares an edge with it. See Figure 4-9.] Divalent iron and magnesium are very close in effective ionic radius, but the slightly larger size of Fe^{2+} suggests the likelihood of some preference for $M(2)$. In addition, $M(2)$ is considerably more distorted than $M(1)$, and crystal field theory (Chapter 11) predicts that a greater crystal field stabilization energy should therefore occur for Fe^{2+} in $M(2)$ that for Fe^{2+} in $M(1)$. These two complementary effects result in a fairly strong tendency for iron to order into $M(2)$, *after* all the calcium has done so.

For calcium-free pyroxenes of the pyroxene quadrilateral, the most common space group is *Pbca*, although monoclinic polymorphs crystallizing in space group $P2_1/c$ are known, and a high-temperature polymorph of enstatite called *protoenstatite* crystallizes in space group *Pbcn*. (See the discussion of polymorphism, p. 179.) In space group *Pbca* there is only one type of $M(1)$ site and one type of $M(2)$ site. For end-member compositions, each is obviously occupied by either Mg or Fe (corresponding to the En or Fs compositions, respectively). Ordering of iron preferentially into $M(2)$ is observed for intermediate members of the solid solution series, however, and the reasons already encountered seem to explain that adequately—slightly greater size of $M(2)$ and greater distortion of that site. Additional insight is provided, however, by calculated site energies [48]. The site energy used in this study was the sum of the electrostatic attractive and repulsive energies, calculated using some admitted but almost unavoidable approximations. They reveal that the combination of Mg in $M(1)$ and Fe in $M(2)$ yields a lower site energy than Fe in

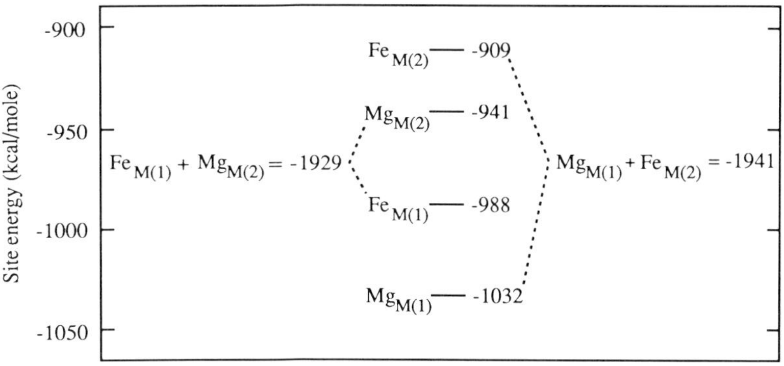

Figure 4-20. Site energies for *M*(1) and *M*(2) occupied by either Mg or Fe^{2+}. Note that Mg in *M*(1) and Fe^{2+} in *M*(2) yields a lower total site energy than does Fe^{2+} in *M*(1) with Mg in *M*(2).

M(1) and Mg in *M*(2), as shown in Figure 4-20. Thus, in addition to considerations of size and crystal-field stabilization, ionic effects are evidently important. The small quantities of $^{[6]}$Al found in some orthopyroxenes is ordered into *M*1.

Along the Di–En edge of the quadrilateral, an interesting phenomenon is observed. When any crystal structure refinement is near completion, it is customary (to say nothing of prudent) to perform a difference Fourier synthesis. This is a calculation, throughout the unit cell, of the difference between observed electron density (determined using the observed structure amplitudes but phase angles assumed from the refined structural model) and calculated electron density (determined solely from the refined model). If observation and model match closely, the difference synthesis will be essentially featureless throughout the unit cell. If an atom that is actually present is missing from the model, or if too light an atom has been assigned to an occupied position in the model, then a positive peak is seen in the difference synthesis at that position. In the event that an atom in the model is not really present in the structure (or if an atom in the structure is represented by too heavy an atom in the model), then a negative "peak" results. In crystal structure refinements of synthetic $Di_{90}En_{10}$ and $Di_{80}En_{20}$ [4] as well as in some natural clinopyroxenes [22;23;55], difference syntheses reveal a peak less than an Ångstrom from *M*(2) along the crystallographic *b* axis. It turns out to be four-coordinated, and very similar to the *M*(2) site in the synthetic monoclinic pyroxene $ZnSiO_3$ [43]; it is interpreted as being occupied solely by magnesium, up to a maximum of only ~7% occupancy, and is called *M*(2)′. Because *M*(2) and *M*(2)′ are too close to be occupied simultaneously, the structure may consist of ordered domains, some of which are diopsidelike in structure and contain (Ca,Mg) in *M*(2) with nothing in *M*(2)′, and some of which are $ZnSiO_3$-like in structure and contain Mg in *M*(2)′ with a vacant *M*(2). Evidence for a similar partially occupied site in Ca-poor hedenbergites has not been found, but the large thermal ellipsoids in those synthetic phases, elongated as they are along the crystallographic **b** direction, suggest positional disorder [49]; it is possible that such a site does exist, but that Fe^{2+} atoms are disordered between it and *M*(2) in such a way that the *M*(2)′ site is not spatially resolved.

The omphacites, which crystallize in space group *P*2/*n*, exhibit a rather complicated ordering scheme. Since the *c* glide is absent, there must be two kinds of *M*(1) polyhedra [formally designated *M*(1) and *M*(11)] and two kinds of *M*(2) polyhedra [*M*(2) and *M*(21)] per octahedral strip. An "ideal" chemical formula for omphacite may be written as $Ca_{0.5}Na_{0.5}(Mg,Fe^{2+})_{0.5}(Al,Fe^{3+})_{0.5}Si_2O_6$. The ionic radii of (Mg,Fe^{2+}) and (Al,Fe^{3+}) are sufficiently different (by about 30% of the latter on average) that they order by alternating along the *M*(1) chain. (This, of course, is what makes two kinds of *M*(1) sites, and leads to the reduction in symmetry from *C*2/*c* to *P*2/*n*.) Because the *M*(2)-type polyhedra share polyhedral elements with three adjacent *M*(1)-type octahedra, one kind of *M*(2)-type site is flanked by the *M*(11)–*M*(1)–*M*(11) sequence $Al(Fe^{3+})$–$Mg(Fe^{2+})$–$Al(Fe^{3+})$ and the other by the *M*(1)–*M*(11)–*M*(1) sequence $Mg(Fe^{2+})$–$Al(Fe^{3+})$–$Mg(Fe^{2+})$. The electrostatic environments are thus significantly different, and the Ca and Na atoms partially order to compensate for the local charge imbalance. Figure 4-21 shows the arrangement, but the atoms in *M*(21) labeled "Ca" are at most 0.75Ca, with the rest being Na, and those in *M*(2) labeled "Na" are at most 0.75Na with the remainder being Ca; likewise, the *M*(1) and *M*(11) sites labeled "Mg" and "Al" contain mostly, but not exclusively, those atomic species [55].

Some *C*2/*c* omphacites of apparent mantle origin have nonstoichiometric chemical formulas that contain fewer than 2.0 nontetrahedral cations per six oxygens. The vacancies are located in the *M*2 sites and typically amount to 0.02–0.08 vacancies per formula unit [39], or 2–8% of the *M*2 occupancy. These omphacites have greater compressibilities and smaller unit-cell volumes than do stoichiometric omphacites of similar jadeite-component concentration, suggesting that they may be stabilized (relative to vacancy-poor omphacites) by high-pressure conditions in the lower part of the upper mantle [40].

Order in the Tetrahedral Chains

Ordering can occur in the tetrahedral chains only if there are two kinds of tetrahedra per chain or two kinds of chains. This is the case in space groups *C*2, $P2_1/c$, *P*2/*n*, and *Pbca*, but in any of these space groups the tetrahedra can be merely geometrically different instead of different because of ordering. As pointed out previously, spodumene crystallizes in space group *C*2 but has been refined in only *C*2/*c*, which averages the two crystallographically distinct tetrahedra; however it contains only Si in tetrahedral coordination, so there is nothing to order. The omphacites (*P*2/*n*) for which crystal structure refinements are available contain nearly 2.00 Si atoms per formula unit (i.e., as in spodumene, there is virtually nothing to order) and statistically identical mean bond lengths for the two independent tetrahedra. Thus, tetrahedral ordering has been confirmed in only pigeonite ($P2_1/c$) and orthopyroxene (*Pbca*).

Both $P2_1/c$ and *Pbca* have two kinds of (100) tetrahedral "layers" (Figures 4-5 and 4-14), one of which contains only Si*A* chains and the other of which contains

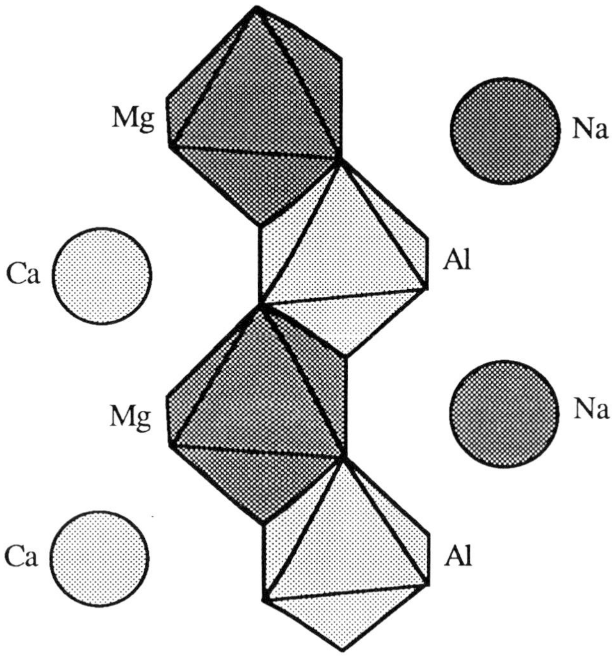

Figure 4-21. The octahedral ordering scheme in omphacite. The darker octahedra are the $M(1)$ sites, and are Mg-rich, and the lighter octahedra [$M(11)$] are Al-rich. The circles represent $M(2)$ (darker) and $M(21)$ (lighter) sites, which are rich in Na and Ca, respectively. The ordering in $M(2)$ and $M(21)$ is not as complete as that in the octahedra; "Na" in the figure represents, in the most ordered cases observed, 0.75Na + 0.25Ca, and "Ca" represents, in those same omphacites, 0.75Ca + 0.25Na.

only Si*B* chains. Each chain possesses just one kind of tetrahedron. The Si*A* chains are characterized by N rotations and the Si*B* chains by P rotations (p. 162). Thus each Si*A* tetrahedron shares an edge with an $M(2)$ polyhedron, and should be inherently smaller than Si*B* due to the shortened shared edge. In fact, where ordering of tetrahedral aluminum has been observed, Al has preferentially ordered into SiB, supporting the proposed difference in inherent tetrahedral size. That is not, however, the only explanation for the size difference. The Si*B* chains are considerably more "kinked" (rotated) than the Si*A* chains, so that the O(3)–O(3)–O(3) chain angles are narrower in order to facilitate proper *M*-site coordination. This naturally results in narrower *T*–O(3)–*T* angles, and the *T*–O bonds must lengthen to increase *T*–*T* separation. (*T* here stands, as usual, for tetrahedral atom, Al or Si). In addition, the larger tetrahedral rotation angle requires larger tetrahedra, so that two of them span the same *c*-axis repeat distance as two tetrahedra of the Si*A* chain span. The larger aluminum should prefer the Si*B* chain whether shared edges, small *T*-O-*T* angles, structural constraints, or a combination of these results in a larger tetrahedron. (In pyroxenes with tetrahedral aluminum some tetrahedra are larger, of course, simply *because* they are occupied by aluminum, and one needs to consider the distinction between cause and effect. In $P2_1/c$ and *Pbca* pyroxenes, however, the Si*B* tetrahedra are the larger ones even if no tetrahedral aluminum is present (for example, [6;42;60], so the tetrahedral size evidently catalyzes the ordering, rather than the other way around.)

Ca-Tschermak's (CaTs) pyroxene, $CaAl(SiAl)O_6$, is an interesting synthetic pyroxene because the Si,Al distribution appears disordered when investigated by x-ray-diffraction methods [52]. Inasmuch as only alternating Si- and Al-containing tetrahedra in a given chain would strictly satisfy the aluminum avoidance principle (p. 78) for a pyroxene of this composition, a disordered tetrahedral chain violates this empirical but generally obeyed rule. The synthesis was carried out at high temperature, so the atoms may actually be disordered but in a metastable "frozen" state. Otherwise, the apparent disorder is an artifact resulting from averaging differently ordered structural elements. These may be (1) individually ordered tetrahedral chains randomly staggered by $\pm c/2$ relative to one another throughout the crystal, or (2) antiphase domains arising from different ordering alternatives in different parts of a crystal that cools from a disordered state at high temperature (analogous to β-quartz, p. 23).

Either of these possibilities would provide an averaged x-ray-diffraction result mimicking disorder, if the numbers of individual structural elements with opposite order were nearly equal; if they were markedly unequal, possibility (1) would result in apparent partial ordering, while possibility (2) would result in diffuse diffraction maxima of the type ($h+k$ odd) that violate the extinction conditions of space group $C2/c$. Were it available, nuclear magnetic resonance spectra of CaTs could at least discriminate between a disordered Si,Al distribution and an ordered one, although it could not distinguish between the two ordered alternatives (1) and (2). We shall later consider antiphase domains and other microstructures in pyroxenes, but to this point transmission electron microscopy of CaTs [20] has revealed no diffraction maxima that violate space group $C2/c$, and thus no reason to believe that $C2/c$ is not the correct space group.

POLYMORPHISM AND PHASE TRANSFORMATIONS IN PYROXENES

The pyroxenes exhibit all of the general types of polymorphic phase transformations that we have discussed previously for the silica polymorphs, the feldspars, and the micas. Some of the transitions are rapid and displacive, and others are sluggish and reconstructive. Some involve a change in only the details of the second coordination sphere, while others involve changes in primary coordination. Some involve the ordering of cations in available polyhedra, and some can be considered polytypic. Some are first order, others are higher order, and many are not yet well enough understood to specify their thermodynamic natures.

Pyroxenes along the enstatite–ferrosilite join (the base of the pyroxene quadrilateral) may exist in more than one structural modification at any given composition; some of these polymorphs are observed in nature and some are known only from experiment. As the polymorphic relationships are not identical at opposite ends of this solid solution series, we shall first consider the polymorphism of enstatite.

Enstatite Polymorphs

Figure 4-22 shows that only three polymorphs of $Mg_2Si_2O_6$ are stable, and that low clinoenstatite (space group $P2_1/c$) is the stable phase at surface conditions. Even so, low clinoenstatite is comparatively uncommon in nature relative to enstatite (space group *Pbca*), and it appears that its occurrence and growth are enhanced by nonhydrostatic (i.e., shear) stresses. The pressure axis on phase diagrams such as the one shown in Figure 4-22 implies only hydrostatic pressure, and yet the experiments on which the low-clinoenstatite–enstatite boundary are based involved some shearing stress; the boundary shown must therefore be considered an upper limit to the stability of low clinoenstatite. It should be pointed out, too, that there is no universal agreement among mineralogists as to whether clinoenstatite or enstatite is the stable low-temperature phase (compare, for example, [36] with [59]). Evidence in favor of the topology of Figure 4-22 is (1) that enstatite has been observed to invert to clinoenstatite in molten $MgCl_2$ [29]—indisputably a hydrostatic environment—indicating that the monoclinic phase is stable at low temperatures; and (2) that the molar volume of clinoenstatite has been found to be very slightly lower

Figure 4-22. Pressure–temperature phase diagram for $Mg_2Si_2O_6$. L + Fo is the field of liquid plus forsterite resulting from the incongruent melting of enstatite. (After [36].)

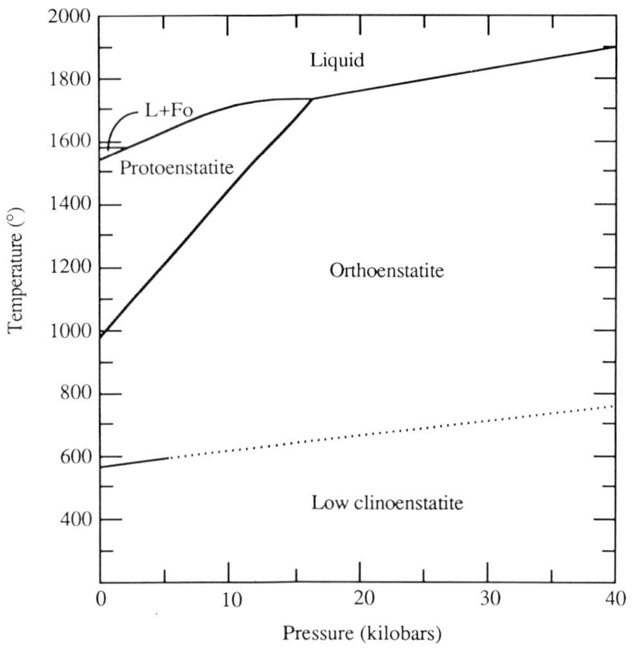

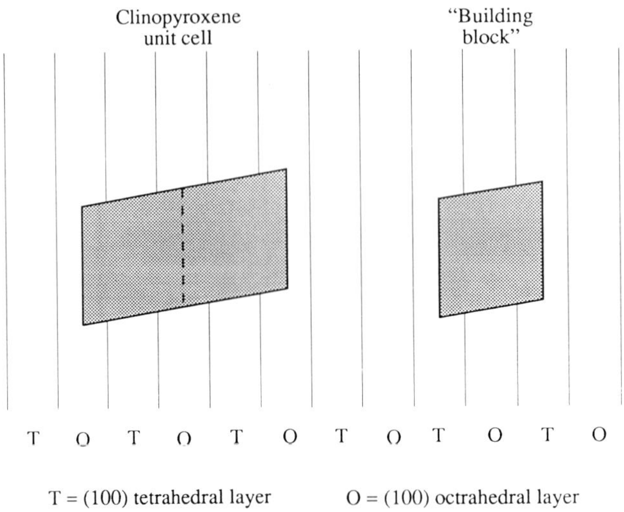

Figure 4-23. The unit cell of clinopyroxene (left), represented as two "pyroxene building blocks" consisting of two layers of tetrahedral chains enclosing the resulting layer of octahedral bands.

than that of enstatite [47], suggesting that clinoenstatite is the high pressure phase. The boundary between enstatite and protoenstatite (space group *Pbcn*) is fairly well established. The small field of forsterite (an olivine) plus liquid represents the incongruent melting of enstatite, but in the dry system it persists only at low pressures. If water is added, the liquidus and solidus boundaries are lowered, and incongruent melting persists to higher pressures [35].

Some understanding of the nature of the phase transformations among the three enstatite polymorphs is gained by modeling them as if constructed of identical slabs stacked differently along (100). Because this model fits the definition of polytypism used in the chapter on micas, the enstatite polymorphs are often considered polytypes. Figure 4-23 depicts the straightforward way in which the "building block" or "unit slab" of the model is derived from the crystal structure of clinoenstatite, and Figure 4-24 shows how building blocks are arranged to yield each polymorph (polytype). If shifts of the building block are considered positive or negative relative to the *c* axis, then the sequence in clinoenstatite is $\cdots|++|++|\cdots$, that in protoenstatite is $\cdots|+-|+-|\cdots$, and in enstatite it is $\cdots|++--|++--|\cdots$. It should be apparent that the plus and minus shifts correspond to the two kinds of skews identified earlier (p. 169 and Figure 4-17). The models of the three enstatite polymorphs may be thought of as derived approximately from low clinoenstatite by precise polysynthetic twinning on the unit-cell level. The model need not be taken too literally, because crystal structure refinements indicate that there are differences in the details of the three structures, and so there are really three nearly identical building blocks, not one (or, alternatively, one that is displacively altered depending on polymorph). For instance, the mean *M*(2)–O bond length in enstatite [47] is slightly longer, and the octahedron is a little more distorted, than that in clinoenstatite [47;51], and there appear to be small differences in chain angles of corresponding Si*A* and Si*B* chains as well; protoenstatite, in space group *Pbcn*, has only one type of tetrahedral chain, and its chain angle of ~167° [45] is considerably greater than any of the chain angles in either low clinoenstatite or enstatite.

An additional polymorph, not appearing in Figure 4-22 because it has no field of stability relative to the phases shown there, is high clinoenstatite, which occurs on heating of low clinoenstatite above ~980°C and has space group symmetry *C*2/*c*. By analogy with the ferrosilite polymorphs to be discussed, it has been suggested by several mineralogists that high clinoenstatite might have a field of stability above that of protoenstatite. Careful single-crystal x-ray investigations of protoenstatite [46;61] up to temperatures at which incongruent melting occurs have failed to confirm this hypothesis, however.

At surface conditions, only two polymorphs of pure $Mg_2Si_2O_6$ are ever found: enstatite and low clinoenstatite. Protoenstatite and high clinoenstatite cannot be quenched from the high temperatures at which they crystallize, and so we conclude

Figure 4-24. Unit cells of the $Mg_2Si_2O_6$ polymorphs, (a) clinoenstatite, (b) *Pbca* enstatite, and (c) *Pbcn* protoenstatite, constructed of the building blocks of Figure 4-23. Solid lines are unit-cell boundaries, and dashed lines are boundaries of building blocks that are not unit-cell boundaries.

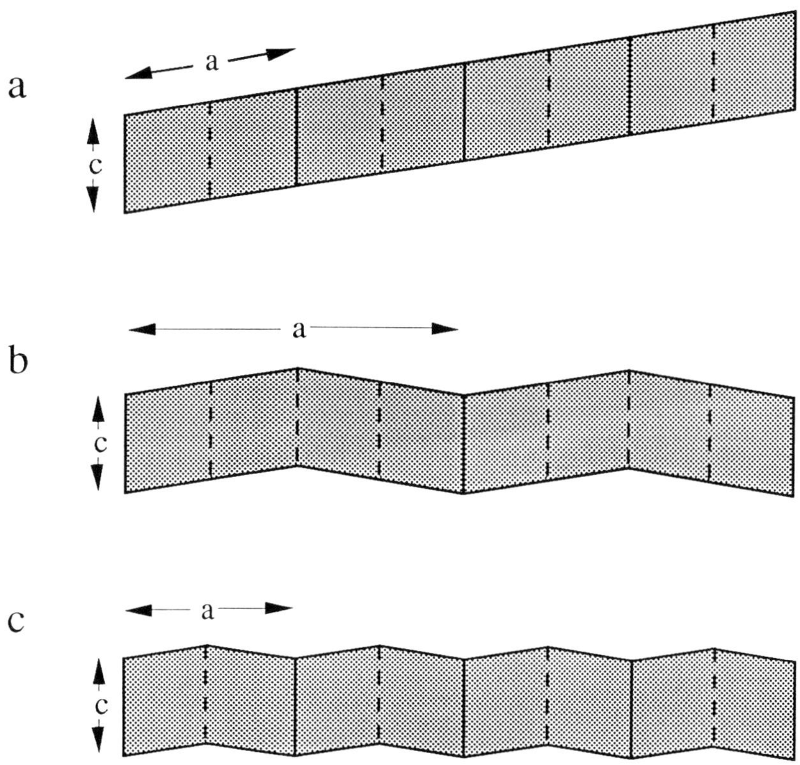

that the transformations they undergo are rapid and probably involve only small differences in free energy. The high–low clinoenstatite transformation is the more straightforward of the two, and is somewhat analogous to the monalbite–analbite transformation (pp. 54–55). At high temperature, the magnesium atoms experience thermal vibrations that make them effectively large, and the result is that the tetrahedral chains are all equivalent and relatively straight [although $M(2)$ is six-coordinated, rather than eight-coordinated as in calcium-rich $C2/c$ clinopyroxenes]. So beginning with low clinoenstatite at a low temperature, as the temperature is increased the effective sizes of the metal cations also increase, and the S-rotated Si*A* chain and O-rotated Si*B* chain straighten to maintain proper coordination of the $M(2)$ cations; this results in the two different types of chains becoming equivalent and the symmetry changing from $P2_1/c$ to $C2/c$. The transition has been more adequately characterized for iron-rich pyroxenes to be discussed in the following, but at present it is not known with surety that the transitions behave identically in iron-rich and iron-free pyroxenes.

When protoenstatite is cooled experimentally to below 700°C, it undergoes a rapid transformation to either enstatite or low clinoenstatite, depending on cooling rate [61]. Very rapid cooling rates yield twinned clinoenstatite, and slower ones yield enstatite. The **a***, **b**, and **c** crystallographic directions are identical before and after each inversion (and thus these are so-called *topotactic inversions*), but the transformations behave in quite different ways. The transformation from protoenstatite to clinoenstatite is reversible, but all of one phase does not invert to the other at once. At any given temperature, transformation occurs in part of a specimen, but proceeds no further regardless of the time allowed. Only after the temperature is raised or lowered (depending on the direction of the reaction being observed) does more of the specimen transform. It appears that there is enough strain energy associated with the coherency of the two oriented lattices that it compensates for the lowering of free energy that would be realized if the reaction proceeded to completion. The transition is thus temperature dependent but time independent, and such transitions are termed *martensitic*. In contrast to these martensitic transitions, the protoenstatite–enstatite, enstatite–protoenstatite, and clinoenstatite–enstatite transformations are sluggish and time-dependent, proceeding further at a given temperature the longer the time allowed.

In the case of the inversion from protoenstatite to low clinoenstatite or vice versa, we are presented with a phenomenon that appears, given our previous discussions of phase transformations in the silica polymorphs and the feldspars, to be enigmatic. This transformation requires that some (100) layers of tetrahedra and octahedra somehow slide past one another, breaking *M*–O bonds in the process, but leaving Si–O bonds intact (refer to Figure 4-24). It must therefore be reconstructive in part, but it is also very rapid and easily reversible, characteristics associated with displacive transitions. Various alternatives have been suggested [3;19;56], but one [61] that seems to require minimum movement and facilitate rapid rearrangement involves a coordinated shift of alternate tetrahedral layers with a concomitant shift of $M(1)$ cations to newly created $M(2)$ positions, and $M(2)$ cations in the same octahedral layer to newly created $M(1)$ positions, as indicated schematically by Figure 4-25. Because the c cell dimension of low clinoenstatite is larger than that of protoenstatite, expansion cracks typical of low clinoenstatite inverted from protoen-

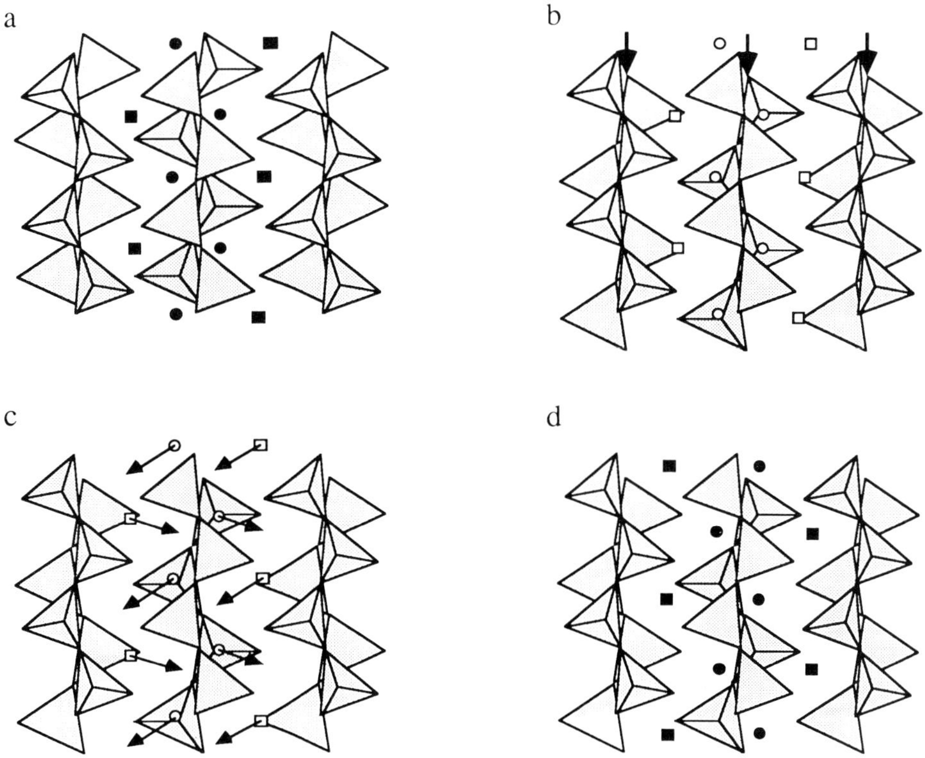

Figure 4-25. A possible mechanism for the phase transformation from protoenstatite to low clinoenstatite. Dots are *M*(1) cations, and filled squares are *M*(2) cations. Open symbols represent former locations of cations, and arrows show the movement of structural elements. Beginning with (a) protoenstatite, (b) the lower tetrahedral layer shifts along **c**, (c) the former *M*(1) Mg moves to a new *M*(2) site and the former *M*(2) Mg moves to a new *M*(1) site, resulting in clinoenstatite (d). (No attempt has been made to show changes in chain geometry, which are necessary during the transformation). (Modified from [61].)

statite develop perpendicular to the **c** direction (see, for example, photomicrographs in [70]).

Ferrosilite Polymorphs

On the ferrosilite side of the En–Fs series, a number of analogies with the polymorphism of enstatite are evident from Figure 4-26, but there are also some differences. A major difference, to begin with, is that no ferrosilite polymorph is stable at surface conditions, but rather the assemblage fayalite plus α-quartz. (A quick mental check will verify that adding the chemical formulas of fayalite and quartz together does indeed yield the formula of ferrosilite. Note that incongruent melting is the rule for all pressures in this system.) Low clinoferrosilite ($P2_1/c$), ferrosilite ($Pbca$), and high clinoferrosilite ($C2/c$) are analogous to, and isostructural with, the enstatite polymorphs with similar names, but ferrosilite-III is a pyroxenoid with a

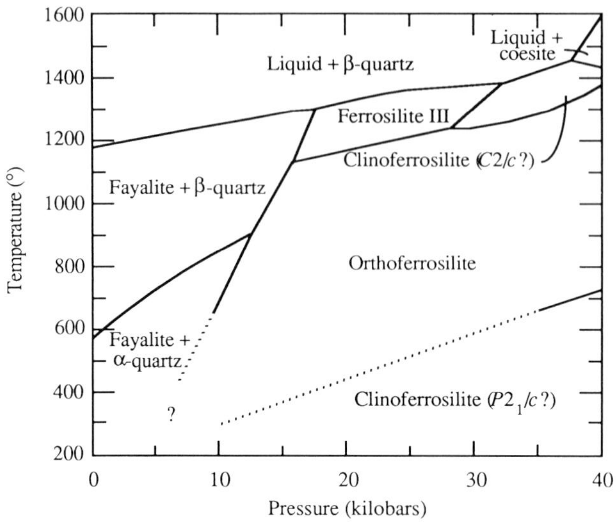

Figure 4-26. Pressure-temperature phase diagram for $Fe_2Si_2O_6$. No ferrosilite polymorph is stable at low pressure, and ferrosilite III is not a pyroxene, but a pyroxenoid. (After [36].)

nine-tetrahedron repeat, and does not correspond structurally to protoenstatite, for which there is no analog in this system.

Phase transitions in the Fs corner of the quadrilateral bear some similarity to those just discussed for the En corner. The transformation from ferrosilite to a twinned high clinoferrosilite occurs at 1025°C (with increasing temperature at 1 atmosphere pressure) [65], and is rapid and reversible. Like the analogous enstatite transition, the two phases share identical **a***, **b**, and **c** crystallographic directions (i.e., the inversion is topotactic). The *c* cell dimension changes continuously in the orthorhombic phase as temperature is increased from 25°C to 1025°C, but discontinuously across the transition temperature, while the other two cell edges appear to change smoothly across the transition temperature, suggesting that the adjustment in chain angle that signals the phase change is abrupt [66]. Figure 4-27, which shows the variation in chain angles, does imply a rather abrupt change at 1025°C. The discontinuity in *c* cell edge suggests that the inversion is first-order, and thus there may be a two-phase ferrosilite–high-clinoferrosilite field of stability; no such field has been delineated to this point.

Transformations at the Fs corner, like those of En, are martensitic [66] and must be partially reconstructive, involving the breaking of *M*–O bonds. It is thus possible (but not yet known) that the same *M*(1)–*M*(2) cation exchange mechanism proposed to explain En transition (Figure 4-25) could apply as well to Fs.

Polymorphs of Intermediate En-Fs Pyroxenes and Pigeonites

Intermediate pyroxenes along the En–Fs join are orthorhombic ferroan enstatite (or magnesian ferrosilite), ferroan low clinoenstatite (magnesian low clinoferrosilite) and ferroan high clinoenstatite (magnesian high clinoferrosilite). The protoenstatite structure appears limited to compositions very near En, containing no more than 2 mol % Wo component nor more than 13 mol % Fs component [61]. One possible explanation for this is seen in Figure 4-28, which compares a portion of the orthopyroxene structure with a similar portion of the protopyroxene structure [45]. In the orthopyroxene structure, the *B* tetrahedral chains are kinked considerably, and oxygen O(3) can thus move away from *M*(2) substantially to allow for larger cations without changing the nature of the structure. In the protopyroxene arrangement, the single type of tetrahedral chain is relatively straight, and the tetrahedral rotation sequence across *M*(2) is NN, producing a small and distorted *M*(2) site; comparatively minor movement of oxygens O(3) away from *M*(2) in order to accommodate larger cations—such as calcium in the hypothetical calcian protoenstatite shown—would not only result in a PP sequence (thus a different structure type), but might also further stretch and distort the *M*(2) site past limits that are structurally reasonable.

The transition from low to high ferroan clinoenstatite or magnesian clinoferrosilite is rapid and displacive, and it takes place at temperatures ranging from ~700°C (Mg-rich compositions) to ~1000°C (Fe-rich compositions). Like the analogous transitions in end-member pyroxenes, it is unquenchable, and involves inversion from $P2_1/c$ to $C2/c$. In addition, the cell dimensions and chain angle change discontinuously at the transition temperature [62;63], making this a first-order transformation. The transformation from ferroan enstatite to high ferroan

Figure 4-27. The variation in the chain angle for high clinoferrosilite, showing an abrupt change at the transformation temperature for ferrosilite. Diamonds represent data points obtained by high temperature crystal structure refinement. (After [66].)

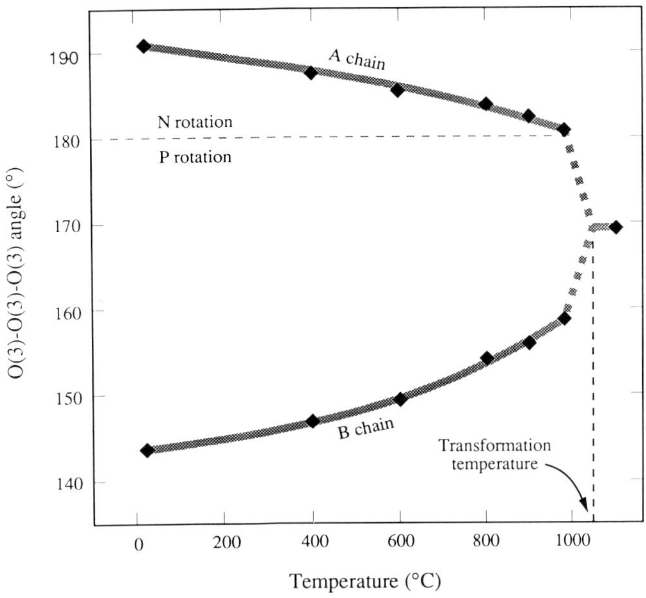

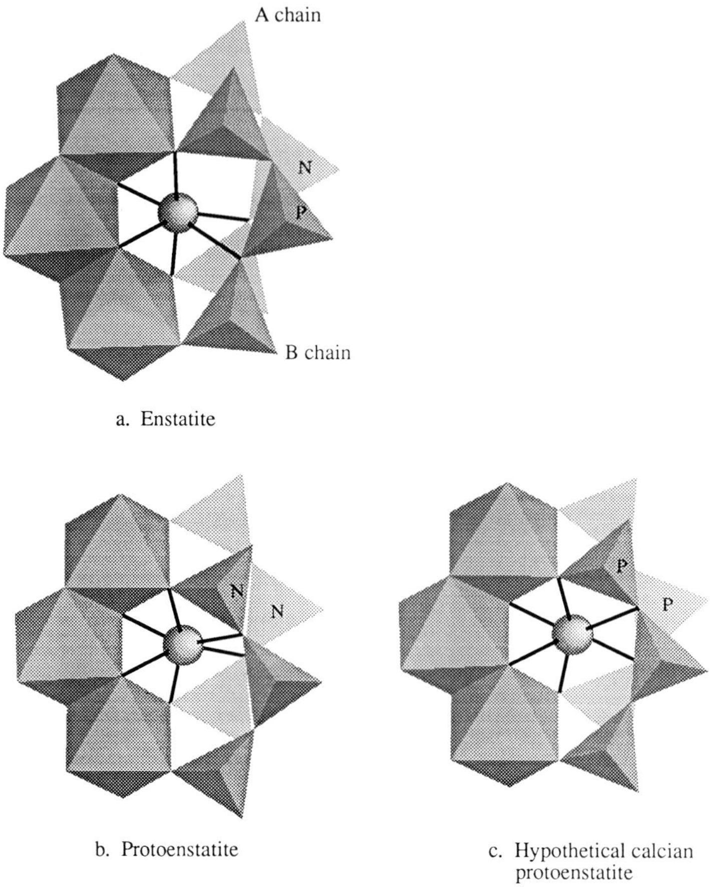

Figure 4-28. Similar portions of the structures of two real and one hypothetical pyroxene. (a) In orthopyroxenes, the Si*B* chain is sufficiently kinked that it can relax to accommodate larger *M*(2) cations without changing the nature of the structure. (b) In protoenstatite, the tetrahedral chain is nearly fully extended, and the rotation sequence is NN. (c) In a hypothetical calcian protoenstatite, the adjustments of the tetrahedral chains to accommodate the larger *M*(2) cations would result in a change from NN to PP rotation sequence and hence a nonprotoenstatite structure type.

clinoenstatite is fairly rapid (slower with increasing Mg) and topotactic if the pyroxene is Ca free, but it is not reversible [67]. Because Fe^{2+} orders preferentially into *M*(2), the *M*(1)–*M*(2) exchange mechanism proposed for transformations in En and Fs polymorphs (Figure 4-25) is unlikely to operate either here or in the pigeonites to be discussed next, owing to distortions that the exchange would require of the crystal structure and the unfavorable crystal field stabilization that would result. If the ferroan enstatite contains as little as 2 mol % Wo, to as much as 5 mol %, then the transformation from orthorhombic to monoclinic forms is neither reversible nor topotactic, but it does proceed rapidly (within a matter of a few minutes in laboratory experiments [67]).

The maximum amount of Ca that can be accommodated in the *Pbca* pyroxene structure appears to be about 5 mol % Wo, so pigeonites have only low ($P2_1/c$) and high ($C2/c$) monoclinic polymorphs. As temperature is lowered, the $C2/c$ high pigeonite undergoes a rapid, displacive, and unquenchable transformation to $P2_1/c$ low pigeonite. Anywhere from ~1000°C to ~200°C, depending again on composition [2], one of the two equivalent tetrahedral chains is caused to rotate more than the other because of the thermal contraction of $M(2)$ with which it shares polyhedral elements, and it becomes the Si*B* chain; the other tetrahedral chain rotates in the opposite direction, and becomes the S-rotated Si*A* chain.

The pigeonite structure, with 5 to 10 mole percent Wo, is stable nearly to compositions on the En–Di boundary (Figure 4-1), but at high temperature and a calcium content less than that of pigeonite (but exceeding that of protoenstatite) it appears that there may be another pyroxene polymorph [8;34;37]. At temperatures between 1300 and 1400°C at 1 atm pressure, a phase that is orthorhombic upon cooling is found, but it does not exhibit the characteristic cracks normal to **c** or polysynthetic twinning found in clinoenstatite inverted from protoenstatite. It does not appear on the Ca-free enstatite *P–T* phase diagram (Figure 4-22), so if that diagram is correct this phase evidently requires a small amount of calcium. Phase equilibrium experiments in the system CaO–MgO–SiO_2 [8] place an upper limit for enstatite stability at 1005°C, so it appears not to be low-Ca enstatite. Its crystal structure, stability, and precise relationships to other Mg-rich polymorphs has not yet been established, and inasmuch as its x-ray powder pattern at room temperature looks like that of normal enstatite, that is what it might be; if that is the case, however, then the thermodynamic models for the magnesian end of the $Mg_2Si_2O_6$–$CaMgSi_2O_6$ system are not at present adequate [9]. On the other hand, this may be an unquenchable phase that is somehow different from enstatite at high temperature.

Omphacite Polymorphs

A final example of polymorphism in pyroxenes involves the transformation between $C2/c$ omphacite, stable at high temperatures, and $P2/n$ omphacite, stable at low temperatures. Unlike the previous examples, it is an ordering transformation involving the movement of cations of different chemical species among *M* sites.

The space group of omphacite is, first of all, composition dependent. In Figure 4-2 the compositional triangle shows a primitive omphacite field and a centered omphacite field. Along the jadeite–diopside join, compositions outside the range $\sim 0.35 < \mathrm{Jd/(Di+Jd)} < \sim 0.65$ yield disordered $C2/c$ omphacites. On heating, the space group remains $C2/c$.

It is within the primitive composition range that the polymorphic transition occurs, and it manifests itself as disordering of the two $M(1)$ and two $M(2)$ sites as temperature is increased. [Recall that in $P2/n$ omphacite, Mg(Fe^{2+}) and Al(Fe^{3+}) are completely ordered in two kinds of $M(1)$ sites—$M(1)$ and $M(11)$, respectively—while Na and Ca are respectively partially ordered on $M(2)$ and $M(21)$.] The maximum ordering so far observed at the two different $M(2)$-type sites is (0.75Ca + 0.25Na) and (0.25Ca + 0.75Na), with a considerably higher degree of order exhib-

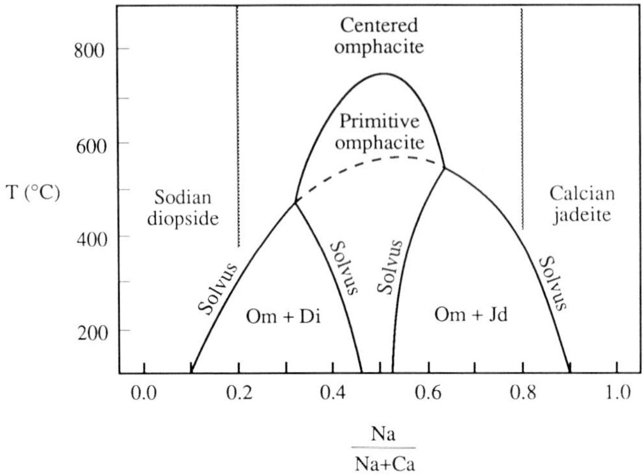

Figure 4-29. A temperature-composition (T–X) phase diagram for the system diopside–jadeite, intermediate members of which are omphacites. The dashed line is a metastable extension of the Di–Jd solvus. The high-temperature curve separating centered from primitive omphacite is not a solvus, but simply the phase boundary across which the transformation occurs. (After [55].)

ited by Mg(Fe^{2+}) and Al(Fe^{3+}) in the $M(1)$ and $M(11)$ sites of the same crystals [55]. At high temperatures (perhaps 650 ± 150°C, depending on composition; the transformation temperature is poorly constrained at present) the Mg(Fe^{2+}) and Al(Fe^{3+}) become disordered, as do the Ca and Na, yielding one kind of $M(1)$ site and one kind of $M(2)$ site, and $C2/c$ symmetry. The inversion is likely second or higher order, because no first-order effects (equilibrium coexistence of ordered and disordered phases of the same composition, and so forth) have been observed, and it is decidedly reconstructive, inasmuch as M–O bonds must be broken to effect the cation exchanges. Figure 4-29 is a tentative T–X (temperature–composition) phase diagram for the diopside–jadeite join, which includes the omphacites, showing what can currently be gleaned from field observations and a few laboratory studies. The arched boundary separating $C2/c$ from $P2/n$ omphacite between the compositions of Jd_{35} and Jd_{65} is not a solvus, but only a phase boundary across which inversion occurs. The two pairs of curves that originate at the ends of that arch do constitute two solvi, however.

MICROSTRUCTURES IN PYROXENES

Antiphase Domain Textures

In three of the phase transformations just discussed, some sites that are equivalent at high temperature become crystallographically distinct at low temperature. One of these transitions is the high–low clinoenstatite (also ferroan clinoenstatite) inversion. Because all silicate chains are identical at high temperature, different parts of the same cooling crystal may effect the *C*2/*c*–*P*2$_1$/*c* collapse to Si*A* and Si*B* chains differently. Any given tetrahedral chain may become an Si*A* chain in one part of the crystal and an Si*B* chain further along *c* in another part. This results in the formation of domains that do not match crystallographically where they meet; they are called *antiphase domains*, and they meet at antiphase boundaries (APBs). (Recall that we encountered a similar phenomenon when discussing plagioclases, p. 85, and described it with the same terminology). As one would expect from the crystallographic similarities between ferroan clinoenstatite and pigeonite, antiphase domains may also develop when high (*C*2/*c*) pigeonite inverts to low (*P*2$_1$/*c*) pigeonite. Note that both the enstatite and pigeonite transformations are displacive.

It appears that the orientations of APBs in pigeonites (and ferroan clinoenstatites?) may be dependent on the calcium content, and it is possible that the domains are oriented so as to optimize the number of presumably larger calcium-containing sites available at the "discordant" boundaries where the structure is presumed to be *C*2/*c* [10;11;44]. (Recall that the regions between α_1 and α_2 domains of low quartz were of necessity β-quartz in structure, and the same reasoning is applied here in postulating *C*2/*c* domain boundaries.) Supporting the hypothesis of calcium-poor domains with calcium-rich boundaries are crystal structure refinements of lunar pigeonites in which the type *a* and type *b* reflections are treated differently [18;50]. In space group *C*2/*c* (or, indeed, any other *C*-centered space group), reflections with diffraction indices of the type $\{h+k \text{ odd}\}$ are forbidden, but they are not forbidden for primitive space groups. In pigeonites, $\{h+k \text{ even}\}$ reflections are called *a* reflections while those with $\{h+k \text{ odd}\}$ are the *b* reflections, and the latter are somewhat diffuse in pigeonites that possess domain textures. In two lunar pigeonites for which the diffuse *b* reflections were used alone in crystal structure refinements, the structures of the primitive domains turned out to resemble those of low-calcium ferroan clinoenstatites; the bulk compositions of both minerals were higher in calcium than low-Ca ferroan clinoenstatites, and it is thus required that the calcium be largely located in the *C*-centered boundaries that do not contribute to *b* reflections. Moreover, refinement with both the *a* and *b* reflections included yielded an average bulk structure (domains and boundaries combined) that was more like a Ca-rich pigeonite.

Another indication that calcium may be concentrated at domain boundaries is the observation of domain "memory" in a kanoite (ideally, $MnMgSi_2O_6$, but containing approximately 0.12 atom Ca for Mg in this case) exsolved in a host manganoan diopside [28]. The kanoite possesses space group *P*2$_1$/*c*, but reverts to *C*2/*c* with no antiphase domains on heating to 350°C. After cooling, however, the domains reappear in the same orientations, many with the same boundaries they had before heating. This suggests that something is concentrated at the boundaries that does not migrate or change with mild heating, and that controls the locations of the boundaries when they reform. The most likely possibility is an elemental concentration (although a concentration of lattice defects cannot be ruled out), and the most likely element is calcium.

Antiphase domains may also develop in omphacites, but for a quite different reason than in Ca–Mg–Fe–(Mn) pyroxenes. During the *C*2/*c*–*P*2/*n* transformation that occurs on cooling, the "choice" of *M*(1) and *M*(11) [thus also *M*(2) and *M*(21)] may be made differently in different parts of the same crystal; octahedral ordering patterns in two such regions would be out of phase with each other, and the octahedral strips would meet at antiphase boundaries.

It is obvious that the mere presence of APBs in a pyroxene reveals something about its previous crystallization history, if no more than that it inverted from a space group of higher symmetry. It has been observed that antiphase domains that arise from the displacive transformation tend to be elongated along c and have irregular boundaries, whereas those resulting from the ordering transformation tend to be more equiaxial with smooth boundaries [14]. With annealing, the elongate domains coarsen rapidly (that is, larger ones grow at the expense of smaller ones), because the energy required to change from one orientation to the other across an APB is very small; domains caused by ordering coarsen more slowly because the energy required to move the octahedral cations around is large.

Attempts to relate domain textures to specific thermal histories have not been altogether successful, largely because the domain texture depends on other variables as well, such as chemical composition and the density of crystal defects. Neither, however, have such attempts been without promise. The disordering of omphacite has been studied, for example, within a framework called time–temperature–transformation (TTT) analysis [13;14] with at least semiquantitative success. Natural ordered omphacites containing antiphase domains were heated at temperatures from 750 to 1260°C at 18 and 30 kbar for times of 1.5 minutes to 16 days, and the presence or absence of the $10\bar{1}$ x-ray reflection was checked on single-crystal photographs. (That reflection is absent for space group $C2/c$, but present for $P2/n$. The absence would thus indicate that a transformation had occurred.) A disordering temperature of 865°C was determined from the experimental results, and the data were extrapolated (using some assumptions and simplifications that need not concern us here) to lower temperatures and a pressure of 8 kbar to yield the TTT diagram of Figure 4-30. The figure shows that infinite time would be required for 95% ordering at 865°C and 8 kbar (i.e., ordering would not occur), but that a small decrease in temperature would result in quite rapid ordering. If the temperature were much lower, however, ordering would be very slow (~50 million years at 350°C). Using a phenomenological model in which time and activation energy are related to antiphase domain size [14] results in the conclusion that maximum antiphase domain sizes are much more dependent on peak metamorphic temperature than on time, and that they may be good estimators of temperature.

Exsolution

It is not uncommon to find pyroxenes whose bulk compositions lie within the miscibility gaps discussed in this chapter. Microscopic or TEM examination of natural pyroxenes with compositions that fall within the miscibility gap of the En–Di–Hd–Fs quadrilateral usually reveals that they consist of exsolution intergrowths of host augite with lamellae of either orthopyroxene or pigeonite or both. This can be crystallographically rationalized by considering how the difference in the effective sizes of Ca and (Mg,Fe) changes with decreasing temperature. As thermal motion subsides, the smaller (Mg,Fe) can be accommodated best in octahedral coordination, but Ca requires a larger eightfold site. The result, for a Ca,Mg,Fe-pyroxene, is con-

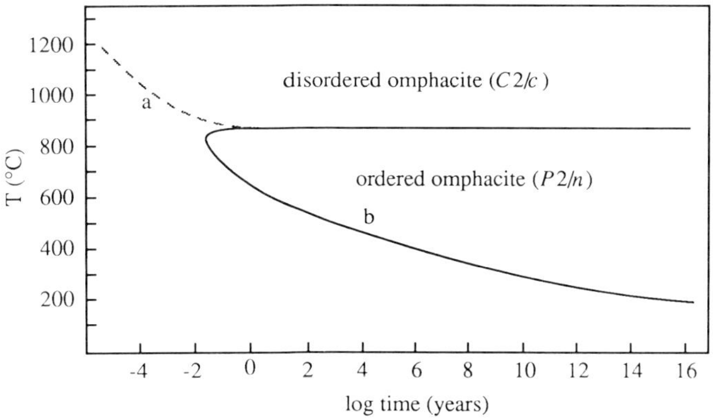

Figure 4-30. A time–temperature–transformation (TTT) diagram for a natural omphacite. Curve a is the experimental *disordering* curve obtained by heating ordered omphacite at 18 kbar until it becomes ordered, as determined by the disappearance of the $10\bar{1}$ x-ray reflection. Assuming that the disordering and ordering are inverse phenomena with respect to rate and mechanism, and adjusting for a pressure of 8 kbar results in curve b, the TTT curve for ordering. The curve shows, for example, that heating at 800°C would result in ordering in about a week; at 400°C, however, the boundary between disordered and ordered omphacite would not be crossed for about a million years. (After [13].)

siderable structural strain. Consequently, a subsolvus migration of ions concentrates the calcium into volumes that become *C*2/*c* augite lamellae, and leaves calcium-poor volumes that become $P2_1/c$ pigeonite or *Pbca* orthopyroxene lamellae. This localizes the strain along boundaries that can be adjusted to achieve the optimum fit of the two lattices. For Mg–Fe–Ca pyroxenes, the optimal phase boundaries yield exsolution lamellae that are nearly parallel to either (001) or (100), but may differ in orientation from those rational planes by up to several degrees [54]. For ~(001) exsolution lamellae, the orientation is governed by misfit of the lamellar and host *a* cell edges, and for ~(100) lamellae, misfit of the *c* dimensions determines orientation. Just as in the feldspars (pp. 69–72), if the lattices are continuous across phase boundaries, then the intergrowth is said to be *coherent*; if the lattices are coherent but the cell dimensions that must match are not exactly the same in the two phases, then *coherency strain* results. Exsolution in pyroxenes is sometimes not lamellar, but rather irregular and patchy (e.g., [53]), suggesting that in such cases it originated by heterogeneous nucleation at crystal defects (p. 70).

Exsolution microstructures in "quadrilateral" pyroxenes may develop either by nucleation and growth (a discontinuous process) or by spinodal decomposition (a continuous process). Differences in textures that may result from these two modes of formation are shown in Figure 4-31—though it must be acknowledged that textural indicators are not necessarily unequivocal [33]. Note that spinodal decomposition sometimes results in compositional modulations in two directions, as was seen in feldspars (Figure 2-11). For spinodal decomposition to occur, cooling must be sufficiently rapid that the spinodal is reached before nucleation can occur near the solvus (refer to Figure 2-27). Spinodal decomposition is therefore more likely in volcanic or kimberlitic pyroxenes than in slowly annealed metamorphic or plu-

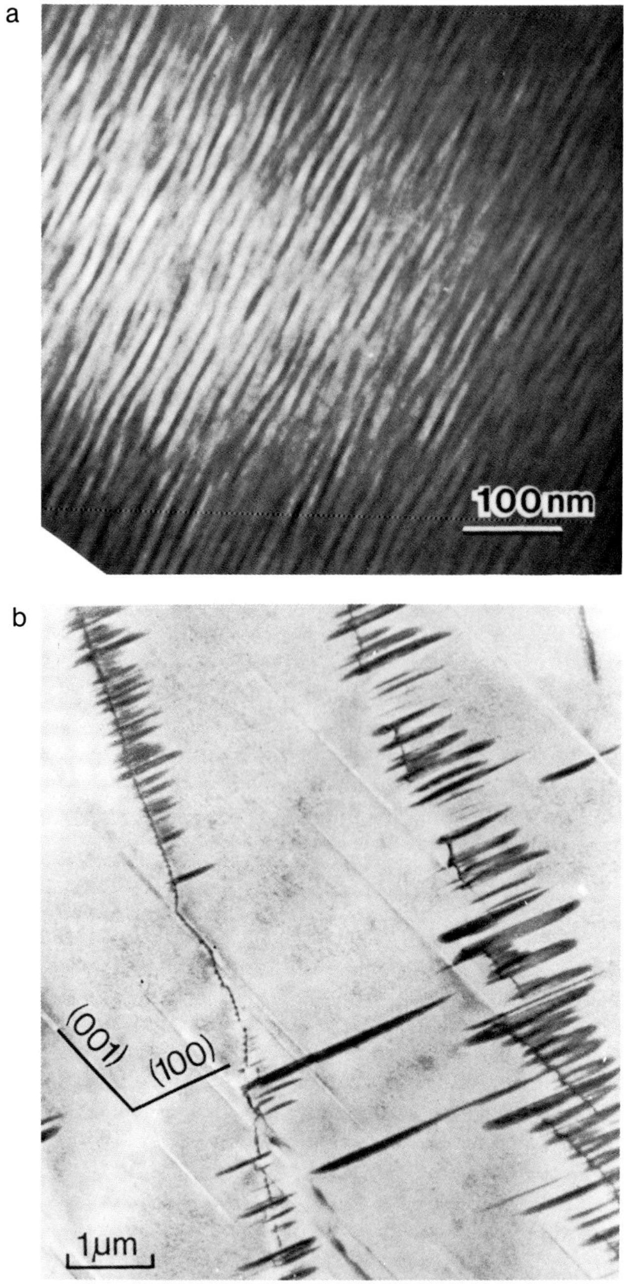

Figure 4-31. (a) An exsolution texture in subcalcic diopside caused by spinodal decomposition to Ca-rich and Ca-poor lamellae. [Photo from R. H. McCallister and G. L. Nord (1976) "Subcalcic Diopsides from Kimberlites: Chemistry, Exsolution Microstructures, and Thermal History." *Contributions to Mineralogy and Petrology* 78, 118–25.] (b) Exsolution of pigeonite lamellae on a dislocation in augite—an example of nucleation and growth. [Photo from G. L. Nord, A. H. Heuer, and J. S. Lally (1976). "Pigeonite Exsolution from Augite." In H.-R. Wenk (ed.), *Electron Microscopy in Mineralogy*, 220–27. Springer-Verlag, Berlin.]

tonic pyroxenes. This suggests that pyroxene microstructures may have the potential for delineating, or at least limiting, pressure–temperature paths for the rocks in which they occur (see, for example, [38], in which pyroxene microstructures are used elucidate the *P–T* path for kimberlites).

Some augite contains orthopyroxene lamellae bounded by thin ferroan clinoenstatite margins, with a high Ca concentration at the augite–clinoenstatite interface [31], suggesting that the lamellae were formed by diffusion of Ca away from the lamellae and toward the augite, and that the orthopyroxene was initially ferroan clinoenstatite; if this is the case, then the orthopyroxene has likely been transformed from ferroan clinoenstatite by glide twinning (see Figures 4-23 and 4-24). It is not yet clear whether all orthopyroxene exsolution lamellae in augite originate in this way; exsolution at high temperatures would result in rapid diffusion rates, and the consequent obliteration of any ferroan enstatite margins that might have persisted at lower temperatures, thus eliminating the evidence for a two-step diffusion-twinning mechanism.

Exsolution also occurs in sodic pyroxenes, in which the intergrowths are jadeite–omphacite or augite–omphacite. In these cases, as seen from Figure 4-29, one of the exsolved phases is ordered *P*2/*n* omphacite, and the other is a disordered *C*2/*c* pyroxene. From a crystallographic point of view, the ordered omphacite acts as an "end member" for the same reason that dolomite, compositionally half-way between $MgCO_3$ and $CaCO_3$, does. In dolomite, the conflicting polyhedral size requirements for accommodating Mg and Ca in the same cation layer make that an energetically much less favorable scheme than simply ordering Mg and Ca into alternating cation layers. In *P*2/*n* omphacite, the conflicting requirements for *M*(1)

Figure 4-32. The intersection of some (low-temperature) isothermal plane with the free energy surface in *G–T–X* space for the system augite–jadeite. The black curve shows how *G* changes with composition for disordered phases, and the gray curve shows how ordering lowers *G* (relative to the disordered state) over an intermediate compositional interval. Because the stable phase is that with the lowest free energy, omphacites within that interval will be ordered. For a bulk composition between jadeite and omphacite at this temperature, the stable assemblage will be a jadeite and an ordered omphacite, as shown by the dashed line.

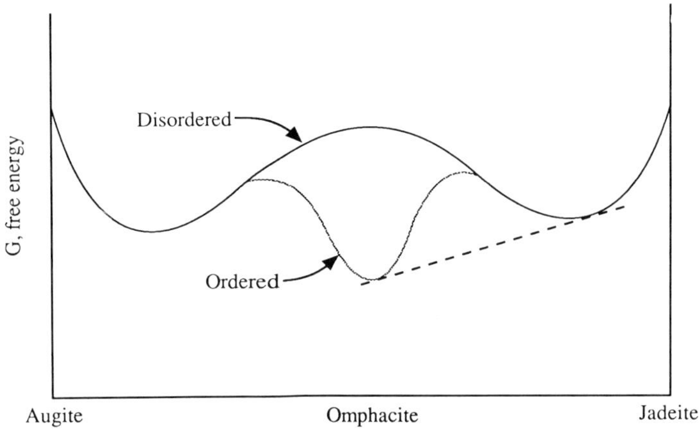

octahedra that need to hold *either* Al or Mg make an ordered, alternating arrangement of Al and Mg much more favorable. From a thermodynamic point of view, the free energy of ordering gives considerable stability to the ordered arrangement; the intersection of the free-energy surface in G–T–X space with some arbitrary (low-temperature) isothermal plane would look something like Figure 4-32. Because the ordering transition is likely second or higher order, the exsolution predicted by Figures 4-29 and 4-32 probably occurs by spinodal decomposition [12]. Evidently, omphacites order from $C2/c$ to $P2/n$ first, then, if their compositions are within either of the two solvi of Figure 4-29, they exsolve to an ordered $P2/n$ omphacite host with disordered $\text{jadeite}_{\text{solid solution}}$ or $\text{augite}_{\text{solid solution}}$ lamellae.

REFERENCES

[1] Bown, M. G. (1965). "Reinvestigation of Clinoferrosilite from Lake Naivasha, Kenya." *Mineralogical Magazine* 34, 66–70.
[2] Brown, G. E., C. T. Prewitt, J. J. Papike, and S. Sueno (1972). "A Comparison of the Structures of Low and High Pigeonite." *Journal of Geophysical Research* 77, 5778–89.
[3] Brown W. L., N. Morimoto, and J. V. Smith (1961). "A Structural Explanation of the Polymorphism and Transitions of $MgSiO_3$." *Journal of Geology* 69, 609–16.
[4] Bruno, E., S. Carbonin, and G. Molin (1982). "Crystal Structures of Ca-Rich Clinopyroxenes on the $CaMgSi_2O_6$–$Mg_2Si_2O_6$ Join." *Tschermaks Mineralogische Petrographische Mitteilungen* 29, 223–40.
[5] Burnham, C. W. (1966). "Ferrosilite." *Carnegie Institution of Washington Year Book* 65, 285–90.
[6] Burnham, C. W., Y. Ohashi, S. S. Hafner, and D. Virgo (1971). "Cation Distribution and Atomic Thermal Vibrations in an Iron-Rich Orthopyroxene." *American Mineralogist* 56, 850–76.
[7] Cameron, M., and J. J. Papike (1981). "Structural and Chemical Variations in Pyroxenes." *American Mineralogist* 66, 1–50.
[8] Carlson, W. D. (1985). "Evidence Against the Stability of Orthoenstatite Above ~1005°C at Atmospheric Pressure in CaO–MgO–SiO_2." *Geophysical Research Letters* 12, 409–11.
[9] Carlson, W. D., J. S. Swinnea, and D. E. Miser (1988). "Stability of Orthoenstatite at High Temperature and Low Pressure." *American Mineralogist* 73, 1255–63.
[10] Carpenter, M. A. (1978). "Nucleation of Augite at Antiphase Boundaries in Pigeonite." *Physics and Chemistry of Minerals* 2, 237–51.
[11] Carpenter, M. A. (1979). "Contrasting Properties and Behaviour of Antiphase Domains in Pyroxenes." *Physics and Chemistry of Minerals* 5, 119–31.
[12] Carpenter, M. A. (1980). "Mechanisms of Exsolution in Sodic Pyroxenes." *Contributions to Mineralogy and Petrology* 71, 289–300.
[13] Carpenter, M. A. (1981a). "Time–Temperature–Transformation Analysis of Cation Disordering in Omphacite." *Contributions to Mineralogy and Petrology* 78, 433–40.
[14] Carpenter, M. A. (1981b). "Omphacite Microstructures as Time-Temperature Indicators of Blueschist- and Eclogite-Facies Metamorphism." *Contributions to Mineralogy and Petrology* 78, 441–51.
[15] Carpenter, M. A., and A. Putnis (1985). "Cation Order and Disorder During Crystal Growth: Some Implications for Natural Mineral Assemblages." In A. B. Thompson and D. C. Rubie (eds.), *Advances in Physical Geochemistry*, Volume 4: *Metamorphic Reactions*. Springer–Verlag, New York.
[16] Carpenter, M. A., and D. C. Smith (1981). "Solid Solution and Cation Ordering Limits in High-Temperature Sodic Pyroxenes from the Nybö Eclogite Pod, Norway." *Mineralogical Magazine* 44, 37–44.
[17] Clark, J. R., D. E. Appleman, and J. J. Papike (1969). "Crystal-Chemical Characterization of Clinopyroxenes Based on Eight New Structure Refinements." *Mineralogical Society of American Special Paper* 2, 31–50.
[18] Clark, J. R., M. Ross, and D. E. Appleman (1971). "Crystal Chemistry of a Lunar Pigeonite." *American Mineralogist* 56, 888–908.

[19] Coe, R. S. (1970). "The Thermodynamic Effects of Shear Stress on the ortho-clino inversion in enstatite and other coherent phase Transitions Characterized by a Finite Simple Shear." *Contributions to Mineralogy and Petrology* 26, 247–64.

[20] Cohen, R. E., and C. W. Burnham (1985). "Energetics of Ordering in Aluminous Pyroxenes." *American Mineralogist* 70, 559–67.

[21] Cosca, M. A., and D. R. Peacor (1987). "Chemistry and Structure of Esseneite ($CaFe^{3+}AlSiO_6$), a New Pyroxene Produced by Pyrometamorphism." *American Mineralogist* 72, 148–56.

[22] Dal Negro, A., S. Carbonin, G. M. Molin, A. Cundari, and E. M. Piccirillo (1982). "Intracrystalline Cation Distribution in Natural Clinopyroxenes of Tholeiitic, Transitional and Alkaline Basaltic Rocks." In S. K. Saxena, (ed.), *Advances in Physical Geochemistry*, Volume 2. Springer–Verlag, New York.

[23] Dal Negro, A., S. Carbonin, G. Salviulo, E. M. Piccirillo, and A. Cundari (1986). "Crystal Chemistry and Site Configuration of the Clinopyroxene from Leucite-Bearing Rocks and Related Genetic Significance: The Sabatini Lavas, Roman Region, Italy." *Journal of Petrology* 26, 1027–40.

[24] Deer, W. A., R. A. Howie, and J. Zussman (1978). *Rock-Forming Minerals*, Volume 2A: *Single-Chain Silicates*. Wiley, New York.

[25] Domeneghetti, M. C., G. M. Molin, and V. Tazzoli (1985). "Crystal-Chemical Implications of the Mg^{2+}–Fe^{2+} Distribution in Orthopyroxenes." *American Mineralogist* 70, 987–95.

[26] Dowty, E., and D. H. Lindsley (1973). "Mössbauer Spectra of Synthetic Hedenbergite–Ferrosilite Pyroxenes. *American Mineralogist* 58, 850–68.

[27] Foit, F. F., Jr., R. L. Hooper, and P. E. Rosenberg (1987). "An Unusual Pyroxene, Melilite, and Iron Oxide Mineral Assemblage in a Coal-Fire Buchite from Buffalo, Wyoming." *American Mineralogist* 72, 137–47.

[28] Gordon, W. A., D. R. Peacor, P. E. Brown, E. J. Essene, and L. F. Allard (1981). "Exsolution Relationships in a Clinopyroxene of Average Composition $Ca_{0.43}Mn_{0.69}Mg_{0.82}Si_2O_6$: X-ray Diffraction and Analytical Electron Microscopy." *American Mineralogist* 66, 127–41.

[29] Grover, J. E. (1972). "The stability of Low-Clinoenstatite in the System $Mg_2Si_2O_6$–$CaMgSi_2O_6$." *EOS* 53, 539.

[30] Huebner, J. S. (1980). "Pyroxene Phase Equilibria at Low Pressure." In C. T. Prewitt (ed.), *Pyroxenes. Reviews in Mineralogy*, Volume 7. Mineralogical Society of America, Washington, D.C.

[31] Isaacs, A. M., and D. R. Peacor (1982). "Orthopyroxene Exsolution in Augite: A Two-Step, Diffusion–Transformation Process." *Science* 218, 152–53.

[32] Jaffe, H. W., P. Robinson, and R. J. Tracy (1978). "Orthoferrosilite and Other Iron-Rich Pyroxenes in Microperthite Gneiss of the Mount Marcy Area, Adirondack Mountains." *American Mineralogist* 63, 1116–36.

[33] Jantzen, C. M. (1984). "On Spinodal Decomposition in Fe-Free Pyroxenes." *American Mineralogist* 69, 277–82.

[34] Jenner, G. A., and D. H. Green (1983). "Equilibria in the Mg-Rich Part of the Pyroxene Quadrilateral." *Mineralogical Magazine* 47, 153–60.

[35] Kushiro, I., H. S. Yoder, Jr., and M. Nishikawa (1968). "Effect of Water on the Melting of Enstatite." *Geological Society of America Bulletin* 79, 1685–92.

[36] Lindsley, D. H. (1980). "Phase Equilibria of Pyroxenes at Pressures $>$ 1 Atmosphere." In C. T. Prewitt, (ed.), *Pyroxenes. Reviews in Mineralogy*, Volume 7. Mineralogical Society of America, Washington, D.C.

[37] Longhi, J., and A. E. Boudreau (1980). "The Orthoenstatite Liquidus Sield in the System Forsterite–Diopside–Silica at One Atmosphere." *American Mineralogist* 65, 563–73.

[38] McCallister, R. H., and G. L. Nord, Jr. (1981). "Subcalcic Diopsides from Kimberlites: Chemistry, Exsolution Microstructures, and Thermal History." *Contributions to Mineralogy and Petrology* 78, 118–25.

[39] McCormick, T. C. (1986). "Crystal-chemical Aspects of Nonstoichiometric Pyroxenes." *American Mineralogist* 71, 1434–40.

[40] McCormick, T. C., R. M. Hazen, and R. J. Angel (1989). "Compressibility of Omphacite to 60 kbar: Role of Vacancies." *American Mineralogist* 74, 1287–92.

[41] Morimoto, N., J. Fabries, A. K. Ferguson, I. V. Ginzburg, M. Ross, F. A. Seifert, J. Zussman, K. Aoki, and G. Gottardi (1988). "Nomenclature of Pyroxenes." *American Mineralogist* 73, 1123–33.

[42] Morimoto, N., and K. Koto (1969). "The Crystal Structure of Orthoenstatite." *Zeitschrift für Kristallographie* 129, 65–83.

[43] Morimoto, N., Y. Nakajima, Y. Syono, S. Akimoto, and Y. Matsui (1975). "Crystal Structures of Pyroxene-Type $ZnSiO_3$ and $ZnMgSi_2O_6$." *Acta Crystallographica* B31, 1041–49.

[44] Morimoto, N., and M. Tokonami (1969). "Domain Structure of Pigeonite and Clinoenstatite." *American Mineralogist* 54, 725–40.

[45] Murakami, T., Y. Takéuchi, and T. Yamanaka (1982). "The Transition of Orthoenstatite to Protoenstatite and the Structure at 1080°C." *Zeitschrift für Kristallographie* 160, 299–312.

[46] Murakami, T., Y. Takéuchi, and T. Yamanaka (1984). "X-ray Studies on Protoenstatite." *Zeitschrift für Kristallographie* 166, 263–75.

[47] Ohashi, Y. (1984). "Polysynthetically-Twinned Structures of Enstatite and Wollastonite." *Physics and Chemistry of Minerals* 10, 217–29.

[48] Ohashi, Y., and C. W. Burnham (1972). "Electrostatic and Repulsive Energies of the *M*1 and *M*2 Cation Sites in Pyroxenes." *Journal of Geophysical Research* 77, 5761–66.

[49] Ohashi, Y., C. W. Burnham, and L. W. Finger (1975). "The Effect of Ca-Fe Substitution on the Clinopyroxene Crystal Structure." *American Mineralogist* 60, 423–34.

[50] Ohashi, Y., and L. W. Finger (1974). "A Lunar Pigeonite: Crystal Structure of Primitive-Cell Domains." *Carnegie Institution of Washington Year Book* 73, 525–31.

[51] Ohashi, Y., and L. W. Finger (1976). "The Effect of Ca Substitution on the Structure of Clinoenstatite. *Carnegie Institution of Washington Year Book* 75, 743–46.

[52] Okamura, F. P., S. Ghose, and H. Ohashi (1974). "Structure and Crystal Chemistry of Calcium Tschermak's Pyroxene, $CaAlAlSiO_6$." *American Mineralogist* 59, 549–57.

[53] Ranson, W. A. (1986). "Complex Exsolution in Inverted Pigeonite: Exsolution Mechanisms and Temperatures of Crystallization and Exsolution." *American Mineralogist* 71, 1322–36.

[54] Robinson, P., H. W. Jaffe, M. Ross, and C. Klein, Jr. (1971). "Orientation of Exsolution Lamellae in Clinopyroxenes and Clinoamphiboles: Consideration of Optimal Phase Boundaries." *American Mineralogist* 56, 909–39.

[55] Rossi, G., D. C. Smith, L. Ungaretti, and M. C. Domeneghetti (1983). "Crystal-Chemistry and Cation Ordering in the System Diopside–Jadeite: A Detailed Study by Crystal Structure Refinement." *Contributions to Mineralogy and Petrology* 83, 247–58.

[56] Sadanaga, R., F. P. Okamura, and H. Takeda (1969). "X-ray Study of the Phase Transformations of Enstatite." *Mineralogical Journal* 6, 110–30.

[57] Sasaki, S. T. Matsumoto, and C. Sawada (1981). "The Influence of Multiple Diffraction on the Space Group Determination of Orthopyroxene, Spodumene, Low Omphacite and Pigeonite." *Physics and Chemistry of Minerals* 7, 260–67.

[58] Sasaki, S., C. T. Prewitt, and G. E. Harlow (1984). "Alternative Interpretation of Diffraction Patterns Attributed to Low ($P2_1ca$) Orthopyroxene." *American Mineralogist* 69, 1082–89.

[59] Smith, J. V. (1969). "Crystal Structure and Stability of the $MgSiO_3$ Polymorphs: Physical Properties and Phase Relations of Mg,Fe Pyroxenes." *Mineralogical Society of America Special Paper* 2, 3–29.

[60] Smyth, J. R. (1973). "An Orthopyroxene Structure up to 850°C." *American Mineralogist* 58, 636–48.

[61] Smyth, J. R. (1974a). Experimental Study on the Polymorphism of Enstatite." *American Mineralogist* 59, 345–52.

[62] Smyth, J. R. (1974b). "The High Temperature Crystal Chemistry of Clinohypersthene." *American Mineralogist* 59, 1069–82.

[63] Smyth, J. R., and C. W. Burnham (1972). "The Crystal Structures of High and Low Clinohypersthene." *Earth and Planetary Science Letters* 14, 183–89.

[64] Sueno, S., M. Cameron, and C. T. Prewitt (1976). "Orthoferrosilite: High Temperature Crystal Chemistry." *American Mineralogist* 61, 38–53.

[65] Sueno, S., and M. Kimata (1981). "On the Ortho–Clino Phase Transition of Rerrosilite ($FeSiO_3$) and Fe-Rich Pyroxenes." *Acta Crystallographica* A37, C-111.

[66] Sueno, S., M. Kimata, and C. T. Prewitt (1984). "The Crystal Structure of High Clinoferrosilite." *American Mineralogist* 69, 264–69.

[67] Sueno, S., M. Kimata, and C. T. Prewitt (1985). "Structural Aspects of Phase Transitions in the Fe–Mg–Ca Pyroxenes." *American Mineralogist* 70, 141–48.

[68] Thompson, J. B. (1970). "Geometrical Possibilities for Amphibole Structures: Model Biopyriboles." *American Mineralogist* 55, 292–93.

[69] Turnock, A. C., D. H. Lindsley, and J. E. Grover (1973). "Synthesis and Unit Cell Parameters of Ca–Mg–Fe Pyroxenes." *American Mineralogist* 58, 50–59.

[70] Yasuda, M., M. Kitamura, and N. Morimoto (1983). "Electron Microscopy of Clinoenstatite from a Boninite and a Chondrite." *Physics and Chemistry of Minerals* 9, 192–96.

5

Amphiboles and Nonclassical Biopyriboles

Lying between pyroxenes and micas, in a crystal structural sense, is the large and diverse assortment of minerals consisting of the amphiboles and the nonclassical biopyriboles. Like the pyroxenes, they are chain silicates, but unlike them, the chains are more than a single tetrahedron wide. On the other hand, these chains *are* of finite width, and so are narrower than the infinitely wide chains (i.e., sheets) of the micas. In our progression from more to less highly polymerized tetrahedra (that is, from framework silicates to orthosilicates), these minerals would logically have been discussed before pyroxenes; that, however, would have obscured their intermediate position between micas and pyroxenes, and that emphasis is worth the slight departure.

The amphiboles are double-chain silicates, whereas the nonclassical biopyriboles have greater chain widths. We shall begin with the amphiboles.

AMPHIBOLES

The most direct way of seeing how amphiboles are related to pyroxenes on the one hand and micas on the other is to look at the crystal structure. Hence we shall consider the general arrangement of atoms in amphiboles before discussing amphibole chemistry.

OVERVIEW OF THE CRYSTAL STRUCTURE

From the point of view of tetrahedral polymerization, the preeminent feature of the amphibole group is the double tetrahedral chain that extends along the **c** direction [Figure 5-1(a)]. Note that it appears to consist of two juxtaposed pyroxene chains, fully extended in the figure but somewhat rotated in real amphiboles. Hydroxyl groups (or sometimes F or Cl) are found at anion position O(3) in the plane of the apical oxygen atoms, just as in the micas. Counting the atoms in one repeat along the chain, and assuming only Si to fill the tetrahedra, one gets $Si_4O_{11}(OH)$ as the formula for the chain. In order to avoid fractions elsewhere in the formula, this part

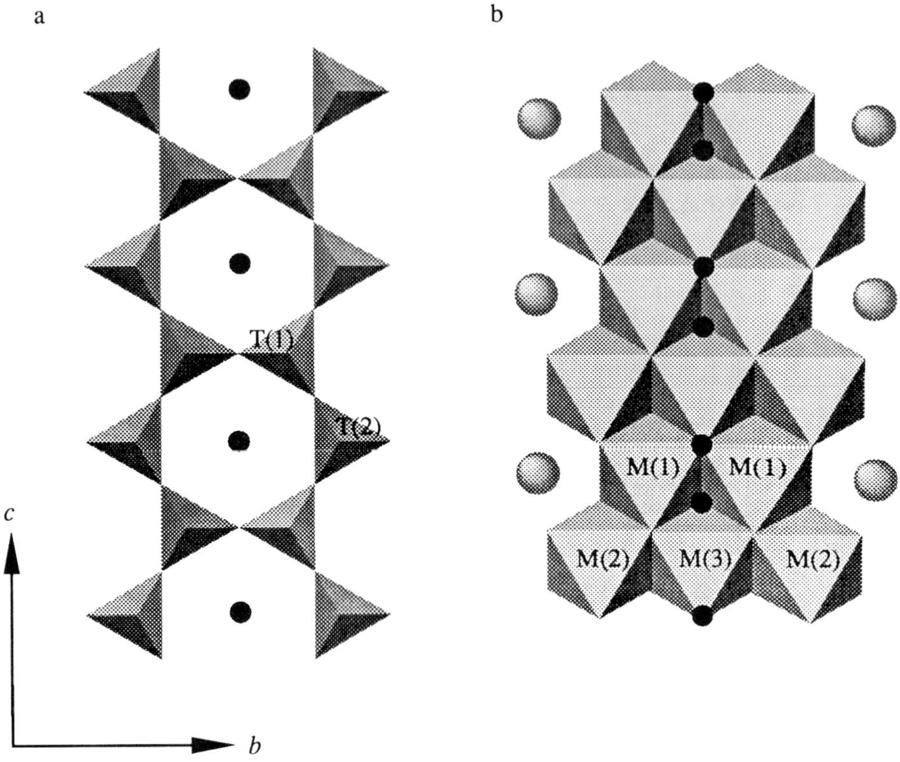

Figure 5-1. (a) The double tetrahedral chain that characterizes the amphiboles, idealized as unrotated regular tetrahedra. (b) The (idealized) octahedral strip, formed by stacking two double chains apex to apex with one offset in the **c** direction by $\sim c/3$. Shaded balls represent the $M(4)$ cations, and small filled circles are O(3) positions, where OH, F, and Cl are located.

is conventionally doubled, and we shall later see that amphibole formulas are characterized by $\cdots Si_8O_{22}(OH)_2$. Not unexpectedly, other atoms, principally Al, may substitute for some Si in the tetrahedral (or T) sites.

As in pyroxenes, the chains in amphiboles are stacked base to base and apex to apex along the **a*** direction, and alternately pointing along +**a*** and −**a*** in the **b** direction (see Figure 5-2, and compare it with Figure 4-4). Successive chains along **a*** are shifted by $\sim c/3$ (that is, in or out of the page in Figure 5-2), as are mica sheets and pyroxene chains. The apical oxygens, along with the hydroxyls, thus create five octahedral sites between the apex-to-apex chains [Figures 5-1(b) and 5-2]. The one on the (010) mirror plane is called $M(3)$, and it has pairs of $M(1)$, and $M(2)$ sites flanking it (although not both in the same plane parallel to the page). With respect to the hydroxyl locations, $M(3)$ is a *trans* site and $M(1)$ is a *cis* site (compare with Figure 3-3). $M(2)$ is not bonded to any hydroxyls, and thus this terminology does not apply to it. $M(4)$ in amphiboles, analogous to $M(2)$ in pyroxenes, is a six- to eight-coordinated site located at the edges of the double chains, and serves to cross-link adjacent chains in the **b** direction. An octahedral band or strip thus comprises one $M(3)$ site, and two each of the $M(1)$, $M(2)$, and $M(4)$ sites.

Some amphiboles contain an atom in a site that is analogous to the interlayer sites of micas. It is known as the A site, and is located between double chains that abut back to back. In the pyroxenes such a site is unnecessary, and in many amphiboles it is also unnecessary, but in other amphiboles it is required for the same reason as it is in the micas: to provide charge balance for the hydroxyls. Thus the atom that occupies the A site is monovalent. For amphiboles that have empty A sites, charge balance is accomplished through the chemistry of the M or T sites.

Four space groups in which amphiboles crystallize are shown in Figure 5-3,

Figure 5-2. The general features of the amphibole atomic arrangement are shown in this amphibole structure, projected down **c**, with **a** vertical and **b** horizontal (the dashed box outlines the clinoamphibole unit cell). Dark filled circles represent hydroxyl groups that occupy positions analogous to those occupied by OH in micas. The small spheres represent Y cations occupying, from left to right in each group, $M(2)$, $M(1)$, $M(3)$, $M(1)$, and $M(2)$. The larger ones represent the X cations, which occupy $M(4)$. The amphibole chains are seen end on. For orthoamphiboles, the repeat along **a** is twice that in monoclinic amphiboles, but the general features of the structure are the same. Compare this figure with Figure 4-4.

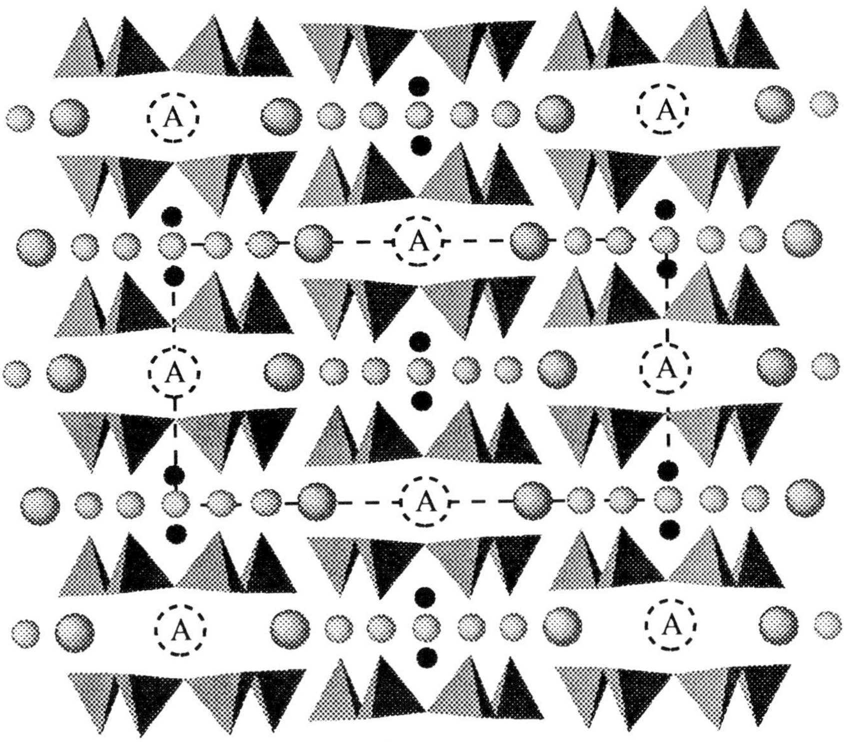

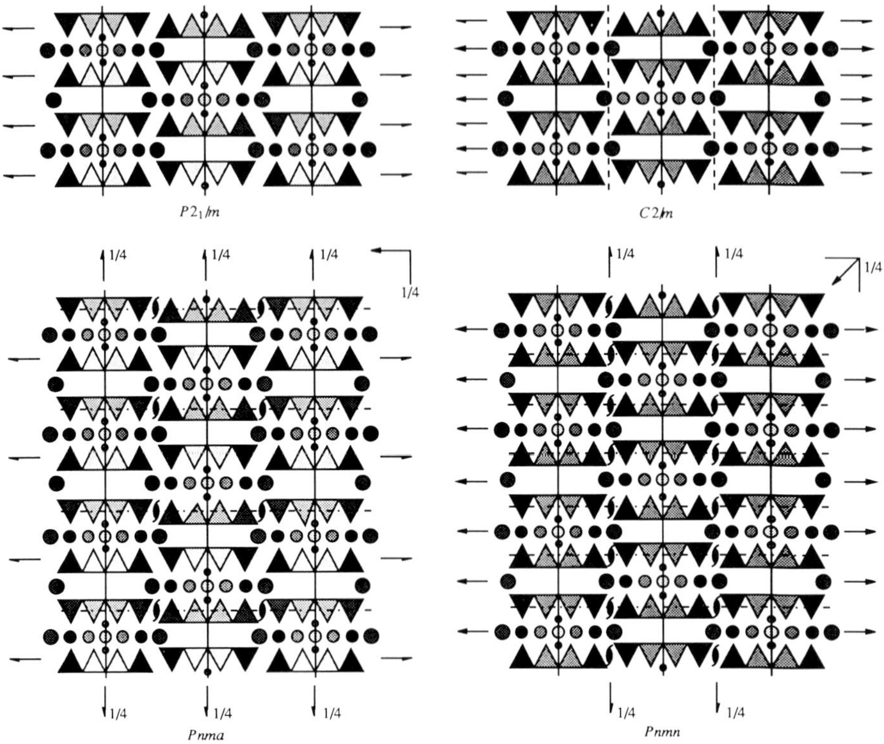

Figure 5-3. Four of the space groups in which amphiboles crystallize. Small shaded circles represent O(3). Standard symbols have been used for symmetry elements, but centers of symmetry have been omitted to avoid confusion. *M*(1), *M*(2), and *M*(3) are represented by shaded circles of intermediate size, and *M*(4) by large shaded circles. As in Figure 4-5, shading indicates equivalence or nonequivalence of cation sites.

which is analogous to Figure 4-5. In each case, *c* comes directly out of the page, and the view in each is directly down the tetrahedral chain. The *b* axis is horizontal in the plane of the page, and the vertical axis in that plane is $a \sin \beta$ (which is, of course, *a* for orthorhombic space groups). As in Figure 4-5, the shadings with which the tetrahedra and the *M*-sites are filled indicate crystallographic equivalence or nonequivalence among those sites in the various space groups. No known natural amphiboles crystallize in *Pnmn*, but the drawing is included because protoamphibole does, and that synthetic phase will come up again when we consider polymorphism in the amphiboles. There is one known amphibole space group not represented in Figure 5-3; *P*2/*a* is the space group of joesmithite, $(CaPb)Ca_2(Mg,Fe^{2+},Fe^{3+})_5Si_6Be_2O_{22}(OH)_2$, a rare mineral that will not be dwelt on here. Unlike all of the space groups in Figure 5-3, there is no mirror plane through *M*(3) in *P*2/*a*, and so there are two types of *M*(1) sites, two types of *M*(2) sites, and two types of *M*(4) sites. As in $P2_1/m$ and *Pnma*, there are two types of *T*(1) and two types of *T*(2) tetrahedra.

There is another way of visualizing how the amphibole structure is related to the structures of pyroxenes and micas [36]. Consider the rather idealized structure projections in Figure 5-4(a); those corresponding to amphibole and pyroxene are projections down **c**, while that for mica is along **a**. Note that the amphibole structure can be thought of as consisting of alternating (010) slabs of mica and pyroxene structure. (To extract the mica and pyroxene portions, the amphibole structure must be cut apart as indicated; the center portion is then a mica slab, and the two outer portions may then be reassembled properly to form a pyroxene slab.) Figure 5-4(b) shows how the arrangement of such slabs, with appropriate translations along amphibole **a**, yields the amphibole structure. This approach may seem a bit obscure, but it will facilitate the description of nonclassical biopyriboles later in this chapter.

Figure 5-4. (a) Ideal representations of mica slabs and pyroxene slabs. Open circles in the mica slab represent the interlayer cation site, and filled circles in the pyroxene slab represent the $M(2)$ site. In the mica structure, $M(1)$ is hidden behind the two $M(2)$ octahedra, and in the amphibole structure $M(3)$ is hidden behind the two $M(1)$ octahedra. (b) The structure of amphibole, as formed by a mica slab and a split pyroxene slab with half on each side of the mica slab. Here the open circles represent the A site and the filled circles represent the $M(4)$ site. The $M(3)$ octahedron is hidden behind the two $M(1)$ octahedra.

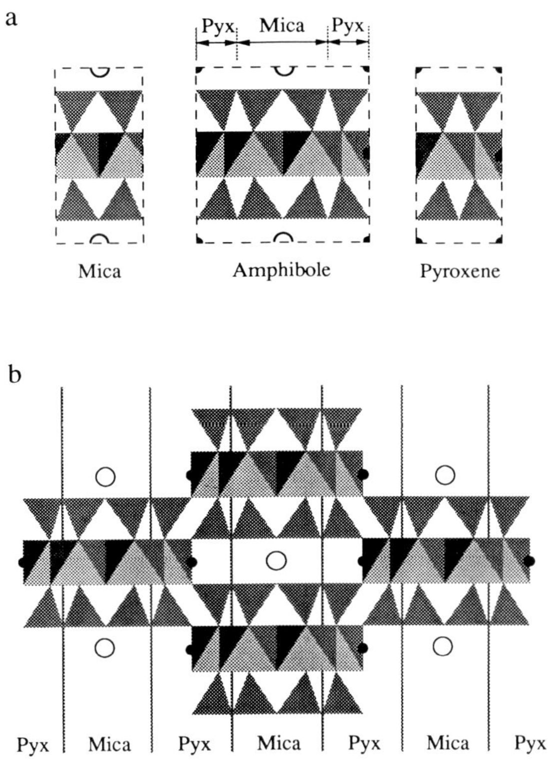

GENERAL CHEMISTRY

Compared with the generalized formulas of minerals we have considered thus far, that of the amphiboles looks complicated:

$$W_{0-1}X_2Y_5Z_8O_{22}(OH)_2.$$

Consistent with notation used in the previous two chapters, Z represents Si, with up to about 3.0 atoms of Al per formula unit (although commonly not this much); these are in tetrahedral coordination in the double chains (the T sites), and there is no compelling direct evidence that any other types of atoms populate these sites in common varieties [14]. The X and Y cations are located in octahedral sites, but not randomly. Y corresponds to Mg, Fe^{2+}, Fe^{3+}, Mn, Al, and Ti, and constitutes the elements found in the $M(1)$, $M(2)$, and $M(3)$ sites; we shall later see that there also are ordering preferences among these sites. The X cations are Na, Ca, Fe^{2+}, Mg, Mn, and Li, and these are found in the large $M(4)$ sites. When present, the W cations occupy the A sites between the backs of the double chains and usually consist of K and Na. Note how similar this distribution of elements is to that found in natural pyroxenes (p. 151). As one would expect, minor amounts of several other elements are occasionally found (for example, Zn, Ni, and perhaps B), unusual amphiboles may contain major amounts of unusual elements (such as joesmithite, in which one-fourth of the Si is replaced by Be), and amphiboles with compositions far from those found in nature have been synthesized.

There is one further site in amphiboles at which substitution commonly occurs. The O(3) site in amphiboles is usually predominantly occupied by OH, but anion substitution is possible. Natural specimens containing considerable F have been analyzed, and synthetic amphiboles containing only F in O(3) are known. No synthetic amphiboles with only Cl in O(3) have yet been made, but there is one published chemical analysis showing O(3) completely occupied by Cl [19], and many analyses show some Cl. In oxy-amphiboles OH at the O(3) site is replaced by O^{2-} (compare with discussion on oxybiotite, p. 106)

Amphibole Nomenclature

The large numbers of chemical substituents possible in amphiboles leads to a bewildering variety of end and intermediate members and chemically related families, accompanied by a rather rigid nomenclature [20;21] designed to minimize confusion about exactly what a given amphibole name implies. The international committee that undertook the task of standardizing amphibole nomenclature recommended the abandonment of over 200 amphibole names found in the literature, but still found that adequate chemical description required over 60 names and a long list of possible modifiers. This is not surprising, given the crystal structure of amphibole as described above. The $M(3)$ and $M(1)$ sites are similar to the octahedral sites of micas, while the $M(2)$ and $M(4)$ sites resemble the sites of the octahedral band in pyroxenes. Thus there are several positions available for metal atoms ranging in size from Al to Ca or Na. In addition, the *A* site resembles the interlayer site in micas, and can accommodate Na or K. It is little wonder, then, that there is a great variety of chemically distinct amphiboles.

Amphiboles are divided into four broad categories, depending on the chemistry of the *M*(4) site, as follows:

1. *Iron–magnesium–manganese amphiboles* contain less than 1.34 atoms of Ca+Na in *M*(4)—that is, in the *X* position in the formula.
2. *Calcic amphiboles* contain $(Ca+Na)_X \geq 1.34$, but $Na_X < 0.67$ atoms.
3. *Sodic-calcic amphiboles* are those with $(Ca+Na)_X \geq 1.34$ and $0.67 < Na_X < 1.34$ atoms.
4. *Alkali amphiboles* have $Na_X \geq 1.34$ atoms.

It should come as no great surprise by this point that, similar to pyroxenes, amphiboles that have relatively small metal atoms (Fe or Mg) in both *M*(1–3) and *M*(4) may occur in either orthorhombic or monoclinic forms. These are, of course, restricted to the iron–magnesium–manganese family. The orthorhombic amphiboles (orthoamphiboles) are anthophyllite, $Na_p(Mg,Mn,Fe^{2+})_{7-q}Al_q(Al_{p+q}Si_{8-p-q})O_{22}(OH,F,Cl)_2$ (where $p+q < 1.0$); gedrite, $Na_p(Mg,Mn,Fe^{2+})_{7-q}Al_q(Al_{p+q}Si_{8-p-q})O_{22}(OH,F,Cl)_2$ (where $p+q \geq 1.0$); and holmquistite, $Li_2(Mg,Fe^{2+})_3(Fe^{3+},Al)_2Si_8O_{22}(OH,F,Cl)_2$. The prefixes sodium-, ferro-, or magnesio- may be attached to the names to specify particular end members, as in sodium-anthophyllite, $Na(Mg,Fe^{2+})_7(AlSi_7)O_{22}(OH)_2$, and sodium-gedrite, $Na(Mg,Fe^{2+})_6Al(Al_2Si_6)O_{22}(OH)_2$. Compositional ranges for other members of the iron–magnesium–manganese family are shown in Figure 5-5, with enough information to construct pure end member formulas.

The monoclinic analog of the anthophyllite–gedrite series is the cummingtonite-grunerite series, $(Mg,Fe^{2+},Mn)_7Si_8O_{22}(OH,F,Cl)_2$, with magnesio-cummingtonite as the Mg end member and grunerite as the Fe end member. The Mn clinoamphiboles are tirodite and dannemorite (Figure 5-5). Not shown in Figure 5-5 is clinoholmquistite, $Li_2(Mg,Fe^{2+},Mn)_3(Fe^{3+},Al)_2Si_8O_{22}(OH,F,Cl)_2$, which has magnesio- and ferro- end members also.

Before considering the other amphibole families, it is worth noting that the guidelines of the Subcommittee on the Amphibole Group of the International Mineralogical Association Commission on New Minerals and Mineral Names [20], which have been strictly followed in the preceding discussion, can lead to confusion in discriminating between sodium-anthophyllite and sodium-gedrite. The formula for pure sodium-anthophyllite given here requires that $p=1$ and $q=0$ in the general formula; but this makes $p + q = 1.0$, and anthophyllites must have $p + q < 1.0$. The solution is that the mineral is an anthophyllite only if the tetrahedral aluminum content is no greater than 0.99 atoms per formula unit, and a sodium-anthophyllite only if $Na \geq 0.50$ atoms per formula unit. Thus the formula given for pure end-member sodium-anthophyllite is essentially, but not strictly, correct; the closest approach to that formula that is actually sodium-anthophyllite is $Na_{0.99}(Mg,Fe^{2+})_7(Al_{0.99}Si_{7.01})O_{22}(OH)_2$.

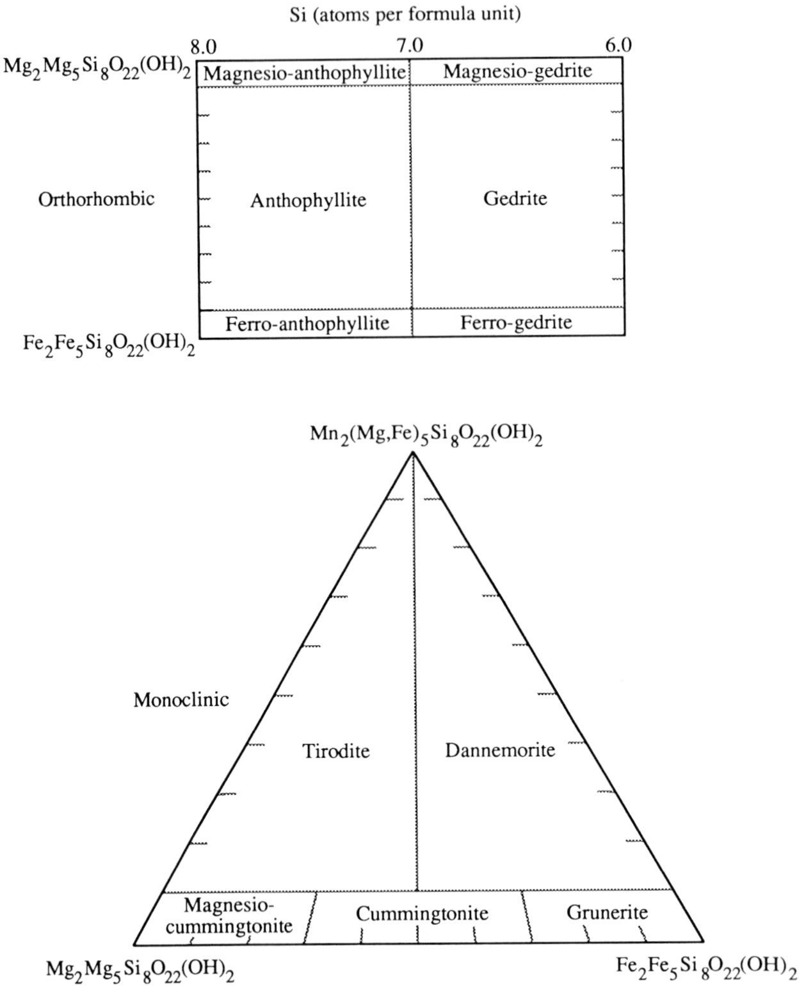

Figure 5-5. Compositional limits for orthorhombic and monoclinic amphiboles of the Fe–Mg–Mn family.

The calcic, sodic–calcic, and alkali amphiboles are all monoclinic and involve large numbers of end members, some of which have clumsy but descriptive prefixes attached to their names. End members of the calcic family are as follows:

	W	*X*	*Y*	*Z*	
Tremolite		Ca_2	Mg_5	Si_8	$O_{22}(OH)_2$
Ferro-actinolite		Ca_2	Fe_5^{2+}	Si_8	$O_{22}(OH)_2$
Edenite	Na	Ca_2	Mg_5	Si_7Al	$O_{22}(OH)_2$
Ferro-edenite	Na	Ca_2	Fe_2^{2+}	Si_7Al	$O_{22}(OH)_2$
Pargasite	Na	Ca_2	Mg_4Al	Si_6Al_2	$O_{22}(OH)_2$
Ferro-pargasite	Na	Ca_2	$Fe_4^{2+}Al$	Si_6Al_2	$O_{22}(OH)_2$
Hastingsite	Na	Ca_2	$Fe_4^{2+}Fe^{3+}$	Si_6Al_2	$O_{22}(OH)_2$
Magnesio-hastingsite	Na	Ca_2	Mg_4Fe^{3+}	Si_6Al_2	$O_{22}(OH)_2$
Alumino-tschermakite		Ca_2	Mg_3Al_2	Si_6Al_2	$O_{22}(OH)_2$
Ferro-alumino-tschermakite		Ca_2	$Fe_3^{2+}Al_2$	Si_6Al_2	$O_{22}(OH)_2$
Ferri-tschermakite		Ca_2	$Mg_3Fe_2^{3+}$	Si_6Al_2	$O_{22}(OH)_2$
Alumino-magnesio-hornblende		Ca_2	Mg_4Al	Si_7Al	$O_{22}(OH)_2$
Alumino-ferro-hornblende		Ca_2	$Fe_4^{2+}Al$	Si_7Al	$O_{22}(OH)_2$
Kaersutite	Na	Ca_2	Mg_4Ti	Si_6Al_2	$(O+OH)_{24}$
Ferro-kaersutite	Na	Ca_2	$Fe_4^{2+}Ti$	Si_6Al_2	$(O+OH)_{24}$

Figure 5-6 shows the compositional ranges over which these end member names and some intermediate calcic amphibole names apply.

The end members of the sodic-calcic group are:

	W	*X*	*Y*	*Z*	
Richterite	Na	CaNa	Mg_5	Si_8	$O_{22}(OH)_2$
Ferro-richterite	Na	CaNa	Fe_5^{2+}	Si_8	$O_{22}(OH)_2$
Ferri-winchite		CaNa	Mg_4Fe^{3+}	Si_8	$O_{22}(OH)_2$
Alumino-winchite		CaNa	Mg_4Al	Si_8	$O_{22}(OH)_2$
Ferro-alumino-winchite		CaNa	$Fe_4^{2+}Al$	Si_8	$O_{22}(OH)_2$
Ferro-ferri-winchite		CaNa	$Fe_4^{4+}Fe^{3+}$	Si_8	$O_{22}(OH)_2$
Alumino-barroisite		CaNa	Mg_3Al_2	Si_7Al	$O_{22}(OH)_2$
Ferro-alumino-barroisite		CaNa	$Fe_3^{2+}Al_2$	Si_7Al	$O_{22}(OH)_2$
Ferri-barroisite		CaNa	$Mg_3Fe_2^{3+}$	Si_7Al	$O_{22}(OH)_2$
Ferro-ferri-barroisite		CaNa	$Fe_3^{2+}Fe_2^{2+}$	Si_7Al	$O_{22}(OH)_2$
Magnesio-ferri-katophorite	Na	CaNa	Mg_4Fe^{3+}	Si_7Al	$O_{22}(OH)_2$
Magnesio-alumino-katophorite	Na	CaNa	Mg_4Al	Si_7Al	$O_{22}(OH)_2$
Ferri-katophorite	Na	CaNa	$Fe_4^{2+}Fe^{3+}$	Si_7Al	$O_{22}(OH)_2$
Alumino-katophorite	Na	CaNa	$Fe_4^{2+}Al$	Si_7Al	$O_{22}(OH)_2$
Ferri-taramite	Na	CaNa	$Fe_3^{2+}Fe_2^{3+}$	Si_6Al_2	$O_{22}(OH)_2$
Magnesio-ferri-taramite	Na	CaNa	$Mg_3Fe_2^{3+}$	Si_6Al_2	$O_{22}(OH)_2$
Alumino-taramite	Na	CaNa	$Fe_3^{2+}Al_2$	Si_6Al_2	$O_{22}(OH)_2$
Magnesio-alumino-taramite	Na	CaNa	Mg_3Al_2	Si_6Al_2	$O_{22}(OH)_2$

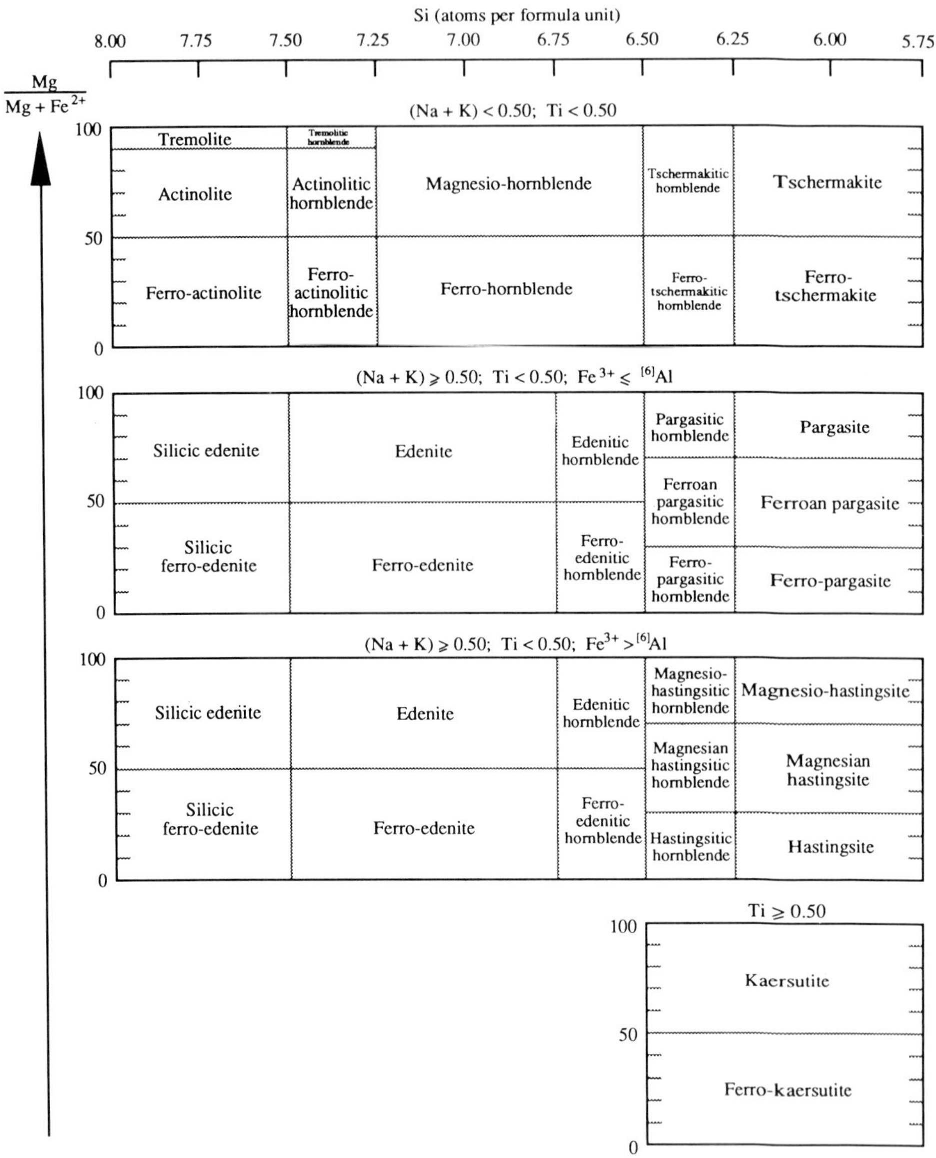

Figure 5-6. Compositional limits for calcic amphiboles. Compositional ratios expressed on the ordinates are ratios of atoms in the standard chemical formulas. The bracketed 6 adjacent to Al indicates coordination number.

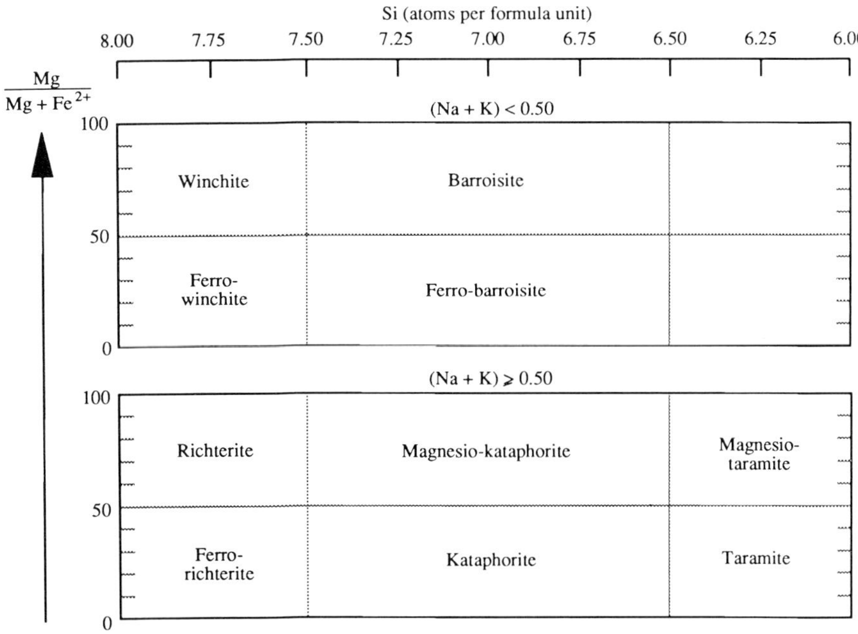

Figure 5-7. Compositional limits for sodic–calcic amphiboles. Compositional ratios expressed on the ordinates are ratios of atoms in the standard chemical formulas.

Compositional ranges over which most of these end member names are valid are shown in Figure 5-7.

Finally, the end members of the alkali amphibole family are as follows:

	W	*X*	*Y*	*Z*	
Glaucophane		Na_2	Mg_3Al_2	Si_8	$O_{22}(OH)_2$
Ferro-glaucophane		Na_2	$Fe_3^{2+}Al_2$	Si_8	$O_{22}(OH)_2$
Magnesio-riebeckite		Na_2	$Mg_3Fe_2^{3+}$	Si_8	$O_{22}(OH)_2$
Riebeckite		Na_2	$Fe_3^{2+}Fe_2^{3+}$	Si_8	$O_{22}(OH)_2$
Eckermannite	Na	Na_2	Mg_4Al	Si_8	$O_{22}(OH)_2$
Ferro-eckermannite	Na	Na_2	$Fe_4^{2+}Al$	Si_8	$O_{22}(OH)_2$
Magnesio-arfvedsonite	Na	Na_2	Mg_4Fe^{3+}	Si_8	$O_{22}(OH)_2$
Arfvedsonite	Na	Na_2	$Fe_4^{2+}Fe^{3+}$	Si_8	$O_{22}(OH)_2$
Kozulite	Na	Na_2	$Mn_4(Fe^{3+},Al)$	Si_8	$O_{22}(OH)_2$

Figure 5-8 clarifies the compositional limits of these names as well as of an intermediate alkali amphibole, crossite. We shall discuss solid solutions later, but at this point it needs to be recognized that the presence of two amphibole names in adjacent spaces on the same figure (5-6, 5-7, or 5-8) does not imply complete solid solu-

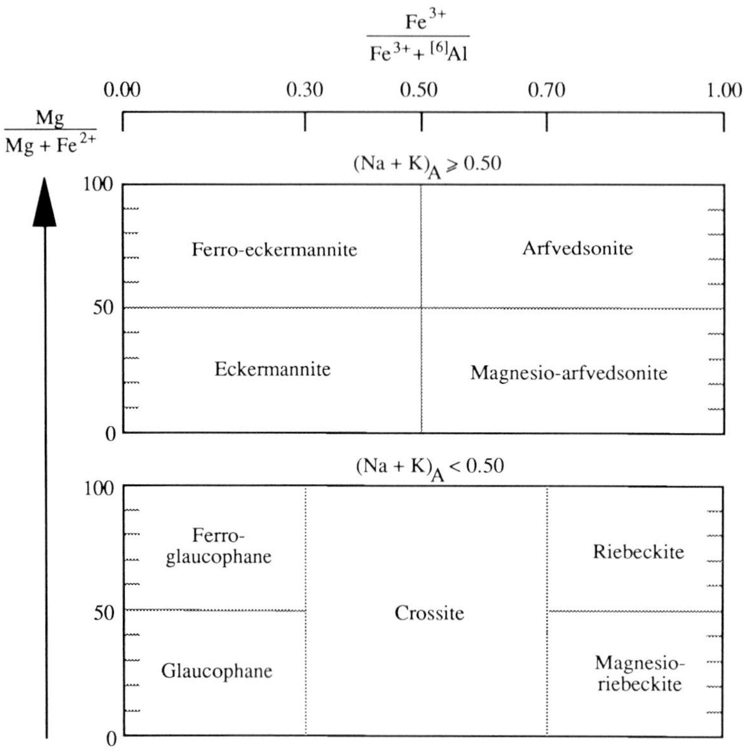

Figure 5-8. Compositional limits for alkali amphiboles. Compositional ratios expressed on the ordinates are ratios of atoms in the standard chemical formulas.

tion between the minerals, or that arbitrary compositions on those figures are necessarily represented by stable minerals.

SOME STEREOCHEMICAL DETAILS

Recall that in micas and pyroxenes the sizes of *M* cations and the geometries of the tetrahedral arrangements are intimately interrelated. Given the similarities of the amphiboles to these groups, it comes as no surprise that analogous relationships are found here.

The Double Tetrahedral Chain

The tetrahedra of amphibole chains are rotated (Figure 5-9), but within narrower limits than in the pyroxenes and micas; there are amphiboles with nearly straight chains, but none with tetrahedra rotated over about 10°. As in pyroxenes, S rotations in amphiboles would result in the sharing of tetrahedral edges with the *M*(4) polyhedron and should be less favorable than O rotations; in fact, only O rotations have been observed to occur. Also as with pyroxenes, the symbols P and N can be used to describe rotations of tetrahedra with respect to the triangular faces of octahedra attached to those tetrahedra through basal oxygen atoms, whereas O and S describe tetrahedral rotations with respect to triangular faces of octahedra linked to the tetrahedra through apical oxygen atoms. Monoclinic amphiboles contain only P rotations, while *Pnmn* amphiboles contain only N rotations, and *Pnma* amphiboles contain both.

The amount of tetrahedral rotation is a factor of the misfit between the *c* repeat of an extended double chain and the *c* repeat of the octahedral strip formed by the apical oxygens of the tetrahedra and the O(3) atoms. This concept is familiar from our discussion of pyroxenes, and we shall not belabor it. Inasmuch as even amphiboles with the smallest tetrahedra—those containing only Si—are observed to have slight tetrahedral rotations, it follows that placing larger atoms into the tetrahedral chain will cause greater rotation, given similar dimensions for the octahedral strip. Figure 5-10 shows two distinct populations of rotation angles, the population with smaller angles corresponding to little or no tetrahedral Al, and the one with larger angles corresponding to significant tetrahedral Al. In addition, the scatter in rotation angles for tetrahedra of nearly the same size reveals that other factors besides the substitution of Al for Si affect tetrahedral rotation. Among these may be sub-

Figure 5-9. Tetrahedral rotation in the double chain of hornblende [24]. The rotation angle may be calculated as one-half the supplement of the O(5)–O(6)–O(5) angle. While this assumes that the tetrahedral bases are parallel to (100), the assumption introduces only a trivial error.

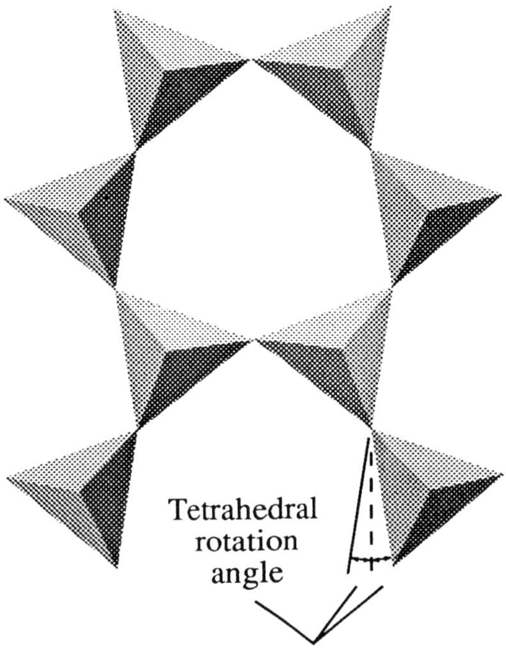

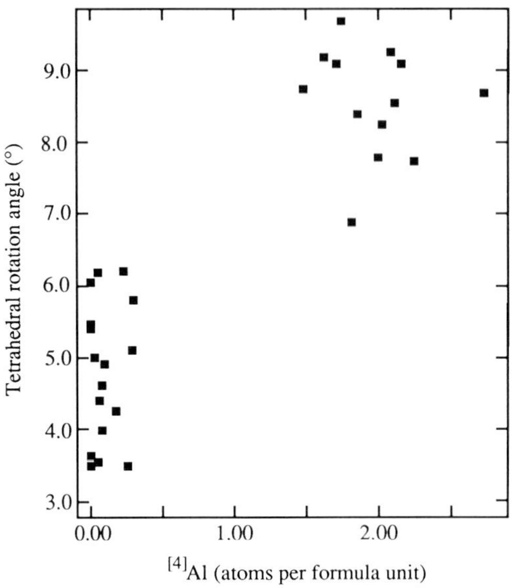

Figure 5-10. Tetrahedral rotation angle for 31 well-refined amphibole structures [15] as a function of tetrahedral Al.

stitutions that affect the dimensions of the octahedral band, the occupancy of the A site, and bonding effects involving either $M(4)$ or A and O(5) or O(6) (see Figures 5-1 and 5-11 for site nomenclature).

Tetrahedral rotation is not the sole mechanism by which the double chain articulates to the octahedral band. As seen in Figure 5-2 the double chain is warped, with its concave side away from the octahedral strip. This is analogous to the tetrahedral tilt perviously discussed with regard to micas (p. 114), and it allows the apical oxygen atoms to span a set of sites that is inherently wider than the double chain. Figures 5-11 and 5-12 show in detail how tetrahedral rotation and warping, respectively, facilitate tetrahedral–octahedral fit in a $C2/m$ amphibole.

Each amphibole double chain contains two symmetrically unique tetrahedra: In $C2/m$ and $Pnma$ they are $T(1)$ and $T(2)$; in $P2_1/m$ and $Pnmn$ there are two distinct chains, and the tetrahedra are $T(1A)$ and $T(2A)$ in the A chain, and $T(1B)$ and $T(2B)$ in the B chain. In nearly all cases wherein the tetrahedral sites contain only Si, the $T(2)$ tetrahedra are larger than the $T(1)$ tetrahedra. For example, in tremolite [24] $\langle T(1)\text{–O}\rangle = 1.620$Å and $\langle T(2)\text{–O}\rangle = 1.632$ Å, and in synthetic fluor-richterite [5] $\langle T(1)\text{–O}\rangle = 1.621$Å and $\langle T(2)\text{–O}\rangle = 1.636$ Å. Clearly ionic size is not the controlling factor here, because both tetrahedra contain Si, but there are at least two other factors involved. First, the mean coordination number of oxygen atoms bonded to $T(1)$ is 2.75, whereas that of oxygen atoms bonded to $T(2)$ is 3.00. It makes qualitative sense that cation-anion bond distances will be larger for anions that must bond to larger numbers of cations, and this has been shown to be quantitatively true also [2]. In addition, the nonbridging T–O bonds in amphiboles become longer with the mean electronegativity of the nontetrahedral cations coor-

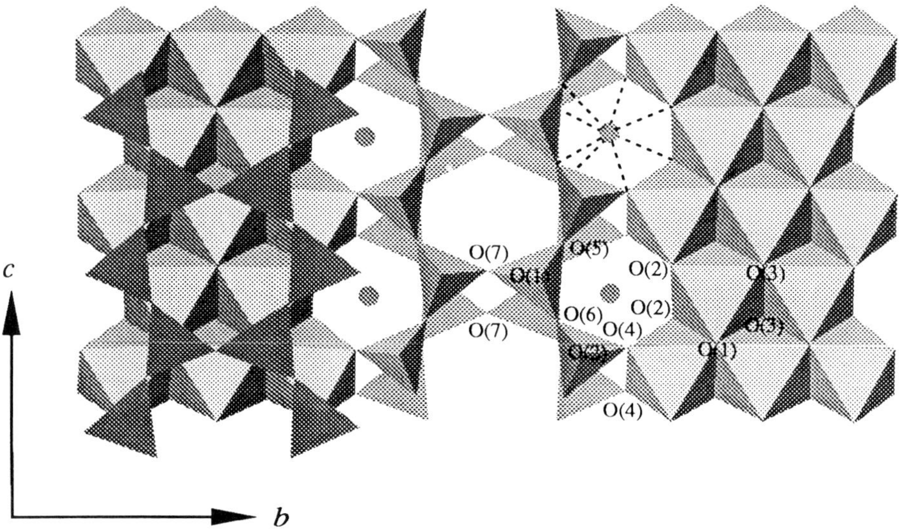

Figure 5-11. (100) projection of a $C2/m$ amphibole structure, showing rotation of the tetrahedra in the double chain. After [15].

dinating the nonbridging oxygens [2]; $T(2)$ has two nonbridging T–O bonds, whereas $T(1)$ has only one.

[The point made here—that cation-anion bond distances will be larger for anions that must bond to larger numbers of cations—may not be self-evident, and a short diversion is appropriate. Another way of stating that is that an increase in coordination number results in longer bond lengths. It seems apparent that an increase in the coordination number of a cation increases the size of the surrounding anion polyhedron, thereby lengthening the vertex-to-center distance, but Pauling's sec-

Figure 5-12. A portion of the unit cell of tremolite viewed down **c**. Note the curvature of the double chains, concave sides away from the octahedral band. (Atomic coordinates from [24].)

ond rule and a conceptual extension of it provide a basis for a more quantitative approach. As given by Pauling [25] the second rule takes no account of variations in bond lengths; that is, in the calculation of Pauling bond strengths it tacitly assumes that all distances between a given type of cation and a given type of anion are equal. Clearly this is not the case, and various methods have been devised to account for the fact that shorter bonds must be inherently stronger than longer ones, all other factors being equal. These methods are empirical, and involve the determination of equations or curves called bond-length–bond-strength curves. (See, for example, [3], [4], and [7].) The sums of bond strengths obtained from such curves generally approach formal ionic charges much more closely than do sums of Pauling bond strengths. (Some prefer to call the numbers obtained *bond valences*, reserving the term *bond strength* for numbers obtained by Pauling's second rule.) With regard to the present issue, bonding of larger numbers of cations to a given anion requires that the bond lengths increase in order to reduce individual bond strengths, leaving their sum near the formal anionic charge.]

Returning now to the tetrahedra of the double chain, we consider the ordering of Al. It is tempting to assign Al to $T(2)$, inasmuch as it is inherently larger than $T(1)$, but crystal structure refinements have shown consistently that Al strongly prefers the $T(1)$ site. The resolution of this apparent dilemma is found by examining the sums of bond strengths reaching the oxygens bonded to tetrahedrally coordinated atoms. The bond-strength sums reaching the oxygen atoms are shown in Table 5-1 (where barred numerals indicate infinitely repeating decimals). Tremolite has been used as an example because of its chemical simplicity. Note that all of the oxygen atoms [and the hydroxyl, O(3)] are reasonably charge-balanced with the exception of O(4), which is seriously underbonded. The use of bond-length–bond-strength curves brings the sum for O(4) somewhat closer to 2.00 (as well as reducing those for O(5) and O(6) somewhat), but it is still the least well balanced anion in the structure. Placing a trivalent atom like Al in $T(2)$ would only make matters worse, reducing the sum of Pauling bond strengths for O(4) by 0.25. Thus in order to maintain a more favorable local balance of charges, Al orders preferentially into $T(1)$, although in amphiboles with less than 6.00 Si per formula unit some Al enters $T(2)$. This is rather like the ordering of Al into T_1O in alkali feldspars (p. 45), which also occurs because of charge-balance considerations rather than size. In orthorhombic

Table 5-1. Pauling Bond Strengths for Tremolite ($C2/m$)

	$T(1)$	$T(2)$	$M(1)$	$M(2)$	$M(3)$	$M(4)$	Σ
O(1)	1.00		$0.3\overline{3}$	$0.3\overline{3}$	$0.3\overline{3}$		2.00
O(2)	1.00		$0.3\overline{3}$	$0.3\overline{3}$		0.25	1.92
O(3)			$0.3\overline{3} \times 2$		$0.3\overline{3}$		1.00
O(4)	1.00		$0.3\overline{3}$		0.25		1.58
O(5)	1.00	1.00				0.25	2.25
O(6)	1.00	1.00				0.25	2.25
O(7)	1.00×2						2.00

amphiboles, where there are four independent tetrahedral sites, Al orders as $T(1B) \geq T(1A) \geq T(2B) > T(2A)$.

Ordering in the Octahedral Band

The octahedral band, shown in Figure 5-1, consists of $M(1)$, $M(2)$, and $M(3)$ octahedra with laterally attached $M(4)$ sites that are eight-coordinated positions in alkali, sodic–calcic, and calcic amphiboles, but that have ambiguous coordination in Fe–Mg–Mn amphiboles. It has been suggested [7] that $M(4)$ is bonded to only four oxygen atoms, all of them on the side of the site away from the double tetrahedral chains, and bond-valence calculations lend some tenuous support to this interpretation. On the other hand, values of the isomer shift from Mössbauer spectra of monoclinic ferromagnesian amphiboles are not too different from isomer shifts for Fe^{2+} in sites with known 6- or 8-coordination. Perhaps the strongest argument for 6- or 8-coordination of the $M(4)$ cation [15] is the observation that Fe–Mg–Mn amphiboles crystallize in the space groups *Pnma*, $P2_1/m$, and $C2/m$, and that the major structural differences among these are found in O(5) and O(6), which are the bridging oxygens of the double tetrahedral chains, in the vicinity of $M(4)$. This suggests that O(5) atoms or O(6) atoms, or both, interact with (that is, are bonded to) $M(4)$, and while it does not discriminate between 6- and 8-coordination, it does mitigate against 4-coordination. In cummingtonite ($C2/m$) [10;11] the two O(5) atoms are farther from $M(4)$ than are $T(2)$, $M(1)$, and $M(2)$, suggesting that 6-coordination is the best model [15]. [Recall that a similar argument was used in connection with the coordination number of $M(2)$ in pigeonite.] In anthophyllite (*Pnma*) the distances from $M(4)$ to O($2A$,$2B$), O($4A$,$4B$), O($5A$,$5B$), and O($6B$) are all less than 2.9 Å, but the distance to O(6A) is 3.5 Å, suggesting that seven might be an appropriate coordination number in this case (compare pigeonite, p. 167).

Certain species of metal atoms exhibit distinct preferences for certain sites of the octahedral band. Na and Ca enter $M(4)$ exclusively (except for the possibility of Na in the A site, which we shall deal with later). Li in holmquistite and clinoholmquistite enters $M(4)$, and where it occurs as a minor constituent of other amphiboles most has been assigned to $M(4)$. Al orders strongly into $M(2)$, as does Fe^{3+}, although the ordering of Fe^{3+} may be less complete than that of Al. In Fe–Mg–Mn amphiboles Fe^{2+} and Mn show a strong preference for ordering into $M(4)$, although the ordering is not generally complete [that is, even when there are more than two Fe^{2+} + Mn atoms per formula unit, they do not exclude all Mg from the M(4) site]. In other amphiboles, Fe^{2+} and Mn preferentially occupy $M(4)$ if Ca, Na, and Li have not completely filled it. With regard to the other three sites, Fe^{2+} seems to prefer $M(1)$ and $M(3)$ about equally, but distinctly more than $M(2)$.

The equilibrium distribution of atoms in a crystal is, of course, the one that results in the lowest energy. Drawing on discussions of cation ordering in previous chapters, we recognize among the factors that may influence site preferences ionic size, local charge balance, crystal-field effects, and electrostatic potential energy. Using any one of these alone implies the assumption that the others are unimportant or essentially constant for alternative sites, but the effects of variations may simply cancel one another. Site preferences may often (and generally probably do) involve the complex interaction of several factors, and this complexity is nowhere better demonstrated than in the amphiboles.

Consider *M*(4), for example. It is the largest of the *M* sites, and in alkali, sodic–calcic, and calcic amphiboles it is occupied by Na and Ca. Certainly these are large atoms, and when we recall that Fe^{2+} and Mn prefer *M*(4) in Fe–Mg–Mn varieties, it is tempting to assume that ionic size is the governing factor. In holmquistite, however, Li orders completely into *M*(4), leaving Fe^{2+} and Mg in *M*(1) and *M*(3), and Fe^{3+} and Al in *M*(2) [17;44]. Reference to any of the tables of ionic radii currently in use among mineralogists (e.g., [31]) reveals that Li is smaller than Fe^{2+} and only slightly larger than Mg, however, so ionic size cannot be the sole factor involved in its preference for *M*(4). On the other hand, there is no question that Na and Ca fit best into *M*(4); so size seems to play a significant role.

In amphiboles containing $^{[6]}Al$ or Fe^{3+}, *M*(2) is distinctly smaller than either *M*(1) or *M*(3). Is it smaller only because it contains these relatively small cations, or do Al and Fe^{3+} prefer *M*(2) because that site is inherently smaller than *M*(1) and *M*(3) to begin with? If we consider amphiboles that contain only Mg in octahedral coordination (tremolite [24], protoamphibole [13], fluor-tremolite [6], and fluor-richterite [5]), we find that *M*(2) is the largest of the three sites in all four. Evidently *M*(2) is not constrained by any steric requirements to be smaller than *M*(1) or *M*(3), and rationalization of the ordering of $^{[6]}Al$ or Fe^{3+} based on ionic size alone is not adequate.

When a monovalent cation occupies *M*(4) charge balance may be maintained by placement of a trivalent cation instead of a divalent cation in one of the other M sites, and of course this is what happens in *M*(2). Just stating this does not make it clear why local charge balance could not be maintained as well by placing the trivalent atom in *M*(1) or *M*(3), but the calculation of electrostatic potential (Madelung) energies sheds some light on the question [45]. (See Chapter 11 for a brief introduction to electrostatic potential energies.) The calculations referred to here were done before very many amphibole crystal structures had been refined precisely, and they made it possible to predict ordering preferences that had not been observed to that point. For example, it was predicted that in richterite Na should occupy the *A* site and CaNa the *M*(4) site [rather than Ca at *A* and Na_2 at *M*(4)]; this was later confirmed. Clinoholmquistite was predicted to have Li at *M*(4), and this was also confirmed. Likewise it was predicted that $^{[6]}Al$ and Fe^{3+} should order into *M*(2) in all amphiboles except oxy-amphiboles [those with O at O(3) instead of OH, F, or Cl], and that edenite, katophorite, and taramite should exhibit preferential ordering of Al into *T*(1), and these predictions have been vindicated. Lest this technique be assumed infallible, we point out that it also predicted that Al would be ordered into *T*(2) in tschermakite and magnesio-hornblende, and this has turned out not to be the case.

Crystal field stabilization energy (CFSE) has been useful in rationalizing the preference of transition-metal ions for certain sites (e.g., p. 174), and we have seen that larger values of CFSE are often associated with preferred occupancy by Fe^{2+}. We have also seen that distortion of sites from ideal geometry leads to greater CFSE. In considering the *M* sites of amphiboles, it is clear that *M*(4) is the most distorted when considered as an octahedron; this is usually followed by $M(1) \approx M(3) > M(2)$ when an angular measure of distortion is used. When a measure of distortion that emphasizes differences in bond lengths is used, then $M(2)$ generally turns out to be more distorted than $M(1)$ and $M(3)$. We have found for other minerals that using a measure of angular distortion often yields considerable success in predicting transition-metal site preferences, and it appears to here, also; Fe^{2+} generally prefers $M(4)$, $M(1)$, and $M(3)$ over $M(2)$. It is difficult to say how important crystal-field effects are compared to size, local charge balance, and so forth. Mn^{2+} also prefers $M(4)$, but its electron configuration provides no crystal field stabilization energy regardless of the distortion of the site, so here size evidently is more important than CFSE, and Fe^{2+} yields to Mn^{2+} in $M(4)$ in tirodite [24]. Li competes successfully for $M(4)$ in holmquistite [17], and Fe^{2+} is relegated to other M sites; evidently local charge balance is more important than crystal field effects here.

The *A* Site

The large site between the backs of the amphibole double chains is the *A* site, and it bears some resemblance to the interlayer site in micas. Like that one, it is surrounded by 12 oxygen atoms, all of which are bridging oxygens, and is essentially on the opposite side of each double chain from a hydroxyl site. Unlike in micas, however, where the back-to-back sheets separated by interlayer cations are stacked so that one eclipses the other when viewed perpendicular to the sheets, the amphibole double chains are offset along *c* (see Figure 5-13). Thus even if there were no tetrahedral rotation, the *A* site would not be equidistant from all 12 oxygens surrounding it. In $C2/m$ amphiboles it would instead have two oxygens relatively close (~2.4 Å), two pairs at intermediate distances (~2.6–3.2 Å), and two so far away (~3.7 Å) that they should probably not be considered as bonded atoms. In *Pnma* amphiboles (specifically, gedrites), two oxygens are ~2.4 Å away from the *A* site, four are at ~2.65 Å, and six are considerably more distant, so there is a clear choice for six-coordination.

The electron density about the *A* site, as determined by x-ray-diffraction data, is often quite anisotropic, leading to the suggestion that there is positional disorder at the site. The situation is analogous to the split Na sites in some feldspars (p. 55). Figure 5-13 reveals that the *A* site can be divided into four subsites: $A(2/m)$ is at the formal A position—the intersection of the diad and the mirror in $C2/m$; $A(2)$ is a set of two sites displaced along the diad in opposite directions on either side of the mirror; $A(m)$ is another set of two that lie on the mirror but displaced from the diad; $A(1)$ is a general site that is neither on the mirror nor on the diad, and thus consists of a set of four. (Because of the *C*-centering symmetry operation, the numbers of subsites contained in the unit cell are twice the numbers given in the previous sentence for a single *A* site.) In some amphiboles that have occupied *A* sites, more than one of the sets of subsites contain some concentration of electron density.

It is very difficult to say with certainty what type of atom or atoms occupy a given *A* subsite, because K, Na, and vacancies may all contribute to the total electron density, and often multiple subsites are involved. Bond-strength–bond-length calculations suggest that Na should prefer $A(2)$ and K should prefer $A(m)$ [16]. These

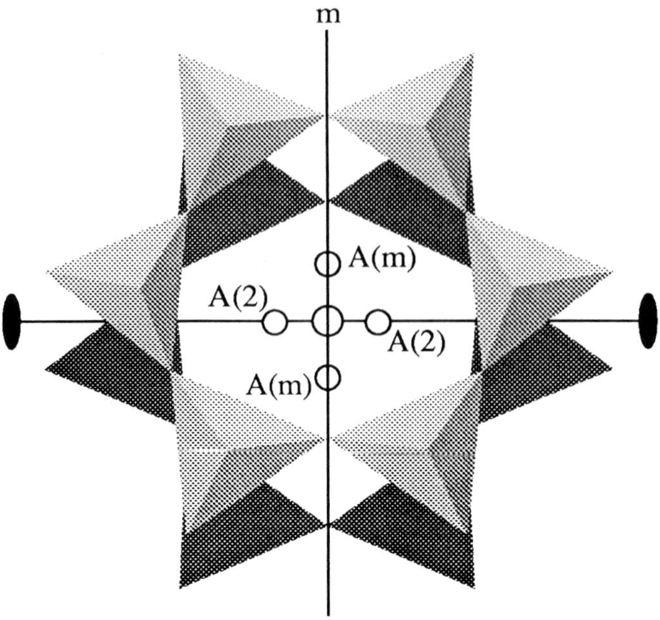

Figure 5-13. The A site in a $C2/m$ amphibole consists of several subsites that may or may not be occupied. The unmarked circle in the center is the $A(2/m)$ site, the notation indicating that it lies at the intersection of the diad and the mirror. The site may be split into two subsites located on the mirror and called $A(m)$ sites, or into two located on the diad and called $A(2)$ sites. The cation may also occupy a general position not on any of the symmetry elements. In this case the site is called $A(1)$ and the symmetry requires that there be four equivalent $A(1)$ sites. [The $A(1)$ sites have been omitted in the figure to avoid excessive clutter.] The separations between $A(2)$ subsites or $A(m)$ subsites are highly exaggerated for clarity in the figure; they are actually only a few tenths of an Ångstrom apart.

predictions are compatible with the observed electron densities at the A subsites of some amphiboles, but there are still too many unknowns and too few equations to derive a unique solution to site occupancy. Besides the identity of the cation(s) occupying the A subsites, the stereochemistry of the tetrahedral double chains, the nature of the anion at O(3), and the details of the $M(4)$ site probably play some role in determining the distribution of cations and vacancies around the A site.

POLYMORPHISM IN THE AMPHIBOLES

Reference to the section above on amphibole nomenclature for the Fe–Mg–Mn family shows that the ideal end-member chemical formulas for anthophyllite–gedrite (*Pnma*) and cummingtonite–grunerite ($C2/m$) are identical, and this suggests polymorphism. It appears that in nature the compositional fields of these two series and manganoan cummingtonite ($P2_1/m$) do overlap [23]. In addition, primitive manganoan cummingtonite ($P2_1/m$) inverts reversibly to $C2/m$ manganoan cummingtonite on heating [26]. It appears that cummingtonite possesses the symmetry of space group $C2/m$ at high temperatures, and attempts to attain *Pnma* symmetry on cooling. This, however, involves a change in stacking sequence with respect to octahedral skew (as defined for pyroxenes). $C2/m$ amphiboles have the sequence $++++$, whereas the sequence $++--$ occurs in *Pnma* amphiboles. The transformation from one to the other is a reconstructive transformation that must have a high activation energy. The transformation from $C2/m$ to $P2_1/m$ is displacive, however, involving decreased rotation of the tetrahedra of the *A* double chain and increased rotation of the tetrahedra of the *B* double chain [34]; moreover, the $M(4)$ cation moves from a position similar to that found in calcic amphiboles to one similar to that in anthophyllite. Thus it may be that the transition from centered monoclinic to primitive monoclinic is alternative metastable behavior.

Comparison of the lattice parameters of glaucophanes synthesized at low pressures and high temperatures with those of glaucophanes synthesized at high pressures and low temperatures has suggested that there may be two forms, glaucophane I and glaucophane II, respectively [9]. Natural glaucophanes are evidently glaucophane II, and they have ordered cation distributions in the *M* sites [22]. It is possible that the difference in the two polymorphs is that glaucophane I has disordered *M* cations, in which case the transformation is one of order-disorder. Verification of this hypothesis must await further work on the synthetic glaucophane I.

SOLID SOLUTION AND EXSOLUTION

Owing to the chemical complexity of amphiboles, we might expect that exsolution from a high-temperature solid solution to chemically simpler low-temperature phases would be common, and indeed it is; and despite the chemical complexity, we find that the same principles that have allowed us to predict exsolution in other minerals also apply here. Essentially, we expect solid solution if the requirements of both charge balance and ionic size are met; if either fails to be satisfied as temperature decreases, we expect exsolution. This means that exsolution is expected between amphiboles that have metal atoms of different sizes in the $M(4)$ site (Ca and Fe–Mg, for example), or that have different charge requirements in the $M(4)$ site (Ca in one, Na in the other), or that have very different occupancies of the *A* site (for instance, filled in one and empty in the other). Such amphibole pairs for which exsolution has been demonstrated include anthophyllite–gedrite, cummingtonite–actinolite, cummingtonite–hornblende, actinolite–hornblende, tremolite–riebeckite, arfvedsonite–cummingtonite, richterite–riebeckite, and cummingtonite–tremolite [12]. Some examples of documented exsolution in amphiboles follow.

Although the ideal formula of cummingtonite does not include Ca, natural high cummingtonites may possess some because their $C2/m$ structural arrangement allows a small amount to fit into $M(4)$. On cooling from high temperature and inverting to low ($P2_1/m$) cummingtonite or to anthophyllite (*Pnma*), the geometry of the $M(4)$ site becomes unfavorable for Ca. At least one specimen consisting largely of low cummingtonite has been observed to have thin lamellae of anthophyllite and a small amount of exsolved tremolite [12]. Evidently the cummingtonite was originally $C2/m$ at high temperature, exsolved tremolite as it displacively inverted to $P2_1/m$, and then partially inverted to anthophyllite, probably aided by the presence of a fluid phase.

A survey of exsolution textures in amphiboles from a variety of geologic environments [30] shows not only that exsolution is very common in these minerals, but that exsolved phases tend to be of the same crystal system as the host. Anthophyllites (*Pnma*) commonly exsolve gedrites (*Pnma*) and vice versa. High cummingtonite ($C2/m$) is often host to lamellae of low cummingtonite ($P2_1/m$) or hornblende ($C2/m$), and hornblende often exsolves lamellae of either high or low cummingtonite. As in other minerals we have considered, exsolution may proceed either by nucleation and growth or by spinodal decomposition, although nucleation and growth seem to constitute the most common mechanism [12].

Having considered exsolution in feldspars and pyroxenes already, we are familiar with the tendency of exsolving phases to assume an orientation relative to the host that results in the least lattice strain for both. The same phenomenon occurs in amphiboles, and careful measurement with a petrographic microscope often reveals that the boundaries between host and lamellae are not rational planes [27]. Even though the *boundaries* are not rational planes, the *lattices* of the two phases are usually parallel or very nearly so [29].

The presence of two or more amphiboles coexisting *in equilibrium* in a rock indicate a miscibility gap between those compositions, even if no exsolution has occurred. This occurs when the crystallization temperature is below the solvus for a given amphibole bulk composition. There are several known examples of multiple-amphibole assemblages involving three or four amphibole phases in equilibrium. For example, the assemblages magnesian hornblende + primitive cummingtonite + anthophyllite (each with exsolution lamellae of different compositions; [28]), and cummingtonite + hornblende + anthophyllite + gedrite [33] have been documented. There is even a report of a single submillimeter-size mineral grain containing five amphiboles (glaucophane, cummingtonite, winchite, actinolite, and hornblende) [32], and although it may not constitute an equilibrium assemblage, the five amphiboles are probably nonetheless mutually immiscible. Interestingly, this assemblage contains members of the Fe–Mg–Mn, alkali, calcic, and sodic–calcic families, emphasizing the control of $M(4)$ chemistry on miscibility.

VARIATIONS IN LATTICE PARAMETERS

The wide variety of chemical substitutions possible in amphiboles limits the applicability of general relationships between composition and lattice dimensions, but some pertinent observations may be made.

We have classified amphiboles on the basis of the type of cation occupying $M(4)$, and that approach has structural as well as chemical implications. Figure 5-14 shows the β angle as a function of the mean (octahedral) radius of the X cation,

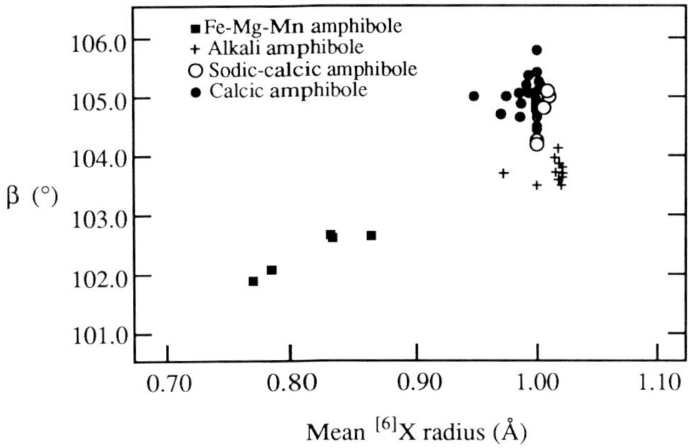

Figure 5-14. The β angle of clinoamphiboles as a function of the mean (octahedral) radius of the X cation, which occupies the $M(4)$ site. Data for about 40 amphiboles are taken from the literature.

which occupies $M(4)$, for a tabulation of about 40 natural clinoamphiboles taken from the literature. For Fe–Mg–Mn varieties there is a clear trend for the β angle to increase with increasing radius of the X cation. Moreover, that trend, when extended, intersects the cluster of data points for calcic, sodic–calcic, and alkali amphiboles. It obviously cannot be said, however, that there is any consistent correlation when all are taken together, although it is clear that Fe–Mg–Mn amphiboles have β angles in the range 101.8–102.8°, alkali amphiboles have β angles in the range 103.3–104.2°, and calcic amphiboles have β angles in the range 104.4–106.1°. The range of β angles for sodic-calcic amphiboles overlaps that for calcic amphiboles considerably.

The nature of the structural connection between the β angle and the size of the $M(4)$ cation can be inferred from Figure 5-15, in which the β angle is plotted as a function of the $M(4)$–O(5) distance. From Figure 5-11, which portrays the orientation of this bond, observe the following: (1) The double chains are offset relative to one another along the **c** direction; (2) the amount of offset depends largely on the geometry of the $M(4)$ site; and (3) the $M(4)$–O(5) bond lengthens as the relative displacement of the two double chains increases in the **c** direction. The β angle decreases as the displacement between double chains increases, and so it also decreases as $M(4)$–O(5) lengthens.

Some other correlations between composition and lattice parameters have been observed for suites of amphiboles of limited chemical variability, but they are not as apparent when all amphiboles are considered. For example, a, b, and c all increase in a very predictable linear way with increasing Fe content of cummingtonites [43]. The same is true for the arfvedsonite–eckermannite series [18] and the glaucophane–riebeckite series [1], but the curves expressing the variations for the three series of minerals are not parallel because the substitutions are not analogous. Figures 5-16(a) and 5-16(b) compare plots of b/c versus mean $M(2)$–O distance [which is governed by $M(2)$ chemistry] for a suite of sodic–calcic to alkali amphi-

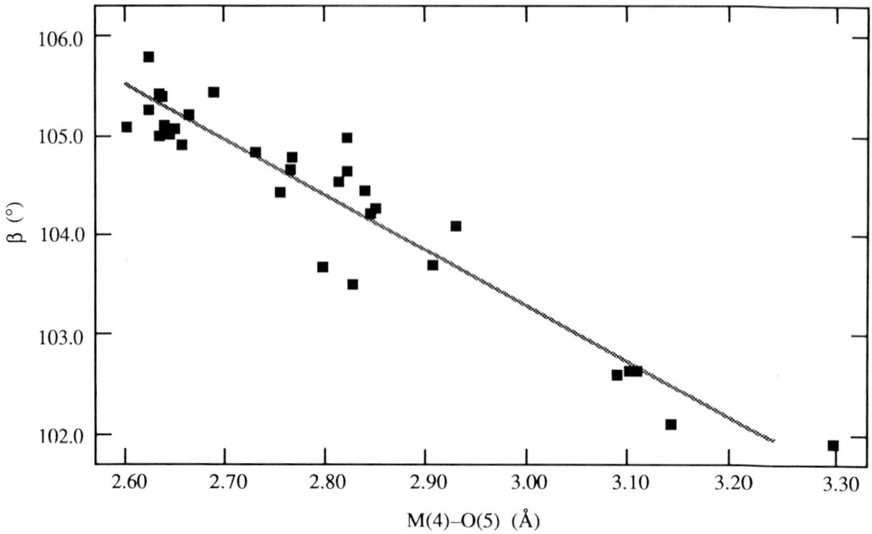

Figure 5-15. The β angle as a function of the M(4)–O(5) distance for 31 amphiboles taken from a compilation in the literature. See text for a structural explanation of the correlation.

boles from a single geologic occurrence [37] and the same parameters for 31 unrelated clinoamphiboles from the literature [15]; it is evident that a correlation that is excellent for a restricted suite of amphiboles is of much more limited use for a sampling of unrelated amphiboles. Using multiple linear regression analyses to determine equations for predicting compositional variables from lattice parameters for suites of unrelated amphiboles has likewise proved of limited use.

While not useful for precise prediction, some of the relationships between lattice parameters and composition provide insight into the structural relationships involved. Figure 5-17 contains some selected plots of mean bond lengths versus lattice parameters for the same 31 unrelated clinoamphiboles as used in Figure 5-16(b). The following observations can be made from these plots:

1. The a cell edges fall into two groups—9.56 ± 0.05 Å and 9.90 ± 0.12 Å. The smaller a cell edges belong to amphiboles of the Fe–Mg–Mn family and to glaucophanes, the latter of which contain high concentrations of Al in the M sites. Despite what thus seems to imply good correlations between the a dimension and the sizes of the M sites (or the atoms they contain), these correlations are very poor for this set of data, ranging from $|r| = 0.06$ to 0.20 for a versus mean bond lengths at the various M sites. No other mean bond lengths show significant correlations with a. [The correlation coefficient r is a measure of "goodness of fit" and varies from $+1.0$ (or -1.0) for perfect positive (or negative) correlation between two variables to 0.0 for no statistical correlation at all. The coefficient of determination, r^2, can be interpreted as the fraction of the variation in the data that is explained by a given correlation.]
2. It is clear from the plots of a versus $\langle T(1)\text{–O}\rangle$ and β versus $\langle T(1)\text{–O}\rangle$ that the

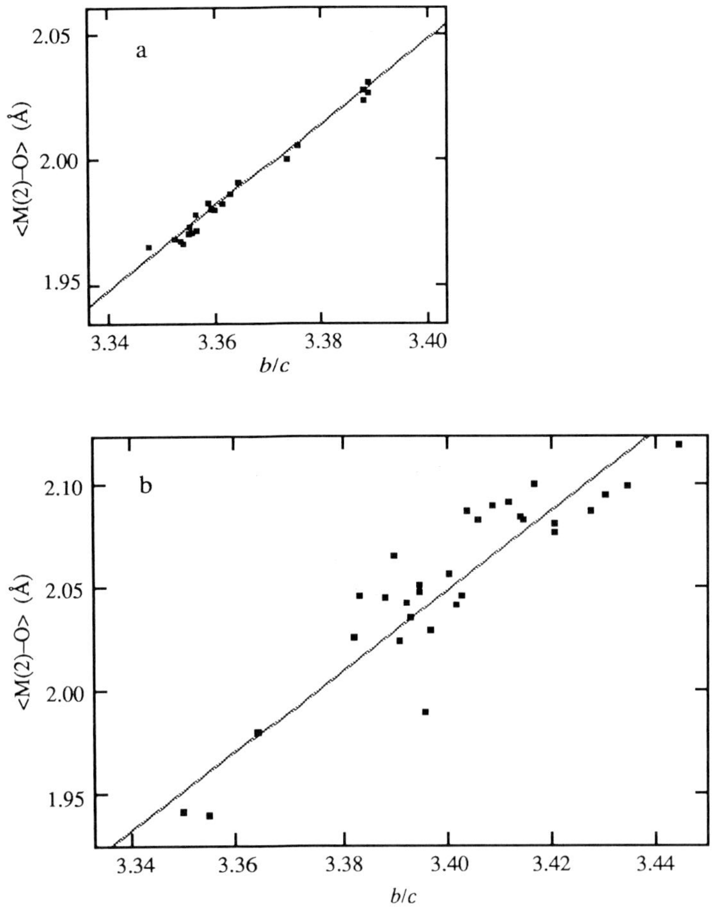

Figure 5-16. (a) Mean $M(2)$–O distance as a function of b/c for a suite of sodic–calcic to alkali amphiboles from a single locality [37]. (b) A similar plot for 31 unrelated amphiboles taken from a compilation in the literature [15].

Figure 5-17. Plots of unit cell parameters of 31 unrelated amphiboles as functions of some mean bond lengths. All units are Å except for β, which is in degrees. See text for discussion of plots.

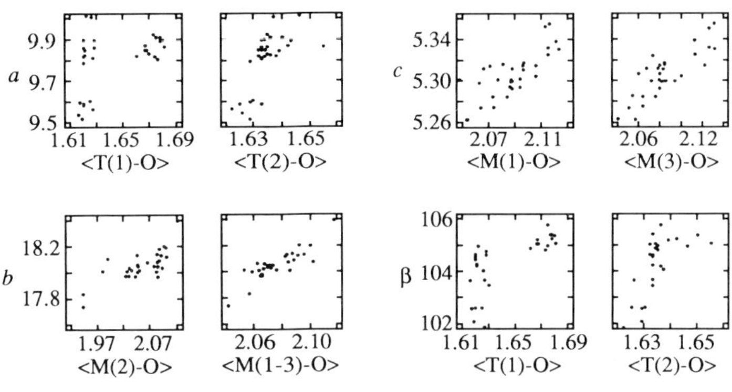

⟨$T(1)$-O⟩ bond lengths are divided into two populations also—a group of smaller values that represent the nonaluminous $T(1)$ sites, and a group of larger values that represent the $T(1)$ sites in $^{[4]}$Al-containing amphiboles.

3. Even though the correlation coefficient for b versus ⟨$M(2)$–O⟩ is not particularly large (0.71, so that $r^2 \approx 0.50$), the trend is clearly visible. Inasmuch as $M(2)$ exhibits the greatest range in mean bond length of any of the octahedra, it is not surprising that the b cell edge (which is essentially a measure of the width of the octahedral strip) is more highly correlated with ⟨$M(2)$–O⟩ than with either ⟨$M(1)$–O⟩ or ⟨$M(3)$–O⟩. However, if the grand mean octahedral bond length [the mean of ⟨$M(1)$–O⟩, ⟨$M(2)$–O⟩, and ⟨$M(3)$–O⟩ = ⟨$M(1$–$3)$–O⟩] is plotted against b, the correlation is much stronger ($r = 0.89$), again reflecting that the b cell dimension is a measure of the width of the octahedral strip.
4. Even though the length of the c repeat is two tetrahedra, measured along either side of the double tetrahedral chain, the c cell dimension is not significantly correlated with either the sizes of the tetrahedra or the tetrahedral rotation angles (not shown in the figure). Instead, the strongest correlations are with ⟨$M(1)$–O⟩ ($r = 0.75$) and ⟨$M(3)$–O⟩ ($r = 0.83$). $M(1)$ and $M(3)$ constitute the mica-like portion of the octahedral strip and, as in micas, the octahedra of amphiboles seem to form the more rigid part of the structure, to which the tetrahedral portion adjusts.
5. The β angle is most strongly correlated with the mean tetrahedral bond lengths, but the correlation coefficients are not particularly high (0.69 and 0.64); for the grand mean tetrahedral bond length, ⟨$T(1,2)$–O⟩, versus β, $r = 0.74$ for this group of amphiboles. β does turn out to show fairly strong correlations with tetrahedral rotation angle ($r = 0.81$) and with a ($r = 0.82$), and these are likely due to (1) the dependence of tetrahedral rotation angle on mean tetrahedral size (that is, $^{[4]}$Al concentration) and (2) the fact that comparison of the a versus ⟨$T(1)$–O⟩ and β versus ⟨$T(1)$–O⟩ plots suggests that a and β respond to both tetrahedral size and $M(4)$ chemistry in very similar ways.

None of the above observations is intended to suggest that very high diagnostic correlations between unit-cell dimensions and structural or chemical parameters do not exist for chemically related suites of amphiboles, but only that it is very difficult to find such correlations that are equally applicable to all amphiboles.

THE NONCLASSICAL BIOPYRIBOLES

Having considered single-chain silicates in the previous chapter and double-chain silicates in this one, it should be apparent that there is no crystallographic reason that minerals with wider silicate chains could not exist, and indeed they do. In fact, in the limit of an infinitely wide chain, we have the micas. For this reason, the broad class of minerals that contains pyroxenes, amphiboles, and micas is often called the *biopyriboles* (from *bio*tites, *pyr*oxenes, and amph*iboles*). Because they are common and have been known for a long time, the micas, amphiboles, and pyroxenes are called the *classical biopyriboles*; minerals with chain widths greater than two single

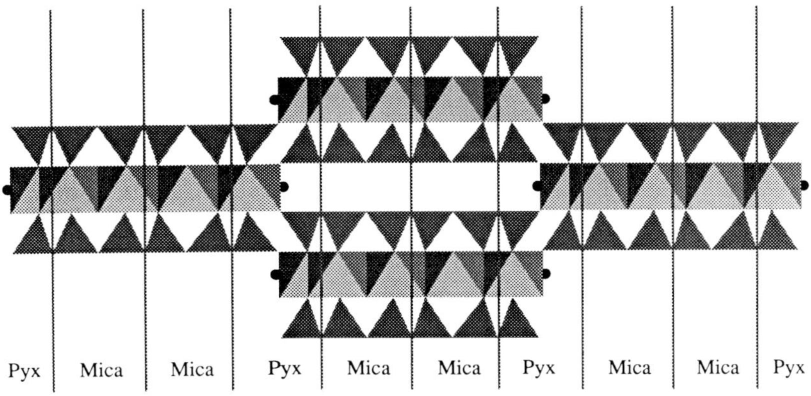

Figure 5-18. A schematic representation of the structure of clinojimthompsonite, a nonclassical biopyribole with a triple silicate chain. It can be visualized as consisting of alternating slabs of mica and pyroxene structure assembled in the sequence · · ·PMMPMMP· · ·

chains (but not infinite) have only recently been recognized and are called the *nonclassical biopyriboles*. These unusual minerals will not be covered exhaustively here, but a brief consideration is worthwhile.

Recall from Figure 5-4 that amphiboles may be visualized as being constructed from alternating slabs of pyroxene structure and mica structure, symbolically represented as · · ·PMPMPM· · ·. Figure 5-18 is a schematic depiction showing these slabs assembled in the sequence · · ·PMMPMMP· · ·, resulting in a triple silicate chain. This is essentially the structure of the nonclassical biopyribole clinojimthompsonite, although in the real structure the chains are convex toward their associated *M*-site bands, as in amphiboles. In principle there could be three, four, or more mica slabs in sequence, producing tetrahedral bands four, five, or more tetrahedra wide. The variety of nonclassical biopyriboles so far recognized in nature turns out to rather restricted, however, and we shall address that in the following.

ORDERED AND DISORDERED NONCLASSICAL BIOPYRIBOLES

It frequently happens in science that new developments are pursued by independent investigators who are initially unaware of one another's work, and in the mid-1970s that was the case for multiple-width silicate chains. On one hand synthesis experiments yielded a triple chain silicate with the possible formula $Na_2Mg_8Si_{12}O_{30}(OH)_6$ [8] and another of composition $Na_2Na_2Mg_8Si_{12}O_{32}(OH)_4$ [35]. On the other hand, three natural minerals with wide chains were found in metamorphosed ultramafic rocks in Vermont [39;40]: jimthompsonite and clinojimthompsonite, both $(Mg,Fe)_{10}Si_{12}O_{32}(OH)_4$, and chesterite, $(Mg,Fe)_{17}Si_{20}O_{54}(OH)_6$. Like the synthetic phases, jimthompsonite and clinojimthompsonite both have triple chains; chesterite has alternating triple and double chains. Because these contain only one type of chain or a rigidly repeating sequence of more than one type, they are called *ordered biopyriboles*. The term refers not to the ordering of cations on alternative sites that we have discussed at length for other minerals, but to the repetition of sequences of M and P slabs. A biopyribole in which the M and P slabs are randomly stacked are said to exhibit chain-width disorder.

CHEMISTRY OF NONCLASSICAL BIOPYRIBOLES

Because the nonclassical biopyriboles can be envisioned as consisting of sequences of M and P slabs, their chemical formulas may be constructed from the formulas of such slabs [38]. The M, or mica, slab has the generalized formula $AM_3T_4O_{10}(OH)_2$, where for example in annite, A is the interlayer K, the three M cations are Mg, and three of the T cations are Si while the other is Al. The P slab has a formula that is twice that of a pyroxene—$M_4T_4O_{12}$—where in enstatite, for instance, the M cations are Mg and the T cations are Si.

Jimthompsonite is a triple-chain biopyribole consisting of the repeated slab sequence · · ·MMP· · ·. Thus the general formula for jimthompsonite would be obtained by adding the formulas for two M slabs—$A_2M_6T_8O_{20}(OH)_4$—and one P slab to yield $A_2M_{10}T_{12}O_{32}(OH)_4$. In this case the A sites are empty (that is, the M slabs are talc slabs), and the formula is $(Mg,Fe)_{10}Si_{12}O_{32}(OH)_4$, allowing for Fe–Mg substitution. Recall that chesterite is characterized by alternation of a triple chain with a double chain: · · ·MMPMP· · ·. The generalized formula is then $A_3M_9T_{12}O_{30}(OH)_6 + M_8T_8O_{24}$, or $A_3M_{17}T_{20}O_{54}(OH)_6$. Here again the M slabs are talc slabs, and the formula is $(Mg,Fe)_{17}Si_{20}O_{54}(OH)_6$.

Other ordered structures have been observed by high-resolution transmission electron microscopy (HRTEM), although it has not been possible to separate them as distinct phases so that they could be classified as new mineral species. HRTEM observations of altered rocks from the Salton Sea Geothermal Field have been observed to contain triple-chain coherent intergrowths in amphiboles and pyroxenes [47]. Analytical electron microscopy (energy-dispersive x-ray analysis with the transmission electron microscope—see Chapter 10) suggests the ideal formula $Ca_2M_8Si_{12}O_{32}(OH)_4$, where M = (Mg,Fe,Mn,Ti), making this another MMP sequence with empty A sites. In the same study some quadruple and quintuple chains were also seen by HRTEM, but compositions were not estimated.

Another interesting observation [41] is that some "long-period" sequences of double, triple, and quadruple chains sometimes occur—that is, sequences of sequences. For example, such complex repetitions as · · ·2233· · ·, · · ·2333· · ·, · · ·43323· · ·, and · · ·43332343332423· · · have been observed. This is reminiscent of the complex polytypes of micas, and the mechanism by which these might have formed is equally mysterious. None of these, however, occurs as discrete, separable mineral grains, and so these intergrowths cannot by properly classed as minerals.

CHAIN-WIDTH DISORDER AND CHAIN TERMINATIONS

HRTEM studies have revealed that many amphiboles and pyroxenes, as well as the nonclassical biopyriboles, display some degree of chain-width disorder. This may range from occasional intercalations of triple chains into an otherwise ordered amphibole to extreme disorder characterized by the presence of many errors with chain widths of 3 or 4 to 8 or 10 (or more), and even regions wide enough to be called volumes of sheet silicate. Local chemical compositions may range from compositions near amphibole to compositions near talc, but do not usually match the stoichiometry of any known ordered biopyribole.

Obviously, a chain-width error must either continue until it meets a grain bound-

Figure 5-19. An *a*-axis view of cooperative chain terminations in a disordered biopyribole. When viewed along **c** at a low angle to the page, the changes from 223 to 322, from 2323 to 3322, and from 23 to 32 are evident. [Photo from D. R. Veblen and P. R. Buseck (1980). "Microstructures and Reaction Mechanisms in Biopyriboles." *American Mineralogist*, 65, 599–623.]

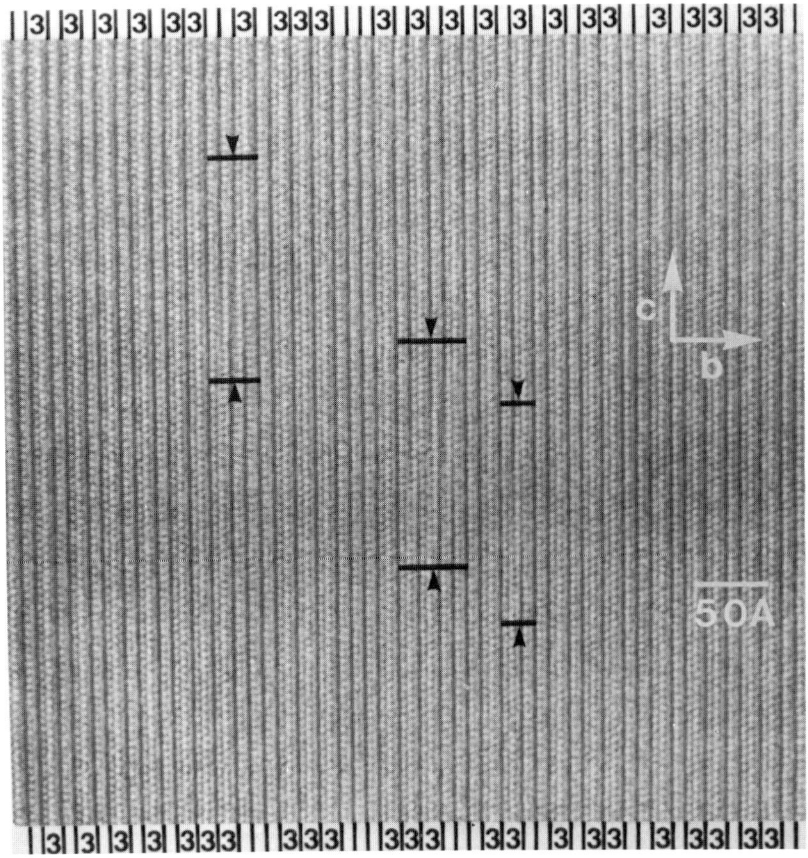

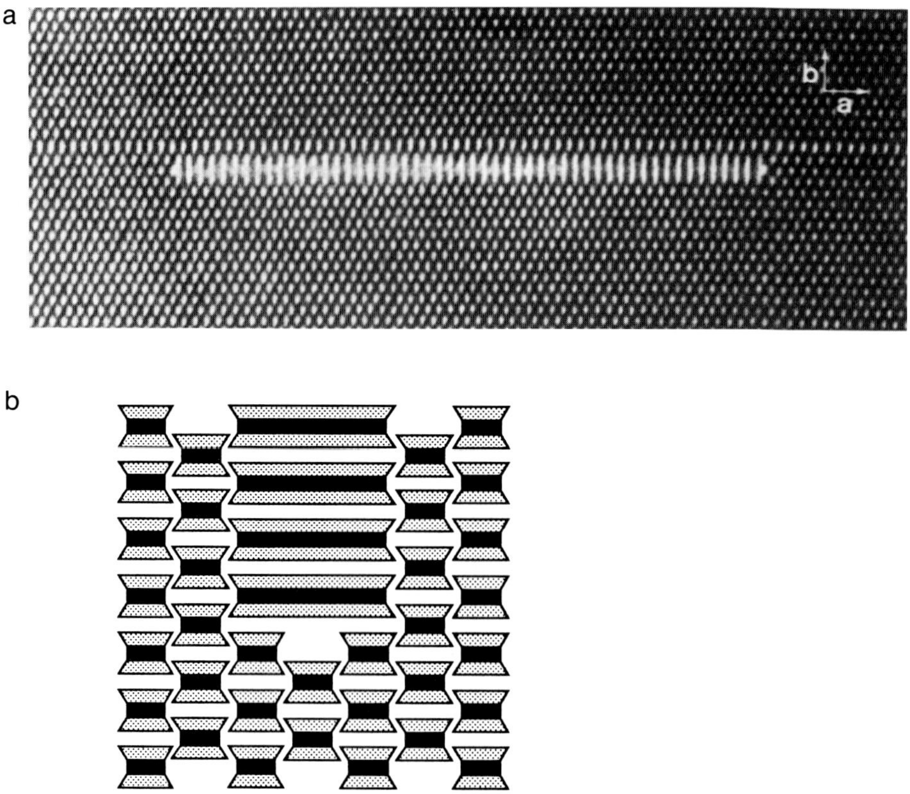

Figure 5-20. (a) HRTEM photomicrograph of a doubly terminated sextuple (6 wide) zipper in an amphibole. Note the triple zipper adjacent to it. [Photo from D. R. Veblen and P. R. Buseck (1980). "Microstructures and Reaction Mechanisms in Biopyriboles." *American Mineralogist*, 65, 599–623.] (b) A schematic representation of the terminations seen in Figure 5-20(a). "I-beam" figures represent two tetrahedral chains of appropriate width sandwiching an octahedral band and viewed down **c**.

ary, or it must terminate by transformation to some other sort of error or to the ordered structure around it. Given that chain-width errors are two-dimensional in extent—that is, they are (010) slabs or lamellae that extend in both the **c** and **a** directions—there are termination mechanisms that arrest the slab along the **c** direction and others that do so along **a**. Because the former terminate [001] chains, they are called *chain terminations*. Viewed along **c** the wide chain errors resemble zippers, so the latter are called *zipper terminations*.

Termination of a wide chain may occur in such a way that the error and the surrounding structure fit together coherently. Figure 5-19 shows an HRTEM photomicrograph in which various sequences of double and triple chains are transformed into different sequences of the same types of chains. Coherency along the **c** direction is maintained as, for example, a 223 sequence becomes a 322 sequence or 2323 becomes 3322. Such "cooperative chain terminations" need not involve simply rearranging the same chain widths; an example in which the sequence 122223 is replaced by 222222 has been documented [48].

Zipper terminations are viewed along **c** and may also either be coherent or incoherent. Figure 5-20(a) shows a sextuple zipper terminated coherently on both ends; note that it is flanked by a triple-wide zipper that spans the entire photograph. Figure 5-20(b) is a schematic structure diagram illustrating one of the terminations.

Two rules govern whether a biopyribole lamella will terminate coherently or

Figure 5-21. (a) A slab of amphibole replaces a triple chain in chesterite. Because the triple chain contains three subchains and the amphibole contains only two, the first rule is violated. Thus the termination is incoherent and would result in a planar fault or distortion of the structure. (b) A violation of the second rule occurs here, where a double and triple slab of the chesterite structure are replaced by a five-wide zipper. Clearly the sharing of corners in the tetrahedra of different chains results in an impossible structure.

a

b

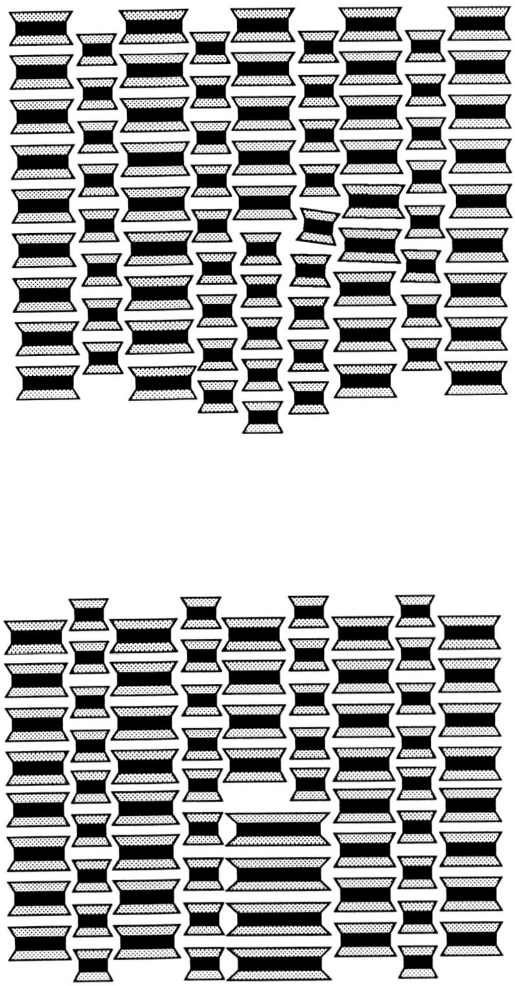

incoherently [42;46]. One rule states that there must be an equal number of subchains in a lamella as in the material it replaces. Here a subchain is essentially a pyroxene chain, so that pyroxenes have one, amphiboles two, and so forth. The second rule requires that the numbers of chains in the lamella and the replacing material be either both even or both odd. In Figure 5-21(a) a slab of amphibole replaces a triple chain in chesterite. Because the triple chain contains three subchains and the amphibole contains only two, the termination is incoherent and would result in a planar fault or distortion of the structure; this is a violation of the first rule. In Figure 5-21(b) a violation of the second rule occurs where a double and triple slab of the chesterite structure are replaced by a five-wide zipper; this structure, of course, makes no sense. Note that in each case the rule not being illustrated is not violated—both must be obeyed for a coherent termination, but the violation of either is sufficient to eliminate the possibility of coherency.

The investigation of nonclassical biopyriboles is still in its infancy. Very sophisticated techniques have been used to characterize those so far discovered, but much more will be learned about these unusual minerals and the metamorphic and alteration reactions that relate pyroxenes, amphiboles, sheet silicates, and nonclassical biopyriboles.

REFERENCES

[1] Borg, I. Y. (1967). "Optical Properties and Cell Parameters in the Glaucophane–Riebeckite Series." *Contributions to Mineralogy and Petrology* 15, 67–92.

[2] Brown, G. E., and G. V. Gibbs (1969). "Oxygen Coordination and the Si–O Bond." *American Mineralogist* 54, 1528–39.

[3] Brown, I. D. (1981). "The Bond–Valence Method: An Empirical Approach to Chemical Structure and Bonding." In M. O'Keeffe and A. Navrotsky (eds.), *Structure and Bonding in Crystals*, Volume 2, pp. 1–30. Academic Press, New York.

[4] Brown, I. D., and R. D. Shannon (1973). "Empirical Bond-Strength–Bond-Length Curves for Oxides." *Acta Crystallographica* A29, 266–82.

[5] Cameron, M., and G. V. Gibbs (1971). "Refinement of the Crystal Structure of Two Synthetic Fluor-Richterites." *Carnegie Institution of Washington Year Book* 70, 150–53.

[6] Cameron, M., and G. V. Gibbs (1973). "The Crystal Structure and Bonding of Fluor-Tremolite: A Comparison with Hydroxyl Tremolite." *American Mineralogist* 58, 879–88.

[7] Donnay, G., and R. Allmann (1970). "How to Recognize O^{2-}, OH^-, and H_2O in Crystal Structures Determined by X-Rays." *American Mineralogist* 55, 1003–15.

[8] Drits, V. A., Y. I. Goncharov, V. A. Aleksandrova, V. E. Khadzhi, and A. L. Dmitrik (1974). "New Type of Strip Silicate." *Kristallografiya* 19, 1186–93.

[9] Ernst, W. G. (1963). "Stability Relations of Glaucophane." *American Journal of Science* 259, 735–65.

[10] Fischer, K. F. (1966). "A Further Refinement of the Crystal Structure of Cummingtonite $(Mg,Fe)_7(Si_4O_{11})(OH)_2$." *American Mineralogist* 51, 814–18.

[11] Ghose, S. (1961). "The Crystal Structure of Cummingtonite." *Acta Crystallographica* 14, 622–27.

[12] Ghose, S. (1981). "Subsolidus Reactions and Microstructures in Amphiboles." In D. R. Veblen (ed.), *Amphiboles and Other Hydrous Pyriboles—Mineralogy. Reviews in Mineralogy*, Volume 9A. Mineralogical Society of America, Washington, D.C.

[13] Gibbs, G. V. (1969). "Crystal Structure of Protoamphibole." *Mineralogical Society of America Special Paper* 2, 101–109.

[14] Hawthorne, F. C. (1981). "Crystal Chemistry of the Amphiboles." In D. R. Veblen (ed.)

Amphiboles and other Hydrous Pyriboles—Mineralogy. Reviews in Mineralogy, Volume 9A, Mineralogical Society of America, Washington, D.C.

[15] Hawthorne, F. C. (1983). "The Crystal Chemistry of the Amphiboles." *Canadian Mineralogist* 21, 173–480.

[16] Hawthorne, F. C., and H. D. Grundy (1978). The Crystal Chemistry of the Amphiboles. VII. "The Crystal Structure and Site-Chemistry of Potassian Ferri-Taramite." *Canadian Mineralogist* 16, 53–62.

[17] Irusteta, M. C., and E. J. W. Whittaker (1975). "A Three-Dimensional Refinement of the Structure of Holmquistite." *Acta Crystallographica* B31, 145–50.

[18] Kempe, D. R. C. (1969). "The Cell Parameters of the Arfvedsonite–Eckermannite Series, with Observations on the MgO and Total Iron Content of Amphiboles." *Mineralogical Magazine* 37, 317–32.

[19] Leake, B. E. (1968). "A Catalogue of Analyzed Calciferous and Subcalciferous Amphiboles Together with their Nomenclature and Associated Minerals." *Geological Society of America Special Paper* 98.

[20] Leake, B. E. (1978). "Nomenclature of Amphiboles." *Mineralogical Magazine* 42, 533–61 (Also *American Mineralogist* 63, 1023–52; *Canadian Mineralogist* 16, 501–20.)

[21] Leake, B. E., and M. H. Hey (1979). "Addendum to the Nomenclature of Amphiboles." *Mineralogical Magazine* 42, 561–63.

[22] Papike, J. J., and J. R. Clark (1968). "The Crystal Structure and Cation Distribution of Glaucophane." *American Mineralogist* 53, 1156–73.

[23] Papike, J. J., and M. Ross (1970). "Gedrites: Crystal Structures and Intracrystalline Cation Distributions." *American Mineralogist* 55, 1945–72.

[24] Papike, J. J., M. Ross, and J. R. Clark (1969). "Crystal-Chemical Characterization of Clinoamphiboles Based on Five New Structure Refinements." *Mineralogical Society of America Special Paper* 2, 117–36.

[25] Pauling, L. (1960). *The Nature of the Chemical Bond.* Cornell University Press, Ithaca, N. Y.

[26] Prewitt, C. T., J. J. Papike, and M. Ross (1970). "Cummingtonite: A Reversible, Nonquenchable Transition from $P2_1/m$ to $C2/m$ Symmetry." *American Mineralogist* 55, 305–06.

[27] Robinson, P., H. W. Jaffe, M. Ross, and C. Klein, Jr. (1971). "Orientation of Exsolution Lamellae in Clinopyroxenes and Clinoamphiboles: Consideration of Optimal Phase Boundaries." *American Mineralogist* 56, 909–39.

[28] Robinson, P., C. Klein, Jr., and M. Ross (1969). "Equilibrium Coexistence of Three Amphiboles." *Contributions to Mineralogy and Petrology* 22, 248–58.

[29] Robinson, P., F. S. Spear, J. C. Schumacher, J. Laird, C. Klein, B. W. Evans, and B. L. Doolan (1982). "Phase Relations of Metamorphic Amphiboles: Natural Occurrence and Theory." In D. R. Veblen and P. H. Ribbe (eds.), *Amphiboles—Petrology and Experimental Phase Relations. Reviews in Mineralogy*, Volume 9B. Mineralogical Society of America, Washington, D.C.

[30] Ross, M., J. J. Papike, and K. W. Shaw (1969). "Exsolution Textures in Amphiboles as Indicators of Subsolidus Thermal Histories." *Mineralogical Society of America Special Paper* 2, 275–99.

[31] Shannon, R. D. (1976). "Revised Effective Ionic Radii and Systematic Studies of Interatomic Distances in Halides and Chalcogenides." *Acta Crystallographica* A32, 751–67.

[32] Smelik, E. A., and D. R. Veblen (1989). "A Five-Amphibole Assemblage from Blueschists in Northern Vermont." *American Mineralogist* 74, 960–64.

[33] Stout, J. H. (1971). "Four Coexisting Amphiboles from Telemark, Norway." *American Mineralogist* 56, 212–24.

[34] Sueno, S., J. J. Papike, C. T. Prewitt, and G. E. Brown (1972). "Crystal Structure of High Cummingtonite." *Journal of Geophysical Research* 77, 5767–77.

[35] Tateyama, H. S., S. Shimoda, and T. Sudo (1978). "Synthesis and Crystal Structure of a Triple Chain Silicate $Na_2Mg_4Si_6O_{16}(OH)_2$." *Contributions to Mineralogy and Petrology* 66, 149–56.

[36] Thompson, J. B., Jr. (1978). "Biopyriboles and Polysomatic Series." *American Mineralogist* 63, 239–49.

[37] Ungaretti, L, D. C. Smith, and G. Rossi (1981). "Crystal-Chemistry by X-Ray Structure Refinement and Electron Microprobe Analysis of a Series of Sodic–Calcic to Alkali–Amphiboles from the Nybö Eclogite Pod, Norway." *Bulletin de la Societe française Minéralogie et Cristallographie* 104, 400–12.

[38] Veblen, D. R. (1981). "Non-Classical Pyriboles and Polysomatic Reactions in Biopyriboles."

In D. R. Veblen (ed.), *Amphiboles and Other Hydrous Pyriboles—Mineralogy. Reviews in Mineralogy*, Vol. 9A. Mineralogical Society of America, Washington, D.C.

[39] Veblen, D. R., and C. W. Burnham (1974a). "New Biopyriboles from Chester, Vermont: I. Descriptive Mineralogy." *American Mineralogist* 63, 1000–09.

[40] Veblen, D. R., and C. W. Burnham (1974b). "New Biopyriboles from Chester, Vermont: II. The Crystal Chemistry of Jimthompsonite, Clinojimthompsonite, and Chesterite, and the Amphibole-Mica Reaction." *American Mineralogist* 63, 1053–73.

[41] Veblen, D. R., and P. R. Buseck (1979). "Chain-Width Order and Disorder in Biopyriboles." *American Mineralogist* 64, 687–700.

[42] Veblen, D. R., and P. R. Buseck (1980)."Microstructures and Reaction Mechanisms in Biophyriboles." *American Mineralogist* 65, 599–623.

[43] Viswanathan, K., and S. Ghose (1965). "The Effect of Mg^{2+}–Fe^{2+} Substitution on the Cell Dimensions of Cummingtonites." *American Mineralogist* 50, 1106–12.

[44] Whittaker, E. J. W. (1969). "The Structure of the Orthorhombic Amphibole Holmquistite." *Acta Crystallographica* B25, 394–97.

[45] Whittaker, E. J. W. (1971). "Madelung Energies and Site Preferences in Amphiboles. I." *American Mineralogist* 56, 980–96.

[46] Whittaker, E. J. W., B. A. Cressey, and J. L. Hutchison (1981). "Terminations of Multiple-Chain Lamellae in Grunerite Asbestos." *Mineralogical Magazine* 44, 27–35.

[47] Yau, Y.-C., D. R. Peacor, and E. J. Essene (1986). "Occurrence of Wide-Chain Ca-Pyriboles as Primary Crystals in the Salton Sea Geothermal Field, California, USA." *Contributions to Mineralogy and Petrology* 94, 127–34.

[48] Zoltai, T. (1981). "Amphibole Asbestos Mineralogy." In D. R. Veblen (ed.), *Amphiboles and Other Hydrous Pyriboles—Mineralogy. Reviews in Mineralogy*, Volume 9A. Mineralogical Society of America, Washington, D.C.

6 The Aluminum Silicate Polymorphs

Although an aluminum silicate can clearly be any chemical compound containing aluminum and silicon bonded to oxygen, the expressions *aluminum silicate polymorphs* and *aluminosilicate polymorphs* are generally understood among mineralogists and petrologists to refer to the three crystalline forms of Al_2SiO_5: kyanite, andalusite, and sillimanite. The designation *aluminum silicate* seems to enjoy wider acceptance than *aluminosilicate* currently, presumably on the grounds that an aluminosilicate may be a compound of aluminum, silicon, oxygen, and one or more other elements (albite is a sodium aluminosilicate, for example), whereas *aluminum silicate* implies a compound consisting of only Al, Si, and O. Mullite is a closely related mineral characterized by a somewhat disordered crystal structure and a compositional range from $3Al_2O_3 \cdot 2SiO_2$ to $2Al_2O_3 \cdot SiO_2$, and we shall include it in this chapter, also.

The transformations among the three aluminum silicate polymorphs are all reconstructive, first order, and involve a change in primary coordination. Indeed, there are many examples of the coexistence of two of the polymorphs, or even all three, in the same rocks. Clearly, these minerals are considered a family not because of structural similarity, as is true of feldspars, micas, and most other recognized mineral families, but because of chemical identity, as with the silica polymorphs.

GENERAL CHEMISTRY

Strictly, the compositions of kyanite, andalusite, and sillimanite are perfectly expressed by the chemical formula Al_2SiO_5, but like the SiO_2 polymorphs, these minerals may contain some "impurity" atoms, substituting largely for Al. Fe^{3+} is the most commonplace substituent in any of the polymorphs (except for Mn in andalusite, discussed below), and is usually the most abundant impurity, not uncommonly ranging up to around 2 wt % Fe_2O_3. Other elements, notably Ti, V, Cr, Mn, B, and Mg, have been reported in many specimens in amounts less than 0.2 wt % oxide [2;15;30;34]. Exceptional concentrations are known, of course—a kyanite with nearly 13 wt % Cr_2O_3 [41], a V-bearing andalusite with over 0.8 wt % V_2O_3 and 0.5 wt % MgO [14], and a sillimanite with 1.3 wt % Fe_2O_3 [18] are examples of exceptions.

The well-known blue or green patchy color distribution of kyanite is largely attributable to the presence of iron, and indicates a concomitantly patchy elemental distribution. In green kyanite essentially all of the iron exists as Fe^{3+}, but in blue specimens there is also a significant proportion of Fe^{2+}, as shown by Mössbauer spectroscopy [31]. The latter requires a charge-compensating substituent, of course, and this is likely Ti^{4+}. The color of blue specimens is therefore caused not only by crystal field electronic transitions within individual iron ions, but also by charge transfer between Fe^{2+} and Fe^{3+} and between Fe^{2+} and Ti^{4+}.

Some sillimanites have been found to contain boron in concentrations exceeding 0.4 wt % B_2O_3 [20;21] but the crystal chemical role of boron in the structure is uncertain. It evidently substitutes for Si, and chemical analyses indicate a concomitant substitution of Mg for Al at an atomic ratio of approximately 0.5 Mg per B. The substitution reaction may look something like $2(B + x\text{Mg}) \rightarrow 2(\text{Si} + x\text{Al}) + (1+x)\text{O}$, where x is around 0.5. Such a reaction would leave the structure of sillimanite charge deficient and would require some local reorganization of the atomic arrangement. Optical spectra yield absorption bands in spectral regions known for absorption by BO_3 groups, so it has been proposed [21] that B in sillimanite exists in either BO_3 planar triangles or in very distorted tetrahedral sites with B coordinated by three close oxygens and one considerably farther away.

A manganese end-member, $MnAlSiO_5$, has the andalusite crystal structure and is called kanonaite [42;43]. Intermediate members were formerly called *viridines* [22], name no longer approved. Intermediate members are now called *manganian andalusites* or *aluminous kanonaites*. The nature of Mn substitution will be discussed below.

CRYSTAL STRUCTURES

Kyanite

Few of the most common minerals are triclinic, but kyanite is one of them; it crystallizes in space group $P\bar{1}$. The oxygen arrangement is a distortion of a cubic-close-packed array, with the close-packed monolayers parallel to (220) in the kyanite unit cell. These layers are apparent in the kyanite structure drawing of Figure 6-1. As with hexagonal close packing, a cubic close packed array of anions contains one octahedral cation site and two tetrahedral cations sites per anion. For Al_2SiO_5, there are thus five octahedral sites and ten tetrahedral sites per formula unit; one of ten available tetrahedral sites is filled with Si and two of five octahedral sites are filled with Al.

The structural formula of kyanite is written $^{[6]}Al_2{}^{[4]}SiO_5$, where the bracketed superscripts indicate coordination numbers. The aluminum-containing octahedra are of four symmetrically independent kinds, two of which form a continuous edge-sharing chain parallel to the c crystallographic axis. These chains of alternating $M1$ and $M2$ octahedra are cross-linked by $M3$ and $M4$ octahedra and by independent

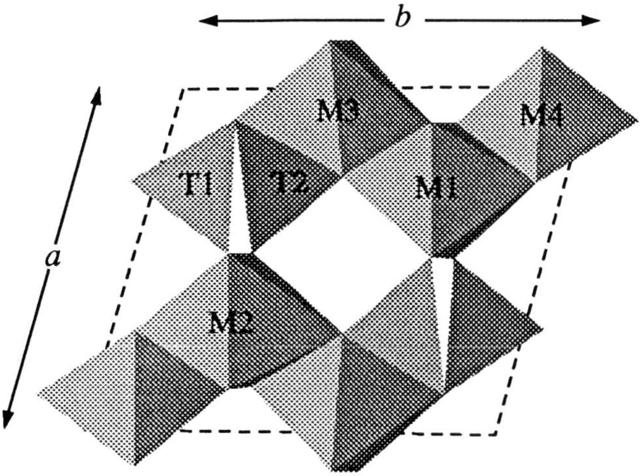

Figure 6-1. The polyhedral arrangement in kyanite. $M1$, $M2$, $M3$, and $M4$ are Al-containing octahedra, and $T1$ and $T2$ are symmetry-independent tetrahedral Si sites. (After [10].)

SiO_4 tetrahedra. Kyanite is consequently an orthosilicate. The crystals are elongated along c, reflecting the structural control of the octahedral chains.

Andalusite

This mineral is orthorhombic, and crystallizes in space group *Pnnm*. The structure (Figure 6-2) possesses, as does kyanite, edge-sharing Al-containing ($M1$) octahedra in chains that extend along c; this accounts for half of the aluminum atoms. Unlike kyanite, however, the other half of the aluminum atoms are in irregular five-coordinated $M2$ voids (approximately trigonal bipyramids) that alternate with Si tetrahedra in double chains that also run parallel to c. The structural formula is therefore $^{[6]}Al^{[5]}Al^{[4]}SiO_5$.

Comparison of x-ray [12] and neutron [18] diffraction structure refinements reveals that the interatomic distances for the neutron-diffraction data are systematically larger than those for the x-ray-diffraction data. This has been interpreted as being due to bonding effects, inasmuch as x-rays are diffracted by the electrons and neutrons are diffracted by the atomic nuclei. Comparison of similar data for sillimanite [18;32] shows no such trend, however, and so this interpretation must remain very tentative.

Substitution of Mn^{3+} in andalusite occurs almost entirely in the octahedrally coordinated $M1$ site [1;42;43]. Two reasons for this have been suggested: First, the $M1$ site in andalusite is inherently large—larger, in fact, than any of the other octahedral sites in the aluminum silicate polymorphs. When only Al is present, the mean $^{[6]}Al–O$ distances at room temperature in kyanite range from 1.896 to 1.919 Å, that in sillimanite is 1.919 Å, and that in andalusite is 1.935 Å [44]; the ionic radii of $^{[6]}Al$ and $^{[6]}Mn^{3+}$ are 0.535 and 0.645 Å, respectively [40]. Second, the $M1$

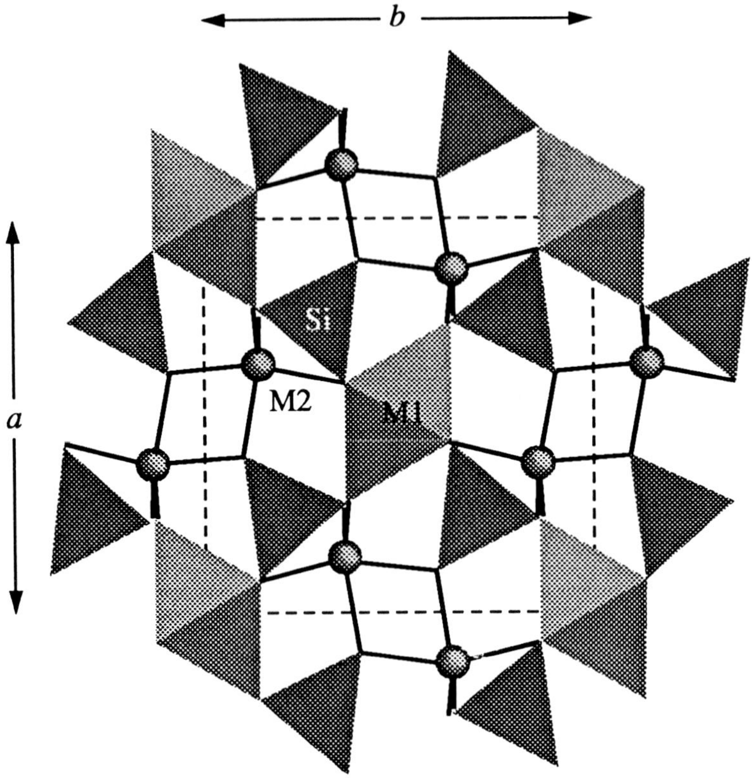

Figure 6-2. The crystal structure of andalusite (and essentially that of kanonaite). *M*1 is the octahedrally coordinated Al site. The spheres represent five-coordinated Al in the *M*2 site. Si is tetrahedrally coordinated. (After [12].)

site in andalusite has approximately $4/mmm$ symmetry, with the predominant distortion from $m\bar{3}m$ symmetry being elongation along one of the fourfold axes (i.e., an extensional tetragonal distortion). This type of distortion is predicted by crystal field theory to provide an energetically favorable environment for Mn^{3+} [23]. Mössbauer spectra indicate that 10–15% of the total Fe^{3+} in aluminian kanonaites may be in *M*2, and this would suggest that some Mn^{3+} might also be there; indeed, a structure refinement of an aluminian kanonaite yields site occupancies of $Mn_{0.74}Al_{0.26}$ for *M*1 and $Al_{0.88}Mn_{0.12}$ for *M*2.

It is noteworthy that manganoan andalusite commonly coexists with Mn-free kyanite or Mn-free sillimanite [34], underscoring the preference of Mn^{3+} for partitioning into the large octahedron.

Sillimanite

The structure of sillimanite (Figure 6-3) is similar in many ways to that of andalusite. It is also orthorhombic, but crystallizes in space group *Pbnm* rather than in

Pnnm. Again, half of the aluminum occupies an edge-sharing octahedral chain that extends along *c*. The other aluminum atoms are tetrahedrally coordinated, as are the silicon atoms. These form chains, as shown in Figure 6-4. Large tetrahedra are occupied by aluminum, small ones by silicon. The structural formula of sillimanite is thus $^{[6]}Al^{[4]}Al^{[6]}SiO_5$. Although sillimanite is not commonly classified as a double chain silicate because of the confusion with the amphibole structure that would result, it is clear from Figure 6-4 that the tetrahedral sites do indeed form a double chain, albeit one of a very different sort from that found in amphiboles.

The completely ordered arrangement of the tetrahedral atoms in sillimanite [9;32] requires perfect alternation of Si and Al tetrahedra. This is another example of the aluminum-avoidance rule, whereby Al-containing tetrahedra do not share corners with each other; we encountered this rule first in the previous discussion of feldspars (p. 78). Two morphologically different varieties of the mineral are known: One, simply called *sillimanite*, has blocky morphology; the other, called *fibrolite*, is a fine-grained fibrous variety. It has been suggested (e.g., [9;45]) that the microscopic factor causing the macroscopic difference between the two might be that fibrolite has a disordered Al–Si distribution. However, preliminary neutron-scattering experiments on powdered fibrolite have indicated that Al and Si are not statistically disordered (less than 5%) [39]. Moreover, the high temperature of Al–Si disordering (over 1700°C) [19] precludes substantial disordering at temperatures

Figure 6-3. The crystal structure of sillimanite. Half of the Al is in the octahedrally coordinated *M*1 site, and the other half is tetrahedrally coordinated. (After [9].)

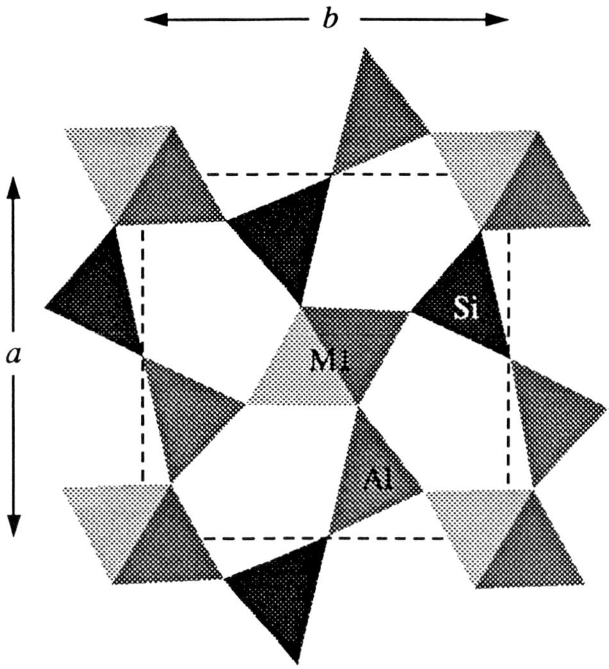

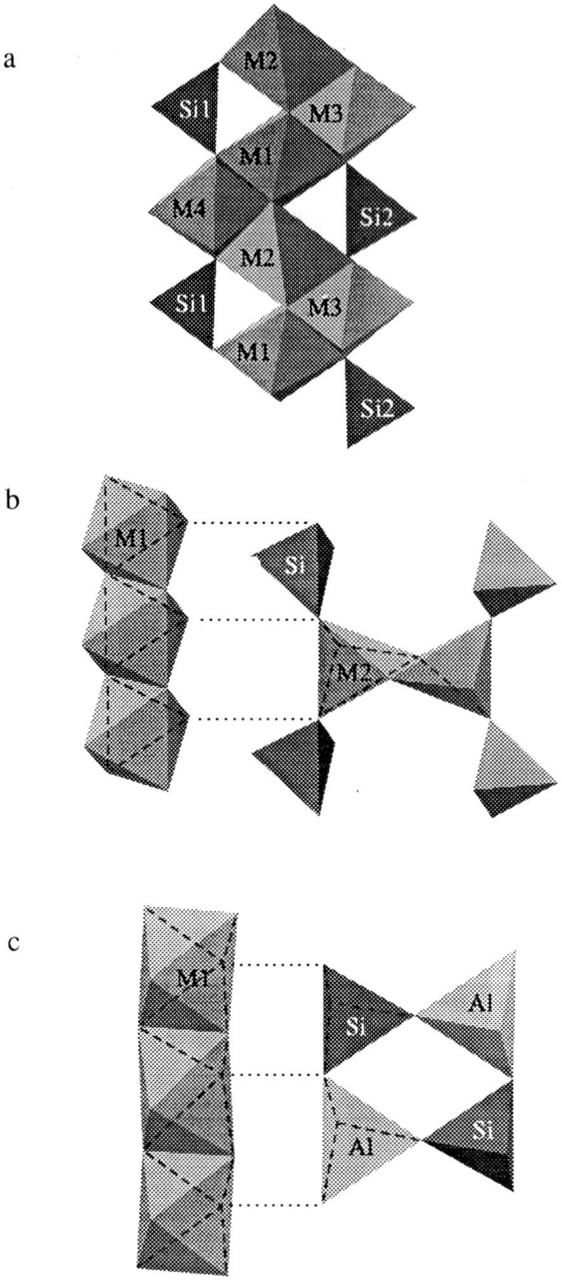

Figure 6-4. (a) In kyanite the *M*1 and *M*2 octahedra form edge-sharing chains parallel to *c*, with *M*3 and *M*4 attached to the sides of the chains and the tetrahedra connected as shown. (After [44].) (b) Dotted lines show how the double chains of Si–*M*2 sites connect to the edge-sharing *M*1 chains, which run along *c*. (After [12].) (c) The double tetrahedral chain in sillimanite articulates to the edge-sharing *M*1 chain as shown by the dotted lines; all chains are parallel to *c*. (After [9].)

appropriate to metamorphism. Transmission electron microscopy (e.g., [17]) has revealed at least two types of stacking faults in sillimanite, and the apparent non-correlation of other characteristics with fibrolitic properties has motivated the proposal that these structural defects are responsible not only for the difference in habit, but also for the anomalous thermodynamic behavior by which experimental determination of the Al_2SiO_5 phase diagram has been plagued (see the following). However, the densities of dislocations typical of the Al_2SiO_5 polymorphs and the energies associated therewith do not lend strong support to this argument either [26].

Experiments carried out with Mössbauer and optical absorption spectroscopy on natural sillimanites containing about 1.5 wt % Fe_2O_3 indicate that both Fe^{2+} and Fe^{3+} are located essentially entirely in the octahedral Al site [24].

Figure 6-4 shows polyhedral linkages for all three polymorphs.

The Effects of Elevated Pressure and Temperature on Crystal Structure

The crystal structures of the aluminum silicate polymorphs have been refined at high temperatures [44] and that of andalusite has also been refined at pressures as high as 37 kbar [33]. Figures 6-5 through 6-8 show how mean bond lengths of the various polyhedra in these minerals are effected by changes in temperature and pressure, and the observed trends are not unexpected: In general, mean bond lengths (that is, polyhedral sizes) tend to increase with temperature and decrease with pressure. It is clear from the slopes of the trends, however, that all polyhedra

Figure 6-5. Changes in mean polyhedral bond lengths with temperature for kyanite. Note that the vertical scale for the octahedral sites is double that for the tetrahedral sites, so the octahedra experience very much more thermal expansion than do the tetrahedra.

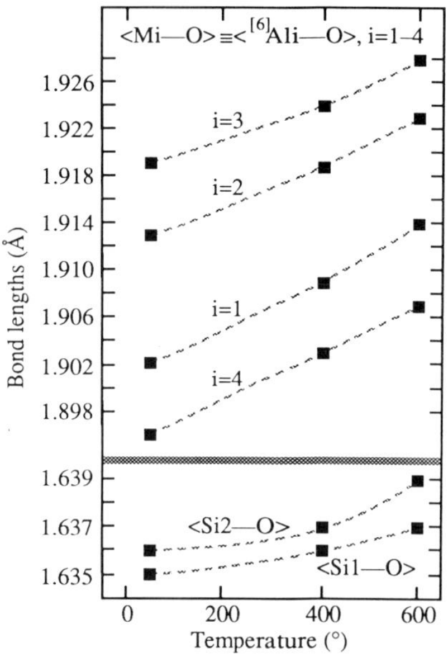

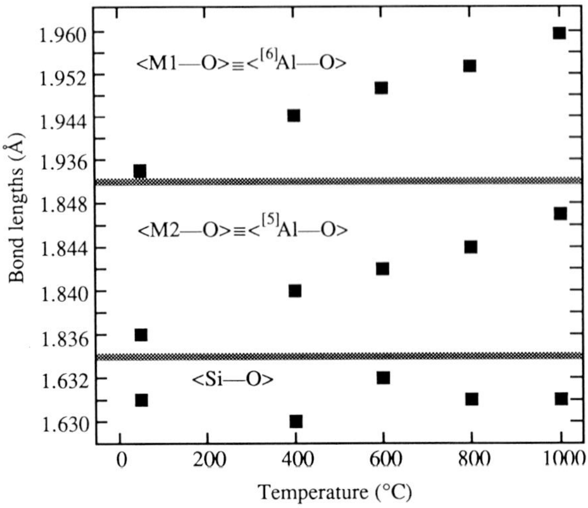

Figure 6-6. Changes in mean polyhedral bond lengths with temperature for andalusite. Note that the vertical scale for the octahedral site is double that for the five-coordinate site, which is in turn double that for the tetrahedral site. Thermal expansion thus increases with coordination number.

are not affected to the same degree. The octahedra of all of the polymorphs have greater thermal expansions than does the trigonal bipyramid of andalusite, which in turn shows greater thermal expansion than any of the tetrahedra. In Figure 6-8 we see that polyhedra respond similarly to pressure, in the sense that the largest ones (i.e., the one with the highest coordination numbers) are most severely effected by increasing pressure: the bulk modulus increases in going from six to five to fourfold coordination in andalusite. These observations may be generalized to other minerals as well—large polyhedra respond to changes in temperature and pressure more than do small ones.

Phase Transitions

The *P–T* phase diagram for the Al_2SiO_5 polymorphs will be discussed in detail later. At this point, however, it has been shown that the structural transformation from kyanite to andalusite to sillimanite involves a change in aluminum coordination that can be represented by

$$^{[6]}Al^{[6]}Al^{[4]}SiO_5 \rightarrow {}^{[6]}Al^{[5]}Al^{[4]}SiO_5 \rightarrow {}^{[6]}Al^{[4]}Al^{[4]}SiO_5.$$

Because the coordination number of the second aluminum atom changes, atomic bonds must be re-formed and the transitions must all be reconstructive. These kinds of changes require a considerable input of energy to effect, and the minerals may exist metastably far outside their thermodynamic stability fields simply because insufficient energy is supplied to generate the transitions. Such was found to be the case with the silica polymorphs related by reconstructive transformations

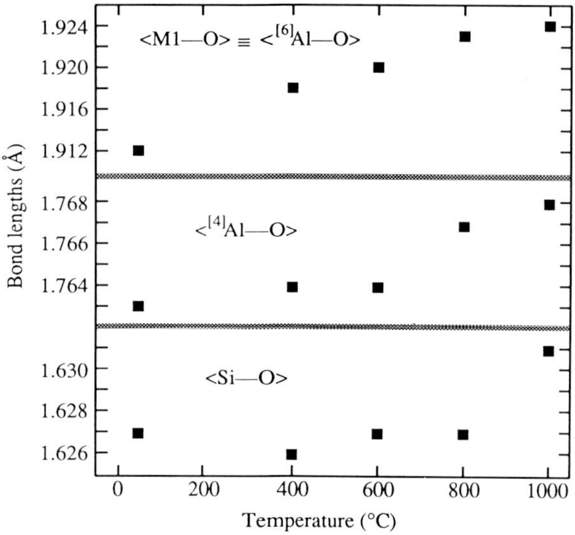

Figure 6-7. Changes in mean polyhedral bond lengths with temperature for sillimanite. Note that the vertical scale for the octahedral site is double that for either tetrahedral site, so the octahedron experiences very much more thermal expansion than do the tetrahedra.

Figure 6-8. The effect of pressure on mean bond lengths for andalusite. Here the vertical scales are the same for all polyhedra. It is apparent that compressibility decreases with coordination number (i.e., bulk modulus increases).

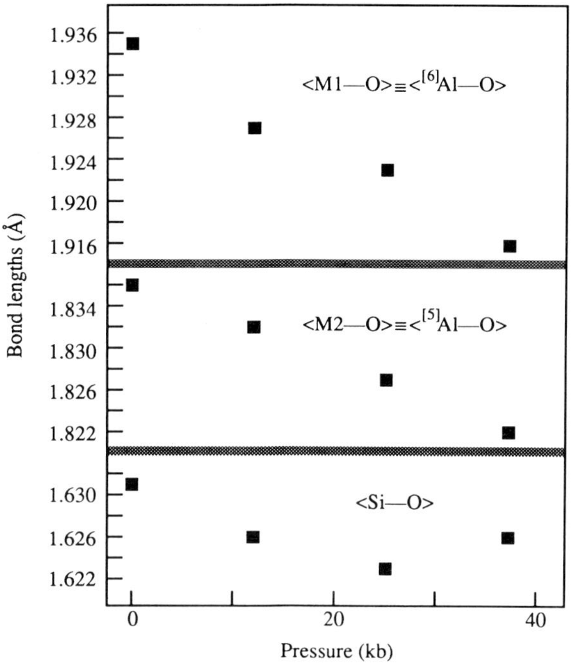

(Chapter 1). Unlike the SiO_2 system, however, no displacive transformations are known to exist in the Al_2SiO_5 system.

Mullite

Mullite is not one of the Al_2SiO_5 polymorphs, but the similarity of its crystal structure to those of sillimanite and andalusite and the fact that it can be produced by raising these two minerals to high temperature (sillimanite → mullite ± quartz; andalusite → mullite + quartz), prompts a brief consideration of it. The compositional range for mullite is generally taken to be $3Al_2O_3 \cdot 2SiO_2$ to $2Al_2O_3 \cdot SiO_2$ [13], whereas Al_2SiO_5 has a 1:1 ratio of Al_2O_3 to SiO_2. Cast in a different way, the (doubled) formula is $Al_2(Al_{2+2x}Si_{2-2x})O_{10-x}$, with x between 0.25 and 0.40, and apparently concentrated around these values for natural specimens. There is no obvious chemical reason that complete solid solution could not exist between Al_2SiO_5 and mullite, but there may be structural reasons. So far no mullites have been reported to have compositions between 3:2 and 1:1 Al_2O_3:SiO_2.

The *average* crystal structures of a 3:2, a 1.92:1, and two 2:1 mullites have been refined in space group *Pbam* ([36];[11];[37]; and [5], respectively) and the average atomic arrangement is shown in Figure 6-9. It is at once obvious that the average mullite structure is similar to that of sillimanite, and has the appearance of somehow being intermediate between the structures of sillimanite and andalusite. The word *average* is stressed here because x-ray-diffraction experiments on mullite reveal both satellite (superstructure) reflections and diffuse streaks, besides the normal sharp diffraction maxima, indicating the presence of some sort of domain structure or modulated (incommensurate) structure (both of which, it will be recalled, were discussed in previous chapters—see particularly Chapter 2 on plagioclase feldspars).

In sillimanite there is an ordered distribution of Al and Si atoms. To envision obtaining the average mullite structure from the sillimanite arrangement, first imagine that the Al–Si distribution in sillimanite becomes completely disordered. (See Figure 6-9.) Next some of the silicon must be replaced by aluminum to obtain the correct Al_2O_3:SiO_2 ratio, and some oxygen atoms must be lost to maintain charge balance. These oxygen atoms are lost from the O_c position; the fraction of vacancies at O_c is $x/2$ in terms of the general formula $Al_2(Al_{2+2x}Si_{2-2x})O_{10-x}$, and thus ranges from 0.125 to 0.20. Removal of 12.5–20% of the O_c atoms means that the same number of Al–Si atoms lose a coordinating oxygen atom. This is a very unstable arrangement, and so these 12.5–20% of (formerly) tetrahedral atoms move to sites designated Al_2^*; these are tetrahedral sites present in sillimanite but not occupied. As suggested by the site name, Al_2^* is occupied almost solely by aluminum. The occupancy of 12.5–20% of the Al_2^* sites requires the adjacent 12.5–20% of the O_c atoms to move off the center of symmetry at (0,1/2,0) to the O_c^* position. We now have the topology of the average structure of mullite. (The new tetrahedral site is designated as Al* or as T^* by some authors.)

The modulated structure consists of two types of ordering patterns with different symmetries, and is presumably stabilized by their presence [6]. One is the ordering scheme described previously for the average structure. The other constitutes a ten-

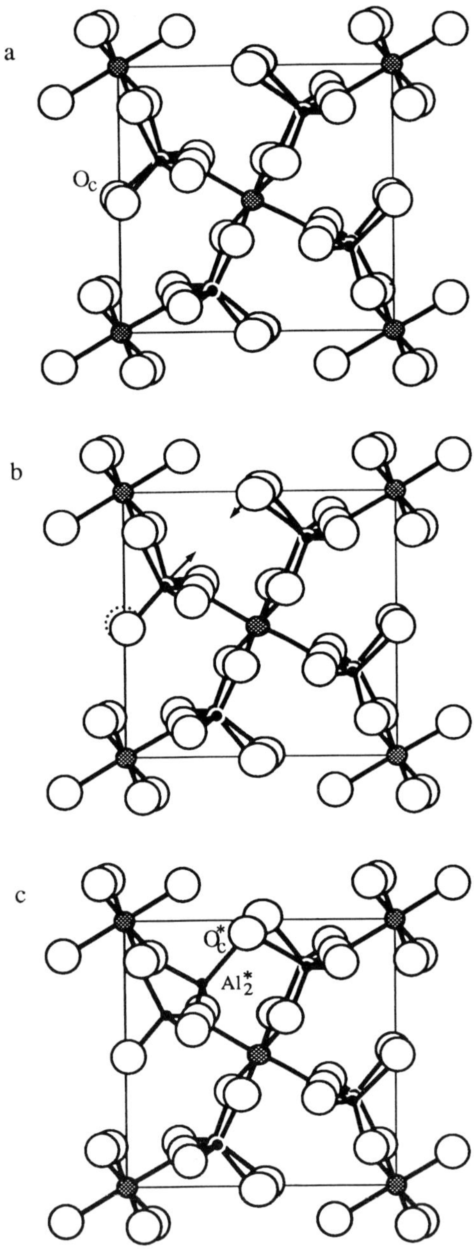

Figure 6-9. (a) The structure of sillimanite (compare Figure 6-3), with *a* directed vertically downward and *b* horizontally to the right. An oxygen O_c that will subsequently be lost for this example is marked. (b) The structural transformation that leads to mullite. An O_c is lost, leaving a cation three-coordinated. The cation moves (long arrow) to a previously unoccupied tetrahedral site, Al_2^*, and an oxygen is pulled off the O_c position at $(0,\frac{1}{2},0)$ to O_c^* (short arrow). (c) The average structure of mullite, showing only one Al_2^* site in this particular unit cell with the rest maintaining essentially the sillimanite topology. A large number of unit cells would include some with alternative Al_2^* sites and many with none, to yield the average *Pbam* structure with about 12.5–20% O_c sites vacant and a like number of tetrahedral Al–Si sites replaced by Al_2^* sites.

dency towards an ordering of Al and Si similar to that found in sillimanite. These two ordering schemes are displaced from one another by $\lambda/4$ along the modulation wave through the structure.

A possible scenario for the transformation of sillimanite to mullite + quartz is as follows [37]: As sillimanite or andalusite is heated, thermal vibration increases to the point at which some atoms may vacate their positions. The octahedral chains seem to be too solidly constructed to allow this, but some of the tetrahedrally coordinated atoms may slip over into the Al_2^* sites, thereby making O_c^* three-coordinated. If a silicon atom does so, then the Pauling electrostatic bond strength received by O_c will be 2.50 to 3.00, which is fairly high and not conducive to stability. If an aluminum atom shifts to Al_2^*, the electrostatic bond strength at O_c will be 2.25 to 2.75, which is not as serious an overbalance. Presumably, then, as silicon moves to Al_2^* and imposes an excessive electrostatic bond strength on O_c, it and corresponding oxygens are exsolved to form silica.

THE Al_2SiO_5 PHASE DIAGRAM

The Al_2SiO_5 polymorphs are metamorphic minerals that occur in regional metamorphic terrains and, less commonly, in contact metamorphic rocks. Nearly all undergraduate geology students encounter the use of these minerals early in their studies, because they are frequently employed to differentiate between various levels of intensity of metamorphism. For example in George Barrow's classic study of the past century [7], two of the mapped isograds were based on the appearance of kyanite and of sillimanite. In other work (e.g., [28]), two pressure regimes have been recognized in some areas: a kyanite–sillimanite (medium-pressure) baric type, and a kyanite–andalusite (low-pressure) baric type. Clearly it is of considerable interest to both mineralogists and petrologists to determine the pressures and temperatures at which the Al_2SiO_5 phase transformations occur. This determination turns out to be no trivial task.

Figure 6-10 shows, superimposed on P–T space, several of the univariant boundaries and invariant points that have been proposed for the Al_2SiO_5 system as a result of high-pressure–high-temperature phase equilibrium studies or solubility measurements. (See Chapter 12 for a review of the methods for determining phase diagrams and the language of equilibrium thermodynamics.) It is clear that there is general agreement about the relative placement of the equilibrium phase fields for the three minerals, but considerable disparity in the exact placement of the boundaries. The reasons for the lack of agreement among the results of various experimentalists fall approximately into three categories: Deficiencies (or at least variations) in experimental procedure, variations in minor element chemistry of the specimens used, and differences in structural defects in the specimens. For example, it has been shown that the reactivity of severely ground mixtures of kyanite and sillimanite is substantially decreased relative to lightly ground mixtures of the same starting materials, presumably because kyanite is preferentially degraded in the mixture [29]; the result is to raise the pressure of the apparent stability limit of sil-

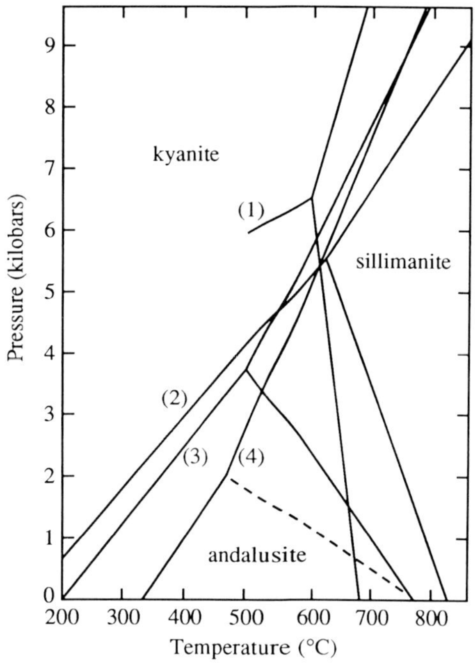

Figure 6-10. Four versions of the aluminum silicate phase diagram. Phase boundaries (1) through (4) are taken from [3], [35], [25], and [8], respectively. The first three were determined by reversal of phase equilibrium at high temperature and high pressure; the fourth was determined by measurement of solubilities of andalusite and kyanite in water.

limanite relative to kyanite, and may explain the high pressure of kyanite–sillimanite boundary number (1) in Figure 6-10. All differences cannot be ascribed to experimental technique, however. A side-by-side run of starting materials use to obtain boundaries (1) and (2), carried out in the same experimental vessel at 750°C and 10 kbar resulted in one kyanite–sillimanite mixture growing kyanite and the other growing sillimanite. This implies that the starting materials themselves were somehow different, and indeed one contained sillimanite with 0.45 wt % Fe_2O_3 and the other contained sillimanite with only 0.03 wt % Fe_2O_3. Likewise, comparison of the starting materials used for boundaries (2) and (3) reveal that the sillimanite used for the former contained much more fibrolite than did the other, and large differences exist in the thermodynamic properties of blocky sillimanite and fibrolite [38]. The cause(s) of these differences is not well understood, but may include variations in minor element chemistry (e.g., [4;27]), may [45] or may not [19] include Al–Si disorder in fibrolite, and may [38] or may not [26] include structural defects in fibrolite.

There have been determinations of the phase diagram from other approaches than traditional experimental mineralogy. For instance, Figure 6-11 shows two sets of boundaries that were determined from (1) measurement of phonon spectra (lat-

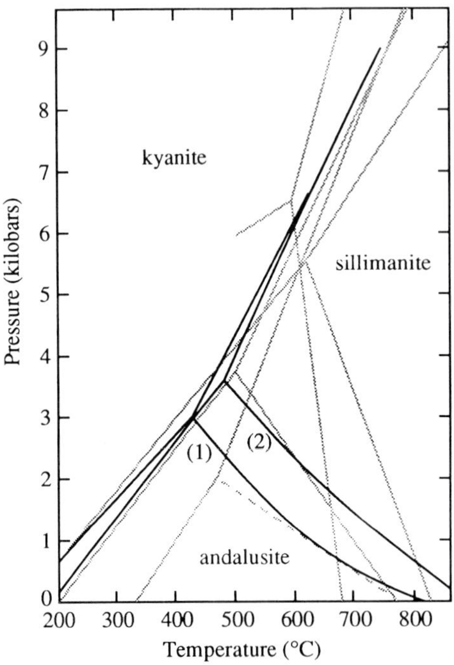

Figure 6-11. The Al_2SiO_5 phase diagram as determined by (1) measurements of lattice vibrations [38], and (2) thermodynamic analysis of experimental data [16]. The boundaries of Figure 6-10 are shown in gray for reference.

tice vibrations) and (2) thermodynamic analysis of experimental data. These appear to agree most closely with the boundary number (3) on Figure 6-10.

What does this mean about the aluminum silicate phase diagram? Is the "best" diagram the one determined using the purest starting materials and only blocky (nonfibrolitic) sillimanite? That depends upon what one wishes to know. Fibrolitic sillimanite occurs in nature, and so do minor element "impurities," and, if these have an effect on laboratory experiments, they will likewise influence the behavior of the Al_2SiO_5 polymorphs in the rocks geologists study in the field. There is both experimental [4] and field [27] evidence that some factor or combination of factors—structural defects and impurities being the most likely candidates—makes the boundaries in the aluminum silicate phase diagram divariant rather than univariant (that is, they are bands of finite width rather than sharp lines). Further, misleading results may be obtained when using sillimanite formed at one temperature to test equilibrium at another temperature [19]. Of greatest use would be a phase diagram determined for pure phases with no structural defects (or Al–Si disorder, if such occurs), together with a quantitative determination of the effects of all factors that perturb the diagram. Boundaries (1) and (2) in Figure 6-11 and (3) in Figure 6-10 may come close to the ideal phase diagram, though there is no certainty about which is closest, but the calibration of perturbing effects is yet a long way off.

REFERENCES

[1] Abs-Wurmbach, I., K. Langer, F. Seifert, and E. Tillmanns (1981). "The Crystal Chemistry of (Mn^{3+},Fe^{3+})-Substituted Andalusites (Viridines and Kanonaite), $(Al_{1-x-y}Mn^{3+}{}_xFe^{3+}{}_y)_2(O|SiO_4)$: Crystal Structure Refinements, Mössbauer, and Polarized Absorption Spectra." *Zeitschrift für Kristallographie* 155, 81–113.

[2] Albee, A. L., and A. A. Chodos (1969). "Minor Element Content of Coexistent Al_2SiO_5 Polymorphs." *American Journal of Science* 267, 310–16.

[3] Althaus, E. (1967). "The Triple Point Andalusite–Sillimanite–Kyanite." *Contributions to Mineralogy and Petrology* 16, 29–44.

[4] Althaus, E. (1969). "Experimental Evidence that the Reaction of Kyanite to Form Sillimanite Is at Least Bivariant." *American Journal of Science* 267, 273–77.

[5] Angel, R. J., and C. T. Prewitt (1986). "Crystal Structure of Mullite: A Re-Examination of the Average Structure." *American Mineralogist* 71, 1476–82.

[6] Angel, R. J., and C. T. Prewitt (1987). "The Incommensurate Structure of Mullite by Patterson Synthesis." *Acta Crystallographica* B43, 116–26.

[7] Barrow, G. (1893). "On an Intrusion of Muscovite-Biotite Gneiss in the South-East Highlands of Scotland." *Geological Society of London Quarterly Journal* 49, 330–58.

[8] Brown, G. C., and W. S. Fyfe (1971). "Kyanite–Andalusite Equilibrium." *Contributions to Mineralogy and Petrology* 33, 227–31.

[9] Burnham, C. W. (1963a). "Refinement of the Crystal Structure of Sillimanite." *Zeitschrift für Kristallographie* 118, 127–48.

[10] Burnham, C. W. (1963b). "Refinement of the Crystal Structure of Kyanite." *Zeitschrift für Kristallographie* 118, 337–60.

[11] Burnham, C. W. (1964). "Crystal Structure of Mullite." *Carnegie Institute of Washington Year Book* 63, 223–27.

[12] Burnham, C. W., and M. J. Buerger (1961). "Refinement of the Crystal Structure of Andalusite." *Zeitschrift für Kristallographie* 115, 269–90.

[13] Cameron, W. E. (1977). "Mullite: A Substituted Alumina." *American Mineralogist* 62, 747–55.

[14] Carlson, W. D., and G. R. Rossman (1988). "Vanadium- and Chromium-Bearing Andalusite: Occurrence and Optical-Absorption Dpectroscopy." *American Mineralogist* 73, 1366–69.

[15] Chinner, G. A., J. V. Smith, and C. R. Knowles (1969). "Transition-metal Contents of Al_2SiO_5 Polymorphs." *American Journal of Science* 267-A, 96–113.

[16] Day, H. W., and H. J. Kumin (1980). "Thermodynamic Analysis of the Aluminum Silicate Triple Point." *American Journal of Science* 280, 265–87.

[17] Doukhan, J. C., N. Doukhan, P. S. Koch, and J. M. Christie (1985). "Transmission Electron Microscopy Investigation of Lattice Defects in Al_2SiO_5 Polymorphs and Plasticity Induced Polymorphic Transformations." *Bulletin de Minéralogie* 108, 81–96.

[18] Finger, L. W., and E. Prince (1972). "Neutron Diffraction Studies: Andalusite and Sillimanite." *Carnegie Institution of Washington Year Book* 71, 496–500.

[19] Greenwood, H. J. (1972). "Al^{IV}-Si^{IV} Disorder in Sillimanite and Its Effect on Phase Relations of the Aluminum Silicate Minerals." *Geological Society of American Memoir* 132, 553–71.

[20] Grew, E. S., and Hinthorne, J. R. (1983). "Boron in Sillimanite." *Science* 221, 547–49.

[21] Grew, E. S., and G. R. Rossman (1985). "Co-Ordination of Boron in Sillimanite." *Mineralogical Magazine* 49, 132–35.

[22] Gunter, M., and F. D. Bloss (1982). "Andalusite–Kanonaite Series: Lattice and Optical Parameters." *American Mineralogist* 67, 1218–28.

[23] Hålenius, U. (1978). "A Spectroscopic Investigation of Manganian Andalusite." *Canadian Mineralogist* 16, 567–75.

[24] Hålenius, U. (1979). "State and Location of Iron in Sillimanite." *Neues Jahrbuch für Mineralogie Monatshefte*, 165–74.

[25] Holdaway, M. J. (1971). "Stability of Andalusite and the Aluminum Silicate Phase Diagram." *American Journal of Science* 271, 97–131.

[26] Kerrick, D. M. (1986). "Dislocation Strain Energy in the Al_2SiO_5 Polymorphs." *Physics and Chemistry of Minerals* 13, 221–26.

[27] Kerrick, D. M., and J. A. Speer (1988). "The Role of Minor Element Solid Solution on the Andalusite–Sillimanite Equilibrium in Metapelites and Peraluminous Granitoids." *American Journal of Science* 288, 152–92.

[28] Miyashiro, A. (1961). "Evolution of Metamorphic Belts." *Journal of Petrology* 2, 277–311.

[29] Newton, R. C. (1969). "Some High-Pressure Hydrothermal Experiments on Severely Ground Kyanite and Sillimanite." *American Journal of Science* 267, 278–84.

[30] Okrusch, M., and B. W. Evans (1970). "Minor Element Relationships in Coexisting Andalusite and Sillimanite." *Lithos* 3, 261–68.

[31] Parkin, K. M., B. M. Loeffler, and R. G. Burns (1977). "Mössbauer Spectra of Kyanite, Aquamarine, and Cordierite Showing Intervalence Charge Transfer." *Physics and Chemistry of Minerals* 1, 301–11.

[32] Peterson, R. C., and R. K McMullan (1986). "Neutron Diffraction Studies of Sillimanite." *American Mineralogist* 71, 742–45.

[33] Ralph, R. L., L. W. Finger, R. M. Hazen, and S. Ghose (1984). "Compressibility and Crystal Structure of Andalusite at High Pressure." *American Mineralogist* 69, 513–19.

[34] Ribbe, P. H. (1980). "Aluminum Silicate Polymorphs and Mullite." In P. H. Ribbe, (ed.), *Orthosilicates. Reviews in Mineralogy*, Volume 5. Mineralogical Society of America, Washington, D.C.

[35] Richardson, S. W., M. C. Gilbert, and P. M. Bell (1969). "Experimental Determination of Kyanite–Andalusite and Andalusite–Sillimanite Equilibria; the Aluminum Silicate Triple Point." *American Journal of Science* 267, 259–72.

[36] Saalfeld, H., and W. Guse (1981). "Structure Refinement of 3:2-Mullite ($3Al_2O_3 \cdot 2SiO_2$)." *Neues Jahrbuch für Mineralogie Monatshefte*, 145–50.

[37] Sadanaga, R., M. Tokonami, and Y. Takéuchi (1962). "The Structure of Mullite, $2Al_2O_3 \cdot SiO_2$, and Relationship with the Structures of Sillimanite and Andalusite." *Acta Crystallographica* 15, 65–68.

[38] Salje E. (1986). "Heat Capacities and Entropies of Andalusite and Sillimanite: The Influence of Fibrolitization on the Phase Diagram of the Al_2SiO_5 Polymorphs." *American Mineralogist* 71, 1366–71.

[39] Salje, E., and C. Werneke (1982). "The Phase Equilibrium Between Sillimanite and Andalusite as Determined from Lattice Vibrations." *Contributions to Mineralogy and Petrology* 79, 56–67.

[40] Shannon, R. D. (1976). "Revised Effective Ionic Radii and Systematic Studies of Interatomic Distances in Halides and Chalcogenides." *Acta Crystallographica* A32, 751–767.

[41] Sobolev, N. V., Jr., I. K. Kuznetsova, and N. I. Zyuzin (1968). "The Petrology of Grospydite Xenoliths from the Zagodochnaya Kimberlite Pipe in Yakutia." *Journal of Petrology* 9, 353–80.

[42] Vrána, S., M. Rieder, and J. Podlaha (1978). "Kanonaite $(Mn_{0.76}Al_{0.23}Fe_{0.01})^{[6]}Al^{[5]}[O|SiO_4]$, a New Mineral Isotypic with Andalusite." *Contributions to Mineralogy and Petrology* 66, 325–32.

[43] Weiss, Z., S. W. Bailey, and M. Rieder (1981). "Refinement of the Crystal Structure of Kanonaite, $(Mn^{3+},Al)^{[6]}(AlMn^{3+})^{[5]}O[SiO_4]$." *American Mineralogist* 66, 561–67.

[44] Winter, J. K., and S. Ghose (1979). "Thermal Expansion and High-Temperature Crystal Chemistry of the Al_2SiO_5 Polymorphs." *American Mineralogist* 64, 573–86.

[45] Zen, E. (1969). "The Stability Relations of the Polymorphs of Aluminum Silicate: A Survey and Some Comments." *American Journal of Science* 267, 297–309.

7
Olivines

GENERAL CHEMISTRY

The general formula for minerals of the olivine family is $XYTO_4$, where the X and Y cations are in octahedral coordination in two crystallographically distinct sites, $M2$ and $M1$, respectively. (Some authors place the X cation in $M1$ and Y in $M2$. In order to maintain consistency with minerals already covered in this text in which sizes of atoms decrease in the order W-X-Y-Z, we shall use the cation placement stated in the first sentence.) The chemical compositions of natural silicate olivines are defined essentially by a tetrahedron whose corners represent Mg_2SiO_4 (forsterite, Fo), Fe_2SiO_4 (fayalite, Fa), Mn_2SiO_4 (tephroite,Te), and Ca_2SiO_4. The last occurs very rarely in nature with the olivine structure [8], and the more common nonolivine structure is called *larnite;* the polymorphism also includes a high-temperature spinel phase and one or two other phases stable at intermediate temperatures. The Greek prefixes are somewhat confusing because they are not used consistently in the literature. The convention is for α to be the prefix for the lowest temperature phase, β for the next-higher-temperature phase, and so forth. The calcio-olivine is the low temperature phase, but is called γ-Ca_2SiO_4 in most mineralogical literature simply because of precedent. The conventional sequence of prefixes is often used in geophysical literature (that is, the high-temperature spinel phase is called γ-Ca_2SiO_4, larnite the β phase, and calcio-olivine the α phase) [9].

Figure 7-1 shows compositional relationships, with shading indicating experimental solid solution ranges at low pressures. The experimental ranges exceed those observed in natural specimens, although it is quite possible that examination of specimens from an ever-wider variety of geologic and chemical environments will push the observed limits closer to the experimental limits. For instance, observed compositions of ternary olivines that would plot on the Fo–Fa–Te face of the tetrahedron suggest a complete solid solution among these three end members [35]. The figure also shows the locations of two solvi—one between forsterite and monticellite [1], and the other between fayalite and kirchsteinite [36]. Note that the empirical rule often taught to introductory students of mineralogy, that solid solution is permitted only if the radii of the substituting atoms differ by 15% or less,

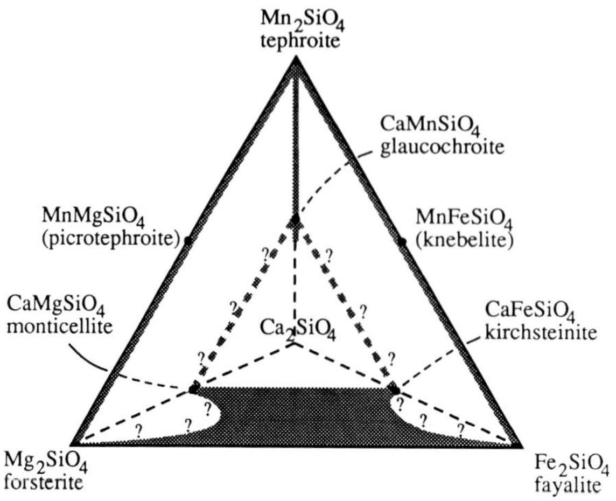

Figure 7-1. A compositional tetrahedron that includes the most common members of the olivine family. Ca_2SiO_4 occurs extremely rarely in the olivine polymorph. *Knebelite* and *picrotephroite* are parenthesized to indicate that these are not preferred mineral names, although they are found in the mineralogical literature. The terms *ferroan tephroite*, *manganiferous fayalite*, *magnesian tephroite*, and *manganiferous forsterite* are preferred. Gray lines indicate known solid solutions, with the dashed ones probable but unconfirmed. Miscibility gaps exist between forsterite and monticellite, and between fayalite and kirchsteinite, but the extent of solid solution within the $(Mg,Fe,Ca)_2SiO_4$ quadrilateral is not known.

applies here; solvi are found only on joins between olivines that contain *X* and *Y* cations differing in radius by more than this amount.

The nomenclature of olivines is complicated by the persistence of varietal names that used to be considered acceptable but are now discouraged. For example, the names *picrotephroite* and *knebelite* were formerly used for $MnMgSiO_4$ and $MnFeSiO_4$, respectively. These, however, are simply intermediate members of a complete solid solution, and the use of the names *tephroite*, *forsterite*, and *fayalite* with the modifiers *manganoan*, *ferroan*, and *magnesian* is now preferred. Similarly the forsterite–fayalite series has, in the past, been divided into forsterite (Fo_{100-90}), chrysolite (Fo_{90-70}), hyalosiderite (Fo_{70-50}), hortonolite (Fo_{50-30}), ferrohortonolite (Fo_{30-10}), and fayalite (Fo_{10-0}), reminiscent of the way in which plagioclases are still sometimes divided. The use of these varietal names is now discouraged.

Chemical components not represented on Figure 7-1 occur to a lesser extent. Olivines that contain significant concentrations of zinc are known, although Zn_2SiO_4 itself is willemite, a hexagonal mineral with a nonolivine crystal structure; at high temperature and atmospheric pressure, concentrations up to about 17 mol % Zn_2SiO_4 in fayalite [13]) can be attained experimentally, and up to 24 mol % Zn_2SiO_4 can be accommodated in forsterite [42]. (Much greater solubilities are attainable at high pressure [3].) Ni, Cr, and a few other elements are sometimes present in very small amounts; in fact, a Ni-containing forsterite, $(Ni,Mg)_2SiO_4$, has been named *liebenberite*. Nonetheless, nearly all silicate olivines encountered in nature are essentially binary solid solutions of the compositions labeled on Figure 7-1, and contain no more than 10 mol % of any other component.

Many synthetic silicate olivines have been produced, including $MgNiSiO_2$, Ni_2SiO_4, Co_2SiO_4, $MgCoSiO_4$, and others. A wide variety of nonsilicates, both natural and synthetic, also crystallize with the olivine structure. Among the mineralogical examples are chrysoberyl (Al_2BeO_4), sinhalite ($AlMgBO_4$), and triphyllite ($LiFePO_4$).

The olivines generally encountered in common rocks are members of the forsterite–fayalite solid solution series, and, in fact, the mineral name *olivine* applies strictly only to members of that series. (In this chapter, however, we shall use the term *olivine* to refer to any member of the family of silicate minerals having the olivine general formula and atomic arrangement.) Ultramafic rocks usually contain olivines with compositions in the range $Fo_{88}Fa_{12}$ to $Fo_{92}Fa_8$ (where Fo stands for mole fraction forsterite, and Fa for mole fraction fayalite). Those found in basaltic rocks range between $Fo_{40}Fa_{60}$ and $Fo_{85}Fa_{15}$. Magmatic fayalite-rich olivines occur in syenites and some silicic volcanic rocks. Fe,Mg olivines are known from contact and regionally metamorphosed rocks of appropriate compositions as well, and it is thought that the upper mantle consists largely of forsteritic olivine. The less common Ca-bearing and Mn-bearing species occur in contact metamorphosed and metasomatized rocks of the proper bulk chemistry, and these often contain Zn as well.

Fayalite rich in ferric iron ("ferrifayalite") has been reported, but has now been found to consist of a mixture of fayalite and laihunite [43]. Laihunite is a fayalite-like mineral with the ideal composition $Fe^{2+}Fe_2^{3+}Si_2O_8$ [25], and a crystal structure like a distorted olivine with every other *M*1 site vacant [18;49]. Note that the ideal formula is a doubled olivine formula with half the required number of *X* cations. It is monoclinic, and occurs in at least three polytypes [26].

THE CRYSTAL STRUCTURE OF OLIVINE

The topology of the anion array in olivine is that of a hexagonal-close-packed arrangement. The mineral does not have hexagonal symmetry, however, for the following reasons: First, the placement of cations within the octahedral and tetrahedral voids is not consistent with hexagonal symmetry; second, this nonhexagonal arrangement of cations introduces distortions in the oxygen array, principally in the form of shortened shared edges, which render it neither truly close packed nor hexagonal; third, given the sizes of the octahedral cations in olivines, the Si–O bond length in a close-packed array would have to be considerably (and unacceptably) longer than ~1.64 Å, the mean value typical of orthosilicates; so structural distortions are required to optimize that distance. The symmetry that results is orthorhombic, and the space group is *Pbnm* (equivalent to the conventional space group *Pnma*, but with a different crystallographic orientation).

The deviations of the oxygen atoms from a close-packed array are not large, and so the structure is topologically identical to a close-packed arrangement, with the same numbers and kinds of sites available for cations. Examination of a model or a drawing of a hexagonal- or cubic-close-packed array of spheres shows that there are two tetrahedral voids and one octahedral void per sphere. (That is, each sphere, or anion, is surrounded by eight tetrahedral voids, of which it "owns" one-fourth of each, and by six octahedral voids, of which its share is one-sixth of each.) Referring to the generalized olivine chemical formula, $XYTO_4$, we see that there are four anions per formula unit. Because there is one octahedral void per anion, and there are two octahedral cations in the formula unit, one-half of the available octahedral sites must be filled. Similarly, because there are eight tetrahedral voids per formula unit (that is, two per oxygen atom), and one tetrahedrally coordinated cation in the formula, one-eighth of the available tetrahedral voids are filled.

The *average* coordination number of all oxygen atoms in a mineral can be computed with the following formula:

$$\Sigma N_C n_C = \langle N_O \rangle n_O,$$

where N_C is the coordination number of a given cation, n_C is is the number of atoms of that cation in the chemical formula, the summation is over all cations, $\langle N_O \rangle$ is the *mean* coordination number of oxygen, and n_O is the number of oxygens in the chemical formula. For olivines, the calculation reveals that the mean coordination number of oxygen is four. Crystal-structure determinations have shown that there are three crystallographically distinct oxygen atoms, each coordinated by one silicon and three octahedral cations.

Although olivines are classified as orthosilicates, their major structural feature is not the silicate tetrahedron, but the chain of edge-sharing octahedra containing the X and Y cations. This chain, extending parallel to the **c** direction, is shown in Figure 7-2 as an ideal close-packed arrangement. Here the dark shaded octahedra form bands at fractional height 0.0 or 1.0 along **a**, and the light shaded octahedral band is at height 0.5. The straight central chain of each band comprises $M1$ sites, while $M2$ sites alternate on either side, forming a serrated chain.

Figure 7-3 shows projections onto (100) of the actual olivine structure, with space-group symmetry elements in the same orientation. The b glide parallel to (100) at $a/4$ relates the octahedral chain shown at $x = 0.5$ to one not shown, at $x = 0.0$. Consequently each octahedral band in one layer (either $x = 0.0$ or 0.5) is flanked on either side by a band of the same shape that contains no octahedral cations, and also by such a band above and below along $\pm a$. It is in these bands with vacant octahedra that the tetrahedra are found, two per c translation in each band, one pointing along $+\mathbf{a}$, and one along $-\mathbf{a}$. This is represented in a very schematic fashion in Figure 7-4. Each tetrahedron shares its three basal edges with one octahedral band, and its apex with another. Figure 7-3 shows that all cation sites lie on special positions. The silicon site and $M2$ are on the mirror plane at $c/4$, and $M1$ is located at a center of symmetry. Thus the site symmetry for Si is m (but very nearly $3mm$); for $M1$ it is $\bar{1}$, and for $M2$ it is m. (See, for example, [6] or [54] for typical structure refinements of ferromagnesian olivines.)

Note the very different kinds of distortions experienced by $M1$ and $M2$ owing to the sharing of different combinations of polyhedral edges with each other and with the silicate tetrahedra. For ferromagnesian olivines and others in which the X and Y atoms are very close in size, the $M1$ site is the quantitatively more distorted of

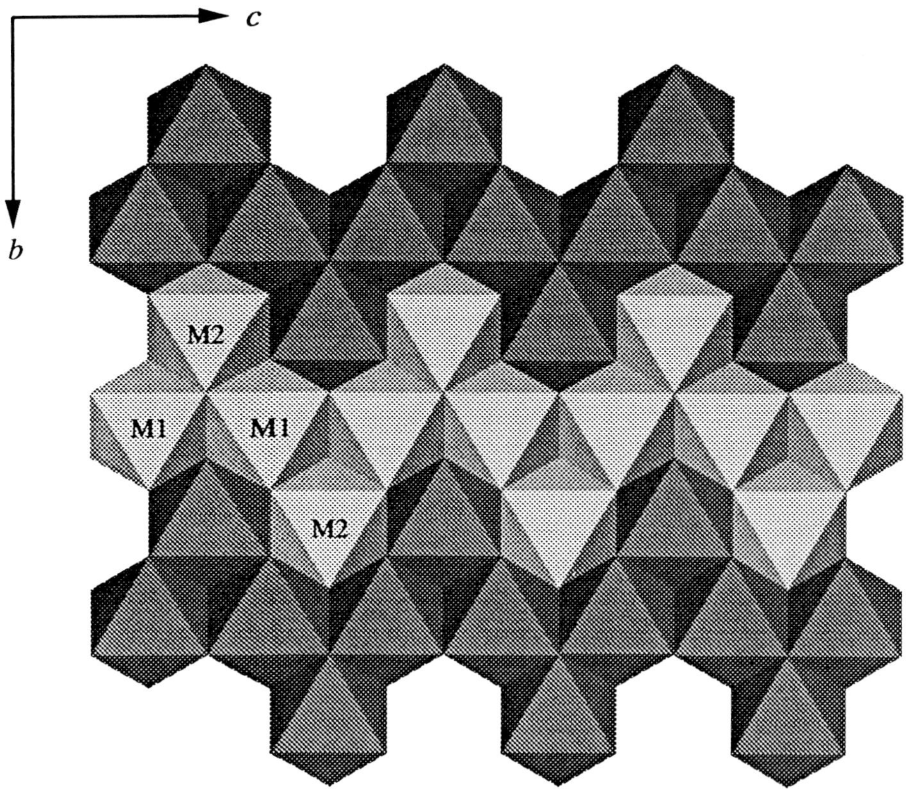

Figure 7-2. The octahedral chains in olivine, drawn as if the mineral were ideally close packed. Identities of $M1$ and $M2$ are indicated. The darker octahedral chains are at height $x = 0$ and the lighter chain is at height $x = 0.5$ along the **a** direction.

the two, as we shall see in the discussion of ordering to follow. Electrostatic energy calculations [7] on observed and hypothetical olivines predict that the geometries of the two M sites are interrelated, however, and for olivines like picrotephroite [17], monticellite [28], and glaucochroite [28] that contain metal atoms of quite different sizes, $M2$ is found to be the more distorted octahedron. As we would generally expect (see Figure 3-13), the edges shared between polyhedra are all shorter than unshared edges in the same polyhedron. Of course, we expect that from a qualitative application of Pauling's familiar rules, but that may not be the only rationale that applies—or even the correct one. A careful analysis of semiempirical molecular-orbital calculations [32] on a cluster of three octahedra sharing basal edges with a tetrahedron (the olivine topology), in which different interactions contributing to the energy of the cluster have been examined independently, has suggested that the shortening of the shared tetrahedral edges is due to a combination of several factors, and that Mg–Si repulsion was not one of the important ones. The fact that the calculations were semiempirical (specifically, they were of the CNDO type—see Chapter 11) means that many approximations were made in reaching the

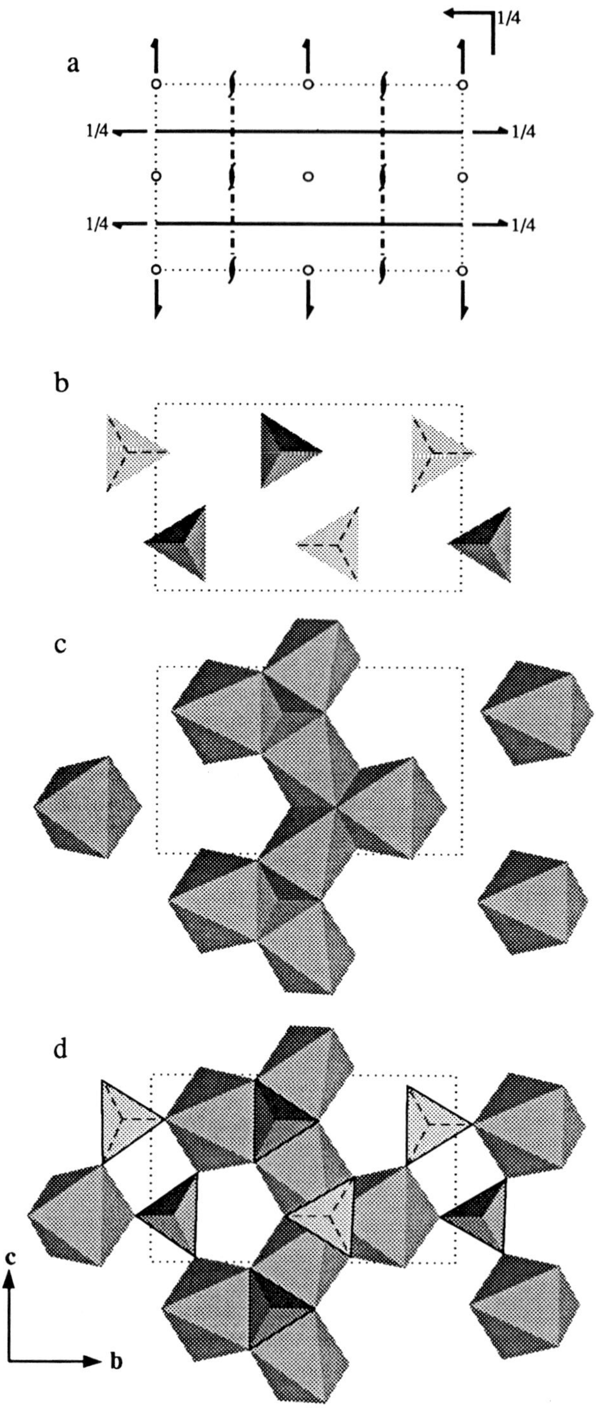
a
1/4
1/4
1/4
1/4
1/4
b
c
d
c
b

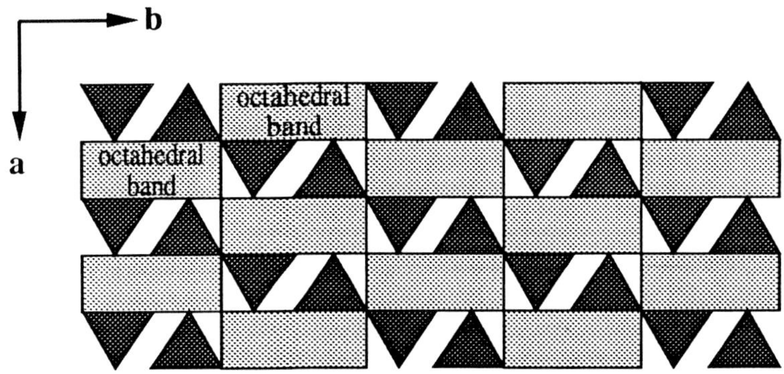

Figure 7-4. A schematic view of the olivine structure projected down the **c** direction. Rectangles represent the edge-sharing octahedral chain and triangles the independent tetrahedra that share edges and corners with octahedra.

results, which should therefore be viewed conservatively. Nonetheless, such results constitute a caution against the categorical acceptance of convention and precedent, and an encouragement constantly to test established ideas against new data.

ORDERING IN OLIVINES

With only one unique tetrahedral site and virtually no substitution for silicon, ordering in olivines involves only the various metal atoms that may occupy the $M1$ and $M2$ sites in the octahedral band. As in pyroxenes, the $M1$ and $M2$ sites in olivines are symmetrically distinct even if they contain the same species of atom, and they occur in symmetrically related layers containing bands of edge-sharing octahedra with two of each kind of octahedron per octahedral band per c repeat. Recall that we have found that ordering is often accompanied by a decrease in symmetry; for example, ordering in monoclinic pyroxenes was often coupled with a change in space group from the high-symmetry $C2/c$ configuration to something lower. This has not been observed to occur in olivines. Although ordering does occur, as we shall discuss, all olivines so far investigated crystallize in the space

Figure 7-3. The structure of olivine. (a) The space group *Pbnm* projected down the **a** direction. The dotted lines are not glide planes, but only outline the unit cell. (b) The positions of tetrahedra at the same scale as the space-group projection. (c) The octahedral band in the same orientation. The smaller octahedra that form the "backbone" of the chain contain $M1$ sites, while the larger ones that alternate on either side of the chain contain $M2$ sites. The octahedra shown contain cations at height $x = 0.5$. Identical chains fill the void on either side of the central chain shown at heights $x = 0$ and 1. (d) The tetrahedra of (b) superimposed on the octahedral chains of (c). The two outer pairs of tetrahedra are in the same layer as the octahedra (the "$x = 0.5$ layer"), while the two tetrahedra above the octahedral chain are in the "$x = 0$ layer" and link that octahedral chain to one directly above in the "$x = 0.5$ layer" of the next unit cell above.

group *Pbnm*, indicating that all *M*1 sites are identical, as are all *M*2 sites and all tetrahedral sites. It is conceivable, however, that an olivine with sufficiently complex chemistry could show loss of symmetry elements relative to space group *Pbnm* and have a different space group.

In other minerals covered so far, we have observed that cation size is a major factor (albeit not the only factor) in determining site preference, and Table 7-1 shows that this is likewise true for olivines. The order of radii, from largest to smallest (see, for example, [45]), is Ca>Mn>Fe>Zn>Mg. (Co, Ni, and other possible substituents are omitted because they have not been found as major constituents in natural olivines.) *M*2 is inherently a larger polyhedron than *M*1, as determined by crystal-structure refinements of olivines with identical atoms in the two octahedra, and in all five cases shown in the table Ca occupies, fully or nearly so, the larger octahedral site. Magnesium, the smallest of the metal atoms considered, is always ordered fully into *M*1 in these olivines. For the other atoms, ordering is only partial, but it is apparent that ionic size is a significant factor influencing the cation distribution; in no case is complete disorder (equal distribution of an element between *M*1 and *M*2) observed. That there are probably other important factors influencing ordering in these olivines is evinced by electrostatic energy calculations [4;7] that yield lower energies for the ordered cation distributions, with the larger cations in *M*2, even though the calculations do not explicitly include ionic size.

Conspicuous in their absence from Table 7-1 are the common olivines of the forsterite–fayalite series, and, in fact, they behave somewhat differently. Until the early 1970s, all published crystal-structure refinements of Fe,Mg olivines showed a completely disordered arrangement of iron and magnesium. While it is true that the ionic radii of Mg and Fe^{2+} are nearly the same, Fe^{2+} is slightly larger, and an extension of the trend for other olivines would suggest that it should exhibit at least a slight preference for *M*2. When lunar olivines were brought back to earth by Apollo astronauts, they were scrutinized with a great deal more care than had previously been exercised for terrestrial olivines, and indications of slight Fe^{2+} enrichment in *M*1 (rather than in *M*2 as size considerations would suggest) were found [14;15]. Because Fe^{2+} is a transition metal with a d^6 electron configuration, crystal field theory seems likely to provide some insight into this behavior. (The tendency

Table 7-1. Site Occupancies for Selected Olivines

		Mg	Fe	Mn	Ca	Other
Zincian magnesian tephroite	*M*1	0.35	0.04	0.43	—	0.18Zn
	*M*2	—	0.08	0.87	—	0.05Zn
Manganoan fayalite	*M*1	0.05	0.66	0.29	—	—
	*M*2	—	0.37	0.63	—	—
Monticellite	*M*1	1.00	—	—	—	—
	*M*2	—	—	—	1.00	—
Kirchsteinite	*M*1	—	0.85	—	0.15	—
	*M*2	—	—	0.02	0.98	—
Glaucochroite	*M*1	0.09	—	0.91	—	—
	*M*2	—	—	0.02	0.98	—

for a larger cation to segregate into a smaller site is sometimes called *antiordering*, and we shall use that term here for clarity.)

Crystal field theory (discussed in Chapter 11) predicts that the crystal field stabilization energy of Fe^{2+} will be increased (that is, the ion will be relatively stabilized) by more distorted coordination and by shorter Fe–O bonds. Octahedral site distortion may be quantified using octahedral angle variance [41],

$$\sigma^2_{oct} = \sum_i [(\theta_i - 90°)^2/11],$$

which is a measure of the deviation of the O–M–O angles (θ_i) from 90°, the value in an undistorted octahedron. For olivines of the Fo–Fa series, the value of σ^2_{oct} for $M1$ is a few percent larger than it is for $M2$. Thus, based on this rationale, Fe^{2+} should exhibit a slight preference for $M1$. In addition, recall that $M1$, which has six shared—and thus short—edges, is inherently smaller than $M2$, which has only three shared edges. Consequently, the mean Fe–O bond length at the $M1$ site is a few percent less than it is at the $M2$ site, providing another reason for Fe^{2+} to prefer the former on the basis of crystal-field stabilization.

A quantitative estimate of the stabilization energy associated with Fe^{2+} in the $M1$ site and the resulting predicted distribution of Fe^{2+} between the two octahedral sites has been made ([53], corrected and modified in this discussion), and the exercise of following the calculation through is worthwhile: Strong tendencies toward transition metal ordering into $M1$ have been observed in the synthetic olivines $Ni_{1.03}Mg_{0.97}SiO_4$ [40] and $Co_{1.10}Mg_{0.90}SiO_4$ [19]. The experimentally measured values of the crystal field splitting, Δ_o, for Ni^{2+} (d^8), Co^{2+} (d^7), and Fe^{2+} (d^6) in undistorted coordination by six H_2O molecules in aqueous solution, are 8200, 8400, and 10 300 cm^{-1} [50], respectively, and it is easily determined (see Chapter 11) that the crystal field stabilization energies for these ions in ideal octahedral coordination are $\frac{6}{5}\Delta_o$, $\frac{4}{5}\Delta_o$, and $\frac{2}{5}\Delta_o$, respectively. Thus the stabilization energies, E_{st}, for geometrically ideal coordination are 9840 cm^{-1} for Ni^{2+}, 6720 cm^{-1} for Co^{2+}, and 4120 cm^{-1} for Fe^{2+}. $M1$ and $M2$ are not geometrically ideal in olivines, however, and the quantity of interest is the *difference* in the stabilization energies experienced at the two sites, $\Delta E_{st} = E_{st}(M1) - E_{st}(M2)$. That has been calculated to be 680 cm^{-1} for the nickel olivine [53], and on the reasonable assumption that the corresponding polyhedra are similar enough from one structure to another that $\Delta E_{st}/E_{st}$ is roughly constant, then ΔE_{st} for Co^{2+} and Fe^{2+} are about 464 and 284 cm^{-1}, respectively. If we define the distribution coefficient, K_D, as $(Mg_{M2}R^{2+}_{M1})/(Mg_{M1}R^{2+}_{M2})$, where R^{2+} is the transition metal, then the observed K_D for the nickel olivine, obtained from the crystal structure refinement, is 9.225. Assuming now that the ratio of the square root of K_D to ΔE_{st} is also constant from one of these structures to another, the predicted values of the distribution coefficients are 4.30 for the cobalt olivine and 1.61 for the iron olivine. The observed K_D for $Co_{1.10}Mg_{0.90}SiO_4$ is 4.60, which is in quite good agreement with the predicted value. For the majority of Fe,Mg olivines so far observed, the distribution coefficients vary from about 1.0 to as high as 1.75, again in reasonable agreement with the prediction. (Note: The square root of K_D was used here because, for olivines in which Mg:R^{2+} = 1:1, it is R^{2+}_{M1}/R^{2+}_{M2} that is linearly related to ΔE_{st}, and that ratio is the *square root* of the distribution coefficient defined previously, as a simple algebraic exercise will show. Even when that ratio is not quite 1:1, as in the cases of the Ni and Co olivines cited here, the square root of K_D is still nearly linearly related to ΔE_{st}.)

Simple crystal field theory thus provides a conceptual framework within which the tendency for the larger Fe^{2+} to exhibit a slight preference for the smaller *M*1 site can be rationalized. Nevertheless, ferromagnesian olivines are not always antiordered in this way; values of K_D for olivines of the forsterite–fayalite series are occasionally observed to be less than unity, indicating a preference of Fe^{2+} for *M*2. For thirty-seven ferromagnesian olivines for which structure refinements (some at both room and high temperatures) were published from 1973 to 1985 [2;5;10;38;39;48;54], the results at room temperature are as follows: 6 were lunar olivines, and all 6 showed significant antiordering of Fe in *M*1. Of the 23 terrestrial volcanic olivines, 10 showed slight to moderate antiordering of Fe in *M*1, 5 showed slight ordering of Fe in *M*2, and the other 8 were disordered. 8 were terrestrial metamorphic olivines; of them, 3 showed antiordering, 4 showed ordering, and 1 was disordered.

The foregoing data suggest a positive, but quite rough, correlation between cooling rate and antiordering of Fe into *M*1; the faster the olivine cools, the greater the tendency for antiordering. The conclusion, of course, is that Fe prefers *M*1 at high temperatures, but that annealing tends to result in disordering, or even ordering (Fe in *M*2). X-ray structure refinements carried out at high temperature confirm this [2;10;48]. There thus exists an apparent potential for using the cation distribution in olivine as a cooling-rate thermometer, but there are problems that make this impractical at present. First, the range of cation distribution observed is not very large, and, given the experimental errors, only the grossest estimates of cooling rate would be possible. Second, there is likely a pressure dependence that has not yet been determined. Third, interpretation of Mössbauer spectra [12;51] has led to generally larger estimates of antiordering than that indicated by x-ray-diffraction techniques, or to results that contradict those we have just reviewed [46]. Finally, other factors may be important in determining Fe,Mg order in olivines, and we shall discuss one of them (oxygen partial pressure) shortly.

It is worth pointing out here that the idea of a mineral becoming more ordered (or antiordered) with increasing temperature violates the intuition we have developed in the earlier chapters of this text. Without exception, all of the minerals that have so far been discussed become more *disordered* with increasing temperature, if disordering occurs at all. In the Fo–Fa series, we are dealing with a mineral in which the differences between the possible combinations of competing atoms and sites are not only very small but also tend to counteract one another, leading to a distribution coefficient that is near unity under *any* conditions. As temperature increases, there are minor changes in polyhedral geometries that evidently cause iron to exhibit a slight preference for *M*1 over *M*2; high-temperature crystal-structure refinements [10;47;48] of ferromagnesian olivines show that the mean M–O distances and the polyhedral volumes of the *M* sites increase with temperature at approximately the same rate for both sites, but that octahedral angle variance increases more rapidly for *M*1 above ~500–600°C. These changes probably affect crystal-field stabilization, but may also involve electronegativity (covalency) or other factors.

Cation distribution in Mg,Fe olivines may be more successful as an indicator of oxygen partial pressure in the environment of crystallization than as a geothermometer. In an experiment on natural forsteritic crystals [38;55] with K_D = 1.09 and 1.06, those subjected to oxygen partial pressures of 10^{-21} bar at 750°C exhibited increased antiordering to $K_D = 1.2$. However, those subjected to oxygen partial pressures of 10^{-16} bar became slightly ordered (Fe in *M*2), with $K_D = 0.8$. Assuming a linear relationship (for lack of data to the contrary), between p_{O_2} and K_D, and including lunar olivines at the generally accepted value of $p_{O_2} = 10^{-20}$ bar for the moon, the straight line shown in Figure 7-5 is obtained. It suggests that the oxygen partial pressure in the environment of formation of a ferromagnesian olivine may be predictable given the observed intracrystalline cation distribution coefficient. The compositions of the crystals were $Fo_{90}Fa_{10}$ and $Fo_{88}Fa_{12}$, so it is apparent that a wider range of compositions needs to be considered, as well as complications arising from differences in cooling rates.

Some final thoughts on ordering in the olivines: We have emphasized the roles of ionic size and crystal-field stabilization in this discussion, but there must be other factors besides these that influence the octahedral cation distribution. Were it not so, then Ni^{2+}, which is smaller than Mg^{2+} and experiences a crystal-field enhancement in addition, would order entirely into *M*1, resulting in an undefined (infinite)

Figure 7-5. The distribution coefficient K_D for olivines of the Fo–Fa series as a function of the partial pressure of oxygen during crystallization. (After [55].)

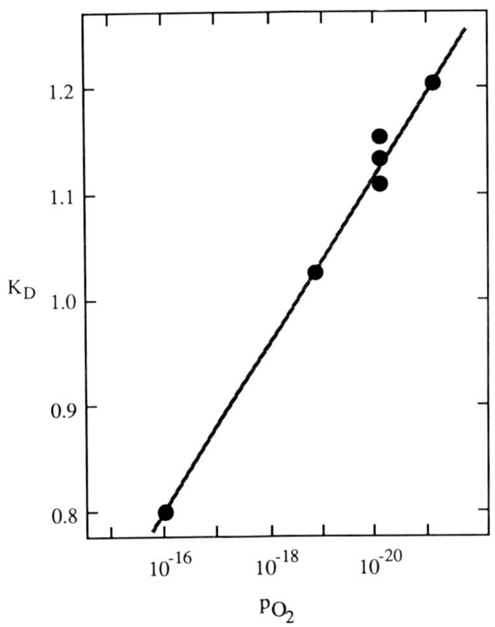

distribution coefficient in $Ni_{1.03}Mg_{0.97}SiO_4$, instead of the 9.225 observed. Similarly, Mn^{2+}, which is larger than Fe^{2+} and experiences no crystal-field effects, would order entirely into *M*2 in $Fe_{1.03}Mn_{0.92}Mg_{0.05}SiO_4$, yielding a distribution coefficient of 0.0 instead of 0.258. We have already seen that oxygen partial pressure affects ordering in at least the ferromagnesian olivines. Among the other effects that must be important in these and the other olivines are the electronegativity of the ordering cations (which is related to the covalency of the metal–oxygen bond) and the need to configure the octahedra so that acceptable Si–O distances result; the relative importance of these and other possible effects are not yet understood in a quantitative way.

LATTICE-PARAMETER VARIATIONS IN NATURAL OLIVINES

The Forsterite–Fayalite Series

Because Fe^{2+} and Mg are not the same size, changes in composition result in changes in unit-cell dimensions. Figure 7-6 provides plots that are useful for estimating compositions of Fe,Mg olivines, to an accuracy of probably 3 or 4 mol %. Of course, the presence of Ca or Mn (or any other less common but possible substituent) decreases the effectiveness of the method, and there is no way of determining the amounts of such "impurities" from the lattice dimensions alone. In addition to unit-cell parameters, the *d* spacings of some lattice planes vary with composition. The one that seems to be most accurate for compositional determinations in this series is d_{130}, which is insensitive to ordering, and for which the determinative curve is shown in Figure 7-7. For those who prefer an equation [44],

$$X_{\mathrm{Fa}} = 7.522 - 14.9071(3.0199 - d_{130})^{1/2}.$$

The advantage in using this curve (or the equation) is that the position of only a single peak need be known; so it is faster than a procedure requiring refined unit cell dimensions; the disadvantage is that any error in the peak position that may still persist after calibration to an internal x-ray standard, will not be "smoothed out" by the usual least-squares refinement treatment by which accurate lattice parameters are obtained (see Chapter 10), and thus an error in estimated composition may result.

Other Olivines

For olivines in the forsterite–tephroite and forsterite–monticellite series, unit cell parameters are very sensitive to both composition and degree of order. The equations

$$a = 0.932r_{M1} + 0.236r_{M2} + 3.918,$$

$$b = 0.505r_{M1} + 3.211r_{M2} + 7.535,$$

calculated by multiple linear regression analysis of data for 29 silicate olivines [30], indicate that the *a* cell edge is much more dependent on the mean radius of

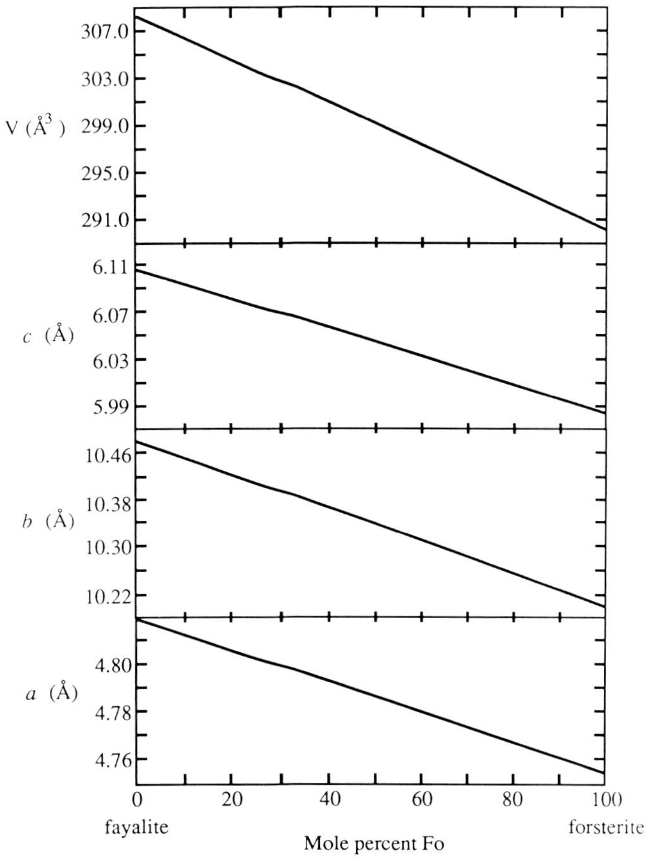

Figure 7-6. Lattice parameters of olivines of the Fo–Fa series as functions of composition. (After [29].)

Figure 7-7. The *d* spacing of the 130 reflection as a function of compositon for the Fo–Fa series. (After [29].)

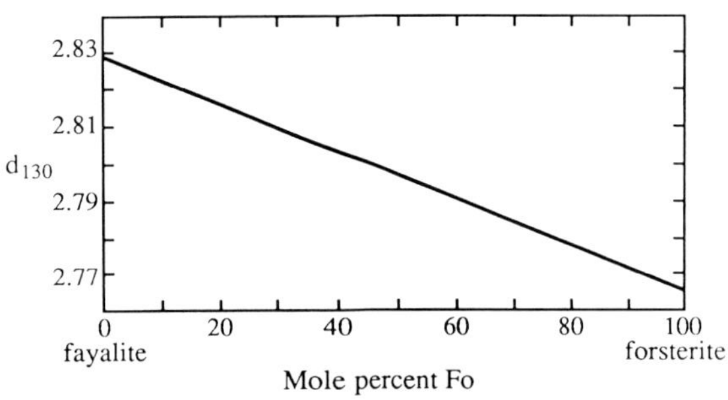

the $M1$ cation than on that of $M2$, while the opposite is true for the b cell edge. An a–b plot, similar in concept to the b–c plot for alkali feldspars (p. 63), may be constructed for the determination of both ordering and composition of olivines in the Fo–Te, Fo–Mo, and Fa–Te series, in which the radii of the octahedral cations are quite different. The plot for the Fo–Te series is given in Figure 7-8. Values of a and b, obtained by least-squares refinement of x-ray powder-diffraction data, locate a unique point on the plot. The nearly vertical lines yield ordering (atoms of Mn^{2+} in $M2$), and oblique lines indicate composition. A partial a–b plot that covers the range of practical importance for the Fo–Mo series is shown in Figure 7-9, and the plot of the Fa–Te series appears in Figure 7-10.

Because the (112) d spacing is insensitive to ordering, it may be used for estimation of composition. For minerals of the Fo–Te series, the appropriate relation is

$$[\text{Mg}](\text{at. }\%)/\Sigma M = 2484.87 - 970.42(d_{112}),$$

where ΣM is the total number of cations in octahedral coordination [31]. [The d spacing of the (130) planes may also be used for this series, but is nonlinear because of ordering.] The equation [16] is

$$[\text{Mg}](\text{at. }\%)/\Sigma M = -31981.250 + 23707.779(d_{130}) - 4378.2695(d_{130})^2.$$

Figure 7-8. The a–b plot for the forsterite-tephroite series. Numbers on the "NW–SE" lines are mol % Fo, and those on the nearly vertical lines are atoms of Mn in the $M2$ site. (After [31].)

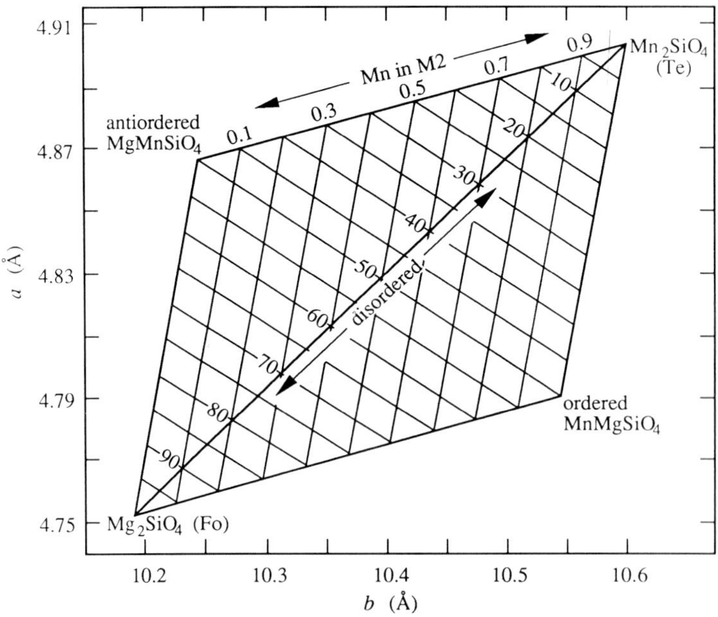

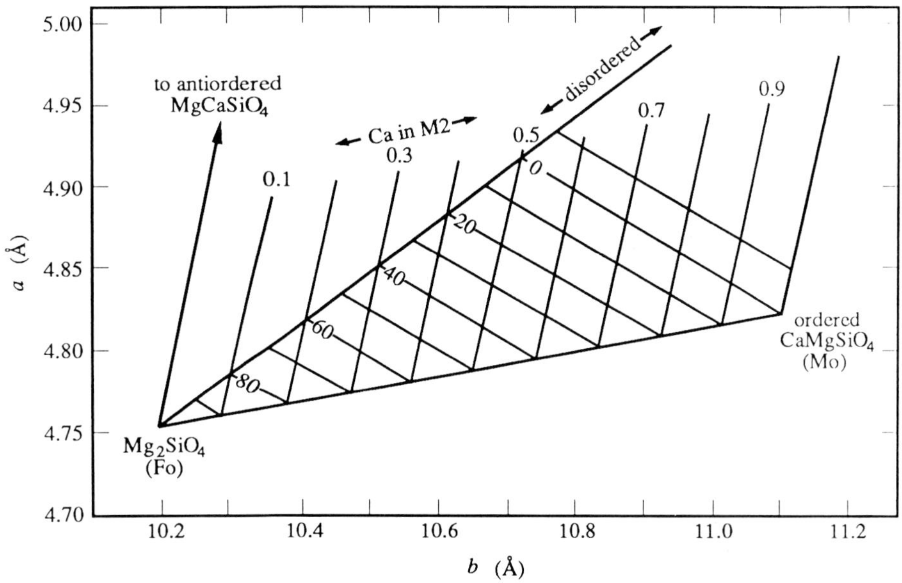

Figure 7-9. A partial *a–b* plot for the forsterite–monticellite series. Numbers on the "NW–SE" lines are mol % Fo in the Fo–Mo series, and those on the nearly vertical lines are atoms of Ca in the *M*2 site. (After [31].)

Figure 7-10. A partial *a–b* plot for the fayalite–tephroite series. Contours labeled in multiples of 10 are mol % Fa in the Fa–Te series, and the others are atoms of Ca in the *M*2 site.

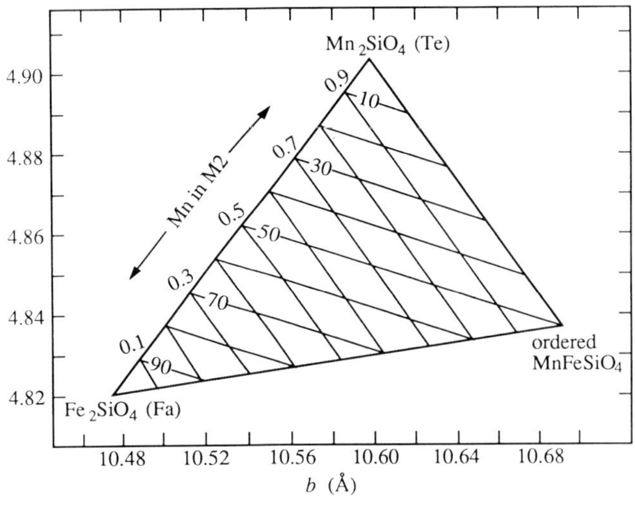

For olivines of the Fo–Mo series, the analogous relationship for d_{112} is

$$[\mathrm{Fo}](\mathrm{mol}\ \%) = 1946.25 - 752.14(d_{112}),$$

where 100[Fo]/([Fo]+[Mo]) is the mole percent forsterite [31]. The *a–b* plots yield composition as accurately as do these equations (around ±2 mol % Fo component), and should be used if both composition and order are desired. If only composition is needed, however, the measurement of 2θ for one peak is certainly a faster procedure than a determination of unit-cell parameters. (It is a good idea to scan enough of the powder pattern to include at least two peaks from an internal standard—one peak higher and one lower in 2θ than the olivine (112) or (130) peak—in order to calibrate the $2\theta_{112}$ or $2\theta_{130}$ measurement.)

POLYMORPHISM AND PHASE TRANSFORMATIONS IN THE Fo–Fa SERIES

A large number of compounds have the A_2BX_4 generalized formula, and the kind of structure adopted depends upon the size of the *A* and *B* atoms. Figure 7-11 is a structure field map for A_2BX_4 compounds. (Note that the radii in common use among mineralogists, such as [45], have not been used in this plot. Rather an older set in which the radius of Si^{4+} was taken to be 0.40 Å has been employed. Recasting the figure in terms of presently used radii would result in no additional insights.) The potential for polymorphism in olivines is evidently related to how the relative sizes of the octahedral and tetrahedral cations in a given species change with pressure and temperature. Under crustal conditions, the polyhedra of ferromagnesian olivines do not experience enough compression or expansion for one of the other structure types to become stable; consequently, polymorphism is not observed in

Figure 7-11. A structure field map for compounds that have the generalized formula A_2BX_4, showing the dependence of the stable structure type on the radii of the *A* and *B* atoms. (After [9] and [37].)

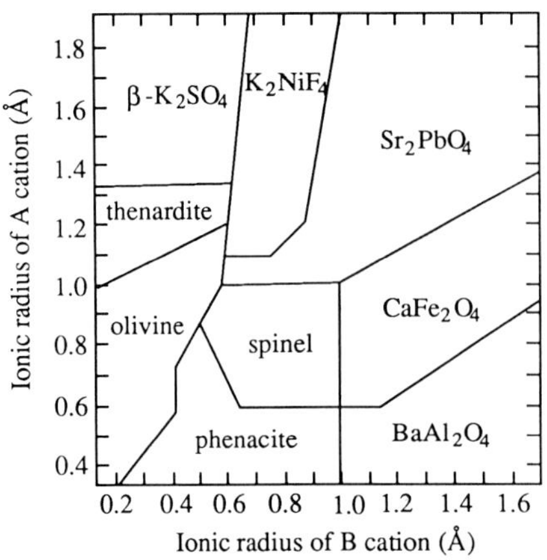

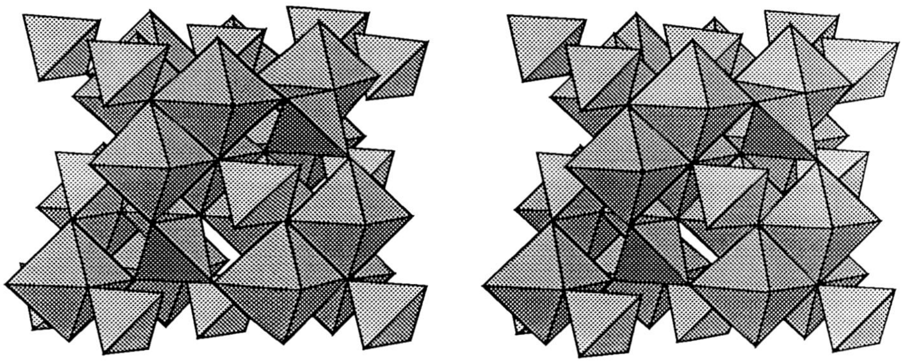

Figure 7-12. Stereoscopic view of the spinel structure. (After [56].)

the forsterite–fayalite series. The same is true of the less common members of the olivine family, with the exception of Ca_2SiO_4, which occurs very rarely in the olivine structure [8] as well as the more common (but still very *un*common) larnite structure.

Under the physical conditions that obtain in the mantle, the situation is different. At very high pressures and temperatures (over 115 kbar and 1000°C) Mg_2SiO_4 is not stable in the olivine structural configuration, but assumes the spinel structure. Because the upper mantle is thought to consist mostly of olivine near forsterite in composition, the olivine–spinel transformation is of great geophysical importance. There are two nonolivine polymorphs of Mg_2SiO_4, and one of them, the γ polymorph, is spinel; Figure 7-12 shows a stereo pair of the that structure. The spinel polymorph is based on cubic close packing of the oxygen atoms, with the silicon and magnesium atoms having the same coordination numbers as in olivine. The β polymorph, which is orthorhombic and is often called *modified* or *distorted spinel,* is also based on cubic close packing, but a crystal structure refinement based on powder diffraction [34] reveals the presence of Si_2O_7 groups. While the spinel and modified spinel structures contain both octahedral and tetrahedral sites, they are disposed differently than in the olivine structure, and the transformation from olivine is reconstructive. It results in a density increase of about 10%, consistent with spinel being the stable polymorph at higher pressures.

Whether the transition involves nucleation and growth or is a martensitic transformation is not yet clear, because there are observations that might support both mechanisms. The transformation in Ni_2SiO_4 shows features that are not satisfactorily explained by either mechanism [20], but some caution must be exercised in applying results for analog systems directly to the near-forsterite composition of olivine in the mantle. Ni_2SiO_4 transforms directly from the olivine (α) form to the spinel (γ) form, with no evidence of a modified spinel (β) polymorph. Thus, observations of the transformation behavior of Ni_2SiO_4 may not be valid for the composition $Fo_{0.9}Fa_{0.1}$.

The general configuration of the P–T phase diagram for Mg_2SiO_4 is reasonably sure, but the details are not. Figure 7-13 shows such a diagram, and the widths of

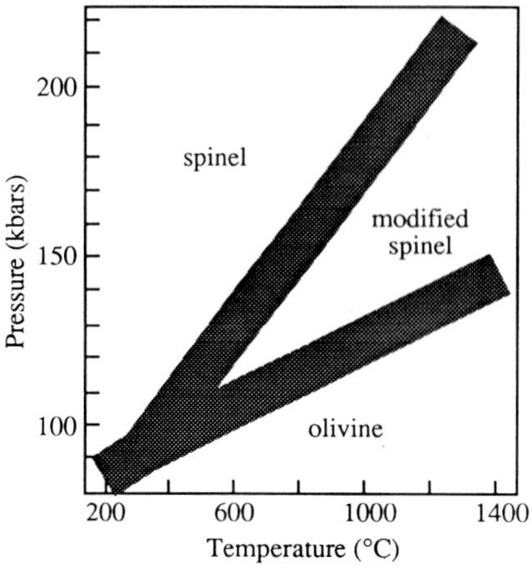

Figure 7-13. A somewhat schematic *P–T* phase diagram for Mg_2SiO_4. The widths of the boundaries understate the discordance among various published versions of the diagram.

the boundaries understate the discordance of the sources from which it was compiled. Note that the modified spinel phase is absent at low temperatures, but is certainly present at temperatures typical of the mantle. Polymorphism as a function of composition and pressure for the $(Mg,Fe)_2SiO_4$ composition range at two temperatures is shown in Figure 7-14. Observe that the stability field of modified spinel shrinks with decreasing temperature, consistent with Figure 7-13. (The phase boundaries on Figure 7-14 were drawn as sharp for clarity, but are obviously not well determined if those in Figure 7-13 are not.) Of significance is the disappearance of the β phase as composition shifts toward Fe_2SiO_4, because it limits the usefulness of "fayalite" spinel (which is much easier to synthesize in large crystals than is "forsterite" spinel) as a crystallochemical analog of the spinel that occurs in the mantle (probably about $Fo_{90}Fa_{10}$). One of the reasons that Fe_2SiO_4 spinel is stable at lower pressures than the magnesium analog is that the free energy change due to differences in crystal field splitting of the olivine and spinel forms (absent, of course, in Mg_2SiO_4) favors the spinel form at all pressures and temperatures [11].

Early high-pressure crystal-structure studies of forsterite suggested that the mean Si–O bond distances did not change with pressure, but that the mean Mg–O distances decreased linearly with increasing pressure [21;22;23]. Subsequent work at higher pressures [27] revealed different trends: The mean tetrahedral bond length does, in fact, decrease linearly with increasing pressure, from 1.634 Å at atmospheric pressure to about 1.58 Å at 149 kbar; the mean *M*2–O distance decreases almost linearly to 149 kbar, but the mean *M*1–O bond length stops decreasing around 80 kbar. At 149 kbar, both mean octahedral bond lengths are 2.05 Å.

Of course, as depth in the Earth increases, temperature goes up as well as pressure, so the influence of temperature on the crystal structure of olivine is also important in understanding the crystallography of the phase transformation. We would generally expect that rising temperature and rising pressure would exert opposite crystallographic effects, and this is largely confirmed. In the range from 25 to 900°C,

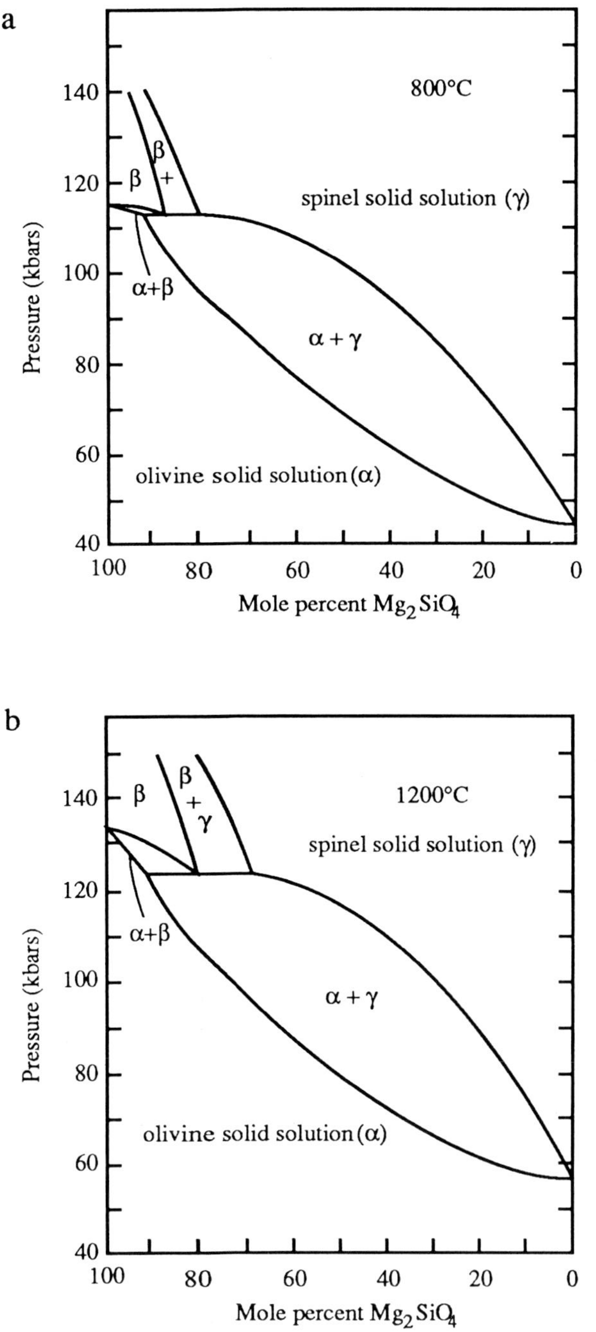

Figure 7-14. *P–X* phase diagrams for the system Mg_3SiO_4–Fe_2SiO_4 at (a) 800°C and (b)1200°C. (After [3].)

the mean $M1$–O and $M2$–O bond lengths increase steadily with temperature, perhaps linearly, but not demonstrably so with data currently available [48]. Contrary to intuition, the dimensions of the silicate tetrahedron remain essentially constant as the temperature is increased. Figure 7-15 shows the effects of both pressure and temperature, independently, on mean bond lengths. Unfortunately, the technology does not yet exist for carrying out single-crystal structure refinements at the pressures and temperatures of interest simultaneously, and it is not obvious that a simple linear combination of the effects of the two variables would yield the correct in situ crystal-chemical quantities.

In the olivine structure, the tetrahedra share three edges with octahedra, and

Figure 7-15. (a) The changes in mean bond lengths in forsterite with pressure (after [27]). (b) Variation in mean bond lengths in forsterite with temperature. (Data from [48].)

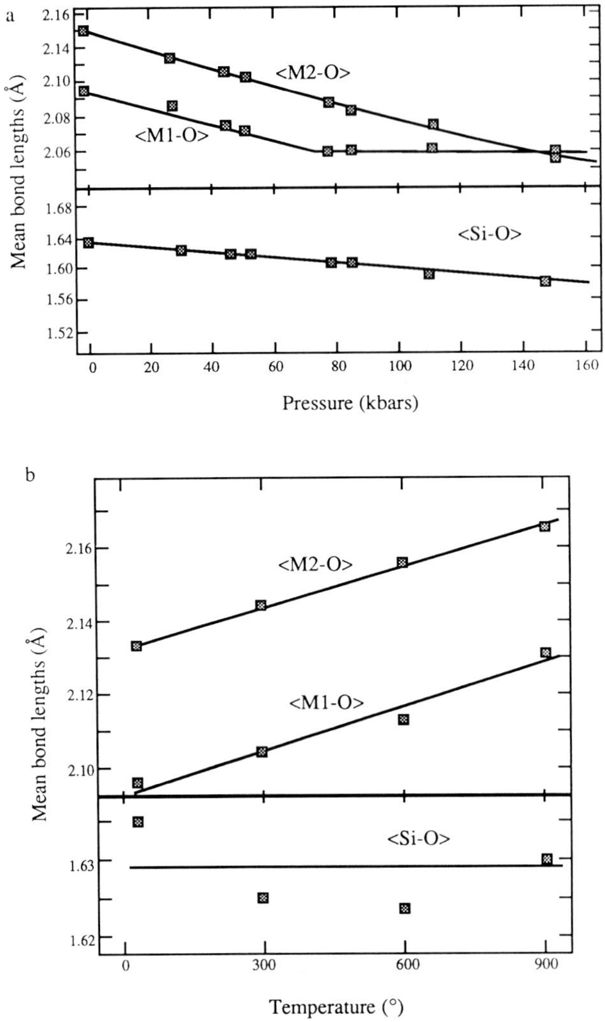

octahedra share three with one another; in the spinel structure no tetrahedral edges are shared with octahedra, but the latter share six with one another. The usual rationalization for the olivine–spinel transformation [24] involves an application of Pauling's rules. It is clear from Pauling's third rule that tetrahedron–octahedron (*T–O*) edge sharing should be less favorable than octahedron–octahedron (*O–O*) edge sharing for two reasons: First, the tetrahedron is smaller than the octahedron, so *T–O* edge sharing places the cations closer to one another than does *O–O* edge sharing. Second, the tetrahedron contains a cation with a formal charge of 4+, and the octahedra contain cations of formal charge 2+, and therefore *T–O* edge sharing would result in greater repulsion even if the polyhedral sizes were the same. The reason, in this argument, that the olivine configuration is favored at low pressure, in apparent contradiction of the third rule, is the shortening of shared edges predicted by Pauling's fourth rule. In olivine, the shared edges are shorter than the unshared ones, allowing greater separation of the cations and interposing some shielding by the electrons of the anions (see Figure 3-10). In order to allow the required Si–O and Mg–O distances in the spinel structure, the shared edges must be longer than the unshared ones, naturally leading to destabilization owing to Mg–Mg repulsion. Only at high pressures are the oxygen atoms presumably forced close enough together to provide adequate shielding between adjacent octahedral cations to stabilize the spinel form.

This explanation, simple and attractive as it is, may obscure other factors that drive the transformation. We have seen previously (p. 251) that molecular-orbital calculations suggest that Si–Mg repulsions are not the primary reason for the shortening of shared *T–O* edges in olivine, and so it may be that the reduction of that repulsion is not as important as Pauling's rules would suggest. Furthermore, the high-pressure crystal-structure refinements of forsterite yield a mean Si–O bond length of only 1.59 Å at 111 kbar (see Figure 7-15)—a very short distance for any silicate, and particularly for an orthosilicate. The driving force for the transformation may well be the unfavorable compression of the silicate tetrahedron, given that high temperature apparently does not counteract that compression. The crystallographic explanations for the phase transition, which almost surely plays some significant role in the ~400 kilometer seismic discontinuity, will likely remain uncertain until new techniques for crystal-structure refinement allow examination of $Fo_{0.9}Fa_{0.1}$ in its olivine, modified spinel, and spinel polymorphs under in situ mantle conditions of simultaneously high temperature and high pressure.

REFERENCES

[1] Adams, G. E., and F. C. Bishop (1985). "An Experimental Investigation of Thermodynamic Mixing Properties and Unit-Cell Parameters of Forsterite-Monticellite Solid Solutions." *American Mineralogist* 70, 714–22.

[2] Aikawa, N., M. Kumazawa, and M. Tokonami (1985). "Temperature Dependence of Intersite Distribution of Mg and Fe in Olivine and the Associated Change of Lattice Parameters." *Physics and Chemistry of Minerals* 12, 1–8.

[3] Akimoto, S., Y. Matsui, and Y. Syono (1976). "High-Pressure Crystal Chemistry of Orthosilicates and the Formation of the Mantle Transition Zone." In R. J. G. Strens (ed.), *The Physics and Chemistry of Minerals and Rocks,* New York, Wiley.

[4] Alberti, A., and G. Vezzalini (1978). "Madelung Energies and Cation Distributions in Olivine-Type Structures." *Zeitschrift für Kristallographie* 147, 167–75.

[5] Basso, R., A. Dal Negro, A. Della Giusta, and G. Rossi (1979). "Fe/Mg Distribution in the Olivine of Ultrafemic Nodules from Assab (Ethiopea)." *Neues Jahrbuch für Mineralogie Monatshefte*, 197–202.

[6] Birle, J. D., G. V. Gibbs, P. B. Moore, and J. V. Smith (1968). "Crystal Structures of Natural Olivines." *American Mineralogist* 53, 807–24.

[7] Bish, D. L., and C. W. Burnham (1984). "Structure Energy Calculations on Optimum Distance Model Structures: Application to the Silicate Olivines." *American Mineralogist*, 69, 1102–09.

[8] Bridge, T. E. (1966). "Bredigite, Larnite, and γ Dicalcium Silicates from Marble Canyon." *American Mineralogist* 51, 1766–74.

[9] Brown, G. E. (1980) "Olivines and Silicate Spinels." In P. H. Ribbe (ed.), *Orthosilicates. Reviews in Mineralogy*, Volume 5. Mineralogical Society of America, Washington, D.C.

[10] Brown, G. E., and C. T. Prewitt (1973). "High-Temperature Crystal Chemistry of Hortonolite." *American Mineralogist* 58, 577–87.

[11] Burns, R. G., and C.-M. Sung (1978). "The Effect of Crystal Field Stabilization on the Olivine-Spinel Transition in the System Mg_2SiO_4–Fe_2SiO_4." *Physics and Chemistry of Minerals* 2, 349–64.

[12] Bush, W. R., S. S. Hafner, and D. Virgo (1970). "Some Ordering of Iron and Magnesium at the Octahedrally Coordinated Sites in a Magnesium-Rich Olivine." *Nature* 227, 1339–41.

[13] Ericsson, T., and A. Filippidis (1986). "Cation Ordering in the Limited Solid Solution Fe_2SiO_4-Zn_2SiO_4." *American Mineralogist* 71, 1502–09.

[14] Finger, L. W. (1970). "Fe/Mg Ordering in Olivines." *Carnegie Institute of Washington Year Book* 69, 302–05.

[15] Finger, L. W., and D. Virgo (1971). "Confirmation of Fe/Mg Ordering in Olivines." *Carnegie Institute of Washington Year Book* 70, 221–25.

[16] Francis, C. A. (1985). "New Data on the Forsterite–Tephroite Series." *American Mineralogist* 70, 568–75.

[17] Francis, C. A., and P. H. Ribbe (1980). "The Forsterite-Tephroite Series: I. Crystal Structure Refinements." *American Mineralogist* 65, 1263–69.

[18] Fu, P., Y. Kong, and L. Zhang (1982). "Domain Twinning of Laihunite and Refinement of its Crystal Structure." *Geochemistry* 1, 115–33.

[19] Ghose, S., and C. Wan (1974). "Strong Site Preference of Co^{2+} in Olivine, $Co_{1.10}Mg_{0.90}SiO_4$." *Contributions to Mineralogy and Petrology* 47, 131–40.

[20] Hamaya, N., and S. Akimoto (1982). "On the Mechanism of the Olivine–Spinel Phase Transformation." *Physics of the Earth and Planetary Interiors* 29, 6–11.

[21] Hazen, R. M. (1976). "Effects of Temperature and Pressure on the Crystal Structure of Forsterite." *American Mineralogist* 61, 1280–93.

[22] Hazen, R. M. (1977). "Effects of Temperature and Pressure on the Crystal Structure of Ferromagnesian Olivine." *American Mineralogist* 62, 286–95.

[23] Hazen, R. M., and L. W. Finger (1980). "Crystal Structure of Forsterite at 40 kbar." *Carnegie Institution of Washington Year Book* 79, 364–67.

[24] Kamb, B. (1968). "Structural Basis of the Olivine–Spinel Stability Relation." *American Mineralogist* 53, 1439–55.

[25] Kan, X., and J. M. D. Coey (1985). "Mössbauer Spectra, Magnetic and Electrical Properties of Laihunite, a Mixed Valence Iron Olivine Mineral." *American Mineralogist* 70, 576–80.

[26] Kondoh, S., M. Kitamura, and N. Morimoto (1985). "Synthetic Laihunite ($\square_xFe^{2+}_{2-3x}Fe^{3+}_{2x}SiO_4$), an Oxidation Product of Olivine." *American Mineralogist* 70, 737–46.

[27] Kudoh, Y., and Y. Takéuchi (1985). "The Crystal Structure of Forsterite Mg_2SiO_4 Under High Pressure up to 149 kb." *Zeitschrift für Kristallographie* 171, 291–302.

[28] Lager, G. A., and E. P. Meagher (1978). "High-Temperature Structural Study of Six Olivines." *American Mineralogist* 63, 365–77.

[29] Louisnathan, S. J., and J. V. Smith (1968). "Cell Dimensions of Olivine." *Mineralogical Magazine* 36, 1123–34.

[30] Lumpkin, G. R., and P. H. Ribbe (1983). "Composition, Order-Disorder and Lattice Parameters of Olivines: Relationships in Silicate, Germanate, Beryllate, Phosphate, and Borate Olivines." *American Mineralogist* 68, 164–76.

[31] Lumpkin, G. R., P. H. Ribbe, and M. E. Lumpkin (1983). "Composition, Order-Disorder and Lattice Parameters of Olivines: Determinative Methods for Mg–Mn and Mg–Ca Silicate Olivines." *American Mineralogist* 68, 1174–82.

[32] McLarnan, T. J., R. J. Hill, and G. V. Gibbs (1979). "A CNDO/2 Molecular Orbital Study of Shared Tetrahedral Edge Conformations in Olivine-Type Compounds." *Australian Journal of Chemistry* 32, 949–59.

[33] Miller, M. L., and P. H. Ribbe (1985). "Methods for Determination of Composition and Intracrystalline Cation Distribution in Fe–Mn and Fe–Ni Silicate Olivines." *American Mineralogist* 70, 723–28.

[34] Moore, P. B., and J. V. Smith (1970). "Crystal Structure of β-Mg_2SiO_4: Crystal-Chemical and Geophysical Implications." *Physics of the Earth and Planetary Interiors* 3, 166–77.

[35] Mossman, D. J., and D. J. Pawson (1976). "X-ray and Optical Characterization of the Forsterite–Fayalite–Fephroite Series with Comments on Knebelite from Bluebell Mine, British Columbia." *Canadian Mineralogist* 14, 479–86.

[36] Mukhopadhyay, D. K., and D. H. Lindsley (1983). "Phase Relations in the Join Kirchsteinite ($CaFeSiO_4$)–Fayalite (Fe_2SiO_4)." *American Mineralogist* 68, 1089–94.

[37] Muller, O., and R. Roy (1974). *The Major Ternary Structure Families.* Springer–Verlag, New York.

[38] Nover, G., and G. Will (1981). "Structure Refinements of Seven Natural Olivine Crystals and the Influence of Oxygen Partial Pressure on Cation Distribution." *Zeitschrift für Kristallographie* 155, 27–45.

[39] Princivalle, F., and L. Secco (1985). "Crystal Structure Refinement of 13 Olivines in the Forsterite–Fayalite Series from Volcanic Rocks and Ultramafic Nodules." *Tschermaks Mineralogische und Petrographische Mitteilungen* 34, 105–15.

[40] Rajamani, V., G. E. Brown, and C. T. Prewitt,(1975). "Cation Ordering in Ni-Mg Olivine." *American Mineralogist* 60, 292–99.

[41] Robinson, K., G. V. Gibbs, and P. H. Ribbe (1971). "Quadratic Elongation: A Quantitative Measure of Distortion in Coordination Polyhedra." *Science* 172, 567–70.

[42] Sarver, J. V., and F. A. Hummel (1962). "Solid Solubility and Eutectic Temperature in the System Zn_2SiO_4-Mg_2SiO_4." *Journal of the American Ceramic Society* 45, 304.

[43] Schaefer, M. W. (1985). "Site Occupancy and Two-Phase Character of 'Ferrifayalite.'" *American Mineralogist* 70, 729–36.

[44] Schwab, R. B., and D. Kustner (1977). "Präzisionsgitterkonstantenbestimmung zur Festlegung Röntgenographischer Bestimmungskurven für Synthetische Olivin der Mischkristallreihe Forsterit–Fayalit." *Neues Jahrbuch für Mineralogie Monatshefte*, 205–15.

[45] Shannon, R. D. (1976). "Revised Effective Ionic Radii and Systematic Studies of Interatomic Distances in Halides and Chalcogenides." *Acta Crystallographica* A32, 751–67.

[46] Shinno, I., M. Hayashi, and Y. Kuroda (1974). "Mössbauer Studies of Natural Olivines." *Mineralogical Journal* 7, 344–58.

[47] Smyth, J. R. (1975). "High-Temperature Crystal Chemistry of Fayalite." *American Mineralogist* 60, 1092–97.

[48] Smyth, J. R., and R. M. Hazen (1973). "The Crystal Structure of Forsterite and Hortonolite at Several Temperatures up to 900°C." *American Mineralogist* 58, 588–93.

[49] Tamada, O., B. Shen, and N. Morimoto (1983). "The crystal structure of Laihunite ($\square_{0.40}Fe^{2+}_{0.80}Fe^{3+}_{0.80}SiO_4$)—nonstoichiometric olivine-type mineral." *Mineralogical Journal* 11, 382–91.

[50] Tanabe, Y, and S. Sugano (1954). "On the Absorption Spectra of Complex Ions. II." *Physical Society of Japan* 9, 766–79.

[51] Virgo, D., and S. S. Hafner (1972). "Temperature-Dependent Mg,Fe Distribution in a Lunar Olivine." *Earth and Planetary Science Letters* 14, 305–12.

[52] Walsh, D., G. Donnay, and J. D. H. Donnay (1974). "Jahn–Teller Effects in Ferromagnesian Minerals: Pyroxenes and Olivines." *Bulletin de la Societé française de Mineralogie et Cristallographie* 97, 170–83.

[53] Walsh, D., G. Donnay, and J. D. H. Donnay (1976). "Ordering of Transition Metal Ions in Olivine." *Canadian Mineralogist* 14, 149–50.

[54] Wenk, H.-R., and K. N. Raymond (1973). "Four New Structure Refinements of Olivine." *Zeitschrift für Kristallographie* 137, 86–105.

[55] Will, G., and G. Nover (1979). "Influence of Oxygen Partial Pressure on the Mg/Fe Distribution in Olivines." *Physics and Chemistry of Minerals* 4, 199–208.

[56] Zoltai, T. (1965). *Steroscopic Drawings of Polyhedral Mineral-Structure Models.* University of Minnesota, Minneapolis.

8
Garnets

GENERAL CHEMISTRY

Garnets are common orthosilicates that occur in a wide variety of rock types and geologic environments, and that are often used—alone or with other minerals—in the estimation of the pressures and temperatures to which metamorphic rocks have been subjected. Although natural garnets with end-member compositions are not very common, most natural specimens can be characterized in terms of six ideal end members:

pyrope, $Mg_3Al_2Si_3O_{12}$	uvarovite, $Ca_3Cr_2Si_3O_{12}$
almandine, $Fe_3^{2+}Al_2Si_3O_{12}$	grossular, $Ca_3Al_2Si_3O_{12}$
spessartine, $Mn_3Al_2Si_3O_{12}$	andradite, $Ca_3Fe_2^{3+}Si_3O_{12}$

The first three garnets listed are collectively known as *pyralspites* (from *py*rope, *al*mandine, *sp*essartine), and the final three are called *ugrandites* (*u*varovite, *gr*ossular, *and*radite). The garnets are grouped in this way because of their rather strongly bimodal chemistry in nature. Garnets with compositions that fall within the pyrope–almandine–spessartine compositional triangle are common, as are those with compositions that lie within the uvarovite–grossular–andradite field. Solid solution within each group is substantial, but field evidence for *complete* solid solution between groups is not compelling, although *substantial* solid solution can be demonstrated in natural occurrences. This topic will be discussed at greater length in the following.

Natural garnets whose compositions fall outside both the pyralspite and ugrandite groups are uncommon, but not unknown. Even though they may occur only rarely as pure minerals, some of them occur as minor components in many garnets of largely pyralspite or ugrandite compositions. Among these are the following:

knorringite, $Mg_3Cr_2Si_3O_{12}$
calderite, $Mn_3Fe_2Si_3O_{12}$
khoharite, $Mg_3Fe_2Si_3O_{12}$
goldmanite, $Ca_3V_2Si_3O_{12}$
kimzeyite, $Ca_3(Zr,Ti)_2(Al,Fe,Si)_3O_{12}$
schorlomite, $Ca_3(Fe,Ti)_2(Si,Fe)_3O_{12}$

Melanite is a Ti-rich andradite (that is, a Ti-poor schorlomite). A series of hydrogarnets also exists with formulas like $Ca_3Al_2(SiO_4)_{3-n}(OH)_{4n}$ (hydrogrossular) and $Ca_3Fe_2(SiO_4)_{3-n}(OH)_{4n}$ (hydroandradite). The end-member hydrogarnet is synthetic $Ca_3Al_2(O_4H_4)_3$, and has not been found to occur naturally. It has essentially the garnet crystal structure (to be discussed), with the substitution $H_4O_4 = SiO_4$. In addition, small amounts of F may replace some oxygen, presumably by the substitution $4F = SiO_4$.

In terms of geologic occurrences, the major end-member garnets are fairly restricted. Uvarovite is rare; indeed, many mineralogists and petrologists ignore it altogether and call the group to which it belongs *grandites*. Grossular and andradite are found principally in contact metamorphic rocks, whereas almandine is a product of regional metamorphism of moderate to high grade. Pyrope is found in rocks of the eclogite facies, suggesting a high-temperature and very high-pressure paragenesis. Spessartine occurs in silicic volcanic rocks, and is of primary magmatic origin. With the exception of pyrope, all of the pyralspite and ugrandite end members can be synthesized under atmospheric pressure (with controlled atmospheres in some cases); the synthesis of pyrope requires high pressure. We shall look more closely at the physical conditions under which various garnets form when we consider solid solutions and stability.

The general chemical formula for the garnet group is often written

$$\{X\}_3[Y]_2(Z)_3O_{12},$$

where $\{X\}$ represents a divalent cation in 8-coordination, $[Y]$ represents an octahedrally coodinated trivalent cation, and (Z) indicates a tetravalent, tetrahedrally coordinated cation. The braces, square brackets, and parentheses are used by convention (after [13]) so that even in a garnet formula containing many chemical elements, it will be immediately obvious which elements fulfill which structural roles. For the garnets of geologic significance, (Z) is silicon, with occasional but minor substitution of aluminum or perhaps iron. The large cation, $\{X\}$, may be Ca, Mn, Fe^{2+}, or Mg, and $[Y]$ may be Al, Fe^{3+}, Cr^{3+}, V^{3+}, and Ti^{4+}. Obviously, the introduction of Ti^{4+} for a trivalent atom requires the loss of charge somewhere else (perhaps Al for Si, or Na for Ca). Mössbauer spectra have been interpreted as indicating that even small amounts of Fe^{2+} and Fe^{3+} in grossular are ordered on the $\{X\}$ and $[Y]$ sites, as in almandine and andradite, respectively [32]. Other elements have been reported, but are not of major significance in the crystal chemistry of the group.

There are many nonsilicate garnets of importance in solid-state technology. Among them are garnets wherein (Z) = Al, Ga Ge, or Fe^{3+}. As above, where (Z) is trivalent, charge balance (i.e., electrostatic neutrality) must be maintained with substitutions or changes in oxidation number at other positions.

THE CRYSTAL STRUCTURE OF GARNETS

Most garnets are cubic (space group $Ia\bar{3}d$), and have an a cell edge between 11.5 and 12.5 Å. The structure consists of a continuous network of alternating corner-sharing octahedra and tetrahedra (i.e., $[Y]O_6$ and $(Z)O_4$ groups). The oxygen framework of this arrangement also encloses large voids shaped like triangular dodecahedra (Figure 8-1). As can be seen in Figure 8-2, these voids can be considered as distorted cubes if the slight noncoplanarity of the four oxygen atoms defining a "cube face" is ignored. These eight anions coordinate the $\{X\}$ atoms.

There are 96 oxygen atoms in the unit cell. If these anions are approximated by spheres of radius 1.3 Å, then they occupy between ~45% and ~60% of the volume of the unit cell, depending upon the cell edge. Since spheres in a closest-packed array occupy ~74% of the total volume, it is clear that garnets are not very efficiently packed structures. Nonetheless, because of the fairly high atomic weights of the cations, garnets are quite dense (ρ = 3.5–4.3 g/cm^3) and rather hard (6.5–7.5), and have high refractive indices (1.70–1.94).

A portion of the garnet structure is shown in Figure 8-3. The number of shared polyhedral edges is quite large; each tetrahedron shares two opposite edges with triangular dodecahedra, each octahedron shares six edges with triangular dodecahedra, and each triangular dodecahedron shares two edges with tetrahedra, four with octahedra, and four with other triangular dodecahedra. We have rather consistently observed that shared polyhedral edges are shorter than unshared ones; the traditional explanation for this involves mutual repulsion of cations across the shared edge, but we have also seen that molecular-orbital calculations tend to predict the observed geometry (at least for the olivines) without invoking cation–cation repulsions (p. 251). Whatever the ultimate cause, all of the shared edges in garnets are shorter than unshared ones, except in the case of the octahedra of Ca-garnets. In these the edges shared between the octahedron and triangular dodecahedron are significantly longer than the other octahedral edges. For garnets in which $\{X\}$ =

Figure 8-1. A triangular dodecahedron (the $\{X\}$ site), an octahedron (the $[Y]$ site) and a tetrahedron [the (Z) site] as they occur in the garnet crystal structure.

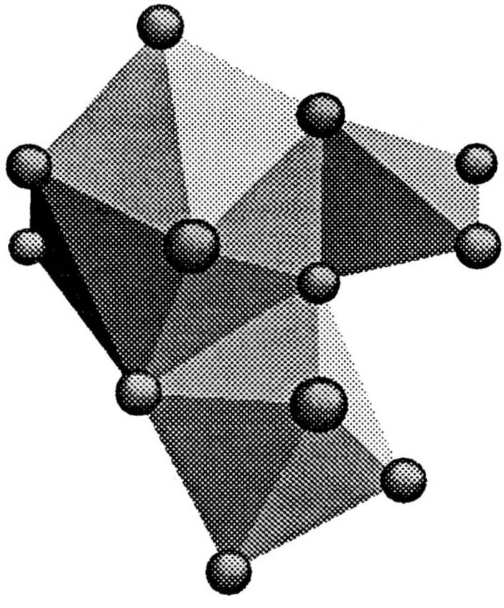

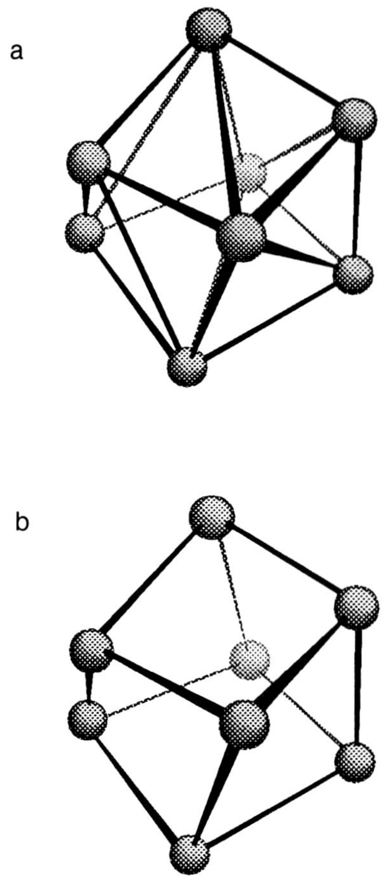

Figure 8-2. (a) The triangular dodecahedron can be visualized as (b) a distorted cube. The distorted site is rather like a cubic antiprism—a figure formed when one face of a cube is rotated 45° with respect to the opposite face and the remaining four faces are skewed as they follow the rotation.

(Ca,Mn), the mean size of the triangular dodecahedron is smaller than in Ca-garnets, and the two types of octahedral edges are statistically equal in length. If the $\{X\}$ cation is Fe^{2+} or Mg, the shared octahedral edges are the shorter ones, as expected. Thus the relative sizes of the polyhedra sharing edges may cause exceptions to the rule. Cations of low formal charge in large polyhedra may not repel adjacent cations enough to shorten the shared edges (if, indeed, repulsion is the effect of importance).

The crystal structure of garnet turns out to be quite simple from a quantitative standpoint. Only the oxygen atoms are on general positions (see Chapter 9). The (Z) atoms occupy the Wyckoff *d* positions with site symmetry $\bar{4}$; the $[Y]$ atoms are at Wyckoff *a* sites ($\bar{3}$), and the $\{X\}$ atoms are at Wyckoff *c* positions (222). As a consequence, the fractional coordinates of the $\{X\}$, $[Y]$, and (Z) atoms are known if the cell edge is known, and the oxygen position (Wyckoff position *h*) is the only one that need be determined by single-crystal diffraction.

This rather specialized crystallography results in simple relationships between chemistry and structural details. The cell edge may be estimated from the following equation, obtained by multiple regression on data from many silicate garnets [41]:

$$a = 9.04(2) + 1.61(4)\langle r_{\{X\}}\rangle + 1.89(8)\langle r_{[Y]}\rangle,$$

where $\langle r_{\{X\}}\rangle$ is the mean radius of the $\{X\}$ cation (as obtained from reference [47]), $\langle r_{[Y]}\rangle$ is the mean radius of the $[Y]$ cation, and the numbers in parentheses are the standard deviations of the regression coefficients. (Note that these standard errors can lead to a predicted cell parameter that is in error by more than 0.1 Å.) Inasmuch as there are 96 oxygen atoms in the unit cell and there are 96 general equivalent positions in the space group, all 96 oxygen atoms must be equivalent and there is but one unique oxygen position (x,y,z). The positional parameters of that oxygen atom may be estimated as follows [41]:

$$x = 0.0059 + 0.022\langle r_{\{X\}}\rangle + 0.014\langle r_{[Y]}\rangle,$$

$$y = 0.0505 - 0.023\langle r_{\{X\}}\rangle + 0.037\langle r_{[Y]}\rangle,$$

$$z = 0.6431 - 0.009\langle r_{\{X\}}\rangle + 0.034\langle r_{[Y]}\rangle.$$

Figure 8-3. A portion of the garnet structure showing the high proportion of shared polyhedral edges. (After [41].)

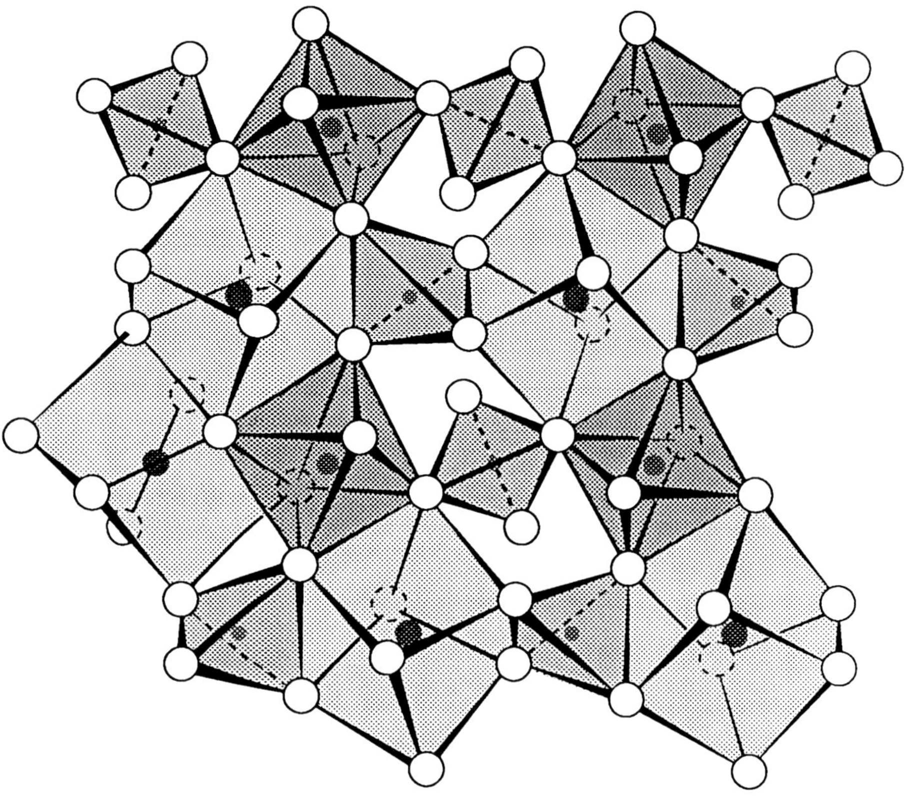

These equations apply only to silicate garnets, because they are independent of the radius of the (Z) atom. Silicate garnets are, of course, the only kind of significance in geology, but there are many synthetic nonsilicate garnets, and these equations are inappropriate for them. For any oxide garnets, including the silicate garnets, the following equations, which include the radius of the tetrahedrally coordinated cation, may be used [19]:

$$x = 0.0141 + 0.0278\langle r_{\{X\}}\rangle + 0.0123\langle r_{[Y]}\rangle - 0.0482\langle r_{(Z)}\rangle,$$

$$y = 0.0523 - 0.0237\langle r_{\{X\}}\rangle + 0.0200\langle r_{[Y]}\rangle + 0.0321\langle r_{(Z)}\rangle,$$

$$z = 0.6519 - 0.0102\langle r_{\{X\}}\rangle + 0.0305\langle r_{[Y]}\rangle - 0.0217\langle r_{(Z)}\rangle.$$

Similarly, an alternative equation for the unit-cell edge of any oxide garnet, including silicate garnets, is [40]:

$$a = 8.44 + 1.71(1)\langle r_{\{X\}}\rangle + 1.78(1)\langle r_{[Y]}\rangle + 2.17(1)\langle r_{(Z)}\rangle.$$

Other equations could be—and have been (e.g., [6])—developed for garnets of various types, but they really do not lead to greater understanding of the crystal chemistry of the group. The fact that the cell edge and oxygen positional parameters are simply related to the chemistry, regardless of the exact form of the relations, is adequately demonstrated.

SOLID SOLUTION AMONG GARNETS

Insight into the solid solutions among members of the garnet family comes both from observations of the chemical compositions of natural garnets that have crystallized under a wide variety of geologic conditions and from phase equilibrium studies on synthetically produced garnets. The joins that have been investigated in laboratory studies are shown in Figure 8-4 for compositions among the six principal end members as well as for knorringite, a less common but petrologically important garnet.

Pyralspite Group

Among natural garnets, complete solid solution is found in the pyrope–almandine–spessartine compositional field. The most abundant pyralspites are solid solutions between pyrope and almandine and between almandine and spessartine, and there are apparently no miscibility gaps in these series. Synthesis experiments along the pyrope–almandine join show that the solubility of pyrope in almandine increases with pressure—for example, the maximum solubility at 2 kbar water pressure is about 22 mol % pyrope (at 648°C) [24], whereas 43 mol % can be accommodated at 10 kbar and 925°C [52]. The requirement of rather extreme pressures and temperatures for the incorporation of large concentrations of Mg is consistent with occurrences of natural garnets.

Natural garnets compositionally intermediate between pyrope and spessartine are uncommon, and known examples do not fill the entire compositional range (see, for example, [46]). While there is no clear evidence for a miscibility gap, com-

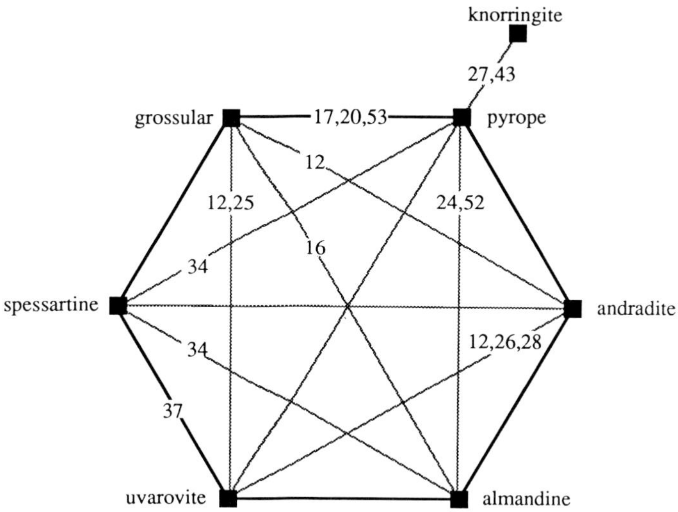

Figure 8-4. Numbers indicate references (in the numerical list of references cited at the end of this chapter) that describe studies of solid solution between the connected garnet end members. (After [35].)

positions more magnesian than $pyr_{60}spe_{40}$ have not been reported. Two crystal chemical observations may bear on this: First, an analysis of the spatial relationships among the atoms of the garnet structure, particularly with respect to the lengths of shared polyhedral edges, has led to the suggestion that Mn is the most stable $\{X\}$ cation and Mg the least stable [8;54]. Second, spessartine may be synthesized in a nonoxidizing atmosphere at a pressure of 1 bar, but the synthesis of pyrope requires very high pressure. These observations suggest a small stability field for pyrope-spessartine solid solutions.

Ugrandite Group

Natural garnets throughout the grossular–andradite (grandite) composition range are very common. Because uvarovite is itself an uncommon end member, it is not surprising that uvarovite–grossular and uvarovite–andradite solid solutions have not been found to any great extent in nature. Nonetheless the complete uvarovite–andradite series has been synthesized at atmospheric pressure and temperatures below about 1140°C [26]. Likewise, the synthetic series uvarovite–grossular has been produced at atmospheric pressure and temperatures up to around 855°C [25]. Thus it appears that, over some pressure range that includes atmospheric pressure, the ugrandites constitute a completely miscible solid solution series among three end members.

Miscibility between Pyralspites and Ugrandites

For many years the existence of a miscibility gap between the pyralspites and the ugrandites was taken for granted. Structurally, however, there is no reason to propose immiscibility between them. Figure 8-5 shows the "structural stability field" of silicate garnets in terms of the radii (taken from [47]) for $\{X\}$ and $[Y]$ cations. The diagram was constructed by using the first set of equations given above for the cell edge and oxygen fractional coordinates to predict structural details for about 200 hypothetical garnets having various combinations of cation radii. Those inside the "stability field" are those that possess crystallochemically reasonable steric details (Si–O bond lengths, nonbonded O···O distances, and so forth). Inasmuch as all of the known silicate garnets fall within this field, it is obvious that all solid solutions between them do also. Thus, from a crystallographic point of view, there is no reason to suspect a miscibility gap. What the "stability" diagram does not convey, however, is that the growth of some of these possible garnets, particularly those near the edges of the "stability field" or those requiring cations of widely disparate sizes to occupy crystallographically similar sites, may demand quite severe temperatures and pressures rarely attained under geologic conditions. Consider, for example, the long-presumed miscibility gap between about $ugr_{35}pyr_{65}$ and $ugr_{65}pyr_{35}$. About 2% of published garnet chemical analyses fall within this "gap," prompting the suggestion that, while these intermediate garnets are produced only under seldom-realized conditions of pressure, temperature, and bulk chemistry, no real gap actually exists [39]. The successful synthesis of a complete solid solution series between uvarovite and spessartine at 27.5 kilobars and 1450°C [37] lends support this conclusion.

The recognition that some garnets occur only in quite restricted geologic environments leads to the idea that garnets may reveal a lot about the physical conditions of formation of the rocks in which they occur. The following observations are

Figure 8-5. A "structural stability field" for silicate garnets, based on the radii of the $\{X\}$ and $[Y]$ cations (as found in [47]). The natural garnets fall within the light shaded rectangle. The darker shaded area outlines cation radii that result in crystal chemically reasonable structural details such as bond lengths and distances between closest nonbonded atoms. (After [41].)

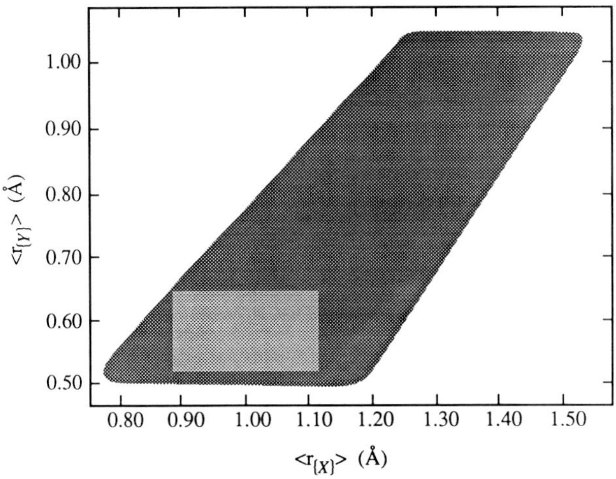

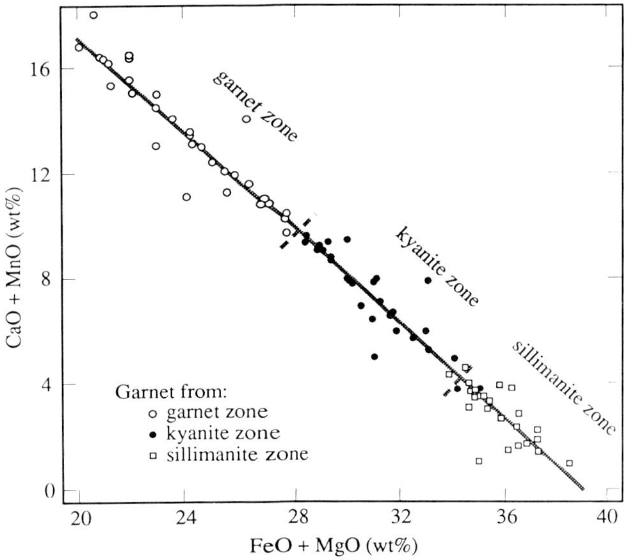

Figure 8-6. Compositions of garnets from regionally metamorphosed rocks as a function of the Barrovian zones in which they occur. After [38].

relevant to garnet paragenesis: (1) A high concentration of the appropriate $\{X\}$ and $[Y]$ cations in the rocks facilitates formation of the intermediate garnets. (2) As the grade of metamorphism increases, more diverse types of garnets become completely miscible. (Diversity here is measured in terms of the ionic radii of the respective cations.) For example, a grossular–spessartine garnet is found in some pyroxene–hornfels skarns, a grossular–almandine garnet is found in some almandine amphibolite and glaucophane schist facies rocks, and grossular–pyrope and grossular–almandine–pyrope garnets occur in the highest grades of the eclogite facies. (3) Pressure is more important than temperature in the formation of grossular–pyralspite garnets.

With this information and the general occurrences of end-member species outlined previously, it is apparent that correlations may exist between chemical–physical variables and the metamorphic grade of garnetiferous rocks. The plots of Figures 8-6 and 8-7 show ([CaO] + [MnO]) versus ([FeO] + [MgO]) and ([FeO] + [MgO])/([CaO] + [MnO]) versus the *a* cell edge for garnets collected from the garnet, kyanite, and sillimanite zones of Barrovian metamorphic terrains. The results suggest that these chemical parameters are useful for characterizing the specific metamorphic grades of garnetiferous rocks in areas of progressive regional metamorphism of the kyanite–sillimanite (that is, medium-pressure) facies series. It should be emphasized that these plots are not intended for use with garnets from contact metamorphic rocks or rocks from high- or low-pressure facies series [36].

Despite the "crystallographic stability field" of Figure 8-5, and observations and syntheses of garnets with compositions intermediate between pyralspites and ugrandites, there is some evidence that exsolution in garnets should and does occur.

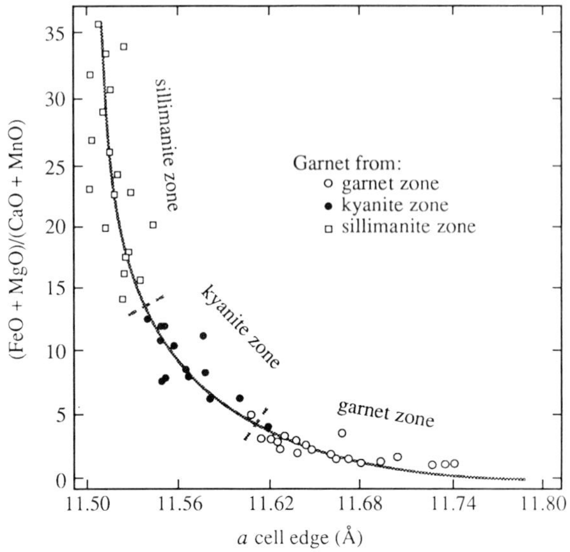

Figure 8-7. The correlation between ([FeO] + [MgO])/([CaO] + [MnO]) and the unit-cell edge for garnets of regionally metamorphosed rocks. After [38].

Calculation of the thermodynamic properties of garnet solutions suggests that a solvus should exist, for example, in the pyrope-grossular system, with exsolution below about 700°C at low pressures [17;21]. Neither optical nor x-ray data have directly confirmed immiscibility between pyralspite and ugrandite garnets.

There is reported exsolution [22;23], however, involving garnets that display birefringence (and which are therefore not cubic; see the following discussion) and are about midway between andradite and grossular in composition—that is, both phases are ugrandites. A tweedlike texture visible in thin section with polarized light has been shown by imaging backscattered electrons in a scanning electron microscope to consist of lamellae of somewhat different compositions (Figure 8-8). The texture is superficially similar to microcline twinning [see Figure 2-10(b)], and an origin by exsolution seems probable. The crystal lattices of these two phases appear only slightly misoriented, as judged from x-ray precession photographs suggesting the (unconfirmed) possibility of origin by spinodal decomposition.

Because nearly all garnets are isotropic (i.e., cubic), exsolution in these would not be visually apparent under the polarizing microscope even if present. Careful observations using electron microscopy, and perhaps X-ray precession photography, may show it to be much more common in isotropic garnets than so far recognized.

BIREFRINGENT GARNETS

In general, garnets crystallize in the isometric space group $Ia\overline{3}d$, and are therefore optically isotropic. Thus under the polarizing microscope with nicols crossed, gar-

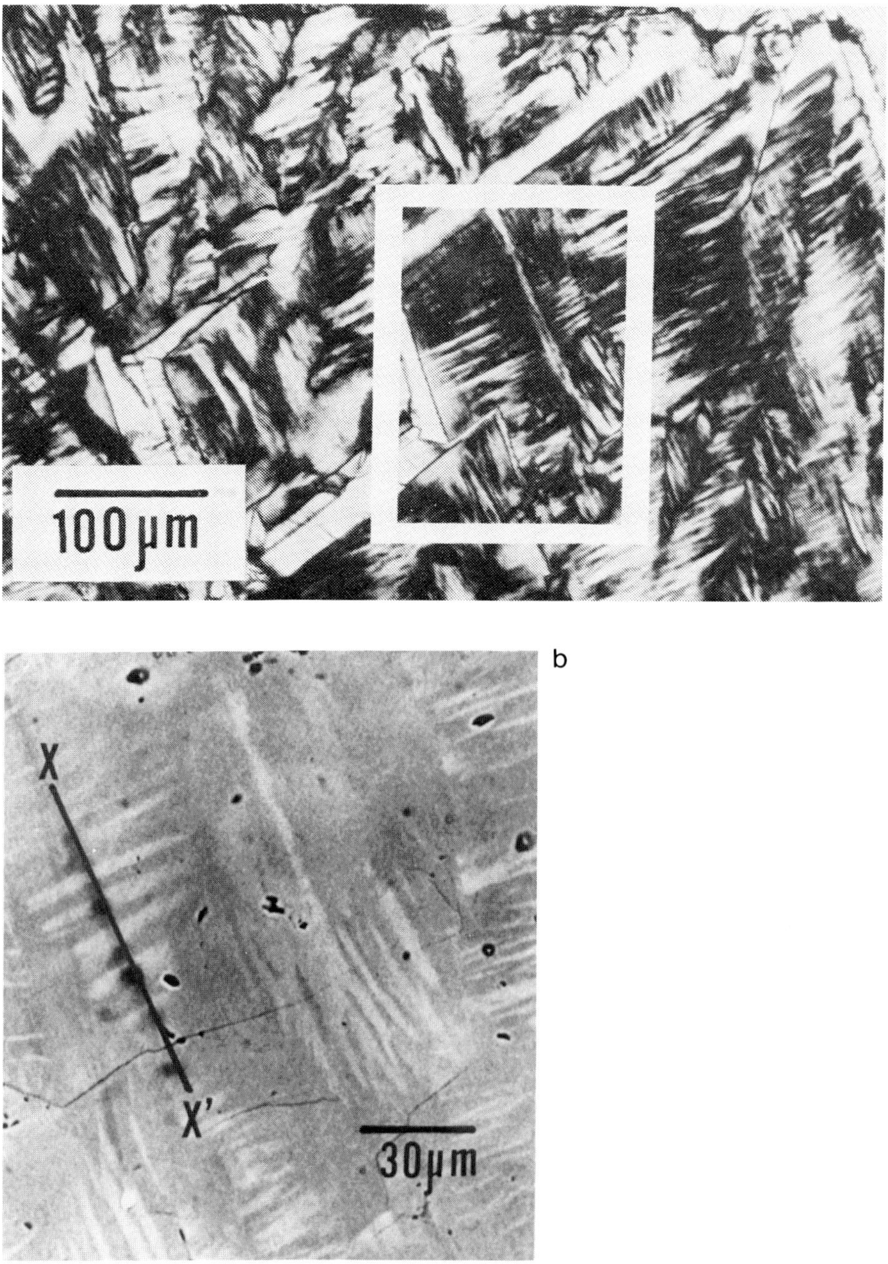

Figure 8-8. (a) A photomicrograph taken in cross-polarized light showing the tweedlike texture in a birefringent $and_{50}gro_{45}spe_{5}$. (b) A back-scattered electron image of the area outlined in (a). Contrast in this micrograph arises from differences in the average atomic numbers of the intergrown lamellae. [Photos from H. Hirai and H. Nakazawa (1986). "Visualizing Low Symmetry of a Grandite Garnet on Precession Photographs." *American Mineralogist*, 71, 1210–13.]

nets are extinguished in all orientations. It has long been known, however, that some garnets are birefringent, albeit weakly, and cannot therefore be truly isometric. This is most common among calcic garnets from contact metamorphic and metasomatic environments. In the past this has been attributed to crystallographic strain (which may well be correct in a relatively few cases), but a different explanation is now apparent for most such garnets.

Recent structure refinements of birefringent garnets [49;50] have revealed that some are orthorhombic, and crystallize in the space group *Fddd*. Figure 8-9 shows the relationship between the cubic and orthorhombic lattices, looking down the orthorhombic *b* axis. The pseudocubic cell edges differ from one another by only ~0.003 Å, and interaxial angles deviate from 90° by only ~0.1°. The origin of the birefringence and noncubic symmetry is to be found in ordering of the [*Y*] cations. In the space group *Fddd*, the 16 [*Y*] sites that are equivalent in $Ia\bar{3}d$ split into two groups of 8 sites each. In two birefringent grossular–andradite (grandite) garnets, the numbers of Fe^{3+} atoms in the two nonequivalent octahedral sites turns out to be 0.255 and 0.145 in one case, and 0.65 and 0.51 in the other [49]. (The remainder in each site is, of course, Al.)

Other birefringent garnets of the grossular–andradite series have been determined to be triclinic, $I\bar{1}$. One with composition $gro_{68}and_{32}$ shows Fe^{3+} occupancies in the eight nonequivalent octahedra of $I\bar{1}$ ranging from 0.15 to 0.49 atom [49]. Even garnets with compositions very close to that of an end member may be noncubic, as evinced by an $I\bar{1}$ garnet of composition $and_{94}gro_5spe_1$ that contains from 88 to 99 percent Fe^{3+} in the eight nonequivalent octahedral sites [30].

Figure 8-9. The relationship between the cubic *a* and *c* (light lattice) and the orthorhombic (pseudocubic) *a* and *c* (dark lattice, boldface labels) for garnets with space group *Fddd*. The orthorhombic unit cell is outlined. Angular differences between the two lattices have been greatly exaggerated for clarity. (After [49].)

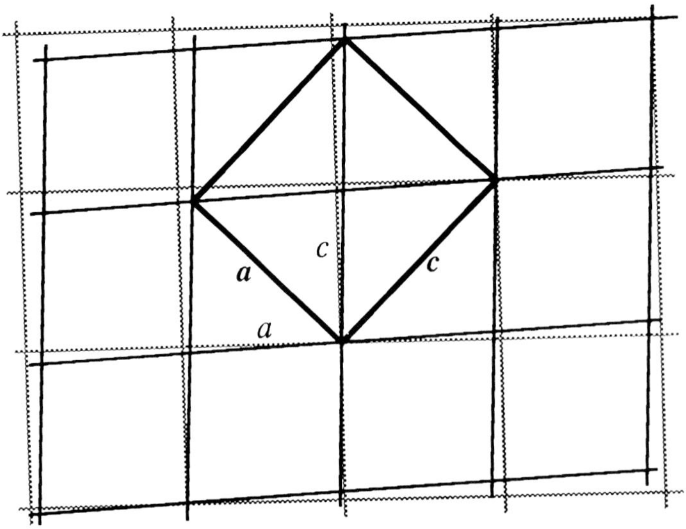

Some garnets may deviate from $Ia\overline{3}d$ symmetry for reasons other than ordering. In at least one case [44], anisotropy of a near end-member grossular has been attributed to a noncubic orientation of hydroxyl groups as revealed by infrared spectroscopy (see the following for a discussion of hydroxyl in garnet). In another garnet, a $gro_{90}and_6alm_4$, infrared and x-ray-diffraction experiments showed that there was ordering of Al and Fe^{3+} on the [Y] sites, ordering of Ca and Fe^{2+} on the {X} sites, and a noncubic distribution of hydroxyl groups [4]; however, loss of water content upon heating to 800°C did not eliminate the birefringence while an increase in temperature to 870°C did, suggesting that hydroxyl orientation may contribute to birefringence but that the primary cause is ordering.

Grandite garnets that owe their birefringence to ordering can often be made isotropic by heating (as mentioned previously). Temperatures of inversion of 800–950°c have been reported, and they decrease with increasing pressure [49]. If birefringence is a measure of disorder among octahedral cations, then it must depend on composition as well as temperature. (It should be obvious that neither a pure grossular nor a pure andradite could be ordered, because there is only one kind of octahedral cation.) In a study of natural garnets from a single locality, ranging in composition from $gro_{72}and_{28}$ to $gro_{33}and_{67}$, the most highly birefringent garnet was found to be $gro_{60}and_{40}$ [33], rather than $gro_{50}and_{50}$ as one might expect.

All of this suggests that contact metamorphic garnets might have significant potential as geothermometers. If the amount of birefringence can be correlated with both the degree of octahedral disorder and composition, and if degree of disorder and composition can be correlated with temperature (and pressure?) of inversion from $Ia\overline{3}d$ symmetry, then birefringent garnets might be used to estimate conditions of metamorphism. A great deal of work in this area will be required before that is possible.

Not all anisotropic garnets are compositionally close to the grossular–andradite join. The crystal-structure of a birefringent garnet with composition $alm_{63}gro_{24}pyr_8spe_3and_1$ (approximately $pyralspite_{75}ugrandite_{25}$) from an ancient subduction zone in Turkey [14] has been refined in space group $I4_1/acd$; slight ordering occurs among the X cations, such that 16 of the 24 cubic {X} sites are preferentially occupied by a small excess of Ca and 8 are preferentially occupied by an excess of (Fe^{2+},Mg,Mn). Some pyralspites with compositions near the almandine–spessartine join are also birefringent; these often contain small amounts of fluorine. Certain garnets may have symmetry lower than $Ia\overline{3}d$, but still be cubic and isotropic. For example, a pyrope with several mole percent grossular component has been reported [9] to show reflections on single-crystal photographs that are forbidden in $Ia\overline{3}d$, but are consistent with $I2_13$; Ca and Mg are the largest and smallest of the {X} cations, respectively, and the likely reason for the reduction in symmetry involves distortions created as the large Ca atom substitutes for the smaller Mg atom, moving some atoms off the special ($Ia\overline{3}d$) positions. Garnets such as these would, of course, be optically isotropic because they are cubic, though of a lower symmetry than $Ia\overline{3}d$.

Silicate garnets that are distinctly noncubic have been produced artificially at very high pressures and temperatures [2;5;10]. At about 1000°C and 125 kbar, $MnSiO_3$ undergoes a polymorphic phase transformation from the clinopyroxene

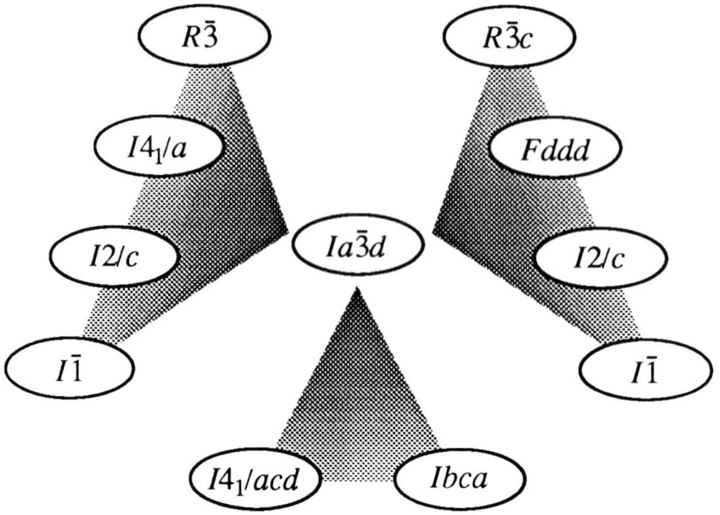

Figure 8-10. Possible transformations from space group $Ia\bar{3}d$ to subgroup symmetry. Transformations between space groups within the same shaded area are permitted, but transformations between space groups in different shaded areas are not. (After [18].)

structure to the garnet structure. Multiplying the formula by four and recasting it to fit the standard garnet model, $Mn_3(MnSi)Si_3O_{12}$, reveals that a Si atom is octahedrally coordinated—a circumstance unknown except in high-pressure phases. The garnet is tetragonal, space group $I4_1/a$, with three unique tetrahedra, two octahedra, and two triangular dodecahedra. Interestingly, it is also reported to be biaxial; if this is due to the intrinsic crystallography of the phase rather than to crystallographic strain induced by the conditions of synthesis, then the symmetry must actually be lower than tetragonal. $MgSiO_3$ also transforms to the tetragonal garnet structure [5;29] at very high pressures; it has been synthesized at 170 kbar and 1800°C from a glass of enstatite composition, and, like $MnSiO_3$ garnet, crystallizes in space group $I4_1/a$. This makes $MgSiO_3$ garnet a phase of great geophysical interest, because magnesium silicate is thought to be an important component of the Earth's mantle above the 670 km seismic discontinuity.

Whether anisotropic garnets experience a phase transformation from $Ia\bar{3}d$ to lower symmetry, or crystallize initially in the observed space groups is a question that has not yet been fully resolved. If the symmetry of some birefringent grandite garnets can be raised to $Ia\bar{3}d$ by heating and then reduced again by annealing [15], then certainly that constitutes an artificially induced phase transformation; on the other hand, it has not been unequivocally shown that such phase transformations occur in nature. Some [3;11] have argued that ordering of Al and Fe^{3+} on {110} growth steps during crystal formation could account for the anisotropy and the observed space groups; such a process would generate domains of symmetry $I\bar{1}$ under any single growth step, the superposition of which could allegedly yield apparent symmetry of any centrosymmetric subgroup of $Ia\bar{3}d$ (which includes $Fddd$ and $I4_1/a$ as well as $I\bar{1}$).

The alternative view (namely, that the observed space groups could arise out of phase transitions) is supported by a group-theoretical analysis of the phase transformations from $Ia\bar{3}d$ allowable under given mechanisms [18]. It turns out that a transition caused solely by ordering of the $[Y]$ cations must result in a phase of symmetry $Fddd$, $I\bar{1}$, $I2/c$, or $R\bar{3}c$, the first two of which have been observed in birefringent grandites. Other transformation mechanisms may result in space groups $R\bar{3}$, $I4_1/a$, $I2/c$, or $I\bar{1}$, the second of which is observed in $MnSiO_3$, in which ordering is very different from that in the grandites. Yet other transformation mechanisms lead to space groups $I4_1/acd$ or $Ibca$, the first of which has been found in a natural pyralsite$_{75}$grandite$_{25}$ formed at relatively high pressure [14] (See Figure 8-10.)

Because phase transformations and crystal growth mechanisms are equally capable of predicting the garnet space groups so far observed, it is not yet clear whether one or the other is correct, or if perhaps both might occur in different crystals.

HYDROXYL IN GARNETS

While the garnet general formula, $\{X\}_3[Y]_2(Z)_3O_{12}$, contains no hydrous component, it turns out that many (perhaps even most [1]) garnets do contain "water" in the form of hydroxyl ions, $(OH)^-$. There exists a complete solid solution from grossular ($Ca_3Al_2Si_3O_{12}$) though $Ca_3Al_2(O_4H_4)_3$, which is known only as a synthetic compound. The whole hydrous series constitutes the hydrogrossulars, and two distinct intermediate members are recognized—*plazolite* [$Ca_3Al_2(SiO_4)_2(OH)_4$] (called *hibschite* or *hibscite* in earlier literature), and *katoite* [$Ca_3Al_2(SiO_4)(OH)_8$]. Hydrogarnets with andraditic chemical affinities also occur, though less commonly. A corresponding hydrous series for pyralspites is not known, but most pyralspites appear to contain from 0.02 to 0.25 wt % H_2O as hydroxyl [1].

For many years it has been recognized that the hydroxyl groups must enter the hydrogarnet structure as substituents for SiO_4 oxyanions, but until quite recently there were no crystal-structure refinements confirming the O_4H_4–SiO_4 substitution. Refinements of plazolite [7], katoite [45], and synthetic $Ca_3Al_2(O_4H_4)_3$ [31] available now not only verify that four hydroxyls replace a silicate tetrahedron, but reveal the configuration of the hydroxyls (Figure 8-11) and some interesting trends along the hydrogrossular series. With increasing hydroxyl content, the $(Z)O_4$ tetrahedron becomes larger on the average, owing to H–H repulsion and the fact that no central cation attracts the oxygens in the H_4O_4 tetrahedra; in the fully hydrous species the distance from the center of the tetrahedron to an oxygen atom is 1.950 Å (as opposed to 1.645 Å in anhydrous grossular). This has an interesting effect on other polyhedra.

Recall that the shared edges of the $[Y]$ octahedron in anhydrous grossular are actually longer than the unshared edges (2.756 Å versus 2.686 Å). As H_4O_4 is substituted for SiO_4, the shared edges shorten and the unshared edges lengthen until the shared edges are 2.608 Å long and the unshared edges are 2.811 Å long. Presumably this derives from the fact that the (Z) tetrahedron shares two edges with the $\{X\}$ triangular dodecahedron: As the tetrahedron expands to accommodate H_4O_4, these shared edges necessarily lengthen, and those shared between the octahedron and

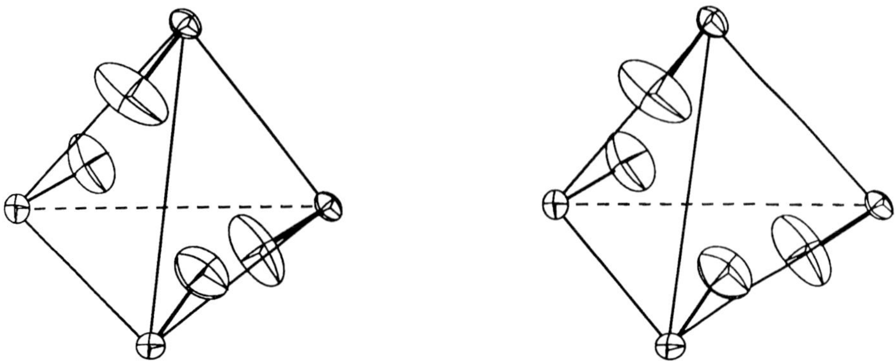

Figure 8-11. Stereographic view of the H_4O_4 tetrahedron in synthetic hydrogarnet. Small thermal ellipsoids at the corners of the tetrahedron represent oxygen atoms, with hydrogens nearly in the tetrahedral faces and showing larger thermal displacements. (After [31].)

the dodecahedron must shorten to maintain a reasonable mean size for the latter; this then drives a lengthening of the unshared octahedral edges for an analogous reason.

Consider now the question of a possible hydrous pyralspite series. In pyralspites, the shared octahedral edges are already shorter than unshared ones (2.617 Å as opposed to 2.716 Å). The chain of related distortions begun by the substitution of H_4O_4 for SiO_4 should lead to unreasonably short shared octahedral edge-lengths after comparatively little hydroxylation [44]. Hence pyralspite garnets with high water content (>1 wt % H_2O) are unknown in nature. Further support for this argument is found in the observation that, in andradite, the difference between the lengths of the (longer) shared octahedral edges and the (shorter) unshared ones is less than that difference in grossular, and indeed hydroandradite is less common than hydrogrossular.

DETERMINATIVE METHODS SHORT OF CHEMICAL ANALYSIS

If the equipment necessary for quantitative chemical analysis is not available, the composition of a garnet can be approximated from the measurements of certain physical properties. The most useful and easily measured properties are the unit-cell dimension a, the refractive index n, and the specific gravity G. Each of these varies linearly with composition, but none varies uniquely with composition due to the large number of end members. Figures 8-12(a) and (b) are nomograms that can be used for the determination of garnet compositions given a, n, and G. Since a and n can be measured more precisely than can G (largely because inclusions that cannot be eliminated affect the measurement of G, but need not affect measurement of the other two variables), these two values constitute the initial input to the diagrams. The value of G aids in determining the "depth" of the composition point relative to the triangular faces. If the garnet is thought (or known) to have no manganese, Figure 8-12(a) is used. Figure 8-12(b) is employed for garnets containing a significant amount of the spessartine component. Compositions are determined by application of the familiar lever rule used with phase diagrams (see Chapter 12). Because three properties are being used to "solve" for five compositional variables, some ambiguity will generally exist. Qualitative chemical tests or knowledge of the rock type and associations will help resolve such ambiguities. Of course, microprobe analysis or some other direct quantitative method (as described in Chapter 11) is always preferable to the techniques just described, but they may not be immediately accessible, and if only approximate compositions are required, use of the nomograms is a relatively fast and inexpensive method.

RECASTING GARNET FORMULAS

The discussion above suggests that many, if not most, garnets should fall within a compositional tetrahedron, or at worst a five-cornered solid if Mn is present. This turns out to be true in a general way—that is, the chemical analyses of most garnets can be partitioned among three to five major common end-members that account for most of the analysis. Even so, garnets often contain Ti, OH, or other chemical components that are not found in the six common end members, and assignment of elements to end members can be done in different ways that result in quite different apportionments. Even if only the six common end members are considered, there are 6!, or 720, different possible sequences in which to calculate the end-mem-

Figure 8-12. Nomograms for estimating garnet composition from index of refraction, *a* cell edge, and density in g/cm^3. (a) Used if MnO is an unimportant or absent component; (b) used if this is not the case. Compositions are in terms of common end-member garnets; no account is taken of Ti, V, Cr or other possible constituents, which may be a source of error. (After [51].)

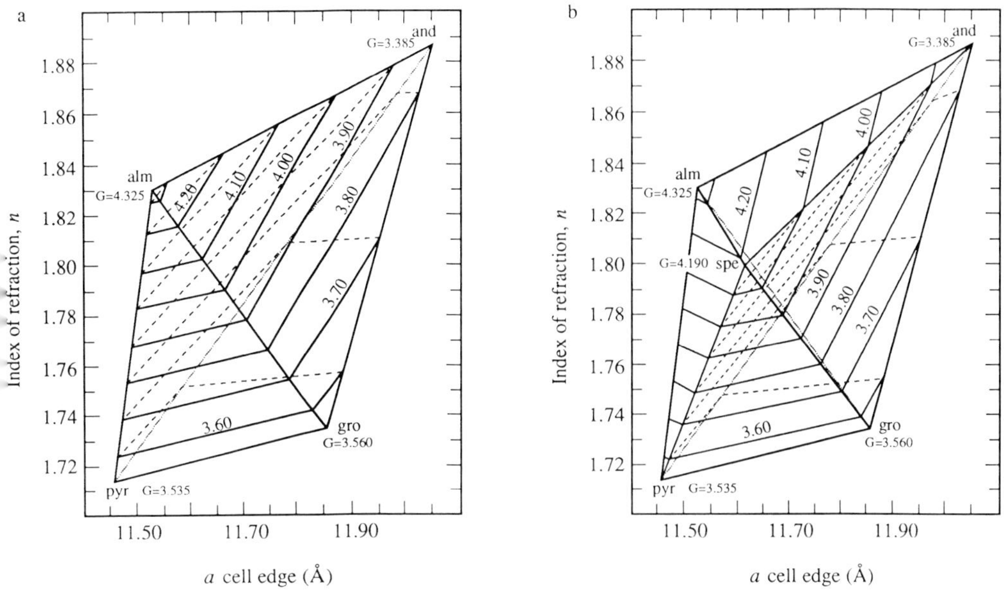

ber components, and these may result in significantly different amounts of each end member, starting with the very same chemical analysis each time; moreover, one cannot say that any of the possible sequences is "correct" or "incorrect"—they are simply different.

In order to standardize the recasting of chemical analyses into percentages of end-member components, a calculation scheme has been proposed [42]. The scheme is not universally followed by mineralogists and petrologists, but it does have certain advantages to recommend it. Among these are the following: (1) It is relatively simple to do, and can be easily programmed on a computer (though it is easy enough to accomplish with a small hand-held calculator); (2) experience has shown that it results in a large percentage of the cations in the analysis entering into one garnet component or another—that is, there is little of the analysis left over that does not fit into some garnet component; and (3) it minimizes the number of unusual garnet components needed, and in fact usually results in only the six common end members and hydrogrossular being utilized to a large degree.

The first step in recasting the analysis is to determine the numbers of each type of cation in a formula based on 12 oxygens. This is done in the normal way: First the oxide weight percentages are divided by their formula weights to obtain molecular proportions of oxides. Next these are multiplied by the number of oxygens in each oxide formula to obtain atomic proportions of oxygen, and these are summed to get the total atomic proportion of oxygen. This number is divided into 12 (in the case of garnet, because the formula is based on 12 oxygens) to get the multiplicative constant, which is then multiplied by each atomic proportion of oxygen to yield the *numbers* of oxygens per 12 oxygens associated with each oxide (i.e., the oxygens are apportioned to the various cations). Finally, the numbers of cations are determined by multiplying the numbers of oxygens (per 12 oxygens) by the numbers of cations per oxygen in the oxide formula unit. The following calculation on a garnet from a skarn illustrates how this is done (chemical analysis from [48]):

Oxide	Wt. %	Mol. prop.	At. prop. O	At. prop. O per 12 O	Cations per 12 O
SiO_2	37.76	0.6284	1.2568	5.8134	2.907 Si
TiO_2	0.14	0.0018	0.0036	0.0167	0.008 Ti
Al_2O_3	19.63	0.1925	1.5775	2.6712	1.781 Al
Fe_2O_3	4.26	0.0267	0.0801	0.3705	0.247 Fe^{3+}
FeO	4.20	0.0585	0.0585	0.2706	0.271 Fe^{2+}
MnO	2.85	0.0402	0.0402	0.1859	0.186 Mn
MgO	0.51	0.0127	0.0127	0.0587	0.059 Mg
CaO	30.31	0.5405	0.5405	2.5001	2.500 Ca
H_2O	0.44	0.0244	0.0244	0.1129	0.226 H
	100.10		2.5943		8.185
			12.0/2.5943 = 4.6255		

Note that the ideal cation total of 8.0 for the garnet structure is considerably exceeded here because of the presence of hydrogen, which substitutes for Si at the rate of 4H for each Si. If we let 4H = 1Si, subtract the 0.226H from the total and add the corresponding amount of Si (0.057), the cation total is then 8.016, which suggests a rather good analysis. The chemical formula is thus approximately

$$\{Ca_{2.500}Fe^{2+}_{0.271}Mn_{0.186}Mg_{0.059}\}[Al_{1.744}Fe^{3+}_{0.248}Ti_{0.008}](Si_{2.907}Al_{0.036}H_{0.226})O_{12},$$

but this does not in itself provide a unique division into end members.

The cations are now apportioned to end-member garnets in the following order:

1. Yttrogarnet	$Y_3Al_2Al_3O_{12}$
2. Yamatoite	$Mn_3V_2Si_3O_{12}$
3. Goldmanite	$Ca_3V_2Si_3O_{12}$
4. Kimzeyite	$Ca_3Zr_2Al_2SiO_{12}$
5. Ferric-kimzeyite	$Ca_3Zr_2Fe_2SiO_{12}$
6. Uvarovite	$Ca_3Cr_2Si_3O_{12}$
7. Andradite	$Ca_3Fe_2Si_3O_{12}$
8. Pyrope	$Mg_3Al_2Si_3O_{12}$
9. Spessartine	$Mn_3Al_2Si_3O_{12}$
10. Hydrogrossular	$Ca_3Al_2H_{12}O_{12}$
11. Grossular	$Ca_3Al_2Si_3O_{12}$
12. Almandine	$Fe_3Al_2Si_3O_{12}$
13. Schorlomite	$Ca_3Ti_2Fe_2TiO_{12}$
14. Hydrogrossular	$Ca_3Al_2H_{12}O_{12}$
15. Hydroandradite	$Ca_3Fe_2H_{12}O_{12}$
16. Repeat from step number 8	
17. Knorringite (=hanléite)	$Mg_3Cr_2Si_3O_{12}$
18. Repeat from step number 9	
19. Khoharite	$Mg_3Fe_2Si_3O_{12}$
20. Skiagite	$Fe_3Fe_2Si_3O_{12}$
21. Calderite	$Mn_3Fe_2Si_3O_{12}$
22. Blythite	$Mn_3Mn_2Si_3O_{12}$

Some comments about the sequence and the procedure are appropriate. First, note that hydrogrossular appears twice in the list, as numbers 10 and 14. This is because of the way in which schorlomite is formed; Ca, Fe^{3+}, and Ti are combined to form schorlomite, and then enough andradite is released to combine with any excess Ti to make additional schorlomite, provided that enough Al and a divalent cation still exist to generate an equal amount of grossular, pyrope, spessartine, or almandine—a process that is repeated as many times as possible. Next, if Mg, Mn, or Fe^{2+} still remains along with extra Fe^{3+} and Ti, then additional schorlomite, with pyrope, spessartine, or almandine, is formed by release of grossular. After the schorlomite is formed by these multiple methods, any grossular that remains is combined with the remaining water to make hydrogrossular, and then any water beyond that is used to make the less common hydroandradite.

Note also that the process is partly iterative. After step number 15, one returns to step number 8 in order to use up Si released in the formation of hydrogrossular and hydroandradite. After step 17, one returns to step 9 to use up Al that was released when some pyrope was released to provide Mg to unite with surplus Cr to make hanléite. It is apparent that the initial steps are intended to use up elements that fit nowhere in the formulas of common garnets, then the common garnets are formed in the bulk of the procedure, and finally, when there is no more Al, the excess common elements are combined in some less common garnet end members. The procedure thus allocates as much of the analysis as possible to the common garnets.

As an example, we continue with the garnet from the skarn. The numbered steps below refer to the steps listed above.

1. There is no yttrogarnet, because there is no Y.
2. There is no yamatoite because there is no V.
3. There is no goldmanite for the same reason.
4. There is no kimzeyite because there is no Zr.
5. For the same reason, there is no ferric-kimzeyite.
6. There is no uvarovite because there is no Cr.
7. Andradite: $Ca_{0.371}\underline{Fe}^{3+}_{0.247}Si_{0.371} = 0.124$ and.
8. Pyrope: $\underline{Mg}_{0.059}Al_{0.039}Si_{0.059} = 0.020$ pyr.
9. Spessartine: $\underline{Mn}_{0.186}Al_{0.124}Si_{0.186} = 0.062$ spe.
10. Hydrogrossular: $Ca_{0.057}Al_{0.038}\underline{H}_{0.226} = 0.019$ hgr.
11. Grossular: $\underline{Ca}_{2.072}Al_{1.382}Si_{2.072} = 0.691$ gro.
12. Almandine: $Fe^{2+}_{0.219}Al_{0.146}\underline{Si}_{0.219} = 0.073$ alm.
13. Schorlomite: $Ca_{0.008}Ti_{0.005}Fe^{3+}_{0.005}Ti_{0.003} = 0.002$ sch. This reduces andradite formed in step 7 by releasing $Ca_{0.008}Fe^{3+}_{0.242}Si_{0.008}$, resulting in a revised andradite amount of $Ca_{0.363}Fe^{3}_{0.242}Si_{0.363} = 0.121$ and.
14. There is no remaining H for the formation of additional hgr.
15. There is no remaining H for the formation of additional hand.
16. Repeat from step 8. At this stage we have 0.052 Fe^{2+}, 0.052 Al, and 0.008 Si remaining.

8. There is no Mg remaining for the formation of additional pyr.
9. There is no Mn remaining for the formation of additional spe.
10. There is no H remaining for the formation of additional hgr.
11. There is no Ca remaining for the formation of additional gro.
12. Almandine: $Fe_{0.008}Al_{0.005}\underline{Si}_{0.008} = 0.003$ alm. Adding this to the 0.073 alm formed in the previous pass through step 12 yields 0.076 alm.

In each step the element whose concentration determines the amount of the particular garnet that is formed is underlined. It was unnecessary to keep track of oxygen, as the oxide analysis guaranteed a sufficiency of that element. The subscripts for the elements not underlined were determined by requiring appropriate ratios of cations as specified in their ideal formulas. The fraction of each end-member was calculated by dividing the $\{X\}$-cation subscript by 3, or the $[Y]$-cation subscript by 2. The total should be 1.000 in the absence of any analytical errors; in this case, it is 0.991.

We now have 0.044 Fe^{2+} and 0.047 Al left over. Because all Si, Ti, and H have been used up, there is no way to form additional garnets, and we have thus gone as far as we can. Moreover, we have accounted for over 99% of the cations, which suggests that this is a very good analysis; because it is a wet chemical analysis rather than a microprobe analysis, one could speculate that the excess Fe^{2+} and Al represent impurities from inclusions in the garnet, but perhaps it is simply minor experimental error. Calculation of end-member percentages yields $pyr_{2.0}alm_{7.7}spe_{6.3}gro_{69.7}$and $_{12.2}sch_{0.2}hgr_{1.9}$. These are listed here with pyralspite components first, followed by ugrandites and then schorlomite and hydrogarnet; the order is obviously inconsequential, and is often chosen as the order of abundance.

REFERENCES

[1] Aines, R. D., and G. R. Rossman (1984). "The Hydrous Component in Garnets: Pyralspites." *American Mineralogist* 69, 1116–26.

[2] Akimoto, S., and Y. Syono (1972). "High Pressure Transformations in $MnSiO_3$." *American Mineralogist* 57, 76–84.

[3] Akizuki, M. (1984). "Origin of Optical Variations in Grossular–Andradite Garnet." *American Mineralogist* 69, 328–38.

[4] Allen, F. R., and P. R. Buseck (1988). "XRD, FTIR, and TEM Studies of Optically Anisotropic Grossular Garnets." *American Mineralogist* 73, 568–84.

[5] Angel, R. J., Finger, L. W., Hazen, R. M., Kanzaki, M., Weidner, D. J., Liebermann, R. C., and Veblen, D. R. (1989). "Structure and Twinning of Single-Crystal $MgSiO_3$ Garnet Synthesized at 17 GPa and 1800°C." *American Mineralogist* 74, 509–12.

[6] Basso, R. (1985). "Crystal Chemical and Crystallographic Properties of Compound with Garnet or Hydrogarnet Structure." *Neues Jahrbuch für Mineralogie Monatshefte*, 108–14.

[7] Basso, R., A. Della Giusta, and L. Zefiro (1983). "Crystal Structure Refinement of Plazolite: a Highly Hydrated Natural Hydrogrossular." *Neues Jahrbuch für Mineralogie Monatshefte*, 251–58.

[8] Born, L., and J. Zemann (1964). "Abstrandsberechnungen und Gitterenergitische Berechnungen an Granaten." *Contributions to Mineralogy and Petrology* 10, 2–23.

[9] Dempsey, M. J. (1980). "Evidence for Structural Changes in Garnet Caused by Calcium Substitution." *Contributions to Mineralogy and Petrology* 71, 281–82.

[10] Fujino, K., H. Momoi, H. Sawamoto, and M. Kumazawa (1986). "Crystal Structure and Chemistry of $MnSiO_3$, Tetragonal Garnet." *American Mineralogist* 71, 781–85.

[11] Gali, S. (1983). "Grandite Garnet Structures in Connection with the Growth Mechanism." *Zeistschrift für Kristallographie* 163, 43–52.

[12] Ganguly, J. (1976). "The Energetics of Natural Garnet Solid Solution. II. Mixing of the Calcium Silicate End-Members." *Contributions to Mineralogy and Petrology* 55, 81–90.

[13] Geller, S. (1967). "Crystal Chemistry of the Garnets." *Zeitschrift für Kristallographie* 125, 1–47.

[14] Griffen, D. T., W. R. Phillips, and D. M. Hatch (1990). "Refinement of the Crystal Structure of a Natural Tetragonal Garnet." *Geological Society of American Program with Abstracts* 22, A215.

[15] Hariya, Y., and M. Kimura (1978). "Optical Anomaly Garnet and Its Stability Field at High Pressures and Temperatures." *Journal of the Faculty of Science, Hokkaido University, Series IV* 18, 611–24.

[16] Hariya, Y. and K. Nakano (1972). "Garnet Solid Solution between Grossular and Almandine." *Journal of the Mineralogical Society of Japan* 10, 373.

[17] Haselton, H. T., and R. C. Newton (1980). "Thermodynamics of Pyrope–Grossular Garnets and their Stabilities at High Temperatures and High Pressures." *Journal of Geophysical Research* 85, 6973–82.

[18] Hatch, D. M., and D. T. Griffen (1989). "Phase Transitions in the Grandite Garnets." *American Mineralogist* 74, 151–59.

[19] Hawthorne, F. C. (1981). "Some Systematics of the Garnet Structure." *Journal of Solid State Chemistry* 37, 157–64.

[20] Hensen, B. J. (1976). "The Stability of Pyrope–Grossular Garnet with Excess Silica." *Contributions to Mineralogy and Petrology* 55, 279–92.
[21] Hensen, B. J., R. Schmid, and B. J. Wood (1975). "Activity–Composition Relationships for Pyrope–Grossular Garnet." *Contributions to Mineralogy and Petrology* 51, 161–66.
[22] Hirai, H., and H. Nakazawa (1986a). "Grandite Garnet from Nevada: Confirmation of Origin of Iridescence by Electron Microscopy and Interpretation of a Moiré-Like Texture." *American Mineralogist* 71, 123–26.
[23] Hirai, H., and H. Nakazawa (1986b). "Visualizing Low Symmetry of a Grandite Garnet on Precession Photographs." *American Mineralogist* 71, 1210–13.
[24] Hsu, L. C., and C. W. Burnham (1969). "Phase Relationships in the System $Fe_3Al_2Si_3O_{12}$–$Mg_3Al_2Si_3O_{12}$–H_2O at 2.0 kilobars." *Geological Society of America Bulletin* 80, 2393–08.
[25] Huckenholz, H. G., and D. Knittel (1975). "Uvarovite: Stability of Uvarovite–Grossular Solid Solution at Low Pressure." *Contributions to Mineralogy and Petrology* 49, 211–32.
[26] Huckenholz, H. G., and D. Knittel (1976). "Uvarovite: Stability of Uvarovite–Andradite Solid Solution at Low Pressure." *Contributions to Mineralogy and Petrology* 56, 61–76.
[27] Irifune, T., and Y. Hariya (1983). "Phase Relationships in the System $Mg_3Al_2Si_3O_{12}$–$Mg_3Cr_2Si_3O_{12}$ at High Pressure and Some Mineralogical Properties of Synthetic Garnet Solid Solutions." *Mineralogical Journal* 11, 269–81.
[28] Isaacs, T. (1965). "A Study of Uvarovite." *Mineralogical Magazine* 35, 38–45.
[29] Kato, T., and M. Kumazawa (1985). "Garnet Phase of $MgSiO_3$ Filling the Pyroxene–Ilmenite Gap at Very High Temperature." *Nature* 316, 803–05.
[30] Kingma, K. J., and J. W. Downs (1989). "Crystal-Structure Analysis of a Birefringent Andradite." *American Mineralogist* 74, 1307–16.
[31] Lager, G. A., Th. Armbruster, and J. Faber (1987). "Neutron and X-ray Diffraction Study of Hydrogarnet $Ca_3Al_2(O_4H_4)_3$." *American Mineralogist* 72, 756–65.
[32] Manning, P. G., and M. J. Tricker (1977). "A Mössbauer Spectral Study of Ferrous and Ferric Iron Distributions in Grossular Crystals: Evidence for local crystal disorder." *Canadian Mineralogist* 15, 81–86.
[33] Mariko, T., and Y. Nagai (1980). "Relations Between Optical Properties and Chemical Composition of Grandite." *Mineralogical Society of Japan Annual Meeting*, Abstract C-26.
[34] Matthes, S. (1961). "Ergebnisse zur Granatsynthese und ihre Beziehungen zur Natürlichen Granatbildung innerhalf der Pyralspit-Gruppe." *Geochimica et Cosmochimica Acta* 23, 233–94.
[35] Meagher, E. P. (1980). "Silicate Garnets." In P. H. Ribbe (Ed.), *Orthosilicates. Reviews in Mineralogy*, Volume 5. Mineralogical Society of America, Washington, D.C.
[36] Miyashiro, A. (1961). "Evolution of Metamorphic Belts." *Journal of Petrology* 2, 277–311.
[37] Naka, S., Y. Suwa, and T. Kameyama (1975). "Solid Solubility between Uvarovite and Spessartite." *American Mineralogist* 60, 418–22.
[38] Nandi, K. (1967). "Garnets as Indices of Progressive Regional Metamorphism." *Mineralogical Magazine* 37, 89–93.
[39] Nemec, D. (1967). "The Miscibility of the Pyralspite and Grandite Molecules." *Mineralogical Magazine* 37, 389–402.
[40] Novak, G. A., and Coleville, A. (1975). "A Linear Regression Analysis of Garnet Chemistries Versus Cell Parameters." (Abstr.) *Geological Society of America Southwest Section Meeting, Los Angeles* 7, 359.
[41] Novak, G. A., and G. V. Gibbs (1971). "The Crystal Chemistry of the Silicate Garnets." *American Mineralogist* 56, 791–825.
[42] Rickwood, P. C. (1968). "On Recasting Analyses of Garnet into End-Member Molecules." *Contributions to Mineralogy and Petrology* 18, 175–98.
[43] Ringwood, A. E. (1977). "Synthesis of Pyrope–Knorringite Solid Solution Series." *Earth and Planetary Science Letters* 36, 443–48.
[44] Rossman, G. R., and R. D. Aines (1986). "Spectroscopy of a Birefringent Grossular from Asbestos, Quebec, Canada." *American Mineralogist* 71, 779–80.
[45] Sacerdoti, M, and E. Passaglia (1985). "The Crystal Structure of Katoite and Implications within the Hydrogrossular Group of Minerals." *Bulletin de Minéralogie* 108, 1–8.
[46] Schmetzer, K., and H. Bank (1981). "Garnets from Umba Valley, Tanzania—Members of the Solid Solution Series Pyrope–Spessartine." *Neues Jahrbuch für Mineralogie Monatshefte*, 349–54.
[47] Shannon, R. D., and C. T. Prewitt (1969). "Effective Ionic Radii in Oxides and Fluorides." *Acta Crystallographica* B25, 925–46.

[48] Shimazaki, H. (1977). "Grossular–Spessartine–Almandine Garnets from Some Japanese Scheelite Skarns." *Canadian Mineralogist* 15, 74–80.
[49] Takéuchi, Y., N. Haga, S. Umizu, and G. Sato (1982). "The Derivative Structure of Silicate Garnets in Grandite." *Zeitschrift für Kristallographie* 158, 53–99.
[50] Takéuchi, Y., and N. Haga (1976). "Optical Anomaly and Structure of Silicate Garnets." *Proceedings of the Japan Academy* 52, 228–231.
[51] Winchell, H. (1958). "The Composition and Physical Properties of Garnet." *American Mineralogist* 43, 595–600.
[52] Yoder, H. S., Jr., and G. A. Chinner (1960a). "Almandite–Pyrope–Water System at 10,000 bars." *Carnegie Institution of Washington Year Book* 59, 81–84.
[53] Yoder, H. S., Jr., and G. A. Chinner (1960b). "Grossularite–Pyrope–Water System at 10,000 bars." *Carnegie Institution of Washington Year Book* 59, 78–81.
[54] Zemann, J., (1962). "Zur Kristallchemie der Granate." *Beiträge zur Mineralogie und Petrologie* 8, 180–88.

PART II
SUPPORTING CONCEPTS

9

Crystal Symmetry

For as long as curious collectors and earnest inquirers have examined crystals, they have been captivated by the symmetry of the external shapes exhibited by well-formed specimens. It is now clearly recognized, of course, that these external shapes are directly related to the arrangements of atoms, and powerful methods exist for revealing these atomic arrangements (that is, the crystal structures). But in the past all that could be learned about the submicroscopic nature of a mineral was gained by careful measurements of the angles between faces of good specimens. Even without any knowledge of atoms and crystal structures, early crystallographers were able not only to determine the symmetries of crystals, but also to completely catalog all of the *possible* external crystalline symmetries (the crystallographic point groups). A complete derivation of all of the possible atomic-level symmetries (space groups) came later, but still well before the advent of x-ray crystallography, which made direct determination of atomic arrangements possible.

Inasmuch as these atomic-level symmetries play a central role in silicate crystal chemistry, it is crucial that the student have at least a working understanding of space groups. We shall begin, however, with a review of point groups, of which the 32 crystallographic point groups, or crystal classes, are a subset, then proceed to the crystal classes and 6 crystal systems, and finally discuss space groups. We shall conclude by briefly considering ways in which crystal structures can be represented tangibly.

REVIEW OF POINT GROUPS

A *point group* is a combination of compatible symmetry elements that intersect in a common point, and that leave that point invariant (that is, stationary) through all of their operations. Those operations consistent with this definition are rotation about an axis, reflection across a "mirror" plane, and inversion through a point. There are two systems of symbology in use for representing the symmetry elements corresponding to these operations. We shall use the *Hermann–Mauguin* (or *international*) system, wherein rotations are represented by Arabic numerals (for example, 4 for a fourfold axis), mirrors by the letter *m*, and enters of inversion by $\overline{1}$ (spoken either "bar one" or "one bar," depending on personal preference). An alternative scheme, called the *Schoenflies system*, uses C_n and S_n for proper and improper rotations, respectively (where the subscript is the "fold" of the rotation, e.g., C_4 for a fourfold axis), the Greek σ for mirrors, and the letter i for centers of inversion.

PROPER AND IMPROPER ROTATIONS

There are two general kinds of rotations—proper ones and improper ones. Proper rotations are those that map a right-handed motif into a symmetry-equivalent right-handed motif; that is, they leave the "handedness" of the motif unchanged. A fourfold axis (4), then, is a proper rotation axis (Figure 9-1).

An improper rotation is one that changes the "handedness" of the motif, which is to say that a right-handed object is mapped into a symmetry-equivalent left-handed version of that same object (Figure 9-2). This may be accomplished in two, and only two, ways. One is to combine the rotation operation with reflection across a plane normal to the rotation axis, and the other is to combine it with inversion through a point on the rotation axis; these new kinds of operations are called *rotoreflections* and *rotoinversions*, respectively. Symbolically, a rotoreflection axis is represented by an Arabic numeral with a tilde over it (for example, $\tilde{4}$), and a rotoinversion is represented by an Arabic numeral with a bar above it (for example, $\bar{4}$). It turns out that either system alone will suffice for the representation of all improper

Figure 9-1. A fourfold axis (4), by which a motif is repeated by rotations of 90°, is an example of a proper rotation. It does not change the handedness of the motif.

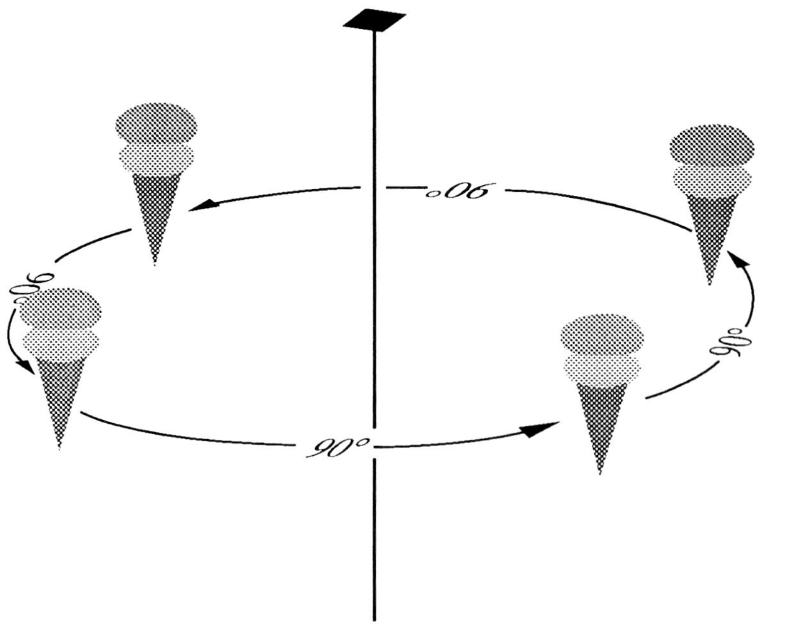

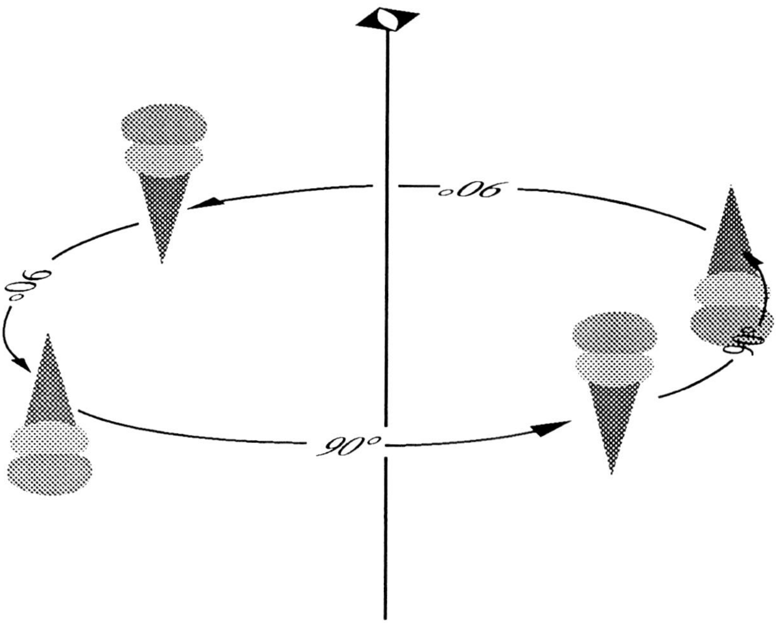

Figure 9-2. A fourfold rotoinversion axis ($\overline{4}$), by which a motif is repeated by rotations of 90° followed by inversion through a point on the axis, is an example of an improper rotation. It changes the handedness of the motif.

point group operations, because they are entirely redundant. Among mineralogists, the rotoinversion axes are used universally by convention. Some of the equivalences are as follows:

$$\overline{2} = \tilde{1}$$

$$\overline{1} = \tilde{2}$$

$$\overline{3} = \tilde{6}$$

$$\overline{4} = \tilde{4}$$

$$\overline{5} = \widetilde{10}$$

$$\overline{6} = \tilde{3}$$

$$\vdots$$

The general rules are,

1. if n is odd, then $\overline{n} = 2\tilde{n}$;
2. if n/2 is an even integer, then $\overline{n} = \tilde{n}$;
3. if n/2 is an odd integer, then $\overline{n} = \tilde{n}/2$.

We have mentioned the symbol $\bar{1}$ previously as a center of inversion; its reappearance here as a onefold rotoinversion axis should suggest that the two are not different (as a little thought about what a onefold axis does will confirm). Another important relationship is the equivalence of a $\bar{2}$ and a mirror plane perpendicular thereto. This is easily illustrated by the use of stereograms, and it is assumed that the reader has had an introduction to stereographic projection elsewhere. (Relatively standard conventions are used: small circles for points projected from the upper hemisphere, crosses for points projected from the lower hemisphere; a solid boundary for the stereogram if a mirror plane exists parallel to the page, otherwise a broken boundary.) Figure 9-3(a) shows an arbitrary crystal face-pole projected from the upper hemisphere, and a $\bar{2}$ axis normal to the page. Figure 9-3(b) shows

Figure 9-3. (a) The stereogram for a single $\bar{2}$ axis, with one arbitrary face pole projected from the upper hemisphere and shown by a small circle. (b) Operation of the $\bar{2}$ on the face pole rotates it 180° and inverts it through the center of the sphere of projection, yielding a symmetrically equivalent face pole in the lower hemisphere. (c) A mirror in the plane of the page operating on the same face pole in the upper hemisphere produces one in the lower hemisphere that is indistinguishable from the one produced by the $\bar{2}$, showing that a twofold rotoinversion axis is equivalent to a mirror normal to that axis.

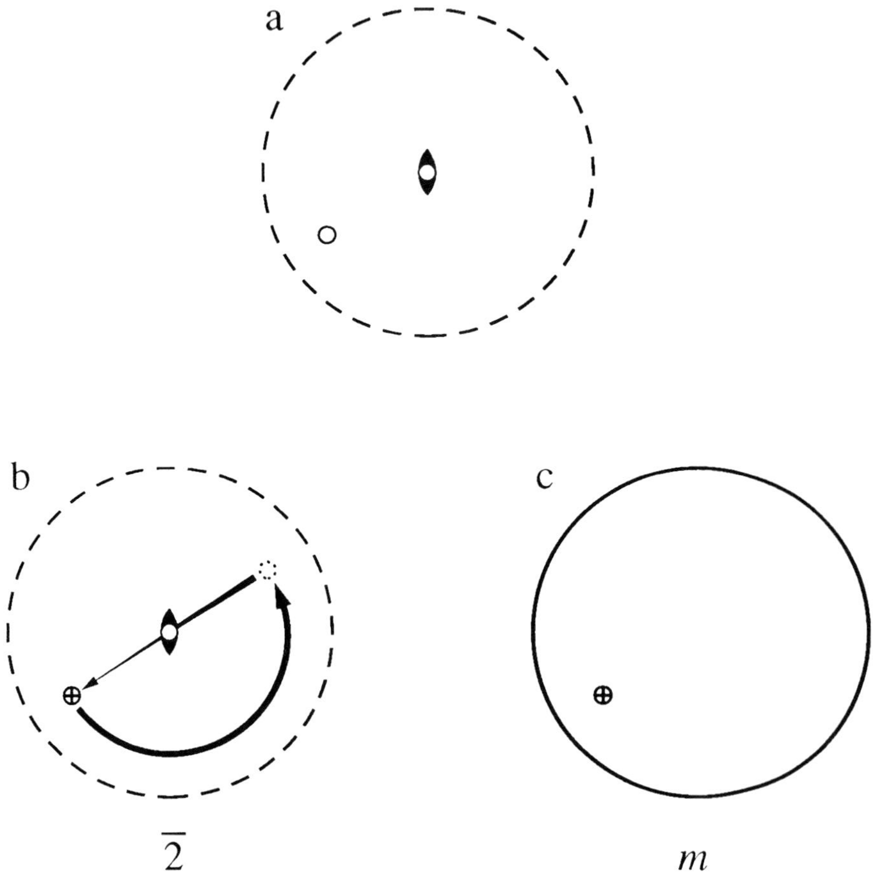

how the $\bar{2}$ axis operates on the face pole to introduce an equivalent one in the lower hemisphere, indistinguishable from the result produced by a mirror perpendicular to the $\bar{2}$ axis [(i.e., in the plane of the page—Figure 9-3(c)]. We conventionally use m rather than $\bar{2}$ to symbolize this operation, but it is important to recognize that *all* point-group symmetry operations—including inversion and reflection—can be considered as some sorts of rotations.

PERMISSIBLE ROTATIONS FOR CRYSTALLOGRAPHIC POINT GROUPS

Though the number of point groups is infinite, because new ones can be found simply by increasing the "fold" of the rotation axis without bound, the number of *crystallographic* point groups is finite. Crystallographic point groups are those whose symmetry elements are consistent with the symmetries of three-dimensional arrays of atoms that make up crystals, and their number is limited by the following constraint: A point group is a crystallographic point group if and only if the set of symmetry elements comprising it is *necessary* for the description of the shape of some crystallographic unit cell that, when repeated indefinitely by translations along three crystallographic axes, will completely fill space. Although the set of symmetry elements must be necessary for the description of an appropriate unit cell, it need not be sufficient. If not sufficient, it will be a subset of the set of symmetry elements that is both necessary and sufficient for such description.

As two intuitive examples, consider point groups consisting solely of (1) a fourfold rotation axis, and (2) an eightfold rotation axis. The unit cell for which a fourfold axis is necessary (though not sufficient) is a square prism. It can easily be visualized [Figure 9-4(a)] that such unit cells could be stacked together along three mutually perpendicular axes so that they occupied all of some arbitrary volume of space. The unit cell containing an eightfold axis would be an octagonal prism [Figure 9-4(b)], but such unit cells cannot be stacked together along any set of axes so as to completely fill space. Thus the point group consisting of a single eightfold axis is not a crystallographic point group. In fact, no point group possessing an eightfold axis could be a crystallographic point group, regardless of other symmetry elements that might be present.

In order to discover the inadmissible rotation axes for crystallographic point groups, one could perform the above exercise for all possible rotations. However, inasmuch as ∞-fold rotation axes are valid symmetry elements (consider, for example, the symmetry of a sphere), one would be a long time in this process. A less empirical approach is available that will identify the limited number of permissible rotations without ambiguity.

Consider an n-fold axis passing through a unit cell in a crystal. This may be either a proper or an improper axis. Because the placement of lattice points in a crystal is arbitrary, so long as the environment of any one of them is identical to that of every other, we shall put a lattice point on the n-fold axis. This means that every other lattice point in the crystal lattice will fall on a similar axis. In Figure 9-5(a), the open circles represent lattice points, each one of which is intersected by an n-fold axis having a rotation angle of α ($= 360°/n$). Our point of view is such that the n-fold axes are perpendicular to the plane of the page. Lattice points 1 through 4 lie on a lattice row that has a unit translation of a. A rotation of α degrees at point 3 moves point 4 to 4′, which is also a lattice point inasmuch as it is symmetrically identical to point 4 by virtue of the n-fold axis. Likewise a rotation of $-\alpha$ (or, equivalently, n-1 rotations of α) degrees at point 2 moves point 1 to 1′. The distance from 1′ to 4′ is b, but since these are lattice points, they must be separated by some integral number of unit cell translations, a. Thus, $b = pa$, where p is some integer. Referring to Figure 9-5b, it can be seen that b is also equal to $a + 2a(\cos \alpha)$. So,

$$pa = a + 2a(\cos \alpha)$$

$$p = 1 + 2(\cos \alpha)$$

$$\cos \alpha = (p-1)/2$$

$$\alpha = \cos^{-1} [(p-1)/2].$$

By substituting for p integers that give permissible values for the cosine function, we can now determine what rotation angles α are allowed in crystallographic point

Figure 9-4. (a) Unit cells with tetragonal (fourfold) symmetry can be stacked together so that space is completely filled. (b) Unit cells with octagonal (eightfold) symmetry cannot be stacked in any way that completely fills space, leaving no gaps.

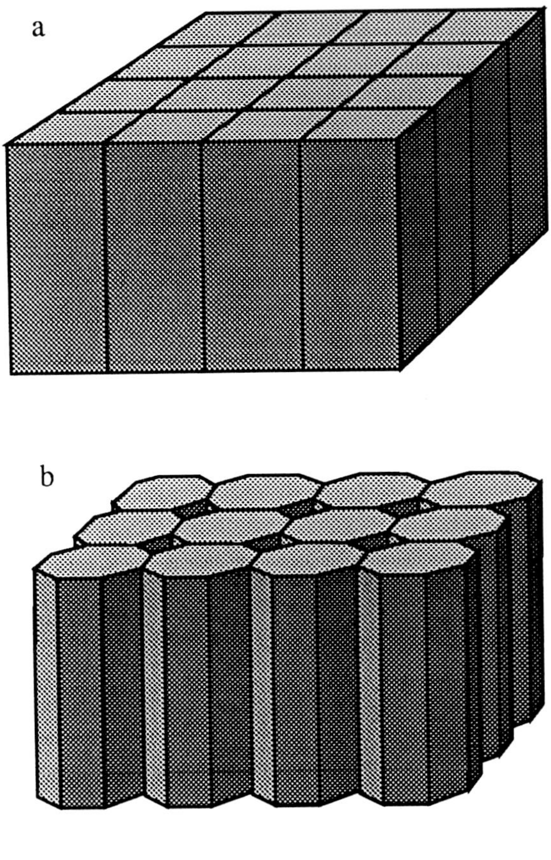

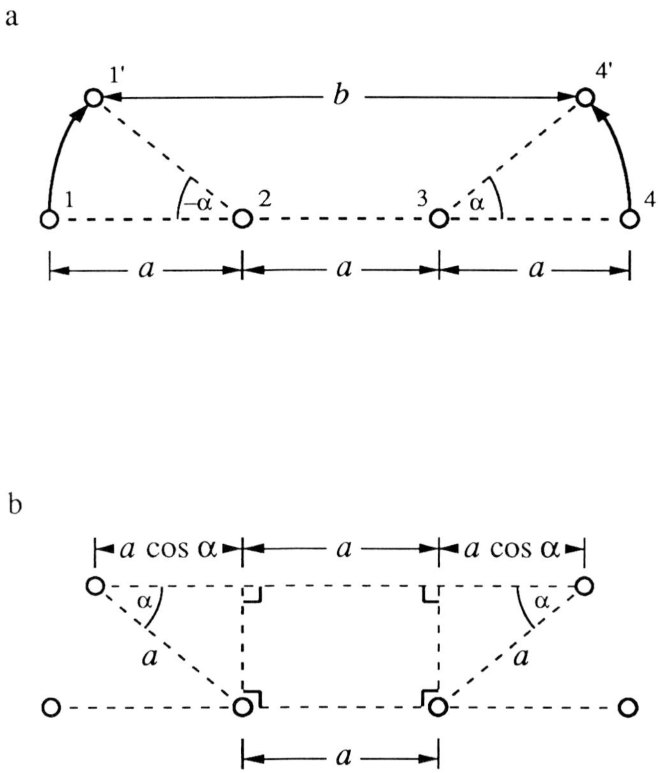

Figure 9-5. (a) Points 1 through 4 are lattice points along a lattice row. Each point is intersected by an n-fold rotation axis normal to the page. The angle for a single rotation of the axis is α. Rotation of point 4 by the axis at point 3 maps it into 4′, and n-1 rotations (or one negative rotation) of point 1 by the axis at point 2 maps it into 1′. The distance between 1′ and 4′ is b, which must be an integral number of lattice translations, pa, since 1′ and 4′ are lattice points on the same lattice row. (b) By constructing perpendiculars at 2 and 3, we see that $b = a + 2a(\cos \alpha)$, and that this equals pa. Hence the angles, α, that yield acceptable separations of points 1′ and 4′ can be found (see text).

groups. The values of p that give cosine values between -1 and $+1$, inclusive, are -1, 0, 1, 2, and 3. The corresponding angles of rotation are 180° (twofold), 120° (threefold), 90° (fourfold), 60° (sixfold), and 360° (onefold). With both proper and improper axes included, then, the only kinds of rotations that can occur in crystallographic point groups are 1, $\bar{1}$, 2, $\bar{2}$ ($\equiv m$), 3, $\bar{3}$, 4, $\bar{4}$, 6, and $\bar{6}$.

EULER'S CONSTRUCTION

Figure 9-6(a) represents a sphere through which pass two rotation axes. One is axis B, which points straight out of the page, and has a rotation angle of β degrees. The other is axis A, with a rotation angle of α degrees. A segment of the great circle through A and B is drawn, as is a segment of the one through A and a point M. The

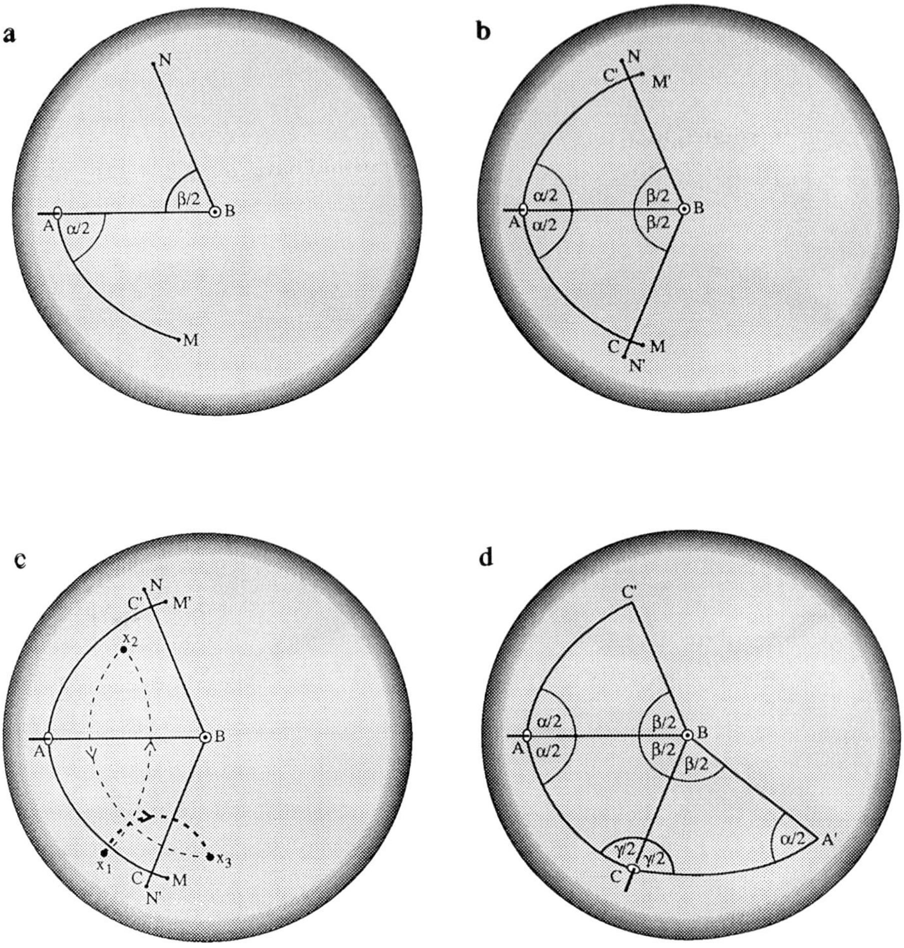

Figure 9-6. Euler's construction. (a) A and B are rotation axes that originate at the center of the shaded sphere and pass through it as shown. A single rotation on A is α degrees, and a single rotation on B is β degrees. The points at which the axes meet the sphere are connected by the great circle segment, AB. AM and BN are great circle segments that make angles of $\alpha/2$ and $\beta/2$, respectively, with AB. (b) Rotation of α about axis A maps AM into AM'; that is followed by rotation of β about axis B, which maps BN into BN'. Point C is the intersection of AM and BN', and C' is the intersection of AM' and BN. Evidently point C existed on AM and C' existed on BN before any rotations occurred, so the rotations mapped C into C', and C' back into C. C has therefore remained invariant throughout. (c) If C has remained invariant, it can only be a location where some new rotation axis intersects the sphere. Where positive rotations of A followed by B move x_1 to x_3, via x_2, a negative rotation of some magnitude maps x_1 directly into x_3. (d) The angle of rotation for axis C is $\gamma/2$.

arc AM is intentionally placed so that it makes an angle of $\alpha/2$ degrees with AB. Another arc, BN, is placed at $\beta/2$ degrees from AB. Now AM is rotated through α degrees about axis A, and is thus mapped into AM', which intersects BN at C' [Figure 9-6(b)]. Next, BN is rotated through β degrees about axis B, and is mapped into BN'. BN' intersects AM at point C, which evidently started on AM, was moved to C' on AM', and has now been returned to its original position. In other words, as x_1 has been rotated to x_2 and thence to x_3, C has effectively been invariant. Therefore, there must be some rotation about C that would have mapped x_1 directly into x_3 without the need for any other rotation.

We can determine the nature of such a rotation. Let the angle at C (and consequently at C') be called $\gamma/2$, and let a rotation of β degrees about axis B be performed on spherical triangle ABC'. The result is shown in Figure 9-6(c). We have already seen that a rotation of α degrees about A followed by a rotation of β degrees about B (symbolically, $A_\alpha B_\beta$) is equivalent to some rotation about C (i.e., $A_\alpha B_\beta = C$). Inasmuch as the effect of $A_\alpha B_\beta$ on point A is to map it into A' ($A_\alpha B_\beta(A) = A'$), then C must be the single rotation that maps A into A' ($C(A) = A'$). From Figure 9-6(d), the angle must be γ degrees, although it is a clockwise rotation whereas the rotations about A and B were counterclockwise. Thus

$$A_\alpha B_\beta = C_{-\gamma}.$$

This is Euler's construction, and its conceptual and quantitative implications are significant. It demonstrates that the combination of two rotation operations will generate a third rotation equivalent to that combination. Inasmuch as we have previously observed that mirror planes and centers of inversion are equivalent to certain improper rotations, and since Euler's construction is valid for improper rotations as well as for proper ones, the following theorem and its corollary can be stated:

Theorem: The combination of any two compatible symmetry elements generates a third symmetry element.

Corollary: The removal of one symmetry element from a point group necessitates the loss of another.

As an example of this theorem, consider two twofold axes lying in the plane of the page and separated by 45°. These are represented by filled ellipses on the stereograms in Figure 9-7(a) through (f). The sequence shows how a single face pole is reproduced by the twofold axes, and reveals that a fourfold axis, oriented vertically out of the plane of the page, has been generated at the center of the stereogram. We did not purposely place it there, but it arose because the combination of two twofold axes at 45° always produces a third symmetry element—a fourfold axis perpendicular to the plane of the twofold axes.

ANGULAR RELATIONSHIPS

If we now consider the spherical triangle ABC of Euler's construction, additional consequences are apparent. In Figure 9-8, u, v, and w are interaxial angles subtended by arcs *BC*, *AC*, and *AB*, respectively. Because the sum of the angles of a spherical triangle must be between 180° and 540°, we have

$$180° < \alpha/2 + \beta/2 + \gamma/2 < 540°.$$

If the letters *A*, *B*, and *C* signify the numbers of rotations necessary to produce an identity operation (that is, the "fold" of each axis), then we may substitute 360°/*A* for α, 360°/*B* for β, and 360°/*C* for γ:

$$180° < 360°/(2A) + 360°/(2B) + 360°/(2C) < 540°$$

$$180° < 180°/A + 180°/B + 180°/C < 540°$$

$$1 < 1/A + 1/B + 1/C < 3.$$

For three non-symmetry-equivalent, noncolinear rotations to be mutually compatible, this inequality must be satisfied. For example, the combination of three twofold axes yields

$$\tfrac{1}{2} + \tfrac{1}{2} + \tfrac{1}{2} = \tfrac{3}{2},$$

which satisfies the inequality and indicates that such a combination is acceptable; but the (noncolinear) combination of a twofold axis, a threefold axis, and a sixfold axis yields

$$\tfrac{1}{2} + \tfrac{1}{3} + \tfrac{1}{6} = 1$$

indicating that these three axes constitute a geometrically impossible combination.

Figure 9-7. If one begins with (a) a general face pole and two diads at 45° to each other, then [(b)–(e)] operates with only the diads, the resultant arrangement of face poles (f) reveals the presence of a fourfold axis normal to the plane of the diads.

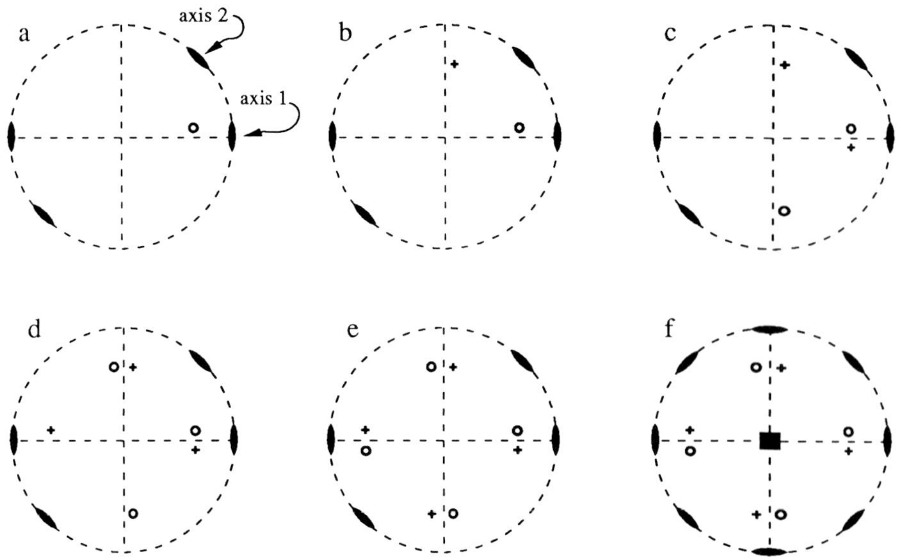

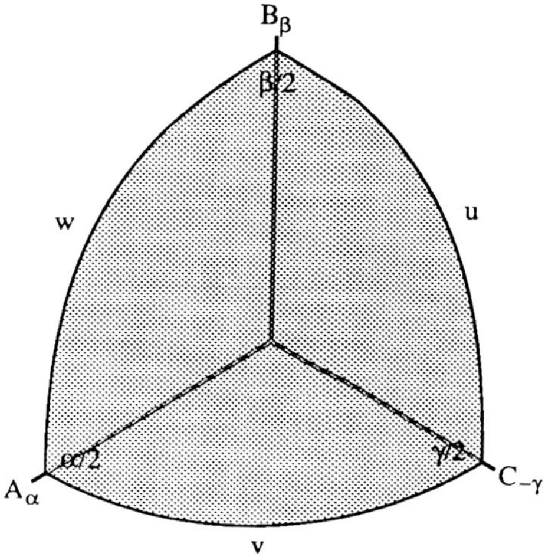

Figure 9-8. The spherical triangle *ABC* of Figure 9-6, showing the half-rotation angles $\alpha/2$, $\beta/2$, and $\gamma/2$, and the interaxial angles u, v, and w.

The following relationships from spherical trigonometry allow us to calculate the interaxial angles, given the rotation angles α, β, and γ for three noncolinear axes found to be compatible by the above inequality:

$$\cos u = [\cos(\alpha/2) + \cos(\beta/2)\cos(\gamma/2)]/[\sin(\beta/2)\sin(\gamma/2)],$$

$$\cos v = [\cos(\beta/2) + \cos(\alpha/2)\cos(\gamma/2)]/[\sin(\alpha/2)\sin(\gamma/2)],$$

$$\cos w = [\cos(\gamma/2) + \cos(\alpha/2)\cos(\beta/2)]/[\sin(\alpha/2)\sin(\beta/2)].$$

Given these three equations and the inequality, we can now summarize the permissible combinations of non-symmetry-equivalent, noncolinear two-, three-, four-, and sixfold axes and the angles between them (Table 9-1).

COMBINATIONS OF PROPER AND IMPROPER ROTATIONS

Consider again the spherical triangle of Euler's construction. The consecutive operation of rotation axes A_α and B_β must bring a motif into a position indistinguishable

Table 9-1. Permissible Crystallographic Combinations of the Rotation of Axes

Axis *A*	Axis *B*	Axis *C*	*u*	*v*	*w*
2	2	2	90°	90°	90°
3	2	2	60°	90°	90°
4	2	2	45°	90°	90°
6	2	2	30°	90°	90°
2	3	3	70°32′	54°44′	54°44′
4	3	2	35°16′	45°	54°44′

from that to which it is brought by rotation $C_{-\gamma}$. Because rotations A_α, B_β, and $C_{-\gamma}$ may be either proper or improper, there exists the possibility that the "handedness" of the motif may be changed by one or more of the operations. If either A_α or B_β (but not both) is improper, then a right-handed motif at point 1 (Figure 9-9) will be mapped into a left-handed one at point 3, and $C_{-\gamma}$ will have to be improper. If both A_α and B_β are proper, or if both are improper, $C_{-\gamma}$ will have to be proper. Thus the possible combinations of proper (P) and improper (I) operations are {PPP}, {PII}, {IPI}, {IIP}, and {P|I P|I P|I}. The last combination is a superposition of the others; the notation P|I denotes a proper rotation with an improper rotation coaxial (that is, along the same direction). Of course, a point group can be monaxial, in which case it may be characterized as {P}, {I}, or {P|I}. By examining systematically these combinations of one-, two-, three-, four-, and sixfold axes, we can now derive all of the crystallographic point groups. It should be stated that, except for point groups possessing no higher symmetry than a center of inversion, onefold axes are not considered, since all objects have an infinite number of them. In Table 9-2 the crystallographic point groups are derived, with the full Hermann–Mauguin symbols for each given in boldface type. The symbols not in boldface type are valid crystallographic point groups that have already appeared elsewhere in the table. Note that a $\overline{2}$ axis is represented as the mirror plane to which it is equivalent, and that an evenfold axis with a coaxial improper axis results in a proper axis with a mirror plane perpendicular to it (n/m).

Observe that there are exactly 32 crystallographic point groups. These are gen-

Figure 9-9. Repetition of an arbitrary motif from position 1 to 2 to 3 by three rotation axes. Note that all three axes may be proper, or two may be improper and the other proper. An odd number of improper operations (1 or 3) would not map a right-handed (or left-handed) motif back into itself.

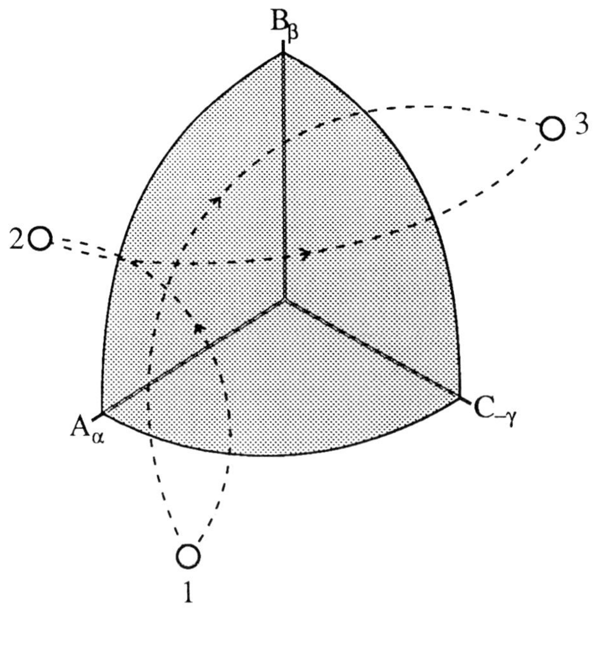

Table 9-2. The 32 Crystallographic Point Groups

P	I	P|I	PPP	PII	IPI	IIP	P|I P|I P|I
1	$\bar{1}$	$1\mid\bar{1}$ = (1)					
2	$\bar{2}$ = **m**	$2\mid\bar{2}$ = **2/m**	**222**	$2\bar{2}\bar{2}$ = **2mm**	$\bar{2}2\bar{2}$ = (m2m)	$\bar{2}\bar{2}2$ = (mm2)	$2\mid\bar{2}\ 2\mid\bar{2}\ 2\mid\bar{2}$ = **2/m 2/m 2/m**
3	$\bar{3}$	$3\mid\bar{3}$ = (3)	**322**	$3\bar{2}\bar{2}$ = **3mm**	$\bar{3}2\bar{2}$ = **$\bar{3}$ 2/m**	$\bar{3}\bar{2}2$ = ($\bar{3}$ 2/m)	$3\mid\bar{3}\ 2\mid\bar{2}\ 2\mid\bar{2}$ = ($\bar{3}$ 2/m)
4	$\bar{4}$	$4\mid\bar{4}$ **4/m**	**422**	$4\bar{2}\bar{2}$ = **4mm**	$\bar{4}2\bar{2}$ = **$\bar{4}$2m**	$\bar{4}\bar{2}2$ = ($\bar{4}$2m)	$4\mid\bar{4}\ 2\mid\bar{2}\ 2\mid\bar{2}$ = **4/m 2/m 2/m**
6	$\bar{6}$	$6\mid\bar{6}$ = **6/m**	**622**	$6\bar{2}\bar{2}$ = **6mm**	$\bar{6}2\bar{2}$ = **$\bar{6}$2m**	$\bar{6}\bar{2}2$ = ($\bar{6}$m2)	$6\mid\bar{6}\ 2\mid\bar{2}\ 2\mid\bar{2}$ = **6/m 2/m 2/m**
			233 = **23**	$2\bar{3}\bar{3}$ = **2/m3**	$\bar{2}3\bar{3}$ = **4/m $\bar{3}$ 2/*m***	$\bar{2}\bar{3}3$ = (4/m $\bar{3}$ 2/m)	$2\mid\bar{2}\ 3\mid\bar{3}\ 3\mid\bar{3}$ = (4/m $\bar{3}$ 2/m)
			432	$4\bar{3}\bar{2}$ = (4/m $\bar{3}$ 2/m)	$\bar{4}3\bar{2}$ = **$\bar{4}$3m**	$\bar{4}\bar{3}2$ = (4/m $\bar{3}$ 2/m)	$4\mid\bar{4}\ 3\mid\bar{3}\ 2\mid\bar{2}$ = (4/m $\bar{3}$ 2/m)

erally called the 32 *crystal classes*, and every crystal must possess the symmetry defined by one of them. Figure 9-10 shows the symmetry elements of all 32 crystal classes, by means of stereographic projections. The classes are arranged by crystal system, and, in cases where the short version of the Hermann–Mauguin symbol differs from the full symbol, the former is given.

TRANSLATIONAL SYMMETRY OPERATIONS

In beginning mineralogy classes, students are taught to determine crystal class by examining faces on well-formed crystals, or on models of ideal crystals. By recognizing those faces that are equivalent by some symmetry operation, the student can determine the presence of rotation axes, mirror planes, and centers of symmetry. The particular combination of symmetry elements identifies the crystal class uniquely and unambiguously.

On the unit-cell level, symmetry elements may be discovered in an analogous way. Here, however, we are dealing not with crystal forms, but with the positions of atoms. The principles involved are similar. Atoms are either symmetrically identical, or the are not. (Chemical identity is a necessary, but not a sufficient, condition for symmetrical identity. Two aluminum atoms in a unit cell, for instance, are not symmetrically identical unless some symmetry operation maps one into the other.) Examination of the atomic positions in a model or drawing of a unit cell reveals the positions of rotation axes, mirror planes, and centers of inversion. The same angular relationships that were derived for crystallographic point groups obtain here. Thus, only kinds of one-, two-, three-, four-, and sixfold axes are possible; symmetry-independent mirror planes that are parallel to a fourfold axis intersect each

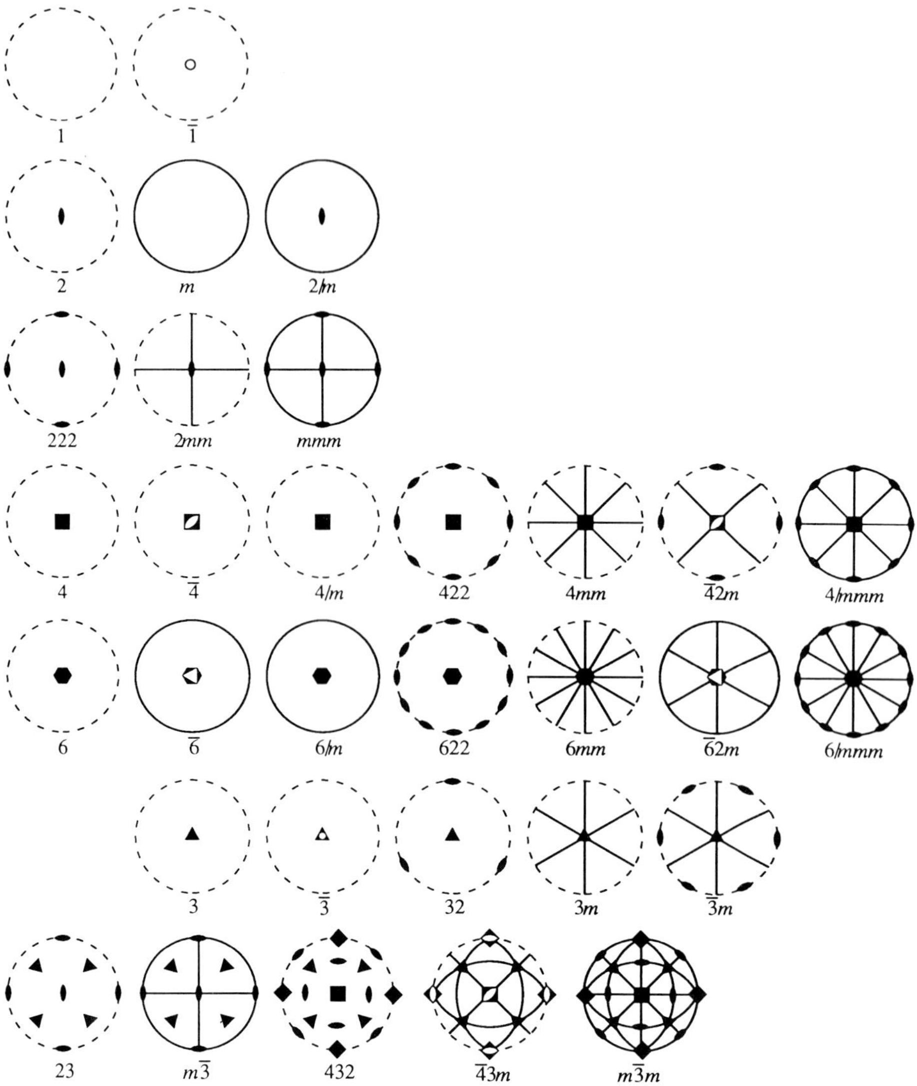

Figure 9-10. Stereograms for each of the 32 crystallographic point groups, or crystal classes. The shortened international Hermann–Mauguin symbols are used. For the full symbols, see Table 9-2.

other at 45°; and so forth. Rotations may, of course, be either proper or improper, but in space groups rotation axes need not intersect.

Unlike point-group symmetry, however, symmetry at the atomic level can involve translation along with rotation. This gives rise to two kinds of symmetry elements not encountered in morphological crystallography—glide planes and screw axes.

Glide Planes

Consider the very simple hypothetical unit cell shown in Figure 9-11. Each sphere represents an atom, spheres of the same size being the same type of atom; all atoms are at the same height along *c*. The solid line parallel to *a* at *b*/2 represents a mirror plane, and pairs of atoms are related by reflection across it parallel to the *b* axis. The dashed lines represent symmetry elements called *glide planes.* If we start at any atom adjacent to the glide plane, reflect across that plane, then translate ("glide") parallel to the *a* axis a distance of $\frac{1}{2}a$, we arrive at a symmetrically equivalent atom. There are several types of glide planes, and they are named for the directions of translation. If the translational component is parallel to a crystallographic axis, then it may be an *a* glide, a *b* glide, or a *c* glide. Diagonal glides (called *n* glides) have translational components of $(0.5a + 0.5b)$, $(0.5b + 0.5c)$, or $(0.5a + 0.5c)$—that is, parallel to, and one-half the length of, diagonals of the unit-cell faces. Diamond glides (*d* glides) have translational components of $(0.25a + 0.25b + 0.25c)$, $(0.25a + 0.25b)$, $(0.25a + 0.25c)$, or $(0.25b + 0.25c)$—parallel to, and one-quarter the length of, the body or face diagonal of the unit cell.

Screw Axes

Measurement of the hypothetical unit cell in Figure 9-12 will confirm that it has a tetragonal shape with the *c* axis perpendicular to the plane of the page. The fractions adjacent to the "atoms" are heights along *c*, so that the elevations from the base of the unit cell, beginning at the lower right and moving counterclockwise, are 0, $0.25c$, $0.5c$, and $0.75c$. If all four atoms were in the same plane, there would be a fourfold axis along *c* at the center of the unit cell (as well as at the corners, which

Figure 9-11. A glide plane is a symmetry element involving translational as well as rotational (mirror) symmetry. Glide planes are represented in this figure by dashed lines, while the solid line is a mirror plane.

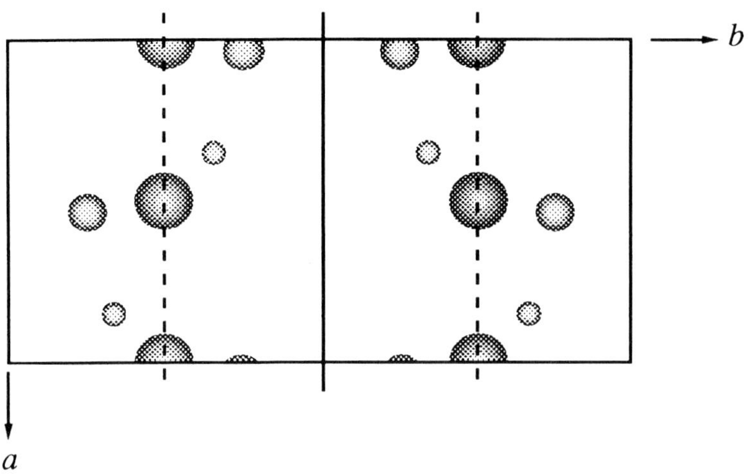

can be verified by adding three more unit cells that share a common corner with this one). Because this is not the case, a fourfold rotation axis is not present, but there is clearly some sort of symmetry here. It is called a *screw axis*, and is the combination of rotation with translation parallel to the rotation axis. The possible types of crystallographic screw axes are shown in Table 9-3; the amount of rotation in each case is the same as for a pure rotation. "Right- handed" and "left-handed" mean that, if the thumb of the corresponding hand is pointed along the positive direction of the screw axis, the fingers curl in the rotational direction for which atom height increases. Thus if the subscript is less than one-half of the "fold" of the axis, the screw axis is right-handed. If greater than one-half, the screw axis is left-handed. If exactly one-half, the screw axis is neutral, because either a right-handed or a left-handed rotation followed by translation will repeat an atom in exactly the same place. The amount of translation, in fractions of the unit-cell dimension in the direction of the axis, is $\frac{1}{2}$ for 2_1, 4_2, and 6_3 axes; $\frac{1}{3}$ for 3_1 and 3_2 axes; $\frac{1}{4}$ for 4_1 and 4_3 axes; $\frac{1}{6}$ for 6_1 and 6_5 axes; and $\frac{1}{3}$ for 6_2 and 6_4 axes. For 4_2, 6_2, 6_3, and 6_4 screw axes, the motif itself must possess symmetry. For 4_2, 6_2, and 6_4 axes, the repeated motif must have twofold symmetry because a full translation occurs after only 180° of rotation; the motif for a 6_3 axis must have trigonal symmetry for a similar reason.

BRAVAIS LATTICES

If, in a crystalline solid, an arbitrary location is chosen and an imaginary point is placed there; and then imaginary points are placed at every identical location throughout the crystal, so that every point has an identical and indistinguishable environment; and then the atoms are removed, leaving only the points; then the array of points is called a *space lattice*. If the unit cell that shows maximum symmetry for a given lattice has lattice points on the corners, but neither on the faces nor in the interior of the cell, the lattice is said to be *primitive* (*P*). If the unit cell

Figure 9-12. Like glide planes, screw axes involve translational symmetry. These atoms are related by rotations of 90° followed by translation along the rotation axis in increments of $\frac{1}{4}$ the repeat distance in that direction.

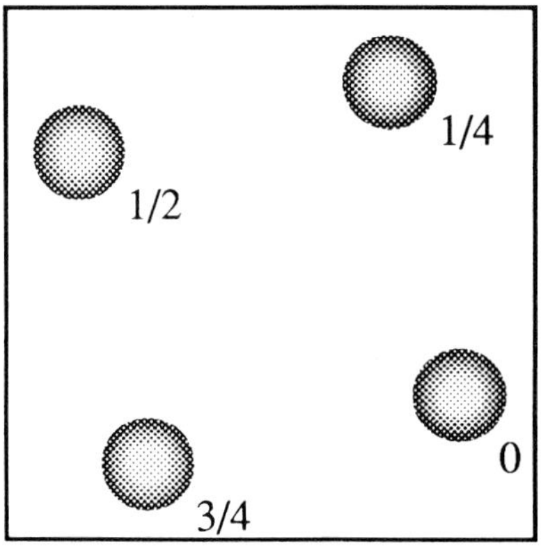

Table 9-3. Types of Screw Axes

Rotation axis	1	2	3	4	6
Right-handed screw axis			3_1	4_1	6_1 6_2
Neutral screw axis		2_1		4_2	6_3
Left-handed screw axis			3_2	4_3	6_4 6_5

has lattice points on faces or inside the cell as well as at the corners, it is said to be *centered*. The following types of centering occur: Body centering (*I*), in which a lattice point is at the geometrical center of the cell; face centering (*F*), in which lattice points are at the centers of all of the faces of the cell; and end centering, in which lattice points are centered on the (100) faces (*A*), the (010) faces (*B*), or the (001) faces (*C*) of the unit cell.

Auguste Bravais (1811–1863) determined that there are only 14 unique kinds of lattices on which crystal structures can be based: *P* triclinic; *P* and *I* monoclinic; *P*, *A*, *F*, and *I* orthorhombic; *P* and *I* tetragonal; *P* hexagonal; *R* rhombohedral; and *P*, *F*, and *I* isometric. Renaming of the crystallographic axes can make the orthorhombic *A* lattice into a *B*- or *C*-lattice; so these 3 are redundant. Redefinition of the *a* or *c* axes in a centered monoclinic lattice can transform an *I*-centered lattice into a *C*-centered (or *A*-centered) one; so these, too, are redundant. The *R* lattice is a rhombohedral lattice with $a = b = c$, and $\alpha = \beta = \gamma \neq 90°$. Figure 9-13 illustrates the geometries of unit cells corresponding to the 14 Bravais lattices.

SPACE GROUPS

When the symmetries of the thirty-two crystal classes are combined with the fourteen Bravais lattices and with possible glide planes and screw axes, the result is 230 unique arrangements of symmetry elements. These arrangements are called space groups, and all crystals, natural or synthetic, belong to one and only one of the 230 space groups. Because of the presence of translational symmetry elements in some of the space groups, no point is necessarily left stationary upon operation of all of the symmetry elements in the group; hence the name space group, as opposed to point group.

Space-Group Symbols

Memorizing the 230 space groups is not a fruitful activity. Lists of them are available in many textbooks on crystallography, and the *International Tables for Crystallography* (Volume A) contains a full description of each one. Nonetheless, space-group symbols can be understood on sight, without recourse to more than trivial memorization beyond the 32 crystal classes.

A space-group symbol consists of the Bravais-lattice symbol followed by either (a) one of the 32 Hermann–Mauguin crystal-class symbols, or (b) one of the crystal-

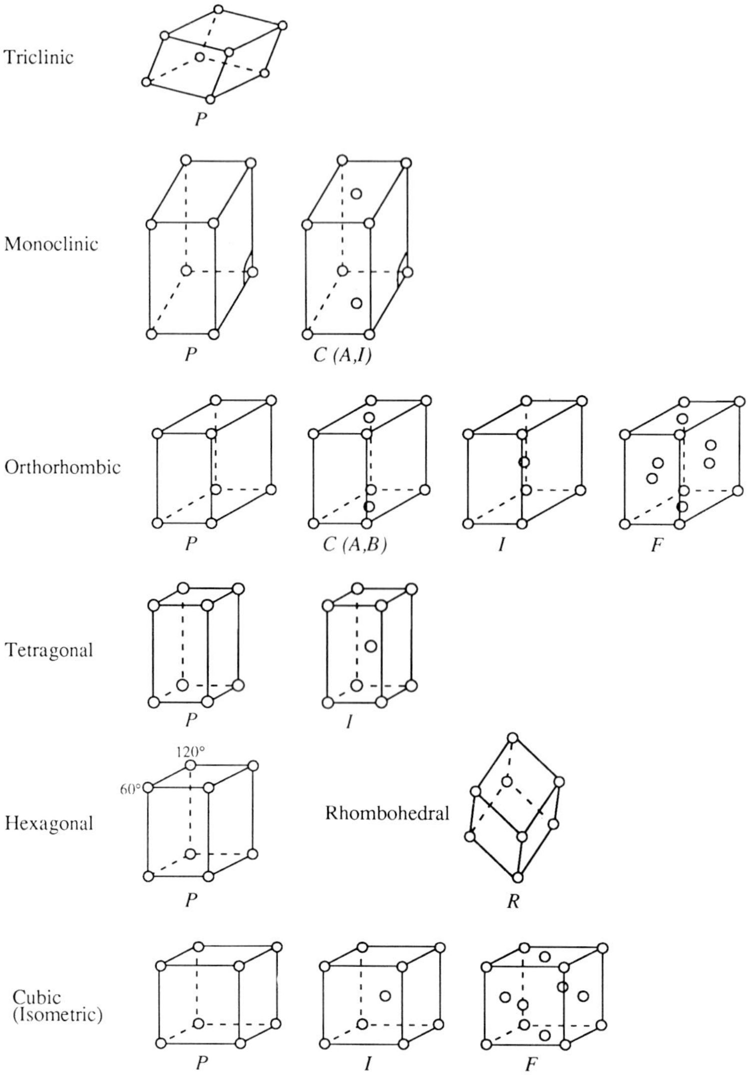

Figure 9-13. The 14 Bravais lattices. The circles represent lattice points, not atoms.

class symbols modified to include translational symmetry elements. The following examples illustrate the ease with which such symbols are interpreted:

1. *P*2/*m*—the lattice is primitive, and the crystal class is 2/*m*.
2. *P*2_1/*m*—the lattice is primitive, and the crystal class, in which translational symmetry elements appear identical to their nontranslational counterparts, is 2/*m*. The diad is replaced, however, by a 2_1 screw axis on the unit-cell level.
3. *C*2/*c*—the lattice is *C* centered, and the crystal class is 2/*m*, with the mirror replaced by a *c* glide in the unit cell.
4. *Pbca*—the lattice is primitive, and the crystal class is *mmm*, which is the short-

ened notation for $2/m\ 2/m\ 2/m$. In the unit cell, the mirror parallel to (100) is replaced by a *b* glide, the mirror parallel to (010) is replaced by a *c* glide, and the mirror parallel to (001) is replaced by an *a* glide. Reference to the *International Tables* (to be discussed) shows that the full space-group symbol is $P2_1/b\ 2_1/c\ 2_1/a$, indicating that all three twofold axes have been replaced by 2_1 screw axes.

5. $I4_1/amd$—the lattice is body centered, and the crystal class is $4/mmm$ ($= 4/m\ 2/m\ 2/m$). At the atomic level, a 4_1 screw axis is present instead of the fourfold axis, the morphological mirror plane perpendicular to *c* is really an *a* glide, the morphological mirror parallel to (100) is in fact a space-group mirror, and the one parallel to (110) is a *d* glide.

While the effective and necessary use of space groups in crystal chemistry does not require a thorough understanding of mathematical group theory (and it is not proposed to convey such understanding here), a brief excursion into the elementary matrix treatment of space-group symmetry is not without substantial reward.

MATRIX REPRESENTATIONS OF SPACE GROUPS

Let $\mathbf{e}_x$ and $\mathbf{e}_y$ be unit vectors that define an orthogonal coordinate system in two dimensional space (Figure 9-14). The vector $\mathbf{r}$ connects the origin and the point (x,y), and makes an angle θ with $\mathbf{e}_x$. We will rotate $\mathbf{r}$ through an angle ρ about an axis normal to $\mathbf{e}_x$ and $\mathbf{e}_y$ so that it terminates at new coordinates (x',y'). From trigonometric definitions and identities we have:

$$\begin{aligned}
\sin\theta &= y/\mathrm{r} \\
\cos\theta &= x/\mathrm{r} \\
\cos(\theta-\rho) &= x'/\mathrm{r} \\
\cos\theta\cos\rho + \sin\theta\sin\rho &= x'/\mathrm{r} \\
(x/r)\cos\rho + (y/r)\sin\rho &= x'/\mathrm{r} \\
x\cos\rho + y\sin\rho &= x' \qquad (1)
\end{aligned}$$

Figure 9-14. In this two-dimensional Cartesian space, **r** is a vector originating at the origin and ending at point (x,y), and making an angle θ with the x axis. It is rotated in a positive sense through an angle ρ.

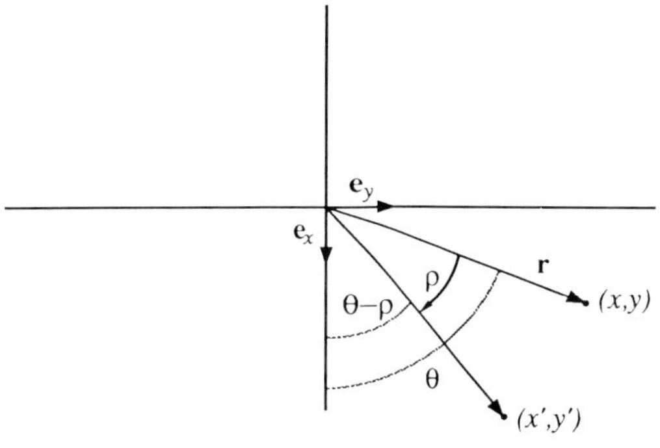

$$\begin{aligned}
&\sin(\theta - \rho) = y'/r \\
&\sin\theta\cos\rho - \cos\theta\sin\rho = y'/r \\
&(y/r)\cos\rho - (x/r)\sin\rho = y'/r \\
&y\cos\rho - x\sin\rho = y' \qquad (2)
\end{aligned}$$

Equations (1) and (2) can conveniently be written in matrix form:

$$\begin{bmatrix} x' \\ y' \end{bmatrix} = \begin{bmatrix} \cos\rho & \sin\rho \\ -\sin\rho & \cos\rho \end{bmatrix} \begin{bmatrix} x \\ y \end{bmatrix}.$$

(It is assumed that the reader is familiar with basic matrix algebra. If this is not the case, reference to the Tutorial at the end of this chapter is advised before proceeding.)

Extending to three dimensions, for a rotation parallel to $\mathbf{e}_z$:

$$\begin{bmatrix} x' \\ y' \\ z' \end{bmatrix} = \begin{bmatrix} \cos\rho & \sin\rho & 0 \\ -\sin\rho & \cos\rho & 0 \\ 0 & 0 & 1 \end{bmatrix} \begin{bmatrix} x \\ y \\ z \end{bmatrix}$$

The 3×3 rotation matrix represents a clockwise rotation of point (x,y,z), which is equivalent to a counter-clockwise rotation of the basis vectors $\mathbf{e}_x$, $\mathbf{e}_y$, $\mathbf{e}_z$. We shall define this as a *positive rotation.* Corresponding positive rotations about $\mathbf{e}_x$ and $\mathbf{e}_y$ are written

$$\begin{bmatrix} x' \\ y' \\ z' \end{bmatrix} = \begin{bmatrix} 1 & 0 & 0 \\ 0 & \cos\rho & \sin\rho \\ 0 & -\sin\rho & \cos\rho \end{bmatrix} \begin{bmatrix} x \\ y \\ z \end{bmatrix}$$

and

$$\begin{bmatrix} x' \\ y' \\ z' \end{bmatrix} = \begin{bmatrix} \cos\rho & 0 & -\sin\rho \\ 0 & 1 & 0 \\ \sin\rho & 0 & \cos\rho \end{bmatrix} \begin{bmatrix} x \\ y \\ z \end{bmatrix},$$

respectively.

The symmetry matrices of practical interest may often be written by inspection of the effect of the symmetry operation on the set of basis vectors. For example, a twofold rotation parallel to $\mathbf{e}_y$ leaves $\mathbf{e}_y$ unchanged, but maps $\mathbf{e}_x$ into $-\mathbf{e}_x$ and $\mathbf{e}_z$ into $-\mathbf{e}_z$. Hence, the corresponding rotation matrix can be written as

$$2 \parallel \mathbf{e}_y = \begin{bmatrix} -1 & 0 & 0 \\ 0 & 1 & 0 \\ 0 & 0 & -1 \end{bmatrix}.$$

Note that $\cos\rho = -1$, so $\rho = 180°$, as it must for a diad. A single fourfold rotation about $\mathbf{e}_z$ maps $\mathbf{e}_x$ into $\mathbf{e}_y$ and $\mathbf{e}_y$ into $-\mathbf{e}_x$, yielding

$$4 \parallel \mathbf{e}_z = \begin{bmatrix} 0 & 1 & 0 \\ -1 & 0 & 0 \\ 0 & 0 & 1 \end{bmatrix}.$$

Here $\cos\rho = 0$ and $\sin\rho = 1$, confirming that $\rho = 90°$.

A reflection across the $\mathbf{e}_x\mathbf{e}_z$ plane leaves $\mathbf{e}_x$ and $\mathbf{e}_z$ invariant, but maps $\mathbf{e}_y$ into $-\mathbf{e}_y$.

$$\text{m} \perp \mathbf{e}_y = \begin{bmatrix} 1 & 0 & 0 \\ 0 & -1 & 0 \\ 0 & 0 & 1 \end{bmatrix}$$

Now recall that $\bar{2}$ axis is a twofold rotation followed by an inversion. We have already seen the matrix for a twofold rotation parallel to $\mathbf{e}_y$, and the matrix for an inversion is the negative of an identity matrix—it maps all basis vectors into their own negatives. Thus

$$\bar{2} \parallel \mathbf{e}_y = [\bar{1}][2 \parallel \mathbf{e}_y]$$

$$= \begin{bmatrix} -1 & 0 & 0 \\ 0 & -1 & 0 \\ 0 & 0 & -1 \end{bmatrix} \begin{bmatrix} -1 & 0 & 0 \\ 0 & 1 & 0 \\ 0 & 0 & -1 \end{bmatrix} = \begin{bmatrix} 1 & 0 & 0 \\ 0 & -1 & 0 \\ 0 & 0 & 1 \end{bmatrix}$$

$$= \text{m} \perp \mathbf{e}_y,$$

and the matrix operations tell us the same thing that the stereograms of Figure 9-3 did: a twofold rotoinversion axis is equivalent to reflection across a plane normal to the axis.

It is possible to write the general form of a matrix for rotation of any angle about any arbitrary direction in space; however, the symmetry operations involving directions not parallel to basis vectors involve other rather special directions, and the corresponding matrices are easily written by inspection. For example, in the tetragonal and cubic systems, some crystal classes have twofold axes parallel to [110]. These map $\mathbf{e}_x$ into $\mathbf{e}_y$, $\mathbf{e}_y$ into $\mathbf{e}_x$, and $\mathbf{e}_z$ into $-\mathbf{e}_z$, so the rotation matrix is

$$2 \parallel [110] = \begin{bmatrix} 0 & 1 & 0 \\ 1 & 0 & 0 \\ 0 & 0 & -1 \end{bmatrix}.$$

Although we have used only orthogonal basis vectors to this point, symmetry matrices are more easily written by inspection if the basis vectors chosen are parallel to the conventional crystallographic axes. For the orthorhombic, tetragonal, and cubic systems they are, of course, orthogonal. For the triclinic system, only a center of symmetry or no symmetry at all is possible, and the inversion and identity matrices look the same for all systems. For the monoclinic system the only possible symmetry axes are a 2 parallel to b or $\bar{2}$ normal to the plane of the other two crystallographic axes, so the nonorthogonality of the basis vectors presents no complications. The hexagonal system presents some complications that cannot be dealt with adequately in this short treatment, but that arise because the direct lattice vectors and the reciprocal lattice vectors are not parallel (except for **c** and **c***) and greater than diad symmetry is possible. In this case the appropriate rotation matrices to represent the transformations of positions (x,y,z) must be obtained by inspection of the reciprocal-lattice basis vectors. For our purposes it is not important to consider the matrix representation of symmetry in the hexagonal system further; this will not prove to be a limitation as we proceed.

Screw Axes and Glide Planes

Both screw axes and glide planes involve translational components which may be represented mathematically by expanding to 4×4 matrices. A twofold screw axis parallel to $\mathbf{e}_y$ is a combination of a diad with a translation of $\frac{1}{2}$ of the b cell repeat along $\mathbf{e}_y$ ($= \mathbf{b}$):

$$\begin{bmatrix} -1 & 0 & 0 & 0 \\ 0 & 1 & 0 & 0 \\ 0 & 0 & -1 & 0 \\ 0 & 0 & 0 & 1 \end{bmatrix} \begin{bmatrix} 1 & 0 & 0 & 0 \\ 0 & 1 & 0 & \frac{1}{2} \\ 0 & 0 & 1 & 0 \\ 0 & 0 & 0 & 1 \end{bmatrix}$$

$$= \begin{bmatrix} -1 & 0 & 0 & 0 \\ 0 & 1 & 0 & \frac{1}{2} \\ 0 & 0 & -1 & 0 \\ 0 & 0 & 0 & 1 \end{bmatrix} \Longrightarrow \left[\begin{array}{c|c} \begin{matrix}\text{rotation}\\ \text{matrix}\end{matrix} & \\ \hline \multicolumn{2}{c}{\text{“dummy row”}} \end{array}\right] \begin{matrix}\text{translational}\\ \text{vector}\end{matrix}$$

Addition of the “dummy row” makes it possible to multiply matrices for symmetry elements together to obtain matrices for other symmetry elements, as we shall do shortly. New coordinates for a point (x,y,z) operated on by this 2_1 axis are obtained by matrix multiplication as follows:

$$\begin{bmatrix} -1 & 0 & 0 & 0 \\ 0 & 1 & 0 & \frac{1}{2} \\ 0 & 0 & -1 & 0 \\ 0 & 0 & 0 & 1 \end{bmatrix} \begin{bmatrix} x \\ y \\ z \\ 1 \end{bmatrix} = \begin{bmatrix} -x \\ y + \frac{1}{2} \\ -z \\ 1 \end{bmatrix}$$

As further examples, the 4×4 matrices representing a c glide normal to $\mathbf{e}_y$, an n glide normal to $\mathbf{e}_z$, and a mirror normal to $\mathbf{e}_x$ are:

$$\begin{bmatrix} 1 & 0 & 0 & 0 \\ 0 & -1 & 0 & 0 \\ 0 & 0 & 1 & \frac{1}{2} \\ 0 & 0 & 0 & 1 \end{bmatrix}, \quad \begin{bmatrix} 1 & 0 & 0 & \frac{1}{2} \\ 0 & 1 & 0 & \frac{1}{2} \\ 0 & 0 & -1 & 0 \\ 0 & 0 & 0 & 1 \end{bmatrix}, \quad \begin{bmatrix} -1 & 0 & 0 & 0 \\ 0 & 1 & 0 & 0 \\ 0 & 0 & 1 & 0 \\ 0 & 0 & 0 & 1 \end{bmatrix},$$

respectively.

To this point we have made no specification of the locations of symmetry operations within the unit cell, which is tantamount to the assumption that the symmetry operations pass through the origin of the unit cell. This is not generally the case, however, and we shall now consider how the locations of symmetry operations are specified in the matrix representations.

Suppose we have a rotation (proper or improper, with or without a translational component) about an axis parallel to $\mathbf{e}_z$, and which intersects the $\mathbf{e}_x\mathbf{e}_y$ plane at coor-

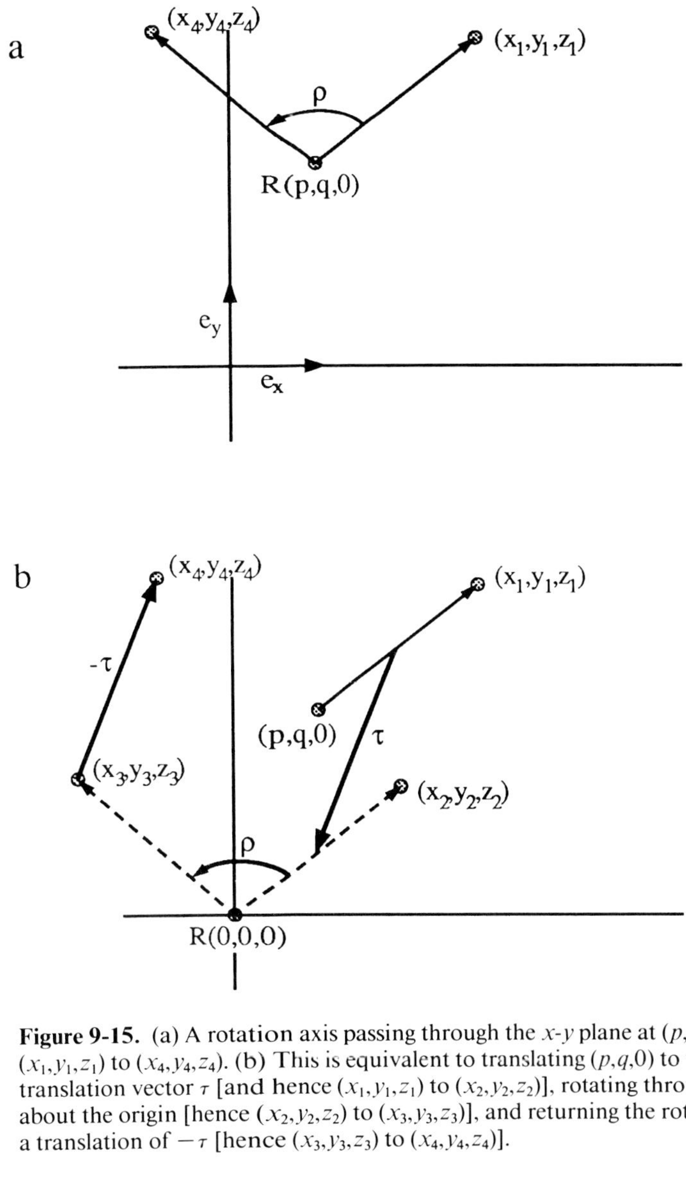

Figure 9-15. (a) A rotation axis passing through the x-y plane at $(p,q,0)$ rotates point (x_1,y_1,z_1) to (x_4,y_4,z_4). (b) This is equivalent to translating $(p,q,0)$ to the origin by the translation vector τ [and hence (x_1,y_1,z_1) to (x_2,y_2,z_2)], rotating through the same angle ρ about the origin [hence (x_2,y_2,z_2) to (x_3,y_3,z_3)], and returning the rotation axis to $(p,q,0)$ by a translation of $-\tau$ [hence (x_3,y_3,z_3) to (x_4,y_4,z_4)].

dinates $(p,q,0)$. Let this rotation operation, which will be designated $R_\rho(p,q,0)$, rotate (x_1,y_1,z_1) through an angle of ρ degrees to a new position (x_4,y_4,z_4), as in Figure 9-15(a). The same result could be obtained by translating $R_\rho(p,q,0)$ to the origin along vector τ, operating on it with rotation $R_\rho(0,0,0)$ to take point (x_2,y_2,z_2) to (x_3,y_3,z_3), and then translating along $-\tau$ back to $(p,q,0)$ by a translation operation τ^{-1} [see Figure 9-15(b)]. In matrix notation, and generalizing to a rotation that may be fixed in the third dimension of the coordinate space as well (such as one involving an inversion center), these operations are represented as follows:

$$\begin{bmatrix} x_2 \\ y_2 \\ z_2 \end{bmatrix} = \begin{bmatrix} x_1 \\ y_1 \\ z_1 \end{bmatrix} - \tau = \begin{bmatrix} x_1 \\ y_1 \\ z_1 \end{bmatrix} - \begin{bmatrix} p \\ q \\ r \end{bmatrix}$$

$$\begin{bmatrix} x_3 \\ y_3 \\ z_3 \end{bmatrix} = \begin{bmatrix} R_\rho(0,0,0) \end{bmatrix} \begin{bmatrix} x_2 \\ y_2 \\ z_2 \end{bmatrix} = \begin{bmatrix} R_\rho(0,0,0) \end{bmatrix} \begin{bmatrix} x_1 - p \\ y_1 - q \\ z_1 - r \end{bmatrix}$$

$$\begin{bmatrix} x_4 \\ y_4 \\ z_4 \end{bmatrix} = \begin{bmatrix} x_3 \\ y_3 \\ z_3 \end{bmatrix} + \tau = \begin{bmatrix} x_3 \\ y_3 \\ z_3 \end{bmatrix} + \begin{bmatrix} p \\ q \\ r \end{bmatrix}$$

$$= \begin{bmatrix} R_\rho(0,0,0) \end{bmatrix} \begin{bmatrix} x_1 - p \\ y_1 - q \\ z_1 - r \end{bmatrix} + \begin{bmatrix} p \\ q \\ r \end{bmatrix}.$$

Omitting intermediate steps in the last line,

$$\begin{bmatrix} x_4 \\ y_4 \\ z_4 \end{bmatrix} = \begin{bmatrix} R_\rho(0,0,0) \end{bmatrix} \begin{bmatrix} x_1 - p \\ y_1 - q \\ z_1 - r \end{bmatrix} + \begin{bmatrix} p \\ q \\ r \end{bmatrix}.$$

Two examples will suffice to demonstrate the usefulness of this relation.

1. Suppose we wish to know the matrix representation of a fourfold rotation axis parallel to c of a tetragonal crystal and intersecting a at 0 and b at $\frac{1}{2}$. $R_\rho(0,0,0)$ is easily written in the same way as we have written rotation matrices by inspection previously:

$$\begin{bmatrix} 0 & 1 & 0 \\ -1 & 0 & 0 \\ 0 & 0 & 1 \end{bmatrix} \begin{bmatrix} x - p \\ y - q \\ z - r \end{bmatrix} + \begin{bmatrix} p \\ q \\ r \end{bmatrix} = \begin{bmatrix} y - q + p \\ -x + p + q \\ z \end{bmatrix}.$$

From the problem specifications it is clear that p is 0, q is $\frac{1}{2}$, and r is 0. Expanding to the 4×4 matrix format and using t_1, t_2, and t_3 for the elements of the translational component,

$$\begin{bmatrix} 0 & 1 & 0 & t_1 \\ -1 & 0 & 0 & t_2 \\ 0 & 0 & 0 & t_3 \\ 0 & 0 & 0 & 1 \end{bmatrix} \begin{bmatrix} x \\ y \\ z \\ 1 \end{bmatrix} = \begin{bmatrix} y - \frac{1}{2} \\ -x + \frac{1}{2} \\ z \\ 1 \end{bmatrix}.$$

Obviously, $t_1 = -\frac{1}{2}$, $t_2 = \frac{1}{2}$, and $t_3 = 0$, so the required matrix is

$$\begin{bmatrix} 0 & 1 & 0 & -\frac{1}{2} \\ -1 & 0 & 0 & \frac{1}{2} \\ 0 & 0 & 1 & 0 \\ 0 & 0 & 0 & 1 \end{bmatrix}$$

Since a translation of $-\frac{1}{2}$ is equivalent to one of $+\frac{1}{2}$, the negative fractions are normally transformed to positive ones, and the matrix for a fourfold rotation parallel to c at $x=0$, $y=\frac{1}{2}$ is

$$\begin{bmatrix} 0 & 1 & 0 & \frac{1}{2} \\ -1 & 0 & 0 & \frac{1}{2} \\ 0 & 0 & 1 & 0 \\ 0 & 0 & 0 & 1 \end{bmatrix}$$

2. Suppose we wish to know what the following matrix represents:

$$\begin{bmatrix} -1 & 0 & 0 & \frac{1}{2} \\ 0 & 1 & 0 & \frac{1}{2} \\ 0 & 0 & -1 & \frac{1}{2} \\ 0 & 0 & 0 & 1 \end{bmatrix}$$

The rotational component specifies a twofold rotation about b. We write

$$\begin{bmatrix} -1 & 0 & 0 \\ 0 & 1 & 0 \\ 0 & 0 & -1 \end{bmatrix} \begin{bmatrix} x-p \\ y-q \\ z-r \end{bmatrix} + \begin{bmatrix} p \\ q \\ r \end{bmatrix} = \begin{bmatrix} -x+2p \\ y \\ -z+2r \end{bmatrix}.$$

Thus $2p = \frac{1}{2}$ and $p = \frac{1}{4}$, and $2r = \frac{1}{2}$; so $r = \frac{1}{4}$. Because $q = 0$, the $\frac{1}{2}$ in the second row of the 4×4 matrix indicates that the twofold axis is a screw axis with translational component $\frac{1}{2}$ along b. Thus the matrix represents a 2_1 axis parallel to b and located at $x=\frac{1}{4}$, $z=\frac{1}{4}$.

One final point is appropriate before we can proceed to illustrate the derivations of a few space groups. Consider the combination of a center of symmetry with a translation along, say, the b axis.

$$\begin{bmatrix} -1 & 0 & 0 & 0 \\ 0 & -1 & 0 & 0 \\ 0 & 0 & -1 & 0 \\ 0 & 0 & 0 & 1 \end{bmatrix} \begin{bmatrix} 1 & 0 & 0 & 0 \\ 0 & 1 & 0 & 1 \\ 0 & 0 & 1 & 0 \\ 0 & 0 & 0 & 1 \end{bmatrix} = \begin{bmatrix} -1 & 0 & 0 & 0 \\ 0 & -1 & 0 & 1 \\ 0 & 0 & -1 & 0 \\ 0 & 0 & 0 & 1 \end{bmatrix}.$$

This represents a center of symmetry at $(0,\frac{1}{2},0)$, as may be easily verified by the method used previously. An analogous result is obtained for a combination of an inversion center with full translations along a, c, [110], or any other crystallographic direction. Thus, if there is an inversion center at the origin, there is also one half-way along each lattice translation of the unit cell. The same is true of twofold axes, twofold screw axes, mirrors, and glide planes.

Derivation of Matrix Representations of Space Groups

Each of the four types of Bravais lattices (P [R], C, I, and F) may be represented by one or more matrices that specify appropriate translation vector(s). For example, a primitive lattice (P), with lattice points separated by one full lattice translation in each crystallographic direction, is simply represented as an identity matrix E:

$$\begin{bmatrix} 1 & 0 & 0 & 1 \\ 0 & 1 & 0 & 1 \\ 0 & 0 & 1 & 1 \\ 0 & 0 & 0 & 1 \end{bmatrix} \equiv \begin{bmatrix} 1 & 0 & 0 & 0 \\ 0 & 1 & 0 & 0 \\ 0 & 0 & 1 & 0 \\ 0 & 0 & 0 & 1 \end{bmatrix}.$$

A C-centered lattice (C) combines the lattice points of a primitive cell with addition lattice points located at $(\frac{1}{2},\frac{1}{2},0)$:

$$\begin{bmatrix} 1 & 0 & 0 & 0 \\ 0 & 1 & 0 & 0 \\ 0 & 0 & 1 & 0 \\ 0 & 0 & 0 & 1 \end{bmatrix} \text{ and } \begin{bmatrix} 1 & 0 & 0 & \frac{1}{2} \\ 0 & 1 & 0 & \frac{1}{2} \\ 0 & 0 & 1 & 0 \\ 0 & 0 & 0 & 1 \end{bmatrix}.$$

The derivation of space groups is best understood by illustration. Four will be given: $C2/c$, Cc, $P4/n$, and $Ibca$.

C2/c

This space group is isomorphic with point group $2/m$, which is the holohedral crystal class of the monoclinic system. There are four faces in the general form of class $2/m$, one corresponding to the identity operation, one to the diad, one to the mirror, and one to the inversion center, as can be seen by constructing a stereogram of $2/m$ showing the repetition of a general face pole. Thus any primitive space group isomorphic with $2/m$ would have four general equivalent positions (GEPs). $C2/c$ is not primitive, however, and the C-centering operation will operate on each of these four "primitive" GEPs to produce four more, a total of eight. We begin by recognizing that a diad and a glide plane normal to it must intersect, so we will pick the point of intersection as a provisional origin for the unit cell.

The rotational components of the 4 × 4 symmetry matrices will be simply the 3 × 3 matrices representing the point group operations, as follows:

$$\mathrm{E} = \begin{bmatrix} 1 & 0 & 0 \\ 0 & 1 & 0 \\ 0 & 0 & 1 \end{bmatrix}, 2 = \begin{bmatrix} -1 & 0 & 0 \\ 0 & 1 & 0 \\ 0 & 0 & -1 \end{bmatrix}, m = \begin{bmatrix} 1 & 0 & 0 \\ 0 & -1 & 0 \\ 0 & 0 & 1 \end{bmatrix},$$

$$\text{and } \bar{1} = \begin{bmatrix} -1 & 0 & 0 \\ 0 & -1 & 0 \\ 0 & 0 & -1 \end{bmatrix}.$$

With the assumption that the diad and glide plane pass through the origin, the positional parts of their translational vectors are zero, and the 4×4 matrices for the identity, diad, and c-glide operations can be written directly:

$$\mathrm{E} = \begin{bmatrix} 1 & 0 & 0 & 0 \\ 0 & 1 & 0 & 0 \\ 0 & 0 & 1 & 0 \\ 0 & 0 & 0 & 1 \end{bmatrix}, \quad 2 = \begin{bmatrix} -1 & 0 & 0 & 0 \\ 0 & 1 & 0 & 0 \\ 0 & 0 & -1 & 0 \\ 0 & 0 & 0 & 1 \end{bmatrix}, \quad c = \begin{bmatrix} 1 & 0 & 0 & 0 \\ 0 & -1 & 0 & 0 \\ 0 & 0 & 1 & \frac{1}{2} \\ 0 & 0 & 0 & 1 \end{bmatrix}.$$

We have no assurance that the center of symmetry is at the origin, but because the presence of two symmetry operations requires a third (Euler's construction, p. 305), we can obtain the matrix for the center of symmetry by multiplying that for the glide plane by that for the diad:

$$\begin{bmatrix} -1 & 0 & 0 & 0 \\ 0 & 1 & 0 & 0 \\ 0 & 0 & -1 & 0 \\ 0 & 0 & 0 & 1 \end{bmatrix} \begin{bmatrix} 1 & 0 & 0 & 0 \\ 0 & -1 & 0 & 0 \\ 0 & 0 & 1 & \frac{1}{2} \\ 0 & 0 & 0 & 1 \end{bmatrix} = \begin{bmatrix} -1 & 0 & 0 & 0 \\ 0 & -1 & 0 & 0 \\ 0 & 0 & -1 & \frac{1}{2} \\ 0 & 0 & 0 & 0 \end{bmatrix}$$

Using the procedure for analyzing location information in the translation vector, it is easily determined that the inversion center is at $(0,0,\frac{1}{4})$.

If this were a primitive space group (e.g., $P2/c$), we would now have all of the matrices and could draw a diagram showing the symmetry elements of the space group in their proper orientations and positions. However, since it is C centered, every symmetry operation that we have found about the lattice point at the origin has a corresponding operation about the lattice point at the center of (001). These can be found by multiplying each of the four matrices already found by the matrix representing the C-centering operation (given previously). The results for the additional matrices are:

$$\begin{bmatrix} 1 & 0 & 0 & \frac{1}{2} \\ 0 & 1 & 0 & \frac{1}{2} \\ 0 & 0 & 1 & 0 \\ 0 & 0 & 0 & 1 \end{bmatrix}, \quad \begin{bmatrix} -1 & 0 & 0 & \frac{1}{2} \\ 0 & 1 & 0 & \frac{1}{2} \\ 0 & 0 & -1 & 0 \\ 0 & 0 & 0 & 1 \end{bmatrix}, \quad \begin{bmatrix} 1 & 0 & 0 & \frac{1}{2} \\ 0 & -1 & 0 & \frac{1}{2} \\ 0 & 0 & 1 & \frac{1}{2} \\ 0 & 0 & 0 & 1 \end{bmatrix},$$

$$\text{and} \quad \begin{bmatrix} -1 & 0 & 0 & \frac{1}{2} \\ 0 & -1 & 0 & \frac{1}{2} \\ 0 & 0 & -1 & \frac{1}{2} \\ 0 & 0 & 0 & 1 \end{bmatrix}.$$

Using what we have already learned about the interpretation of matrices, we see that these four represent, respectively, (1) the centering operation, (2) a twofold screw axis parallel to b and located at $x = \frac{1}{4}$, $z = 0$, (3) an n glide parallel to (010) at $y = \frac{1}{4}$, and (4) an inversion center at $x = \frac{1}{4}$, $y = \frac{1}{4}$, $z = \frac{1}{4}$.

With these eight matrices we can now draw the space-group diagram, using standard international symbology and orientation (Figure 9-16). The standard projection is looking down on the ab plane (or its projection) from the positive c direction, with a (or its projection) pointing downward in the plane of the page, b to the right

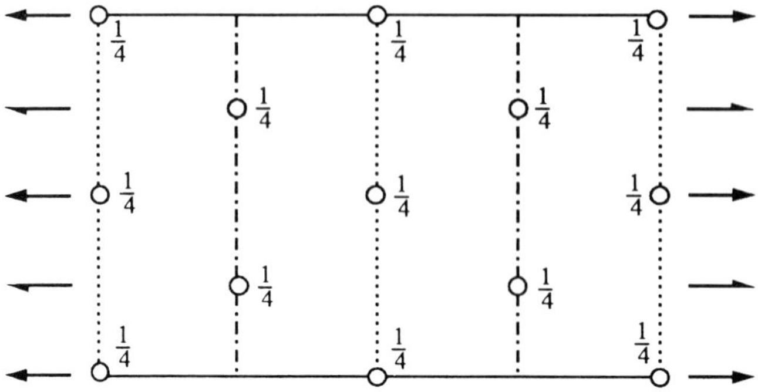

Figure 9-16. The space group $C2/c$ in an orientation that is nonstandard because a center of symmetry does not lie at the origin of the unit cell.

in the plane of the page, and c coming out of the page. The $\frac{1}{4}$ next to each inversion center indicates that it is located at $z=\frac{1}{4}$.

This drawing is technically correct, but the orientation is nonstandard for the following reason: If a space group possesses one or more inversion centers, one of them is placed at the origin by convention. Consequently, to obtain the standard orientation, we must shift all symmetry elements by $z=-\frac{1}{4}$. This will not affect the glides, but it will move the inversion centers and rotation axes, yielding the standard orientation shown in Figure 9-17. The matrices may now be rewritten to conform to this orientation:

$$E = \begin{bmatrix} 1 & 0 & 0 & 0 \\ 0 & 1 & 0 & 0 \\ 0 & 0 & 1 & 0 \\ 0 & 0 & 0 & 1 \end{bmatrix}, \quad c = \begin{bmatrix} 1 & 0 & 0 & 0 \\ 0 & -1 & 0 & 0 \\ 0 & 0 & 1 & \frac{1}{2} \\ 0 & 0 & 0 & 1 \end{bmatrix},$$

$$\bar{1} = \begin{bmatrix} -1 & 0 & 0 & 0 \\ 0 & -1 & 0 & 0 \\ 0 & 0 & -1 & 0 \\ 0 & 0 & 0 & 1 \end{bmatrix}, \quad 2 = \begin{bmatrix} -1 & 0 & 0 & 0 \\ 0 & 1 & 0 & 0 \\ 0 & 0 & -1 & \frac{1}{2} \\ 0 & 0 & 0 & 1 \end{bmatrix}$$

$$C = \begin{bmatrix} 1 & 0 & 0 & \frac{1}{2} \\ 0 & 1 & 0 & \frac{1}{2} \\ 0 & 0 & 1 & 0 \\ 0 & 0 & 0 & 1 \end{bmatrix}, \quad 2_1 = \begin{bmatrix} -1 & 0 & 0 & \frac{1}{2} \\ 0 & 1 & 0 & \frac{1}{2} \\ 0 & 0 & -1 & \frac{1}{2} \\ 0 & 0 & 0 & 1 \end{bmatrix},$$

$$\bar{1} = \begin{bmatrix} -1 & 0 & 0 & \frac{1}{2} \\ 0 & -1 & 0 & \frac{1}{2} \\ 0 & 0 & -1 & 0 \\ 0 & 0 & 0 & 1 \end{bmatrix}, \quad n = \begin{bmatrix} 1 & 0 & 0 & \frac{1}{2} \\ 0 & -1 & 0 & \frac{1}{2} \\ 0 & 0 & 1 & \frac{1}{2} \\ 0 & 0 & 0 & 1 \end{bmatrix}.$$

The general equivalent positions for $C2/c$ may be "read" directly from these matrices as x,y,z; $x,-y,z+\frac{1}{2}$; $-x,-y,-z$; $-x,y,-z+\frac{1}{2}$; $x+\frac{1}{2},y+\frac{1}{2},z$; $-x+\frac{1}{2},y+\frac{1}{2},$

$-z+\frac{1}{2}$; $-x+\frac{1}{2},-y+\frac{1}{2},-z$; and $x+\frac{1}{2},-y+\frac{1}{2},z+\frac{1}{2}$. Thus if an atom is located at coordinates (x,y,z) in the unit cell, a symmetrically identical atom is located at each of the seven other general equivalent positions.

Cc

This space group is isomorphic with point group *m*, a monoclinic crystal class. If the space group were *Pm*, there would be two GEPs: one arising from the identity operation, and one from the mirror operation. Because *Cc* is an end-centered lattice, there are two additional GEPs, obtained by operating on the first two with the *C*-centering operation. We have just seen the matrix for the *C*-centering operation; so the four matrices representing space group *Cc* are:

$$\begin{bmatrix} 1 & 0 & 0 & 0 \\ 0 & 1 & 0 & 0 \\ 0 & 0 & 1 & 0 \\ 0 & 0 & 0 & 1 \end{bmatrix}, \quad \begin{bmatrix} 1 & 0 & 0 & 0 \\ 0 & -1 & 0 & 0 \\ 0 & 0 & 1 & \frac{1}{2} \\ 0 & 0 & 0 & 1 \end{bmatrix}, \quad \begin{bmatrix} 1 & 0 & 0 & \frac{1}{2} \\ 0 & 1 & 0 & \frac{1}{2} \\ 0 & 0 & 1 & 0 \\ 0 & 0 & 0 & 1 \end{bmatrix},$$

$$\text{and} \begin{bmatrix} 1 & 0 & 0 & \frac{1}{2} \\ 0 & -1 & 0 & \frac{1}{2} \\ 0 & 0 & 1 & \frac{1}{2} \\ 0 & 0 & 0 & 1 \end{bmatrix}$$

The first represents the identity operation, the second is the *c* glide, the third is the centering operation, and the fourth is an *n* glide located at $y=\frac{1}{4}$. The GEPs are x,y,z; $x,-y,z+\frac{1}{2}$; $x+\frac{1}{2},y+\frac{1}{2},z$; and $x+\frac{1}{2},-y+\frac{1}{2},z+\frac{1}{2}$, and the space group drawing is given in Figure 9-18. Note that unlike *C*2/*c*, *Cc* has no center of symmetry. Because of the way we wrote the matrix for the *c* glide, it is constrained to lie somewhere in the glide plane (i.e., at $y = 0$), but the location with respect to *x* and *z* is not specified.

*P*4/*n*

*P*4/*n* is a tetragonal space group isomorphic with crystal class 4/*m*, in which the general form contains eight faces. Because this is a primitive space group, there are

Figure 9-17. *C*2/*c*, reoriented so that an inversion center is at the origin.

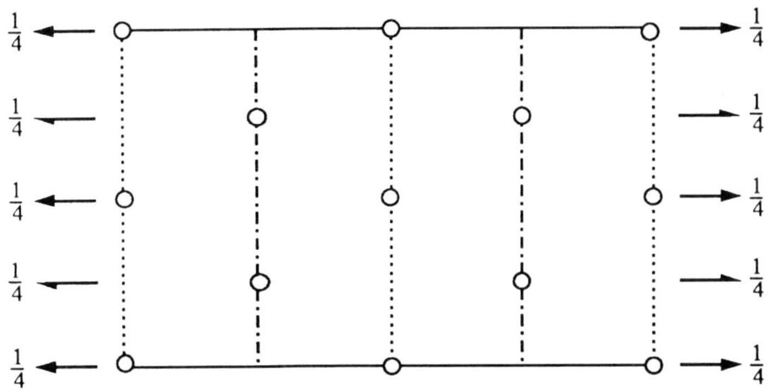

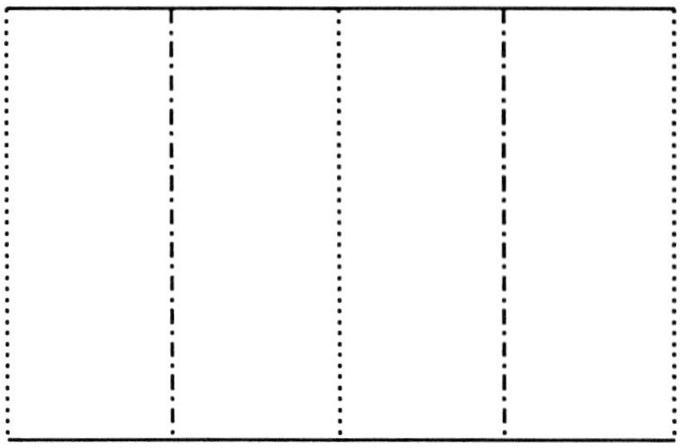

Figure 9-18. Space group *Cc*.

also eight GEPs representing eight unique symmetry operations. It is not immediately apparent what all of the eight are, but we can determine some of them from the space group symbol. There is a fourfold axis that includes a $4^1=4$, $4^2=2$, $4^3=4^{-1}$, and $4^4=E$. (This is to say that a fourfold axis, which rotates objects 90° at a time, inherently includes rotations of 90°, 180°, 270°, and 360°, each of which is represented by a separated matrix. The matrices for the last three may be determined by multiplying that for the 4 by itself the number of times indicated by the exponent.) There is an *n* glide, and because point group $4/m$ contains a center of symmetry, the space group must also. Evidently the other two symmetry operations will appear in the derivation as products of some of these. Recognizing that the rotation axis and the glide plane must intersect, we shall initially assume their intersection to be the origin and write matrices for them as if they passed through the origin.

$$4 = \begin{bmatrix} 0 & 1 & 0 & 0 \\ -1 & 0 & 0 & 0 \\ 0 & 0 & 1 & 0 \\ 0 & 0 & 0 & 1 \end{bmatrix}, \quad 4^2 = \begin{bmatrix} -1 & 0 & 0 & 0 \\ 0 & -1 & 0 & 0 \\ 0 & 0 & 1 & 0 \\ 0 & 0 & 0 & 1 \end{bmatrix},$$

$$4^3 = \begin{bmatrix} 0 & -1 & 0 & 0 \\ 1 & 0 & 0 & 0 \\ 0 & 0 & 1 & 0 \\ 0 & 0 & 0 & 1 \end{bmatrix}$$

$$4^4 = \begin{bmatrix} 1 & 0 & 0 & 0 \\ 0 & 1 & 0 & 0 \\ 0 & 0 & 1 & 0 \\ 0 & 0 & 0 & 1 \end{bmatrix}, \quad n = \begin{bmatrix} 1 & 0 & 0 & \frac{1}{2} \\ 0 & 1 & 0 & \frac{1}{2} \\ 0 & 0 & -1 & 0 \\ 0 & 0 & 0 & 1 \end{bmatrix}.$$

The 4^1, 4^2, 4^3, and 4^4 constitute a subgroup, so that multiplying any combinations of them together will just continue to yield other members of the subgroup. Multiplication of the 4^1, 4^2, and 4^3 by n, however, will yield three new matrices:

$$n\cdot 4^1 = \begin{bmatrix} 0 & 1 & 0 & \frac{1}{2} \\ -1 & 0 & 0 & \frac{1}{2} \\ 0 & 0 & -1 & 0 \\ 0 & 0 & 0 & 1 \end{bmatrix} \quad n\cdot 4^2 = \begin{bmatrix} -1 & 0 & 0 & \frac{1}{2} \\ 0 & -1 & 0 & \frac{1}{2} \\ 0 & 0 & -1 & 0 \\ 0 & 0 & 0 & 1 \end{bmatrix}$$

$$n\cdot 4^3 = \begin{bmatrix} 0 & -1 & 0 & \frac{1}{2} \\ 1 & 0 & 0 & \frac{1}{2} \\ 0 & 0 & -1 & 0 \\ 0 & 0 & 0 & 1 \end{bmatrix}$$

The first is a $\bar{4}$ rotoinversion axis parallel to c and located at $x=\frac{1}{2}$ and $y=0$, the second is a center of symmetry at $(\frac{1}{4},\frac{1}{4},0)$, and the third is a $\bar{4}$ rotoinversion axis parallel to c and located at $x=0$ and $y=\frac{1}{2}$. We can therefore draw the space group as shown in Figure 9-19(a). Here, as in our derivation of $C2/c$, there are two centers of symmetry but neither is at the origin. One center is marked by the small circles at $(\frac{1}{2},\frac{1}{2},0)$, and the other is inherent in the rotoinversion axis and is at $(\frac{1}{2},0,0)$. Which is placed at the origin is an arbitrary decision, so we shall place the one at $(\frac{1}{2},0,0)$ there, resulting in the conventional orientation shown in Figure 9-19b. The GEPs, obtained from matrices rewritten after the change of origin, are x,y,z; $-x,-y,z$; $-y+\frac{1}{2},x+\frac{1}{2},z$; $y+\frac{1}{2},-x+\frac{1}{2},z$; $-x+\frac{1}{2},-y+\frac{1}{2},-z$; $x+\frac{1}{2},y+\frac{1}{2},-z$; $y,-x,-z$; and $-y,x,-z$.

Ibca

The space group *Ibca* is isomorphic with point group *mmm* ($\equiv 2/m\ 2/m\ 2/m$). We know that the b, c, and a glides must intersect in a point, so we will consider their

Figure 9-19. Space group $P4/n$ (a) before and (b) after shifting to a standard orientation with an inversion at the origin. The center of symmetry at $(\frac{1}{4},\frac{1}{4},0)$ could just as well have served as the origin.

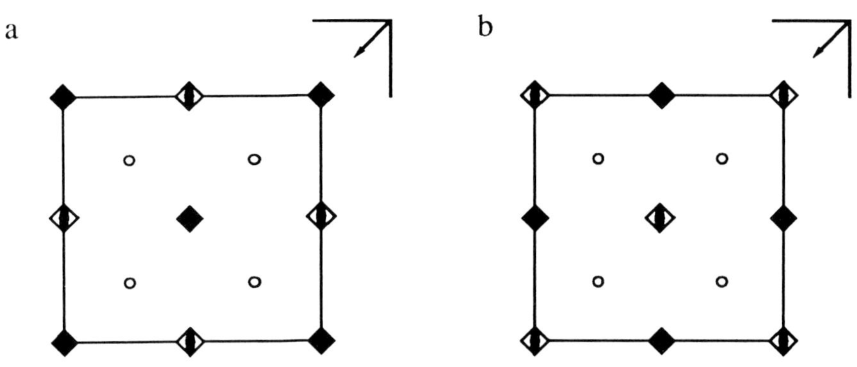

intersection to be the unit-cell origin initially. The body-centering (I) operation is represented by

$$\begin{bmatrix} 1 & 0 & 0 & \frac{1}{2} \\ 0 & 1 & 0 & \frac{1}{2} \\ 0 & 0 & 1 & \frac{1}{2} \\ 0 & 0 & 0 & 1 \end{bmatrix}$$

The matrices for b, c, and a glides passing through the origin are, respectively,

$$b = \begin{bmatrix} -1 & 0 & 0 & 0 \\ 0 & 1 & 0 & \frac{1}{2} \\ 0 & 0 & 1 & 0 \\ 0 & 0 & 0 & 1 \end{bmatrix}, c = \begin{bmatrix} 1 & 0 & 0 & 0 \\ 0 & -1 & 0 & 0 \\ 0 & 0 & 1 & \frac{1}{2} \\ 0 & 0 & 0 & 1 \end{bmatrix}, a = \begin{bmatrix} 1 & 0 & 0 & \frac{1}{2} \\ 0 & 1 & 0 & 0 \\ 0 & 0 & -1 & 0 \\ 0 & 0 & 0 & 1 \end{bmatrix}.$$

From these we get

$$b \cdot c = \begin{bmatrix} -1 & 0 & 0 & 0 \\ 0 & -1 & 0 & \frac{1}{2} \\ 0 & 0 & 1 & \frac{1}{2} \\ 0 & 0 & 0 & 1 \end{bmatrix} = 2_1 \parallel c \text{ at } x = 0, y = \tfrac{1}{4}$$

$$b \cdot a = \begin{bmatrix} -1 & 0 & 0 & \frac{1}{2} \\ 0 & 1 & 0 & \frac{1}{2} \\ 0 & 0 & -1 & 0 \\ 0 & 0 & 0 & 1 \end{bmatrix} = 2_1 \parallel b \text{ at } x = \tfrac{1}{4}, z = 0$$

$$c \cdot a = \begin{bmatrix} 1 & 0 & 0 & \frac{1}{2} \\ 0 & -1 & 0 & 0 \\ 0 & 0 & -1 & \frac{1}{2} \\ 0 & 0 & 0 & 1 \end{bmatrix} = 2_1 \parallel a \text{ at } y = 0, z = \tfrac{1}{4}$$

$$b \cdot c \cdot a = \begin{bmatrix} -1 & 0 & 0 & \frac{1}{2} \\ 0 & -1 & 0 & \frac{1}{2} \\ 0 & 0 & -1 & \frac{1}{2} \\ 0 & 0 & 0 & 1 \end{bmatrix} = \bar{1} \text{ at } x = \tfrac{1}{4}, y = \tfrac{1}{4}, z = \tfrac{1}{4}$$

The remaining matrices are found my multiplying these eight (the seven above plus the identity matrix) by the body-centering matrix. The results are:

$$\begin{bmatrix} -1 & 0 & 0 & \frac{1}{2} \\ 0 & -1 & 0 & 0 \\ 0 & 0 & 1 & 0 \\ 0 & 0 & 0 & 1 \end{bmatrix} = 2 \parallel c \text{ at } x = \tfrac{1}{4}, y = 0$$

$$\begin{bmatrix} -1 & 0 & 0 & 0 \\ 0 & 1 & 0 & 0 \\ 0 & 0 & -1 & \frac{1}{2} \\ 0 & 0 & 0 & 1 \end{bmatrix} = 2 \parallel b \text{ at } x = 0, z = \tfrac{1}{4}$$

$$\begin{bmatrix} 1 & 0 & 0 & 0 \\ 0 & -1 & 0 & \frac{1}{2} \\ 0 & 0 & -1 & 0 \\ 0 & 0 & 0 & 1 \end{bmatrix} = 2 \parallel a \text{ at } y = \tfrac{1}{4}, z = 0$$

$$\begin{bmatrix} -1 & 0 & 0 & 0 \\ 0 & -1 & 0 & 0 \\ 0 & 0 & -1 & 0 \\ 0 & 0 & 0 & 1 \end{bmatrix} = \bar{1} \text{ at } x = 0, y = 0, z = 0$$

$$\begin{bmatrix} -1 & 0 & 0 & \frac{1}{2} \\ 0 & 1 & 0 & 0 \\ 0 & 0 & 1 & \frac{1}{2} \\ 0 & 0 & 0 & 1 \end{bmatrix} = c \text{ glide} \perp a \text{ at } x = \tfrac{1}{4}$$

$$\begin{bmatrix} 1 & 0 & 0 & 0 \\ 0 & 1 & 0 & \frac{1}{2} \\ 0 & 0 & -1 & \frac{1}{2} \\ 0 & 0 & 0 & 1 \end{bmatrix} = b \text{ glide} \perp c \text{ at } z = \tfrac{1}{4}$$

$$\begin{bmatrix} 1 & 0 & 0 & \frac{1}{2} \\ 0 & -1 & 0 & \frac{1}{2} \\ 0 & 0 & 1 & 0 \\ 0 & 0 & 0 & 1 \end{bmatrix} = a \text{ glide} \perp b \text{ at } y = \tfrac{1}{4}$$

$$\begin{bmatrix} 1 & 0 & 0 & \frac{1}{2} \\ 0 & 1 & 0 & \frac{1}{2} \\ 0 & 0 & 1 & \frac{1}{2} \\ 0 & 0 & 0 & 1 \end{bmatrix} = I.$$

The space-group drawing in Figure 9-20 may now be constructed. Note that this time a center of symmetry occurs on the origin, so our initial assumption that the three glide planes intersected at the origin was correct. The center at ($\frac{1}{4}, \frac{1}{4}, \frac{1}{4}$) could also serve as origin, of course. The sixteen GEPs are

x,y,z	$-x,y+\frac{1}{2},z$	$x,-y,z+\frac{1}{2}$	$x,-y+\frac{1}{2},-z$
$x,y+\frac{1}{2},-z+\frac{1}{2}$	$-x,-y+\frac{1}{2},z+\frac{1}{2}$	$-x+\frac{1}{2},y+\frac{1}{2},-z$	$-x,-y,-z$
$x+\frac{1}{2},-y,-z+\frac{1}{2}$	$-x+\frac{1}{2},-y+\frac{1}{2},-z+\frac{1}{2}$	$x+\frac{1}{2},y,-z$	$-x+\frac{1}{2},-y,z$
$x+\frac{1}{2},y+\frac{1}{2},z+\frac{1}{2}$	$-x+\frac{1}{2},y,z+\frac{1}{2}$	$x+\frac{1}{2},-y+\frac{1}{2},z$	$-x,y,-z+\frac{1}{2}$

The International Tables for Crystallography

Because there is a large but finite number of space groups, a comprehensive compilation of the sort of data we have just derived could be made for each one, and it

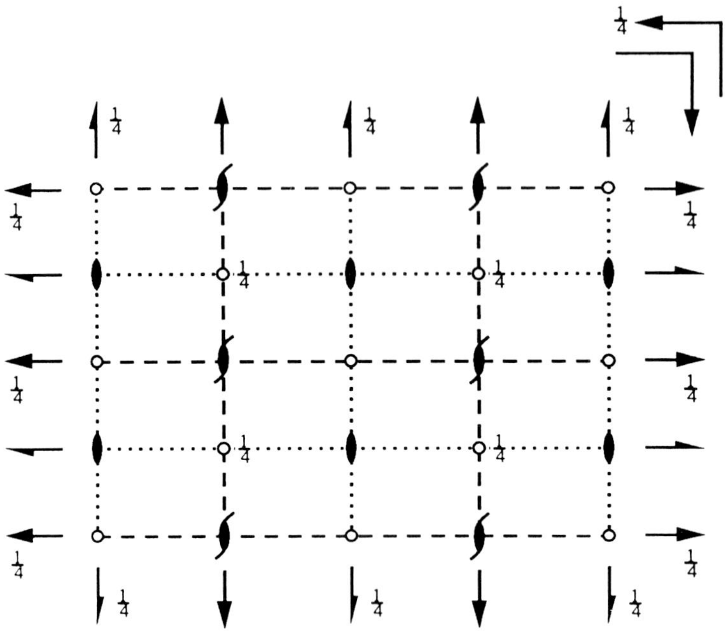

Figure 9-20. Space group *Ibca*.

would save enormous amounts of labor on the part of individuals who routinely deal with space groups. Such a compilation exists in the form of the *International Tables for Crystallography*, Volume A [Th. Hahn (ed.), 1987, D. Reidel Publishing Company, Dordrecht, Holland]. Figure 9-21 is an illustration of part of the information for space group $I4_1/amd$. The data for this space group are presented on four pages in the *Tables*—two each for two choices of origin—but it is condensed here by omission of material that is not necessary for an understanding of the subjects covered in this text. Even with these omissions, a great deal of information is shown, and we shall consider it piecemeal.

1. At the top of the left-hand page (shown at the top here) is information about the symmetry of the space group. Both the shortened symbol and the full Hermann–Mauguin symbol are given. This space group is number 141 out of the total of 230 space groups. The crystal class of crystals having this space group is $4/mmm$, and they belong to the tetragonal crystal system. The Schoenflies symbol for this space group is D_{4h}^{19}. The Patterson symmetry is the symmetry of the Patterson function, used in x-ray crystal-structure determination.
2. Space group drawings are provided. The view is down the *c* axis. The drawing on the left shows how a motif is repeated by the space group symmetry. If a circle represents some right-handed motif, then a circle enclosing a comma represents the left-handed version of that motif. Plus and minus signs indicate relative positions above or below the plane of the page, and the fractions are fractions of the *c* cell edge.

   The right-hand drawing shows the locations of symmetry elements in the unit cell. Fractions, again, are fractions of the c cell edge above the plane of the page, and refer to the symmetry element to which they are adjacent. A key to these symbols is given in Table 9-4. Space-group drawings are conventionally oriented so that a is positive down the page, b is positive to the right, and c is positive out of the page toward the viewer. The origin is in the upper left-hand corner of the unit cell and in the plane of the page.

Figure 9-21. The appearance of parts of two pages from the *International Tables for Crystallography*, Volume A (1987).

$I\,4_1/amd$ D_{4h}^{19} $4/mmm$ Tetragonal

No. 141 $I\,4_1/a\,2/m\,2/d$ Patterson symmetry $I4/m\,m\,m$

ORIGIN CHOICE 1

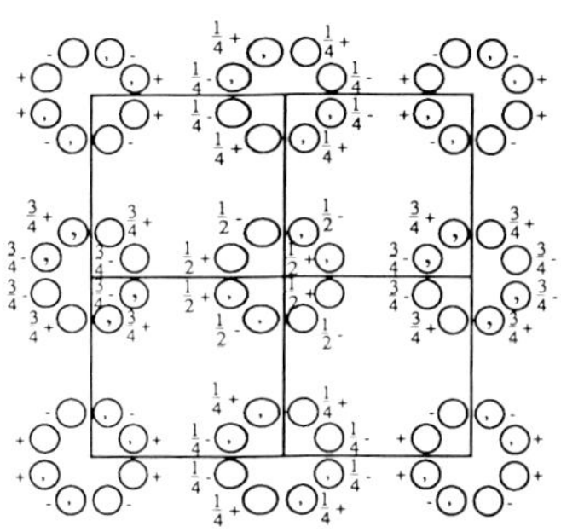

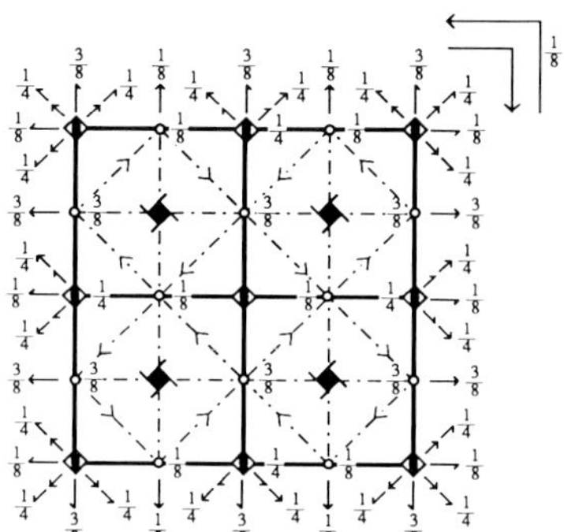

Origin at $\bar{4}m2$, at $0, \frac{1}{4}, -\frac{1}{8}$ from centre ($2/m$)

Asymmetric unit $0 \le x \le \frac{1}{2};\ 0 \le y \le \frac{1}{2};\ 0 \le z \le \frac{1}{8}$

Positions

Multiplicity, Wyckoff letter, Site symmetry — Coordinates

$(0,0,0)+ \quad (\frac{1}{2},\frac{1}{2},\frac{1}{2})+$

Multiplicity	Wyckoff letter	Site symmetry	Coordinates			
32	i	1	(1) x, y, z	(2) $\bar{x}+\frac12, \bar{y}+\frac12, z+\frac12$	(3) $\bar{y}, x+\frac12, z+\frac14$	(4) $y+\frac12, \bar{x}, z+\frac34$
			(5) $\bar{x}+\frac12, y, \bar{z}+\frac34$	(6) $x, \bar{y}+\frac12, \bar{z}+\frac14$	(7) $y+\frac12, x+\frac12, \bar{z}+\frac12$	(8) $\bar{y}, \bar{x}, \bar{z}$
			(9) $\bar{x}, \bar{y}+\frac12, \bar{z}+\frac14$	(10) $x+\frac12, y, \bar{z}+\frac34$	(11) $y, \bar{x}, \bar{z}$	(12) $\bar{y}+\frac12, x+\frac12, \bar{z}+\frac12$
			(13) $x+\frac12, \bar{y}+\frac12, z+\frac12$	(14) $\bar{x}, y, z$	(15) $\bar{y}+\frac12, \bar{x}, z+\frac34$	(16) $y, x+\frac12, z+\frac14$
16	h	.m.	$0, y, z$	$\frac12, \bar{y}+\frac12, z+\frac12$	$\bar{y}, \frac12, z+\frac14$	$y+\frac12, 0, z+\frac34$
			$\frac12, y, \bar{z}+\frac34$	$0, \bar{y}+\frac12, \bar{z}+\frac14$	$y+\frac12, \frac12, \bar{z}+\frac12$	$\bar{y}, 0, \bar{z}$
16	g	..2	$x, x, 0$	$\bar{x}+\frac12, \bar{x}+\frac12, \frac12$	$\bar{x}, x+\frac12, \frac14$	$x+\frac12, \bar{x}, \frac34$
			$\bar{x}, \bar{x}+\frac12, \frac14$	$x+\frac12, x, \frac34$	$x, \bar{x}, 0$	$\bar{x}+\frac12, x+\frac12, \frac12$
16	f	.2.	$x, \frac14, \frac18$	$x+\frac12, \frac14, \frac58$	$\frac34, x+\frac12, \frac38$	$\frac34, \bar{x}, \frac78$
			$\bar{x}, \frac14, \frac18$	$\bar{x}+\frac12, \frac14, \frac58$	$\frac14, \bar{x}, \frac78$	$\frac14, x+\frac12, \frac38$
8	e	2mm.	$0, 0, z$	$0, \frac12, z+\frac14$	$\frac12, 0, \bar{z}+\frac34$	$\frac12, \frac12, z+\frac12$
8	d	.2/m.	$0, \frac14, \frac58$	$\frac12, \frac14, \frac18$	$\frac34, \frac12, \frac78$	$\frac34, 0, \frac38$
8	c	.2/m.	$0, \frac14, \frac18$	$\frac12, \frac14, \frac58$	$\frac34, \frac12, \frac38$	$\frac34, 0, \frac78$
4	b	$\bar{4}m2$	$0, 0, \frac12$	$0, \frac12, \frac34$		
4	a	$\bar{4}m2$	$0, 0, 0$	$0, \frac12, \frac14$		

Table 9-4. A Key to Symbols Used in the *International Tables for Crystallography*

Symbol	Meaning	Symbol	Meaning
	mirror perpendicular to the plane of the page	o	center of symmetry
	axial glide perpendicular to the plane of the page		diad in the plane of the page
	diagonal (n) glide perpendicular to the plane of the page		twofold screw axis in the plane of the page
	diamond (d) glide perpendicular to the plane of the page		
			Axes normal to the page:
	mirror parallel to the plane of the page		2 2_1
	axial glide plane parallel to the plane of the page		3 3_1 3_2
	diagonal (n) glide plane parallel to the plane of the page		4 4_1 4_2 4_3
$\frac{1}{8}$	diamond (d) glide plane parallel to the plane of the page		6 6_1 6_2 6_3 6_4 6_5
			$\bar{3}$ $\bar{4}$ $\bar{6}$

3. The origin of the unit cell is specified in terms of its location with respect to symmetry elements. In this case, it is at the intersection of the $\bar{4}$ axis, mirror plane, and twofold axis, at 0, $\frac{1}{4}$, $-\frac{1}{8}$ from the center of inversion at the intersection of a twofold axis and a mirror plane. Because the origin could have been chosen at $2/m$ as well as at $\bar{4}2m$, a similar set of data with that origin is also given in the Tables.
4. The asymmetric unit, that volume of the unit cell that is unique, and from which the rest of the cell may be obtained by operation of the symmetry elements, is given.
5. On the right-hand page (shown here on the bottom) is a table of equivalent positions. This table shows positions of atoms that are symmetrically equivalent to one another. There are five columns. The first gives the number of atoms that must be present in the unit cell if one such atom is present (the so-called multiplicity). The second gives alphabetical symbols for the positions. The third shows the point symmetry that exists at the particular position. The fourth gives the equivalent positions in terms of fractional coordinates x, y, and z; fractional coordinates are fractions of the unit cell edges. The last column, omitted here, shows conditions for diffraction indices that limit the occurrence of x-ray-diffraction maxima ("reflections"). If the lattice is centered, then appropriate fractions to be added to each position appear under the column heading "coordinates;" in this case, because the lattice is body centered, "(0,0,0)+ $(\frac{1}{2},\frac{1}{2},\frac{1}{2})$+"

means that $(\frac{1}{2},\frac{1}{2},\frac{1}{2})$ must be added to all coordinates shown—the list given is thus only one-half of the total list of coordinates.

Consider Wyckoff position *a* at the bottom of the figure. There are four such positions in the unit cell (column 1). The are 0,0,0; $0,\frac{1}{2},\frac{1}{4}$; $\frac{1}{2},\frac{1}{2},\frac{1}{2}$; and $\frac{1}{2},0,\frac{3}{4}$.(This last position could be written $\frac{1}{2},1,\frac{3}{4}$, inasmuch as that is what one obtains by adding $\frac{1}{2},\frac{1}{2},\frac{1}{2}$ to $0,\frac{1}{2},\frac{1}{4}$, but a translation of 1 unit-cell dimension is the same as no translation at all, since unit cells are identical by definition. Thus we write 0 instead of 1.) If an atom is located at 0,0,0 (the origin), then identical atoms must be found at 0, $b/2,c/4$; $a/2,b/2,c/2$; and $a/2,0,3c/4$. Note that the presence of the atom at 0,0,0 also implies its presence at all other corners of the unit cell (by full translations), because every corner is the origin for some adjacent unit cell. The point symmetry at the origin is $\bar{4}2m$; that is, an observer at the origin would see that symmetry about him.

At Wyckoff position *i*, the point symmetry is 1, which means that no symmetry elements pass through such a position, and that makes this the *general* position. The fractional coordinates x, y, and z may have any values that do not yield one of the other eight kinds of Wyckoff positions. Note that these other eight positions are all on symmetry elements, and so the are called *special* positions. For any atom that occupies a general position, there must be 31 other identical atoms in the unit cell, for a total of 32 such atoms. If there is an atom at x,y,z, then identical atoms are also located at $-x,y,z$; $x,\frac{1}{2}+y,\frac{1}{4}-z$; and so on through the 16 listed; then also at $(\frac{1}{2},\frac{1}{2},\frac{1}{2})+$ each of the 16 listed, for a total of 32. If the constraints required of any special position are imposed on the general position, the latter degenerates to the special positions. For example, requiring x, y, and z all to be zero in the general positions results in the Wyckoff *a* positions.

GRAPHICAL REPRESENTATION OF CRYSTAL STRUCTURES

The information tabulated in the *International Tables* is obviously of great value in crystal chemistry, but in itself it does not provide a visual representation of the arrangement of atoms in any particular mineral. In fact many space groups are common to more than one mineral (often to *many* minerals), and the symmetry relations alone do not yield pictures of the crystal structures. Besides space-group information, it is necessary to have a set of atomic positions, and these are found generally by x-ray- or neutron-diffraction experiments (see Chapter 10). They are usually expressed as fractional coordinates (fractions of *a*, *b*, and *c*) for atoms in the *asymmetric unit*, that group of atoms that constitutes the symmetrically unique atoms, all of the others in the unit cell being obtained by symmetry from these.

Consider, as an example, the mineral zircon ($ZrSiO_4$), which crystallizes in space group $I4_1/amd$. The asymmetric unit contains one Zr atom, one Si atom, and one O atom, with fractional coordinates as shown in Table 9-5. Thus, with reference to Figure 9-21, the Zr atom is on an *a* Wyckoff position, the Si is on a *b* position, and O is on an *h* position. The numbers of positions require unit-cell contents of $Zr_4Si_4O_{16}$, which reduces to the chemical formula given above.

One of the popular ways of representing crystal structures in graphical form is by

Table 9-5. Fractional Coordinates for Zircon

Atom	*x*	*y*	*z*
Zr	0.0	0.0	0.0
Si	0.0	0.0	0.5
O	0.0	0.8161	0.3203

projection of the atomic positions down one of the crystallographic axes. Figure 9-22 shows such a projection of the zircon structure down the *c* axis, with atom heights shown as hundredths of the unit-cell translation. For clarity, atoms are generally pictured with their relative sizes approximately correct, but with all of them uniformly reduced in size relative to the size of the unit cell. The drawing is made very easily by using the fractional coordinates in Table 9-5 and the equivalent positions from the International Tables (Figure 9-21 here). The crystal structure of zircon looks simple enough, but just how the atoms are bonded together is not obvious when only the atomic positions are shown.

To make the picture clearer, we shall connect atoms that are chemically bonded to each other by lines—a "ball-and-stick" model of the structure. Some false but helpful perspective can even be added by tapering the bonds as they recede into the background. Even so, enough bonds and atoms superimpose in the *c*-axis projection of zircon that the structure is still difficult to appreciate. In such a case, changing the axis of projection is generally the best option, and Figure 9-23, which is an *a*-axis projection, shows the structure more clearly than a *c*-axis projection does. The edge-sharing chains of Si and Zr coordination polyhedra running along *c* are evident, but the natures of the polyhedra themselves are somewhat obscure.

Often minerals are considered as arrays of linked polyhedra, formed by connecting the anions that surround a cation. In polyhedral drawings, only the polyhedra are shown, with the anions shrunk to points at the polyhedral corners, and with the cations that are at the centers of the polyhedra usually only implied. The polyhedra in zircon consist of a tetrahedron of oxygen atoms about the silicon, and a triangular dodecahedron about the zircon. The latter is shown, in *c*-axis projection, in Figure 9-24, with heights along *c* given to facilitate comparison with Figure 9-22. Figure 9-25 shows a polyhedral drawing of zircon, projected down *a*.

Packing diagrams, which show atoms not only in their correct relative positions but also occupying their correct volumes of the unit cell, are used infrequently in mineralogy. Figure 9-26 is a packing diagram for zircon, projected again down the *c* axis. It is obvious that nothing is gained in this case, and that other types of representations give a clearer picture of the structure. Nevertheless, there are occasions when a packing diagram may illustrate a point better than a polyhedral or ball-and-stick drawing.

With the widespread availability of computers, high-quality printers and plotters, and software specifically written for making crystallographic drawings, it is no longer difficult to make illustrations of crystal structures, in either orthographic projection or perspective, from any desired direction. This can often be done interactively, and illustrations can be rotated on the screen of a computer terminal until a desirable view is obtained. Figure 9-27 shows the ZrO_8 triangular dodecahedron in zircon rotated to a view that shows it well. There are two features to note in this drawing: (1) The atoms are not depicted as spheres, but as thermal ellipsoids with sizes and shapes proportional to the thermal vibrations of the atoms. (2) Perspective has been added to enhance the three-dimensional appearance of the drawing.

A final enhancement, shown in Figure 9-28, is stereographic projection. Here two drawings, made by projecting the structure from slightly different angles to simulate parallax due to eye separation, are placed side by side. One may either use a stereoscopic viewer to see an apparently three-dimensional picture, or, if able to coordinate movements of the eyes adequately, may be able to "see in stereo" without the aid of a viewer. Stereographic projections are sometimes more confusing

Figure 9-22. The atomic arrangement of zircon, $ZrSiO_4$, viewed down the *c* axis with *a* pointing down in the plane of the page and *b* pointing to the right in the plane of the page. The origin is the upper left-hand corner. Numbers indicate heights of atoms above the plane of the page ($z = 0$) in hundredths of the *c* unit-cell dimension. The unit cell is outlined by the light square. In order of increasing size, the spheres represent Si, Zr, and O.

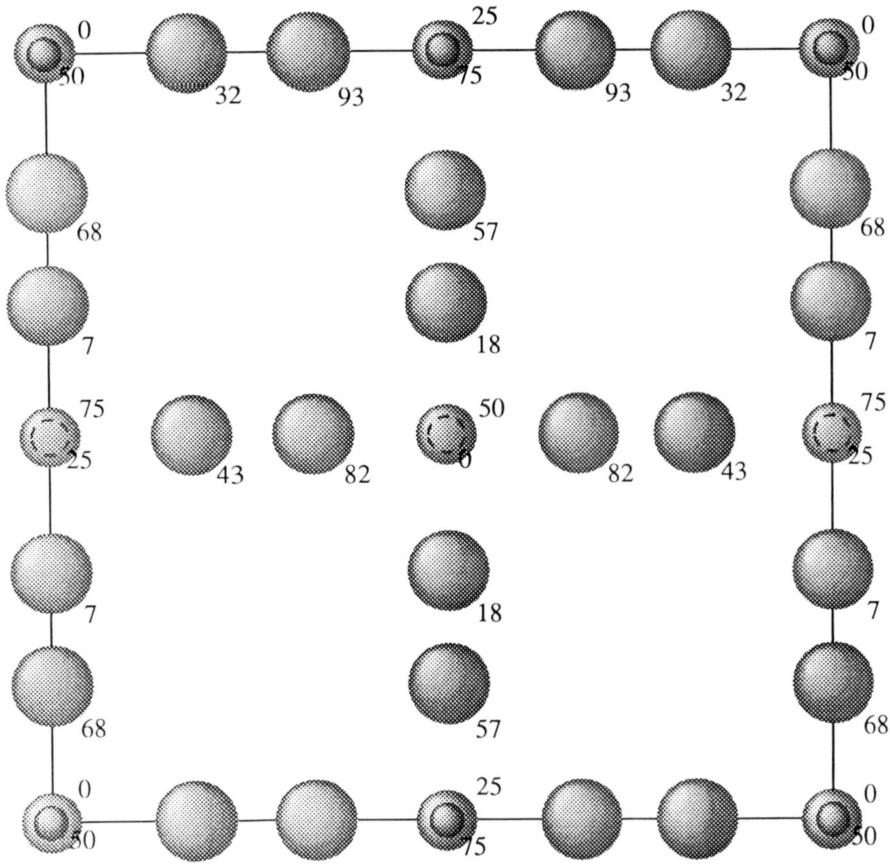

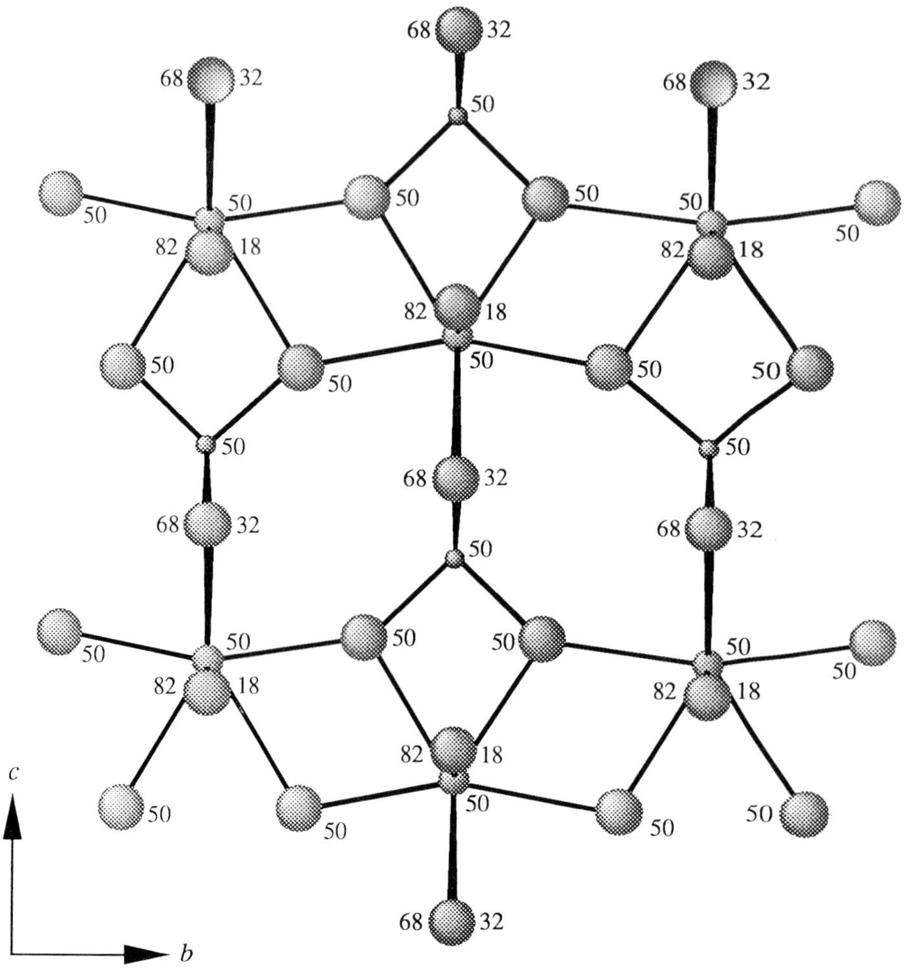

Figure 9-23. A "ball-and-stick" figure of the crystal structure of zircon, viewed down the *a* axis. Only a slab of the full unit cell is depicted; a similar but potentially confusing slab lies behind the one shown. Relative sizes of atom types are as in Figure 9-22.

than helpful—particularly if the unit cell has a large number of atoms and the drawing is cluttered with atom labels or other information; but with a judicious choice of viewing direction, they may provide the best way of portraying some crystal structures in two dimensions.

TUTORIAL ON MATRICES AND MATRIX OPERATIONS

This very brief review instructs those unacquainted with matrix algebra how to carry out the operations of addition, subtraction, and multiplication of matrices.

These are the only ways in which we shall manipulated matrices in this text. The use of matrix algebra is in general so prevalent in scientific computation, however, that a formal course or rigorous self-instruction in the subject should be acquired by anyone pursuing research in the physical sciences. The treatment presented here is no substitute for that.

A matrix is a rectangular array of entries which, for our purposes, will be real numbers. Each entry is known as a matrix *element*, and the *order* of the matrix is the number of rows and columns it contains, designated $m \times n$, where m is the number of rows and n the number of columns. Thus the matrix

$$\begin{bmatrix} 2 & 3 & 1 \\ 4 & 1 & 1 \\ 6 & -1 & 0 \\ 0 & 0 & 8 \end{bmatrix}$$

is of order 4×3 and contains only integral matrix elements.

Two matrices are said to be equal if and only if they are of the same order and every element of one is identical to the corresponding element in the other. The matrices

$$\begin{bmatrix} 2 & 3 \\ 4 & 1 \end{bmatrix} \text{ and } \begin{bmatrix} 2 & 3 & 0 \\ 4 & 1 & 0 \end{bmatrix}$$

are not equal, but would be if the third column were removed from the 2×3 matrix to make it a 2×2 matrix.

Matrices may be added or subtracted only if they are of the same order. The oper-

Figure 9-24. The ZrO_8 polyhedron in zircon. Corners are the locations of oxygen atoms, and the zirconium atom is at the center, but not shown. The orientation is the same as in Figure 9-22.

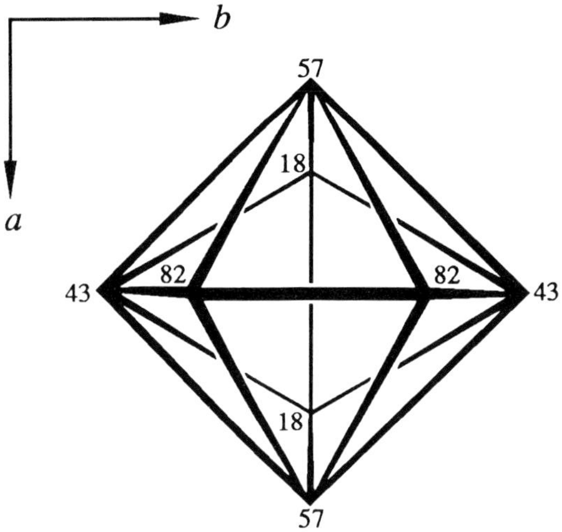

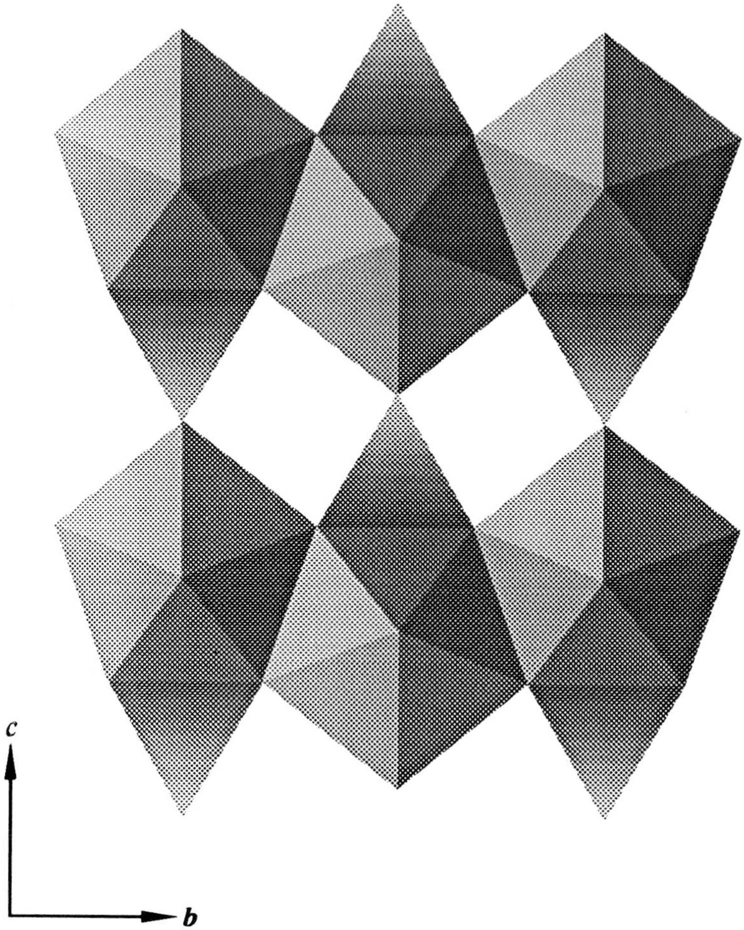

Figure 9-25. A shaded polyhedral drawing of the same slab of the zircon structure that appears in Figure 9-23. As in Figure 9-24, oxygens are at the corners of the polyhedra.

ations are done by adding or subtracting each of the corresponding matrix elements to produce a matrix of the same order:

$$\begin{bmatrix} a_{11} & a_{12} & \cdots & a_{1n} \\ a_{21} & a_{22} & \cdots & a_{2n} \\ & & \vdots & \\ a_{m1} & a_{m2} & \cdots & a_{mn} \end{bmatrix} + \begin{bmatrix} b_{11} & b_{12} & \cdots & b_{1n} \\ b_{21} & b_{22} & \cdots & b_{2n} \\ & & \vdots & \\ b_{m1} & b_{m2} & \cdots & b_{mn} \end{bmatrix}$$

$$= \begin{bmatrix} (a_{11} + b_{11}) & (a_{21} + b_{21}) & \cdots & (a_{1n} + b_{1n}) \\ (a_{21} + b_{21}) & (a_{22} + b_{22}) & \cdots & (a_{2n} + b_{2n}) \\ & & \vdots & \\ (a_{m1} + b_{m1}) & (a_{m2} + b_{m2}) & \cdots & (a_{mn} + b_{mn}) \end{bmatrix}$$

Multiplication of two matrices is possible only if the first has the same number of columns as the second has rows: $[n \times m][m \times p] = [n \times p]$. It is carried out as follows:

$$\begin{bmatrix} a_{11} & a_{12} & \cdots & a_{1n} \\ a_{21} & a_{22} & \cdots & a_{2n} \\ & & \vdots & \\ a_{m1} & a_{m2} & \cdots & a_{mn} \end{bmatrix} \begin{bmatrix} b_{11} & b_{21} & \cdots & b_{1p} \\ b_{21} & b_{22} & \cdots & b_{2p} \\ & & \vdots & \\ b_{n1} & bn2 & \cdots & b_{np} \end{bmatrix}$$

$$= \begin{bmatrix} \sum_{j=1}^{n} a_{1j}b_{j1} & \sum_{j=1}^{n} a_{1j}b_{j2} & \cdots & \sum_{j=1}^{n} a_{1j}b_{jp} \\ \sum_{j=1}^{n} a_{2j}b_{j1} & \sum_{j=1}^{n} a_{2j}b_{j2} & \cdots & \sum_{j=1}^{n} a_{2j}b_{jp} \\ & & \vdots & \\ \sum_{j=1}^{n} a_{mj}b_{j1} & \sum_{j=1}^{n} a_{mj}b_{j2} & \cdots & \sum_{j=1}^{n} a_{mj}b_{jp} \end{bmatrix}$$

Figure 9-26. A packing diagram of zircon, in which the atoms are represented in their approximate true sizes with respect to the unit cell dimension.

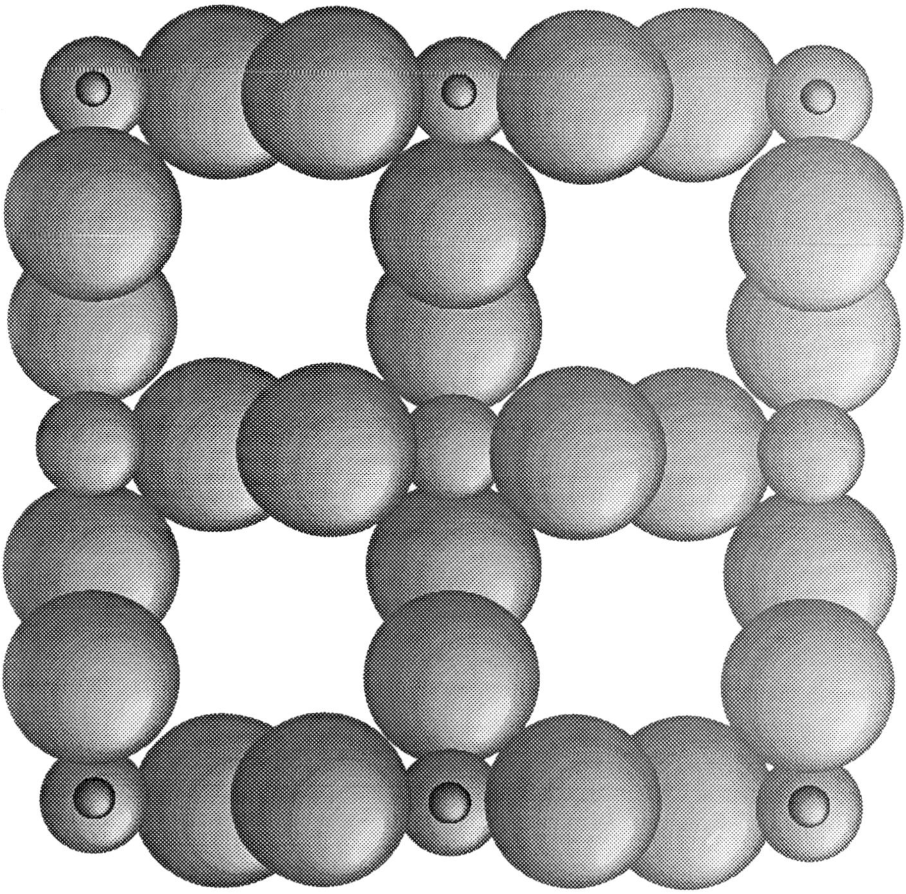

For example,

$$\begin{bmatrix} 2 & 3 & 1 \\ 1 & 2 & 1 \end{bmatrix} \begin{bmatrix} 2 & 3 \\ 1 & 0 \\ 4 & 2 \end{bmatrix} = \begin{bmatrix} (2)(2)+(3)(1)+(1)(4) & (2)(3)+(3)(0)+(1)(2) \\ (1)(2)+(2)(1)+(1)(4) & (1)(3)+(2)(0)+(1)(2) \end{bmatrix}$$

$$= \begin{bmatrix} 11 & 8 \\ 8 & 5 \end{bmatrix}.$$

It is apparent from this example that matrix multiplication is not, in general, commutative: $[A][B] \neq [B][A]$. If we reverse the order of the two matrices above, the result is a matrix of order 3×3:

$$\begin{bmatrix} 7 & 12 & 5 \\ 2 & 3 & 1 \\ 10 & 16 & 6 \end{bmatrix}$$

In general, reversing the order of attempted multiplication would result in an impossible operation. For instance, while it is possible to premultiply a 4×2 matrix by a 3×4 matrix, a 3×4 matrix cannot be premultiplied by a 4×2 matrix.

Figure 9-27. A thermal ellipsoid plot of the ZrO_8 polyhedron, in which the atoms are represented as probability ellipsoids depicting thermal motion, as determined by x-ray crystal-structure refinement. Zr is the central atom. Orientation differs from that of Figure 9-24.

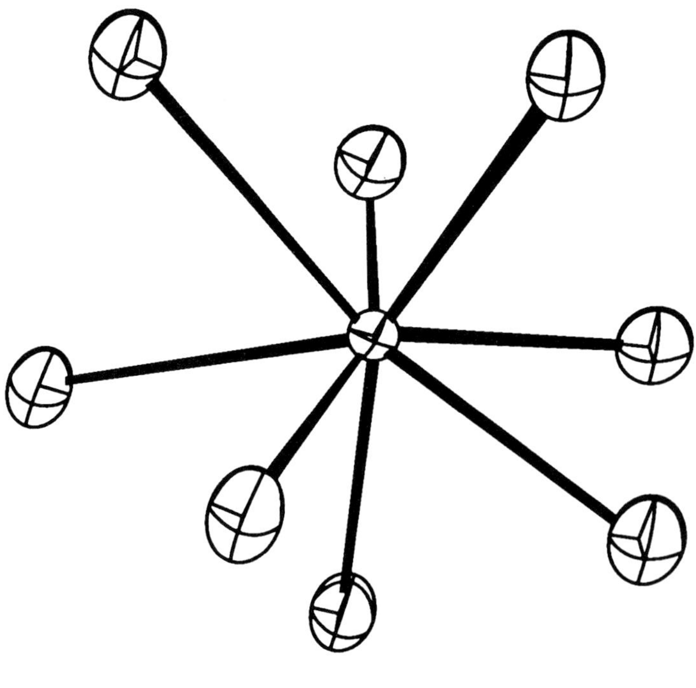

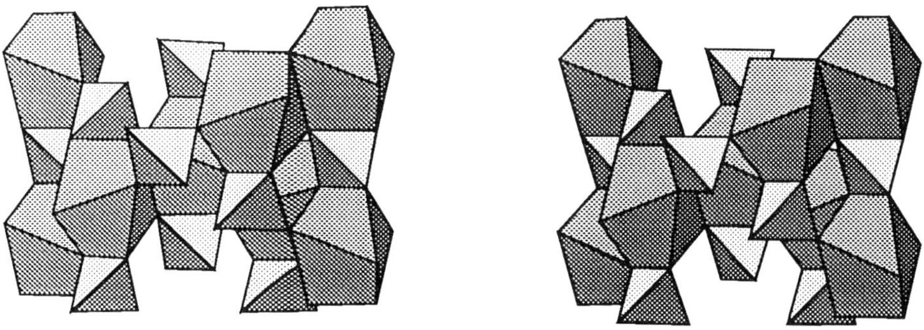

Figure 9-28. A stereoscopic view of the zircon structure. The ZrO_8 polyhedra have been represented as distorted cubes by omitting one edge through each "cube" face. The view is normal to *c*, and somewhat to the left of *a*. [Redrawn from T. Zoltai (1965) *Stereoscopic Drawings of Polyhedral Mineral-Structure Models*, University of Minnesota, Minneapolis.]

In crystallography, as well as in many other common applications of matrices in science, those of greatest use are either square matrices (the number of columns equals the number of rows), column matrices ($n \times 1$ matrices), or row matrices ($1 \times n$ matrices). Row and column matrices are often called row and column *vectors*, respectively. It should be pointed out here that the square matrices used to represent symmetry operations belong to a class called *orthogonal matrices*, and multiplication of two orthogonal matrices *is* commutative (that is, the result of any two consecutive symmetry operations is independent of the order in which they are carried out).

In scalar arithmetic, multiplying a number by 1 leaves it unchanged, as does adding zero to a number. Analogous operations exist in matrix algebra. Multiplication of any square matrix by a matrix of the same order that consists of ones down the main diagonal (upper left to lower right) and zeros everywhere else yields the original matrix; the matrix

$$\begin{bmatrix} 1 & 0 & 0 \cdots\cdots & 0 \\ 0 & 1 & 0 \cdots\cdots & 0 \\ 0 & 0 & 1 \cdots\cdots & 0 \\ & & \cdot & \\ & & & \cdot \\ 0 & 0 & 0 \cdots\cdots & 1 \end{bmatrix}$$

is called the *identity matrix*. Addition of a matrix whose elements are all zero to a matrix of the same order also leaves that matrix unchanged. The matrix

$$\begin{bmatrix} 0 & 0 & 0 \cdots\cdots & 0 \\ 0 & 0 & 0 \cdots\cdots & 0 \\ 0 & 0 & 0 \cdots\cdots & 0 \\ & & \cdot & \\ & & & \cdot \\ 0 & 0 & 0 \cdots\cdots & 0 \end{bmatrix}$$

is the *null matrix*.

10

Instrumental Methods

As in all of the physical and biological sciences, the very rapid advances in mineralogy and crystallography in the past several decades have been made possible by the invention and development of a host of instrumental techniques for gathering and processing data of various types. In Part I of this text, frequent reference has been made to crystal structures and space groups, chemical variations, ordering of cations, and microstructures, all of which can be determined by one or a combination of these methods. While this text is not intended to serve as a handbook of instrumental techniques and does not purport to provide instruction in using these methods, conceptual familiarity with some of them is needed for an appreciation of the material in Part I and in current mineralogical research literature. In this chapter we shall consider seven techniques that have played major roles in the development of modern mineralogy.

A list of only seven instrumental methods, among all that are used by mineralogists, is obviously not exhaustive. Omitted are infrared spectroscopy, optical spectroscopy, calorimetry, atomic absorption spectrophotometry, and a number of other methods available to mineralogists and used to one extent or another. Their omission does not necessarily imply lesser importance, except perhaps in the context of the approach to crystal chemistry we have taken in this text and the methods that have provided most of the data on which it is based.

Some of the methods that will be described in the following pages are based, in part, on similar phenomena. For example, the generation of x-rays is important in several of them. Rather than discuss such topics with every method, they will be considered the first time they are encountered and then assumed thereafter.

X-RAY DIFFRACTION

Within a few years of Max von Laue's important x-ray diffraction experiments in 1912 the atomic arrangements of several simple ionic solids had been solved. Attempts at some more complicated minerals met with similar successes, and by the 1930s the crystal structures of several major rock-forming silicates were known.

X-ray diffraction has remained, to the present, the most widely-used technique for the precise determination of crystal structures.

PRODUCTION OF X-RAYS

X-rays are photons, not different from visible light except that they have higher energies (high frequencies, short wavelengths), and are thus not in the range of wavelengths perceptible by the human eye. They are produced during certain electronic transitions in atoms.

Figure 10-1 shows a schematic energy-level diagram for the electrons in an atom of copper. If a vacancy were to occur in the *s* orbital of the *K* shell, an electron from one of the outer shells would fill it. Because each orbital is characterized by a specific amount of energy, the substituting electron would need to give up some energy as it dropped in from an outer shell to the *K* shell; the quantum of energy emitted

Figure 10-1. Schematic diagram of the electron energy levels for a copper atom (or any atom containing electrons in the *M* shell). Arrows indicate electron transitions responsible for the labeled x-ray wavelengths. All transitions are not shown.

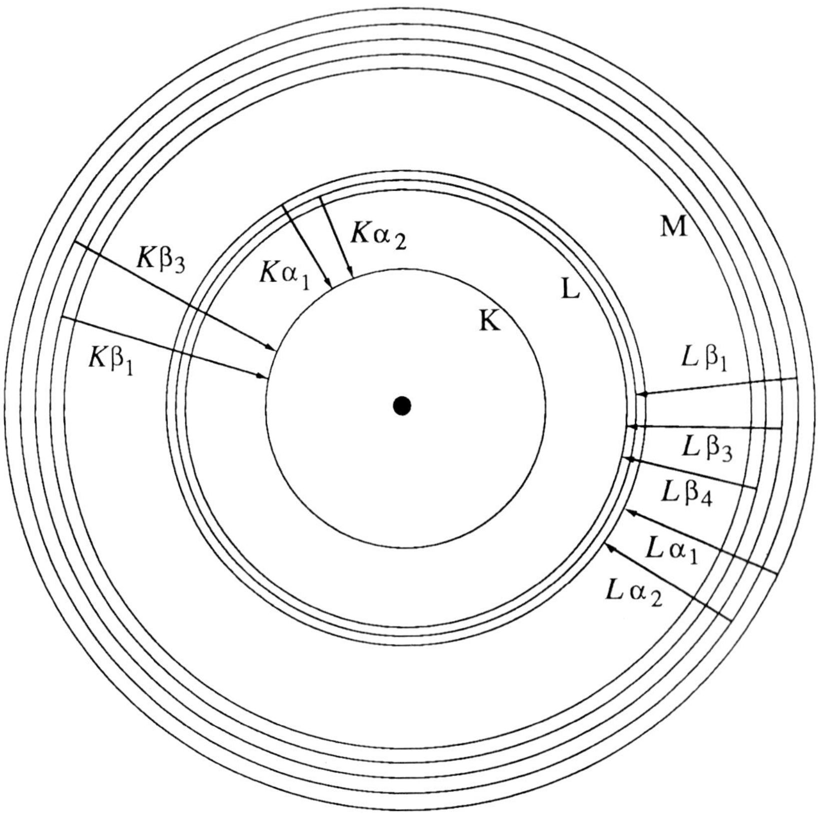

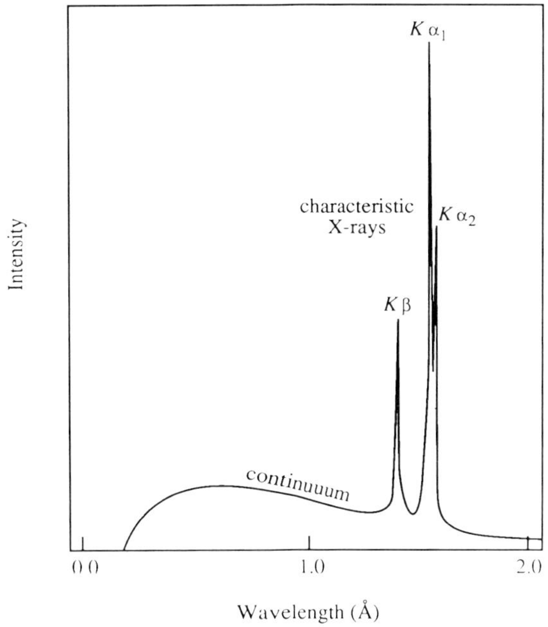

Figure 10-2. The continuous and characteristic spectra of copper.

would be an x-ray. Moreover, for a given element the amount of energy released would depend only on the particular orbital from which the electron had come, and would be constant for any given "donor" orbital; an x-ray produced in this way is called a characteristic x-ray, because it is characteristic of the element involved. Copper is capable of producing a spectrum of x-rays, depending on where the vacancy occurs and where the electron that fills it comes from. In the present example, the vacancy is in the K shell, so the characteristic photon is called a CuK x-ray; depending on whether the substituting electron comes from the L or M shell, it is called a Cu$K\alpha$ or Cu$K\beta$ x-ray, respectively. In fact, if it comes from the L shell, then it is a Cu$K\alpha_1$ or Cu$K\alpha_2$ x-ray, depending on whether it came from a $2p_{x,y}$ or a $2p_z$ orbital, respectively. Each of these types of characteristic radiation has a different, but constant, wavelength. X-rays of the L spectrum are produced when an electron from the M shell fills a vacancy in the L spectrum, and so forth.

The production of x-rays for use in diffraction experiments, then, requires a means of creating vacancies in the inner electron shells of some element—copper in this example. This is done by heating a cathode (usually made of tungsten) to free electrons, and accelerating them toward a copper anode at high speed. (Typically, accelerating voltages of tens of thousands of volts are used.) These cathode electrons hit some of the electrons in the copper target, ejecting them, and creating the required vacancies. Figure 10-2 shows the x-ray spectrum produced by a copper target. Note that not only are characteristic x-rays produced, but also a background called the *continuum* (sometimes referred to as "white radiation," because it contains a continuum of wavelengths, as does white light). The continuum arises because many electrons from the cathode do not hit and eject target electrons, but are deflected by the electric fields around the copper atoms. Such deflection constitutes an acceleration, and accelerated charged particles always emit radiation. The wavelength of the radiation thus emitted depends on the magnitude of the acceleration experienced, but the very sharp cutoff of the continuum at the low-wavelength end of the spectrum represents the acceleration experienced by a cathode electron that hits a target electron. This is known as the *low wavelength limit*, and of course corresponds to the highest acceleration, and thus highest energy emission, of cathode electrons.

Because electrons interact strongly with matter, x-ray tubes must be evacuated. In addition, inasmuch as x-rays are dangerous, x-ray tubes must be properly shielded and provided with devices to prevent inadvertent exposure. Modern x-ray tubes are encased in lead-lined shields containing spring-loaded or electronically controlled ports that permit x-rays to emerge only in certain directions through windows in the tube. The windows are made of material that is essentially transparent to x-rays—generally beryllium metal. (Caution: The windows are fragile, and should never be touched, even if the x-ray tube is not in operation.) The x-rays are admitted from the x-ray tube into shielded collimators that absorb radiation in all but the desired direction, resulting in a narrow beam aimed toward the specimen.

For most x-ray diffraction applications, it is advantageous to have as monochromatic a source of x-rays as possible. (*Monochromatic* here refers to the property of consisting of a single wavelength.) In addition, the range of wavelengths suitable for x-ray diffraction includes only the *K* spectrum of copper (and of other metals commonly used as targets in x-ray tubes). Thus a means must be provided for eliminating both the continuum and all but one peak of the characteristic spectrum. This may be accomplished in either of two ways. The more effective but also more expensive method is to use a device called a *monochromator*. The other is to use a material that filters the radiation by physically absorbing undesirable wavelengths. We shall consider filters first, and then monochromators.

Figure 10-3 shows the absorption spectrum of nickel superimposed on the x-ray spectrum of copper. Note first (from Figure 10-2) that the *L* lines of copper are at such a high wavelength that they are not diffracted, and cause no interference; and that the continuum is so low in intensity in that region of the spectrum that it is not a hindrance to diffraction experiments. Second, observe that the absorption edge of nickel falls in between the Cu$K\alpha$ and Cu$K\beta$ peaks in the spectrum. Thus a piece of nickel foil placed between the x-ray tube and the detector will filter most of the spectrum out except for the Cu$K\alpha_1$ and Cu$K\alpha_2$ peaks. (This is only possible, of course, at the expense of some of the Cu$K\alpha$ x-ray intensity.) Table 10-1 shows some of the more commonly employed target metals in x-ray tubes used in diffraction work, along with the appropriate filters. The best filters are generally metals lower in atomic number by 1 or 2 than the target metal. The $K\alpha_1$ and $K\alpha_2$ peaks are so close in wavelength for the target metals that inability to filter out one or the other is not a hindrance.

A monochromator is itself a device for causing x-ray diffraction, and so a full understanding of how it functions requires diffraction theory, which is to be dis-

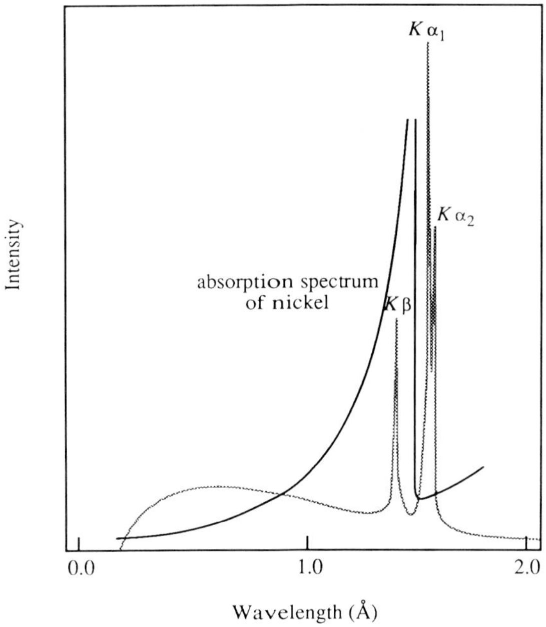

Figure 10-3. The absorption edge of Ni falls between the Cu$K\beta$ and Cu$K\alpha$ wavelengths, and so Ni is effective as a filter for Cu radiation.

cussed below. For now, it suffices to state that there is a specific relationship between the wavelength of the radiation, the perpendicular spacing between the planes of atoms responsible for any given diffraction, and the angle made by the incident beam with those planes. Given any two of those variables, the third is uniquely determined. Thus if a crystal of known interplanar spacing is set in the path of the incident x-ray beam, and aligned so that the appropriate atomic planes make an angle with the incident beam that allows diffraction of only $K\alpha$ radiation, then a beam of essentially monochromatic $K\alpha$ radiation will be diffracted onto the specimen. Like the metal filter, such a device is not sensitive enough to discriminate between $K\alpha_1$ and $K\alpha_2$ and eliminate one of them, but that turns out to be unimportant in practice.

Table 10-1. Some X-ray Targets Commonly Used in Diffraction Work

Target	Wavelength ($K\alpha$)	Filter
Cr	2.2909 Å	V
Fe	1.9373	Mn
Co	1.7902	Fe
Cu	1.5418	Ni
Mo	0.7107	Zr

DIFFRACTION GEOMETRY

The Laue Equations

Figure 10-4 represents a row of point scatterers—approximations for atoms whose electrons scatter x-rays—being illuminated by a beam of x-rays. The "atoms" are separated by a distance a in the one-dimensional lattice, and $\bar{\mu}$ and $\bar{\nu}$ are the angles between the lattice vector **a** and the incident and diffracted x-ray beams, respectively. The incident beam is coherent, which means that all of its individual rays are in phase, and of course the diffracted beam must also be coherent or the large number of individual rays (x-ray photons) would mutually cancel. Specifically, rays 1 and 2 in the incident beam must be in phase at wavefront AB, and their diffracted counterparts must be in phase at wavefront CD. Thus the difference between path AD, traveled by ray 1, and path BC, traveled by ray 2, must be an integral number of wavelengths:

$$AD - BC = h\lambda,$$

where h is an integer and λ is the wavelength of the x-rays. From the figure, we have

$$BC = |\mathbf{a}| \cos \bar{\mu}, \text{ and}$$

$$AD = |\mathbf{a}| \cos \bar{\nu}, \text{ so}$$

$$|\mathbf{a}|(\cos \bar{\nu} - \cos \bar{\mu}) = h\lambda.$$

For a one-dimensional lattice of atoms, this is the only equation that needs to be satisfied for x-ray diffraction to occur. Since the rays are scattered in all directions, coherent scattering is not restricted to the plane of the incident beam and row of

Figure 10-4. A row of point scatterers, with points separated by distance a. Incident x-rays make the angle $\bar{\mu}$ with the row, and the diffracted beam makes the angle $\bar{\nu}$ with it.

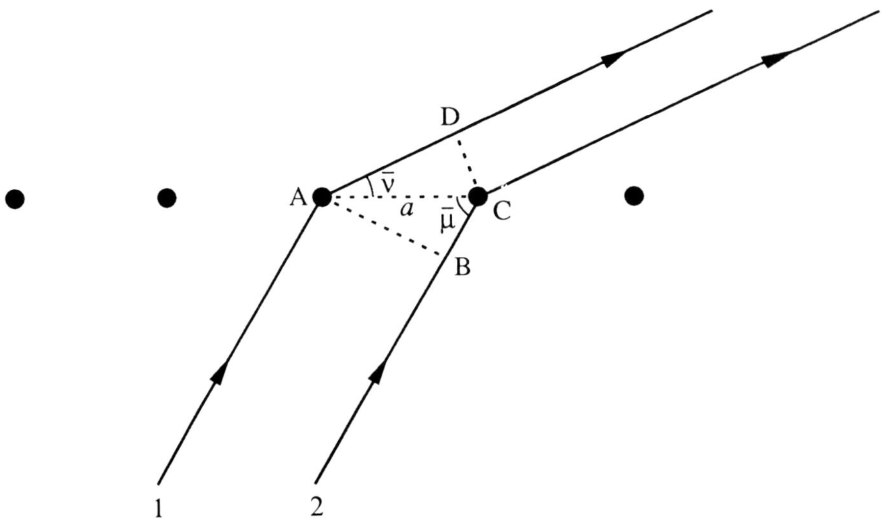

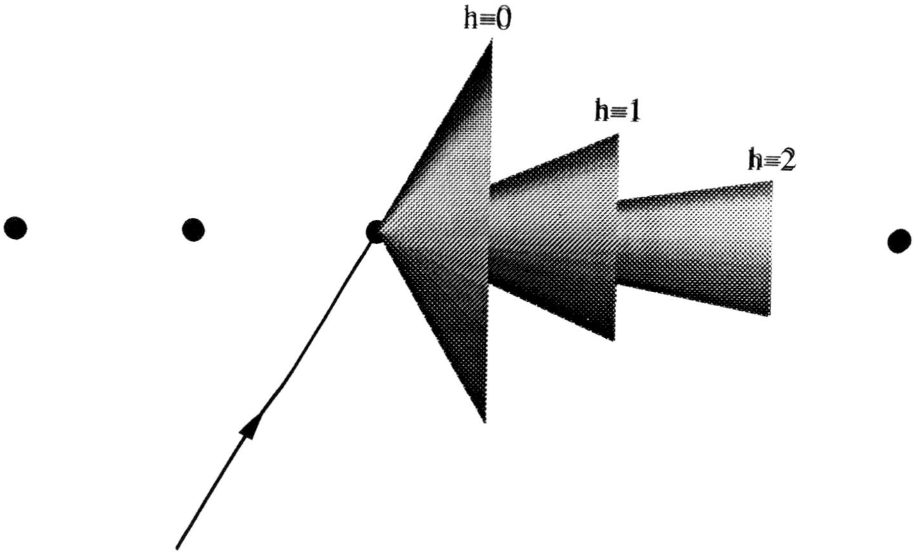

Figure 10-5. Because the diffracted x-rays are not restricted to the plane of the page, a cone of diffraction for each different path-length difference h is produced from the row of scatterers.

scatterers. As indicated in Figure 10-5, coherent diffraction would occur in cones coaxial with the lattice row, one cone for each value of h.

For diffraction from a hypothetical two-dimensional lattice of scatterers, path differences in two directions must be equal to integral multiples of the wavelength, though the two integers need not be the same. Thus two equations must be satisfied simultaneously:

$$|\mathbf{a}|(\cos \bar{\nu}_1 - \cos \bar{\mu}_1) = h\lambda, \text{ and}$$

$$|\mathbf{b}|(\cos \bar{\nu}_2 - \cos \bar{\mu}_2) = k\lambda.$$

The two-dimensional lattice gives rise to two sets of diffraction cones, one coaxial with the lattice translation **a** and one with **b**. Constructive interference occurs where the two cones, one of each set, intersect—that is, in at most two lines emanating from each scatterer for each hk pair.

For a three-dimensional array of scatterers (and this corresponds, of course, to the real case), three cones must intersect. Inasmuch as three cones may only intersect in a single line, a single diffracted beam results. In order for this to happen, three equations must be satisfied simultaneously:

$$|\mathbf{a}|(\cos \bar{\nu}_1 - \cos \bar{\mu}_1) = h\lambda,$$

$$|\mathbf{b}|(\cos \bar{\nu}_2 - \cos \bar{\mu}_2) = k\lambda, \text{ and}$$

$$|\mathbf{c}|(\cos \bar{\nu}_3 - \cos \bar{\mu}_3) = l\lambda.$$

These are called the Laue equations, after the same Max von Laue who discovered x-ray diffraction. The integers h, k, and l are called *diffraction indices*, and they are related to the Miller indices for the atomic plane causing a given diffraction.

The Bragg Equation

While the Laue equations provide a perfectly accurate geometrical description of x-ray diffraction, they are not very convenient from an experimental point of view. It was the desire to find a simpler way to describe the geometry of x-ray diffraction that led Sir Lawrence Bragg to a single, remarkably simple expression. The key was the recognition that diffraction resembles reflection in that the angle made by the incident beam with the diffracting planes of atoms is always equal to the angle made by the diffracted beam with that set of planes; but that it is unlike reflection in that reflection occurs at any incident angle, whereas diffraction occurs only at specific incident angles. Bragg reasoned that reflection from a mirror was always coherent, because the path-length differences were always zero. If the lattice planes causing a given diffraction were considered mirrors all stacked together, then each of the two-dimensional lattice planes would satisfy two of the three Laue equations. Whether the third was satisfied at the same time, resulting in diffraction, depended both on the spacing between the stacked planes and on the angle between those planes and the incident beam; coherent scattering would only occur if that angle were such that rays "reflected" from successive stacked planes differed in their path lengths by some integral number of wavelengths.

Consider Figure 10-6. The diffracting planes are seen edge-on, with ray 1 being scattered from the top plane and ray 2 from the second one in the stack. (In reality, both rays also penetrate further into the stack of planes and are diffracted by each, but the derivation with which we are involved rests only on these two rays and these two planes.) At wavefront AA', the two rays are in phase, and both strike the planes at an angle θ. Both also reach point C, and so the path-length difference between them is

$$A'B'C - ABC = n\lambda, \text{ where } n \text{ is some integer.}$$

Because triangle $B'BD$ is isosceles, $B'B = B'D$, so

$$(A'B'D + BC) - (AB + BC) = n\lambda,$$

$$A'B'D - AB = n\lambda.$$

Now we construct the line segment BN normal to $B'D$. Because the sides of angle NBD are perpendicular to the sides of the angle ABE right-to-right and left-to-left, the angle NBD is also equal to θ. Further resolving line segment $A'D$ into components,

$$(A'B' + B'N + ND) - AB = n\lambda.$$

Note that $A'ABN$ is a rectangle, so that $A'N (= A'B' + B'N)$ is equal to AB, and so

$$(AB + ND) - AB = n\lambda, \text{ and}$$

$$ND = n\lambda.$$

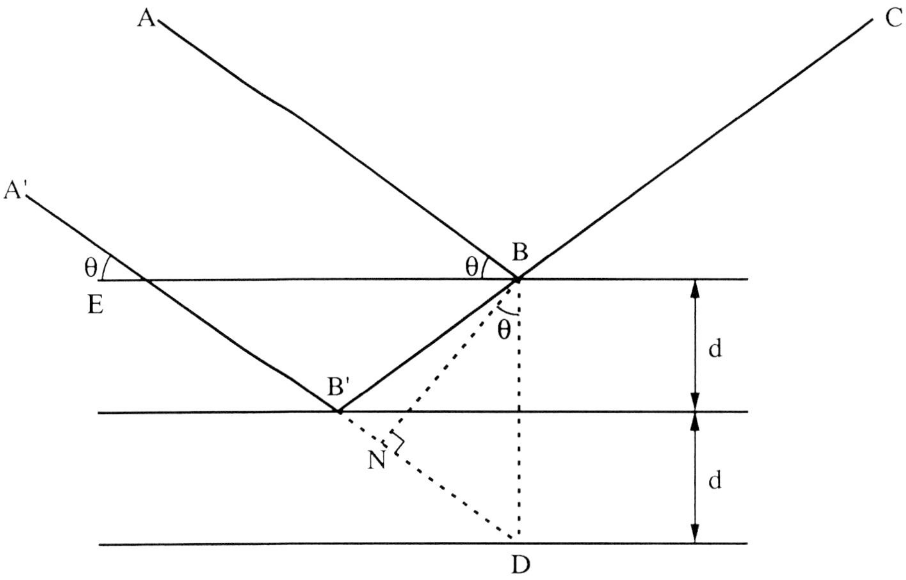

Figure 10-6. Incident x-rays *AB* and *A′B′* are in phase, and are diffracted at the angle θ so that both travel to *C*. The perpendicular distance between the diffracting planes is *d*. See text for further explanation.

But $ND = BD \sin \theta$, and $BD = 2d$, where d is the perpendicular spacing between lattice planes. Thus,

$$n\lambda = 2d \sin \theta,$$

which is known as *Bragg's law*, or the *Bragg equation*. The integer n is the path-length difference for rays diffracting from successive planes, and is not generally an immediately known number; it is called the *order of the diffraction*. The wavelength is known, however, because it is fixed by the type of x-ray tube being used, and the diffraction angle θ is the experimentally measured variable. The interplanar spacing, commonly known as the *d-spacing*, is usually the variable whose value is to be determined after measurement of θ.

The various orders (n) of diffraction for a given set of planes occur at different diffraction angles. We now divide both sides of Bragg's equation by n to yield

$$\lambda = 2(d/n) \sin \theta$$

If the first-order planes have spacing d and Miller indices *hkl*, then the planes with spacing d/n have indices *nh,nk,nl*. Diffraction from the planes *nh,nk,nl* may now be considered first-order, rather than n^{th}-order, diffraction from the *hkl* planes. For example, a third order diffraction from the 111 planes of some mineral is geometrically indistinguishable from a first-order diffraction from the 333 planes. Because these indices (*nh,nk,nl*) contain a common factor, they are not called Miller indices, but rather diffraction indices. Thus, we now rewrite the Bragg equation as

$$\lambda = 2d_{hkl} \sin \theta,$$

and consider all diffractions to be first-order diffractions from planes with diffraction indices *hkl*, where *h*, *k*, and *l* may contain a common factor. Given some experimental way to find the diffraction angle, and knowing the wavelength of x-rays being employed, we now have a single, simple equation in one unknown, which completely describes the diffraction geometry. Because the geometry is so much like reflection, the diffracted beams are often called *reflections*, even in formal literature.

X-RAY DIFFRACTION EXPERIMENTATION

The most common purposes for x-ray diffraction in mineralogy are the identification of minerals or determination of unit-cell geometry, on the one hand, and determination of the crystal structures (atomic arrangements) of minerals, on the other. The kinds of data usually acquired for these separate objectives are different enough that different experimental arrangements are generally used. For determining precise unit-cell geometry or identifying an unknown mineral, powder diffraction techniques are employed, and for determining the crystal structure single-crystal diffraction is used. (While it is true that precise unit-cell geometry can be determined by single-crystal methods—and often is, as part of the structure-determination problem—and that powder methods can be used for solving crystal structures under favorable conditions, the foregoing generalization nonetheless reflects the most common applications of the methods.)

In either case, the geometrical problem is the same: The lattice planes of a crystal must be oriented with respect to the incident x-ray beam so that, for the wavelength being used and the d spacing of the particular planes, Bragg's law is satisfied and diffraction takes place. Figure 10-7 shows, in a schematic way, the geometry required; although the Bragg equation contains the diffraction angle, θ, note that it is usually 2θ, the angle between the incident and the diffracted beams, that is measured experimentally. The specimens used, as implied by the names of the methods, are either powders or single crystals. The detector may be a device for counting x-ray photons, such as a scintillation counter, or it may be a piece of film sensitive to x-rays.

Powder Diffraction

In powder-diffraction experiments, the problem of crystal orientation is solved simply. The mineral is powdered by grinding with a mortar and pestle until the particles are fine enough to pass through a fine (~300-mesh) screen. The powder is then packed into an appropriate specimen holder so that the crystallites have random orientation, and it is placed in the x-ray beam. With this very large number of crystallites in random orientation, a significant (and presumably statistically equal) number of them will satisfy the Bragg equation for each set of lattice planes capable of diffracting.

The type of specimen holder used depends on the nature of the detector. If a cylindrical powder camera is used, the powder is packed into a thin glass capillary or rolled into a cylinder with some amorphous binding medium, and is then placed

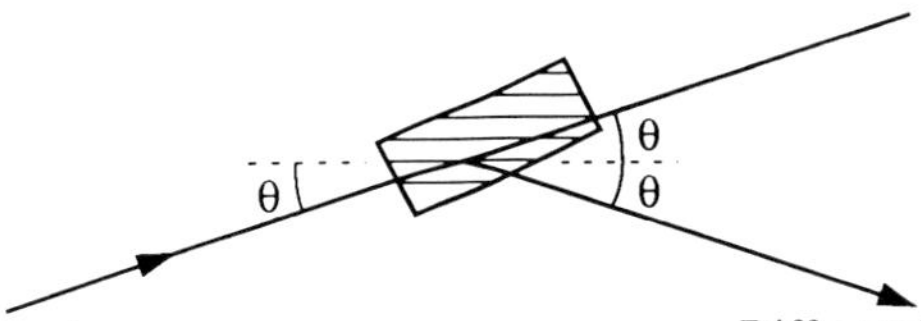

Figure 10-7. The geometry of x-ray diffraction is much like that of reflection, in that incident and diffracted beams make the same angle, θ, with diffracting planes. It is unlike reflection in that θ cannot be an arbitrary angle, but must satisfy the Bragg equation. In experimental work the angle 2θ, rather than θ, is usually measured.

in the incident beam along the axis of the camera [Figure 10-8(a),(b)]. Because the lattice planes that cause a given diffraction must make the Bragg angle with the incident beam, a cone of diffracted radiation results; the angle between the cone and its axis is 2θ. Rotation of the specimen about its axis ensures that the diffraction conditions will be satisfied for all possible diffracting planes. The strip of film wrapped around the cylindrical camera intercepts the cones of diffracted radiation, and, with knowledge of the camera dimensions, the diffraction angles can be easily found and the Bragg equation solved for *d* spacings.

Alternatively (and usually more conveniently) a scintillation counter or similar device may be used for detection of the diffracted beam. In this case, the powder is mounted in a flat specimen holder or mixed with a binder and smeared on an glass slide, and is then placed in a mount that rotates so that the angle between the incident beam and the specimen holder can be changed. The detector is mechanically coupled to the specimen mount so that, when the angle between the incident beam and the specimen holder is 0°, the angle between the detector axis and the incident beam is also 0°. As the specimen is rotated with angular speed ω, however, the detector is rotated at 2ω, so that it always makes twice the angle with the incident beam as is made by the specimen holder [Figure 10-8(c)]. In this way, when the Bragg angle, θ, for some set of lattice planes in the mineral is reached by rotation of the specimen, the detector is at 2θ from the incident beam, and in position to detect the diffracted beam. For a given set of lattice planes (hkl) (which have a specific *d* spacing and therefore a specific diffraction angle), the only crystallites capable of diffracting a beam into the detector are those for which that set of planes is parallel to the flat specimen holder. There are so many crystallites in the specimen, however, that there are many properly oriented ones for any potential diffraction, and diffraction occurs from each possible set of planes. The signal from the detector is processed electronically and displayed as a strip-chart recording, with 2θ read directly from the chart [Figure 10-8(d)].

Mineral Identification

A list of *d* spacings, together with the relative intensities of the diffracted beams (either taken from peak heights on a strip chart, or estimated from line darkness on a photographic negative) constitutes a "fingerprint," as it were, of a mineral (or of any crystalline substance, for that matter). The *d* spacings depend only on the

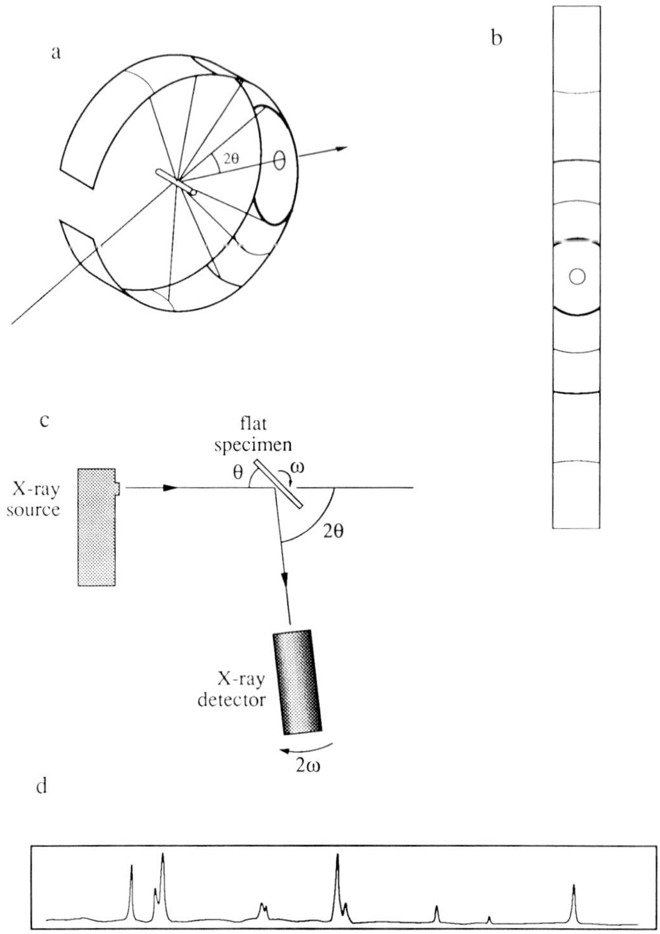

Figure 10-8. (a) Powder cameras use cylindrical specimens with the film wrapped about the camera so that it intercepts cones of diffraction from the specimen. (b) When developed and laid flat, the film shows the diffraction maxima as arcs, for which 2θ can be determined if the camera radius is known. (c) Powder diffractometers usually use specimens mounted in flat-plate holders. The holder is rotated through the range of 2θ values at a rate ω, while the detector is rotated at the rate 2ω so that it is always in position to receive diffracted beams from planes at the proper angle 2θ. (d) The output of a diffractometer is a strip chart showing diffraction maxima as peaks, with 2θ read directly from the chart.

geometry of the unit cell, and the relative intensities depend on the electron densities on the various lattice planes; the latter, of course, depend on the kinds of atoms present and their arrangement in the unit cell. Inasmuch as any mineral is defined by its chemical composition and atomic arrangement, its powder pattern (the combination of *d* spacings and relative intensities) is characteristic, and may be used to identify it.

Comparing the powder pattern of an unknown mineral against those of all known minerals would obviously be a time-consuming task that no one would want to undertake—and certainly not on a routine basis. Instead, the powder patterns of minerals have been compiled into a catalog called the X-ray Powder Diffraction File (XPDF) which can be searched efficiently either manually or by computer. The XPDF is provided with two kinds of indexes, the Hanawalt Index and the Fink Index, both of which index the *d* spacings of the most intense diffraction maxima, but with different schemes for ordering them. The most intense peaks or lines in the pattern of an unknown mineral are generally found to correlate reasonably well in both *d* spacings and relative intensities with only a few—perhaps one to five—patterns in either index, and then the full powder patterns of these possible minerals are checked in the XPDF to determine the correct identification. Volumes of the Powder Diffraction File and the indexes are published that contain only organic compounds, inorganic compounds (including minerals), minerals only, and some other specialized listings. Obviously, the most convenient one for mineralogists are the volume and indexes that contain only minerals.

In cases where the unknown mineral belongs to a family of minerals that exhibits complicated solid solution, identification of the specific mineral in the family may not be possible without additional chemical data. For example, extensive solid solution occurs among members of the garnet family, all of which have rather similar unit-cell dimensions but different chemical compositions. Moreover, the differences in intensities due to different compositions within the garnet family may not be more than those that can be ascribed to normal experimental error or to differences caused by sample preparation. Thus, one may be sure that the unknown mineral is a garnet of some sort, and may be confident that it is not one of the extreme end members whose *d* spacings establish low or high limits for the series, if that is the case; but without further information (partial chemical analysis, index of refraction, density—even color helps), the specific kind of garnet remains unknown.

As just implied, intensities are less reliable than *d* spacings in identifying minerals, because substantial differences may exist between the reference powder patterns and those produced by other workers, even for the same minerals. The reasons for this are many: differences in instrumental conditions or diffractometer design; incomplete grinding of the specimen; improper mounting of the specimen; intensity of background radiation; and preferred orientation of the specimen. Preferred orientation occurs in minerals that show excellent cleavage. In grinding, the crystallites tend to end up as cleavage fragments. Then, when they are placed on a flat specimen holder or packed into a glass capillary, they become aligned so that the orientation is not random; the result is that the peaks corresponding to the cleavage planes are unusually intense. Finer grinding generally alleviates this problem to some extent, but for minerals such as micas, preferred orientation is significant no matter how well the powder is ground. It can often be partially overcome by using careful mounting techniques, such as sprinkling the powder on a glass slide coated with petroleum jelly.

Determination of Unit-Cell Geometry

Because the *d* spacings depend solely on the geometry of the unit cell, they are routinely used for determining that geometry. The reason for doing this may be to establish the unit cell for a new mineral never before described, but more often it is done to establish relationships between chemical composition and unit-cell parameters—or to use established relationships to estimate chemical composition *from* unit-cell parameters. Crude estimates of cell dimensions can be made by examination of the indexed powder pattern—for example, the 200 *d* spacing measured for a tetragonal crystal is half the *a* cell dimension, if there is no experimental error in the measurement. (Indexing a powder pattern, which means assigning diffraction indices to each peak, is usually done for known minerals by comparison with the reference pattern in the XPDF, or with some other previously indexed pattern. Obviously, unit cells and diffraction indices for new minerals can be found from powder patterns, but the methods for doing this are not relevant to the subject of this text.)

If the 100, 200, and 300 peaks of a tetragonal mineral were used for estimation of the *a* cell edge, three slightly different answers would likely be obtained because of random experimental errors. Precise determination of unit-cell parameters requires some way of minimizing such errors, which are always present regardless of the care taken in experimental procedures and measurements. The most common method is to use a mathematical procedure known as *least-squares refinement,* which belongs to a family of techniques that statisticians call *linear regression.* The task in unit-cell refinement is a multiple linear-regression problem, because there is more than one independent variable, but the general nature of the method can be conveyed most easily in terms of simple linear regression. Figure 10-

Figure 10-9. Hypothetical experimental data that appear to reveal an approximately linear relation between the dependent variable, *Y*, and the independent variable, *X*.

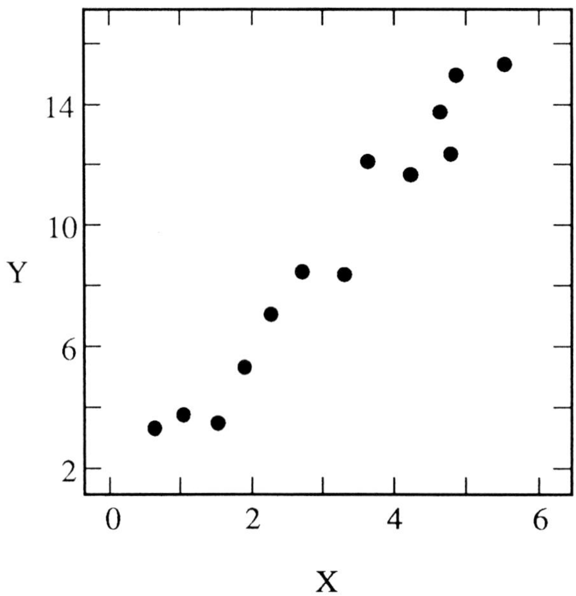

9 shows some hypothetical experimental data relating an observed dependent variable Y to an experimental variable X, which is set at various values by the experimenter. Suppose that there is reason to expect that X and Y are linearly related by an equation of the form $Y = aX + b$. The task is to find a straight line that, in some sense, is a "best fit" of the data.

Because least-squares procedures are used so widely in many sciences for a great variety of problems, it is worth taking a short digression into the details of the method. The plot shows an obvious linear relationship between X and Y, but just where the "best-fit" line should be drawn is not so obvious. It is assumed that all of the error resides in the measured values of Y (that is, that the experiment allowed X to be chosen exactly). The best-fit (least-squares) line is taken to be the one that results in the smallest sum of the squares of the errors in Y, with the errors being defined as the differences between the calculated and observed values in Y.

Let measurements of Y be denoted y_i, and values of X be x_i. The equation relating the two variables will have the form $y_i = ax_i + b + \epsilon_i$, where ϵ_i is the error in the measured y_i. Add and subtract $a\langle x\rangle$ from the right-hand side of the equation, where $\langle x\rangle$ is the mean value of the x_i: $y_i = ax_i + b + a\langle x\rangle - a\langle x\rangle + \epsilon_i$. Now let $b' = b + a\langle x\rangle$. Then $y_i = a(x_i - \langle x\rangle) + b' + \epsilon_i$, and solving for the error yields $\epsilon_i = y_i - b' - a(x_i - \langle x\rangle)$. We shall now successively square the error, sum it over all n observations of (x_i, y_i), and then set the first partial derivatives of that sum equal to zero to find the minima with respect to b' and to a.

$$\epsilon_i^2 = [y_i - b' - a(x_i - \langle x\rangle)]^2$$

$$\sum_{i=1}^{n} \epsilon_i^2 = \sum_{i=1}^{n} [y_i - b' - a(x_i - \langle x\rangle)]^2 \equiv L$$

$$\partial L/\partial a = -2\sum_{i=1}^{n} [(y_i - b' - a(x_i - \langle x\rangle)](x_i - \langle x\rangle) = 0$$

Dividing by -2 and distributing both the multiplication and the summation yields

$$\sum_{i=1}^{n} y_i(x_i - \langle x\rangle) - b'\sum_{i=1}^{n} (x_i - \langle x\rangle) - a\sum_{i=1}^{n} (x_i - \langle x\rangle)^2 = 0$$

But $\Sigma_{i=1}^{n} (x_i - \langle x\rangle) = 0$, so the middle term on the left drops out, and

$$a = \sum_{i=1}^{n} y_i(x_i - \langle x\rangle) \Big/ \sum_{i=1}^{n} (x_i - \langle x\rangle)^2.$$

Next, $$\partial L/\partial b' = -2\sum_{i=1}^{n} [y_i - b' - a(x_i - \langle x\rangle)] = 0,$$

so $$\sum_{i=1}^{n} y_i - nb' - a\sum_{i=1}^{n} (x - \langle x\rangle) = 0$$

Again $$\sum_{i=1}^{n} (x_i - \langle x \rangle) = 0,$$

so $$\sum_{i=1}^{n} y_i = nb'$$

$$b' = \sum_{i=1}^{n} y_i/n = \langle y \rangle,$$

and thus $$b = \langle y \rangle - a\,\langle x \rangle.$$

By substituting the value of a obtained into the equation for b, we guarantee simultaneous solution of both equations, and are thus assured that the solution corresponds to a minimum in $\Sigma_{i=1}^{n} \epsilon_i^2$.

For the determination of precise lattice parameters, we must extend the least-squares procedure to more than one independent variable. This extension is reasonably straightforward, although the algebraic manipulations are a bit different; the equations become unwieldy, however, and are most easily handled with matrix algebra. One begins with some reasonably close guesses for the lattice parameters, though they need not be so close that this becomes an obstacle. The dependent variables are functions of the observed 2θ angles, and the independent variables are functions of the unit-cell parameters. The coefficients that are refined are the differences (corrections) between estimated and "true" cell parameters. The true cell parameters are, of course, unknown, but they are successively approximated during each cycle of refinement by adding the corrections to the old estimates to obtain new estimates. When the corrections are sufficiently small, the unit-cell parameters are said to be *refined.*

The initial, or starting, parameters may be obtained by examination of axial reflections ($h00$, $0k0$, $00l$), although in nonorthogonal crystal systems some estimate of interaxial angles must also be found, or cell parameters from the XPDF or some other source for the same species of mineral may be used. The least-squares calculations are tedious, and several convenient computer programs for carrying them out are in general use by mineralogists throughout the world. They not only refine the starting estimates to yield precise values for the lattice parameters for the particular specimen in question, but they also provide estimates of the precision of the refinement in the form of standard deviations in the refined parameters. Precision on the order of one or two parts in ten thousand is not at all uncommon; given uniform data quality, precision usually improves as the number of peaks included in the refinement increases. Inasmuch as accurate values of 2θ or d spacings are needed if accurate lattice parameters are to be derived, some standard material with well-known d spacings is generally mixed with the mineral of interest, and its peaks are used to correct any systematic errors in the data.

Single Crystal X-ray Diffraction

As implied by the name, the diffraction methods that fall within this category involve illuminating a single crystal in an incident x-ray beam and detecting the diffraction maxima produced by it. Specimen preparation consists merely in selecting a suitable crystal; there is no requirement that it have any crystal form, and in fact, small fragments chipped or cut from larger crystals or fragments are generally used. For most purposes the optimum crystal size is 0.1 to 0.4 mm on a side, and roughly equidimensional specimens are preferred.

Determination of Crystal Symmetry

Data acquisition is again by either film or electronic counter. For the most part, film is used when crystal symmetry is being investigated and quantitative measurements of intensities are unnecessary, and a counter is used when diffraction intensities are needed in order to determine crystal structure. There are several film methods available, and each one requires a different sort of geometrical arrangement of crystal, film, and x-ray beam. The most widely used technique, and the one that is the most straightforward to interpret, is called the *precession method.* Figure 10-10 shows a typical precession photograph, and examination of it reveals two important differences between single-crystal and powder methods.

First, notice the neat geometric arrangement of the spots. (The streaks are caused by radiation from the continuum that is not completely filtered out, even though a filter was used in taking the photograph.) The pattern is related to the symmetry of the crystal along some direction, and that implies that the crystal must be intentionally oriented in a known way. Unlike the powder method, in which a random arrangement of crystallites was not only acceptable but desirable, photographic single-crystal methods require careful orientation of the crystal in the x-ray beam. This is not done visually, but by making adjustments on trial orientation photographs. Only when the crystal is properly oriented are photographs obtained on which crystal symmetry is apparent.

Second, observe that the photograph contains not lines, as do powder photographs, but individual spots. Each spot represents diffraction from a single set of lattice planes, whereas the lines on powder photographs represent diffraction from all symmetrically equivalent sets of planes, and perhaps superposition of some that are not symmetrically equivalent as well. To illustrate, consider a tetragonal crystal. The (100), (010), ($\bar{1}$00), and (0$\bar{1}$0) planes are equivalent by symmetry, and thus have identical *d* spacings; they thus diffract at the same 2θ angle. While some crystallites in a powder are diffracting from the (100) planes, others are simultaneously diffracting from the (0$\bar{1}$0), ($\bar{1}$00), or (010) planes, and all four planes are contributing to the same diffraction line on the film, or to the same peak on a strip chart recording. In addition, if there are other spacings in the crystal lattice that are fortuitously very close to the (100) *d* spacing, their diffractions will be superimposed in the same location. In single-crystal photographs of the same tetragonal crystal, the (100), ($\bar{1}$00), (010), and (0$\bar{1}$0) planes will still diffract at identical values of 2θ, but the diffracted beams will be in different directions, and so will appear as separate spots on the photograph. Even if some planes do have nearly the same *d* spacing, the directions of beams diffracted from them will necessarily be different, and they will also result in separate spots. Each photograph contains spots representing one layer of the lattice.

With precession photographs (and other types of single-crystal photographs as

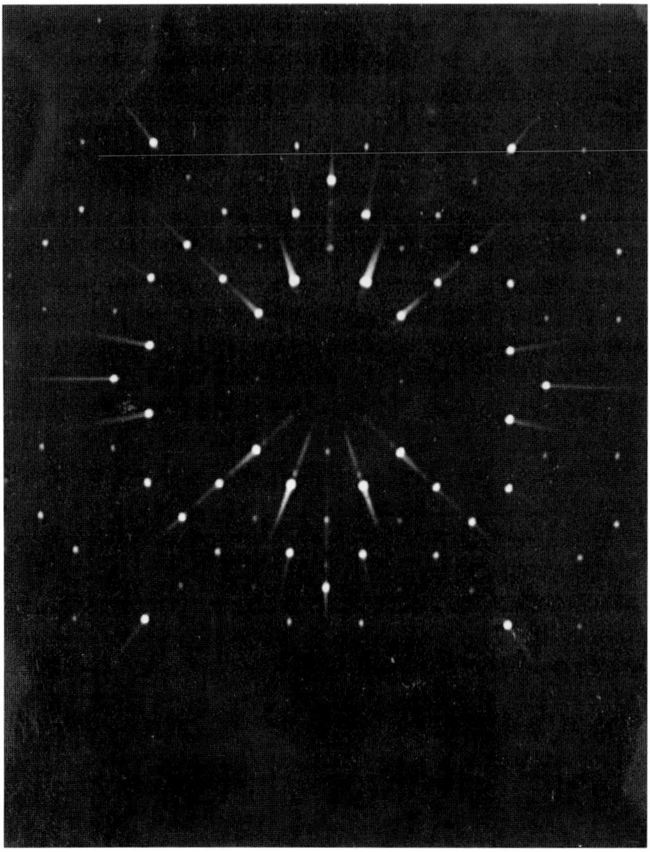

Figure 10-10. A precession photo of a monoclinic pyroxene. Each spot represents a beam diffracted from a different set of planes *hkl*.

well), then, it is possible to determine something about crystal symmetry, including space group. Unfortunately, crystal class cannot be assessed unambiguously, because, by Friedel's Law, diffraction intensities *hkl* and $\overline{hkl}$ are equal, even if the crystal does not possess a center of symmetry. By adding a center of symmetry to the 21 crystal classes that do not possess one, the 32 crystal classes reduce to eleven *Laue classes*, and the Laue class of a crystal can be determined without ambiguity. There are other methods by which the presence or absence of a center of symmetry can be established.

Space group is evaluated by noting the reflections that are systematically absent. (Some reflections will be absent just because their intensities happen to be small; here we are considering whole classes of reflections that are absent.) If there are no systematic absences among general reflections of the type *hkl*, then the crystal has a primitive Bravais lattice. If, however, all reflections are missing for which $h+k+l$ is an odd number, then the lattice is body centered; if h, k, and l are either all even or all odd for all reflections, the lattice is face centered; and if $h+k$ (or $h+l$, or $k+l$) is always an even number, then the lattice is C (or B, or A) centered. If, for reflections of the type $00l$, l is always an even integer, then there is a 2_1 screw axis along the **c** direction. If examination of the $hk0$ reflections shows that k is always even, then there is a b glide parallel to (001). There are other rules governing the interpretation of systematic absences, and the reasons for any of them are not necessarily obvious without additional discussion. Inasmuch as our purpose is simply to demonstrate that single-crystal diffraction provides the data necessary for determination of the space group, we shall leave such a discussion to texts on x-ray crystallography, and provide just two examples.

Suppose a precession photograph of the $h0l$ layer (that is, a photograph on which the value of k is 0 for all spots) of some monoclinic mineral reveals that h is never an odd number, but no other systematic absences are apparent; in addition, on the $h1l$ photograph, h is always odd, but no other systematic absences occur. The conclusion is that, for reflections of the general type hkl, $h+k$ is always an even number, and the lattice is therefore C centered. The space group is $C2/m$ if the crystal is centrosymmetric, but it is either $C2$ or Cm if it is not. As a second example, suppose that the same general conditions held as in the first example, but in addition, the value of l everywhere on the $h0l$ photograph were even. This would indicate the presence of a c glide parallel to (010), and we would conclude that the space group was $C2/c$ (centrosymmetric) or Cc (noncentrosymmetric). Even without knowing the crystal structure, then, the space group can be either determined or narrowed to a small number of possibilities.

Crystal Structure Determination and Refinement

The determination of crystal structures can be done with single-crystal photographs, but because it is necessary to measure the intensities of the diffracted beams, this is usually done now with single-crystal diffractometers equipped with counters, instead of by visual estimates or densitometer measurements of photographic intensities. Because the detector receives only one diffracted beam at a time, it must be precisely positioned, and the location of the diffracted beam must be accurately known. The instrument is constructed so that the detector may be rotated about a vertical 2θ axis, and the crystal may be rotated about a horizontal axis and at least one independent vertical axis. This permits any set of diffracting planes to be oriented in a vertical plane so that they make an angle θ with the incident beam, and it allows the detector to be rotated to the proper 2θ angle. Such maneuvers obviously require explicit knowledge of crystal orientation, either by the experimenter who manipulates the instrument, or by the computer software that controls it. (The most convenient software does not require the experimenter to know the crystal orientation in advance, or to orient the crystal intentionally in any special way.)

Although the Bragg equation again describes the diffraction geometry, it is neither 2θ nor d spacing that is of interest in crystal-structure work. The experimentally measured quantities are diffracted intensities, which are measured only in a relative sense, and which are related to the real quantities of interest, the structure factors:

$$I = K|F_{hkl}|^2,$$

where F_{hkl} is the structure factor for the hkl planes, and K is a constant of proportionality that is largely dependent on instrumental factors. The constant K is not determined in practice, but is replaced by an empirical scale factor that makes I and

$|F_{hkl}|^2$ comparable on the average. The quality $|F_{hkl}|^2$ is not a simple square, but the complex conjugate FF^*, because F_{hkl} is a complex number.

$$F_{hkl} = \sum_n f_n^0 \exp(-B_n \sin^2 \theta/\lambda^2) \cos[2\pi(hx_n + ky_n + lz_n)]$$
$$+ i \sum_n f_n^0 \exp(-B_n \sin^2\theta/\lambda^2) \sin[2\pi(hx_n + ky_n + lz_n)].$$

The use of complex numbers is a mathematical convenience for the sole purpose of keeping the cosine and sine terms computationally separate; there is nothing imaginary about the diffracted beams, however. Here the summation is over all n atoms in the unit cell; f_n^0 is the scattering factor (a measure of diffraction efficiency as a function of atom type, diffraction angle, and x-ray wavelength) of atom n at a temperature of absolute zero; the exponential expression is the temperature factor, which corrects the scattering factor to the temperature of the experiment; and the trigonometric expressions are cosines and sines of the phase angles of the diffracted beams. The latter are functions of the positions of the n atoms in the unit cell, and if the phase angles were observable, the determination of crystal structures would be straightforward. Unfortunately, it is the intensities of the reflections that are observable, and the phases are lost—giving rise to the "phase problem" in x-ray crystallography. The consequence is that crystal structures have usually had to be solved by indirect methods; a class of solution techniques called *direct methods* has been developed relatively recently, by which phases can be assigned to reflections for structures with favorable characteristics. Structures not suitable for direct methods are often quite well suited to indirect ones, however. Such methods make use of the fact that the Fourier transform of the structure factor is the electron density in the unit cell. If the location of one very heavy atom that is responsible for most of the scattering, or of a few moderately heavy atoms, can be determined by indirect methods, then the difference between the Fourier transforms of the observed structure factors (that is, of the square roots of the observed intensities, with phase angles determined from the positions of the heavy atom or atoms) and calculated structure factors (based on the working model of the crystal structure) will generally reveal the positions of the other atoms as peaks not accounted for by atoms in the working model.

Prior to attempts to solve the structure by any of these methods, the intensities must be corrected for background by subtracting the average intensities of x-rays detected just above and just below the 2θ for each peak. They are generally also adjusted for some geometrical effects of diffraction by a correction known as the Lorentz-polarization (LP) correction, and often also for the effects of absorption of x-rays. The absorption correction is an attempt to compensate for the differences in path lengths through the crystal for incident and diffracted beams; often these differences are due to inequant crystal shape, but even a spherical crystal experiences variation in absorption with 2θ. This correction is unnecessary if the absorption coefficient of the crystal is small, but the absorption coefficient should always be calculated to make sure.

Once the structure is solved, it remains to refine the positions and temperature factors of the atoms. (The latter give an indication of the amount of thermal vibration experienced by an atom.) Refinement is carried out by a least-squares procedure, as with powder pattern refinement. The dependent variables are functions of the observed intensities, and the independent variables are functions of the "true" atomic positions and thermal parameters. Each cycle of refinement results in corrections to the positions and temperature factors used in the previous cycle. At some stage in the refinement, the temperature factors are made anisotropic, so that the atoms are approximated as vibrating in an ellipsoidal volume of space. Often the more reliable reflections—as judged by counting statistics—are given higher weights than the others, resulting in a so-called *weighted refinement.* When the corrections become small enough, the refinement is complete. With each new set of estimates, values of F^2 (or F) are calculated and compared with I(or $|F_{obs}|$) to yield an agreement index called the R factor. Refinements are considered good if the R factors are below about 0.08 (an obviously arbitrary and subjective value), and refinements of very high quality may have R factors as low as 0.015 to 0.020.

A great deal more detail could be provided on the use of x-ray diffraction for determining and refining crystal structures, but several very good texts describing both the theory and practice already exist. This brief outline will at least give the reader who is unfamiliar with the field of x-ray crystallography sufficient background to appreciate the data used in Part I of this text.

ELECTRON PROBE MICROANALYSIS

The electron probe microanalyzer, usually referred to as the *microprobe,* is the most widely used instrument for determining the chemical compositions of minerals. The microprobe enjoys several advantages over other methods of compositional analysis for minerals. It requires only minute amounts of specimen, and, regardless of the quantity of specimen available, it is possible to analyze only a few hundred cubic micrometers at a time. This means that inclusions of "foreign" minerals can be avoided; it also means that compositional variations over very small distances can be documented, and that the compositions of individual tiny single crystals used for crystal-structure work can be determined. The requirements for specimen preparation are fairly rigid, but are not particularly difficult to meet. It is not required that the specimen be dissolved, powdered, or fused; moreover, the method is largely nondestructive, so that a substantial portion of a specimen can be recovered intact after analysis if necessary.

To be sure, the method has some disadvantages, and may be inadequate in some instances. The equipment is quite expensive, and may require substantial electronic and mechanical maintenance support. It cannot be used to distinguish among different oxidation states of elements. It is not routinely useful for quantitative analysis of elements lighter than fluorine (atomic number 9) unless the instrument is equipped (at considerable expense) with special windowless detectors. For these last two disadvantages in particular, other methods can be used to overcome the specific deficiency without needing to use them for the entire analysis; for example, Mössbauer spectroscopy (which is incapable of providing a complete analysis) can be used to obtain the $[Fe^{2+}]/[Fe^{3+}]$ ratio, and atomic absorption spectrophotometry (for which specimen preparation for silicates is troublesome) can be used for very light elements.

Principles and Instrumentation for Wavelength-Dispersive Analysis

A microprobe is somewhat like a big x-ray tube. Like the x-ray tube, it is evacuated, and it has a metal cathode at one end that generates electrons that are driven by a potential to the other end where they strike an "anode" and generate characteristic x-rays. In the case of the microprobe, however, the "anode" is the mineral specimen whose composition is desired, and instead of generating characteristic x-rays from just one element, the electron beam generates characteristic x-rays from all of the elements ($Z \geq 9$) that constitute the mineral. Provided that there is some way to measure the intensities of the separate characteristic emissions and compare them with the intensities of similar emissions from specimens of known compositions, it is then possible to determine the concentrations of each element present. Given this brief overview, we begin again at the cathode filament and consider some of the details.

Figure 10-11 shows, in a very schematic fashion, the components of an electron microprobe that pertain to generation of the primary x-rays. At the top is the electron gun, which produces the electrons by heating a tungsten wire filament. The electrons are then accelerated down the body of the instrument by a potential applied to the anode plate, which contains a central hole through which the electron beam passes. On the way they pass through an electromagnetic condensing lens that focuses the beam, and several apertures that collimate it. Most instruments are also equipped with a beam-current monitor that allows for maintenance of beam stability during operation. The objective lens focuses the beam to a cross-section as little as a fraction of a micrometer in diameter precisely at the specimen surface. Wider analytical spots can be used by defocusing the beam, and this is sometimes

Figure 10-11. A schematic diagram of the essential features of an electron microprobe.

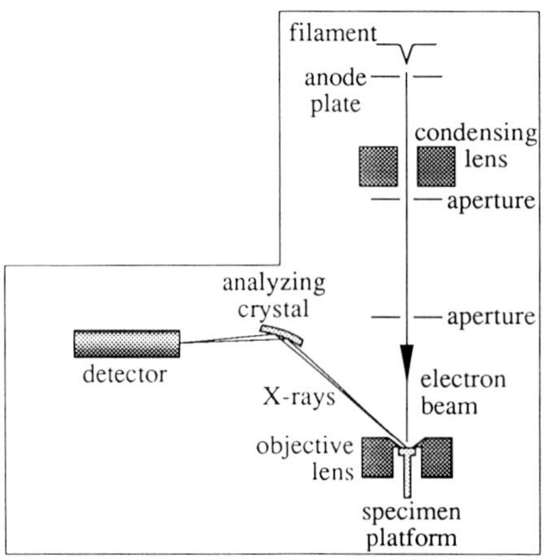

advisable to avoid loss of volatile elements or intentionally to average compositional variations over small distances.

Many of the beam electrons collide with inner-shell electrons of the elements in the specimen, producing characteristic x-rays in the same way that an x-ray tube does. Because the electrons penetrate some small distance into the specimen, the x-rays are produced from a volume of material that, at its widest, is slightly wider than the diameter of the electron beam itself. Some other types of radiation are produced in the process—some of which are not desirable and some of which cause no problems. The electron beam produces visible light (cathodoluminescence) in some specimens, for example, and this allows for determining the correct alignment and focus of the beam. Some of the characteristic x-rays produced in the specimen may, however, be capable of ejecting inner-shell electrons in other elements in the specimen; when these fluorescent x-rays are generated, they increase the apparent concentration of the fluorescing element and decrease that of the exciting element. We shall consider how to correct for such effects when data reduction is covered.

Once the x-rays are produced, the problem is to discriminate among the different wavelengths (or, equivalently, energies) of x-ray photons generated by different elements and to measure their intensities. Discrimination can be approached in two quite different ways—one called *wavelength-dispersive analysis* and one called *energy-dispersive analysis*. We shall postpone discussion of energy-dispersive analysis until later in the chapter.

In wavelength-dispersive analysis, which is generally considered to be the more accurate of the two methods, a curved "analyzing crystal" is used to isolate the x-rays from any particular element and focus them at the detector. According to the Bragg equation (p. 351), specification of both the wavelength of diffracting radiation and the d spacing of the crystal involved uniquely determines the diffraction angle θ at which diffraction occurs. Thus, if it is desired to allow only characteristic x-rays from silicon atoms to reach the detector, one need only place an appropriate analyzing crystal in position to intercept the generated x-rays (which are from all atoms in the sample volume illuminated by the electron beam, of course), adjust its orientation so that its diffracting planes make the required angle θ with the generated x-rays (which are analogous to the incident beam in x-ray diffraction), and then place the detector at the appropriate angle 2θ with the generated x-rays. An appropriate analyzing crystal is one with d spacings that allow for diffraction angles that are mechanically accessible to the detector. Typical microprobes have several x-ray detectors, and with each of them more than one kind of analyzing crystal may be used, so that it is possible to analyze several elements simultaneously.

Unlike in x-ray diffraction, where only the $K\alpha$ characteristic peak is used, any characteristic x-ray emission is potentially useful for analysis with the microprobe. With several elements present in the unknown and a wide variety of characteristic x-rays being unavoidably generated, there is the possibility that the analyzing crystal may diffract both the first-order characteristic radiation of the element of interest and some higher-order characteristic radiation of another element at essentially the same diffraction angle. Even though the interfering radiation is diffracted to the detector by the analyzing crystal, it is of a different wavelength than the radiation of interest, because some multiple of the wavelength actually satisfies the Bragg equation. (That is, it is not λ, but $n\lambda$, where $n \neq 1$, that satisfies Bragg's law.) The detector is therefore equipped with a pulse-height selector that allows x-ray photons to be counted only if they fall within a preselected energy range. (A numerical example of pulse-height selection is provided in the discussion of x-ray fluorescence spectrometry, p. 373.)

Determining the concentration of an element is done by comparing the x-ray intensity for that element in the "unknown" with the x-ray intensity for the same element in a standard of precisely known composition. The unknown and the standard must be subjected to exactly the same instrumental conditions—operating voltage, beam current, specimen geometry, and so forth—for effective analysis. Even then, one does not simply take a ratio of total unknown counts per second to standard counts per second for the element of interest; a substantial amount of data reduction must occur before unknown and standard are compared.

Specimen Preparation and Data Reduction

Specimens may be polished thin sections, polished grain mounts, or polished billets, but in any case the polishing is of prime importance. Figure 10-12 shows that failure to prepare specimens to a high polish leads to enhancement or absorption effects for which there is no way to correct. After polishing, the specimens are coated with a very thin layer of carbon in an evaporative sputterer. The carbon coating does not interfere with analysis, inasmuch as carbon is not analyzed with the microprobe anyway, but it does provide a way to prevent the buildup of sample current caused from bombardment by the electron beam. The specimen discharges continuously through a conducting paint that is applied between the carbon-coated surface and the metal specimen platform.

After the analyzing crystals and detectors are properly set for the elements desired, but before any data are taken from the unknown, counts are obtained from the standards. Often a different standard must be used for each element, although

Figure 10-12. If the specimen surface is inadequately polished, unpredictable absorption effects result during microprobe analysis.

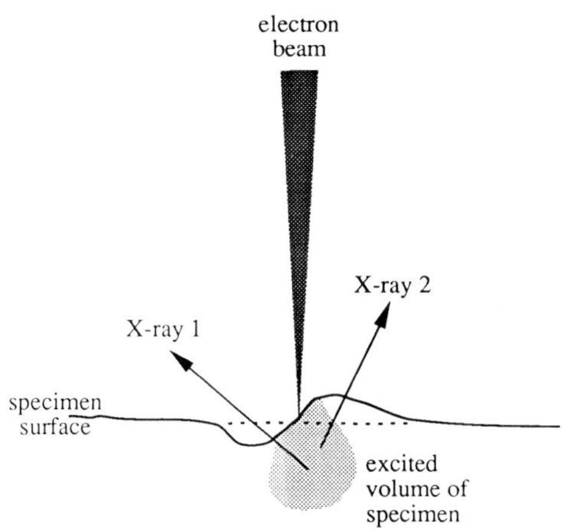

sometimes a standard is available that has suitable concentrations of more than one relevant element. As a rule, it is helpful to choose standards in which the concentrations of the desired elements are fairly close to the expected concentrations in the unknown, although this is not always practical and need not be an obstacle if such standards are not available. Usually several 10- or 20-sec counting intervals are used, with the standard being moved slightly in between to avoid beam damage. These are averaged during data reduction to obtain counts per second for each element in the standard, and a similar procedure is used for the unknowns. After the unknowns have been counted, the standard is counted again to check for electronic drift.

Either at the beginning or the end of the data collection for a given set of elements, backgrounds must be measured for them. This can be done by changing the detector settings so as to measure the counts above and below the characteristic x-ray peak, and then averaging the two values. It can also be done by leaving the detector set for the element currently being analyzed, and taking counts on two or three substances of different mean atomic numbers that contain none of that element. Background depends on the position of the detector relative to the specimen (the closer it is, the more stray background radiation is detected) and on the mean atomic number of the specimen (specimens with higher mean atomic numbers are more efficient generators of x-rays, including background radiation). For a given element, the relative position of specimen and detector is constant, so that a plot of background against mean atomic number for a few specimens that contain none of the element of interest allows one to find the background for the mean atomic number of the unknown specimen. Of course, the mean atomic number of the unknown specimen is not accurately known at this stage, but it can generally be approximated with sufficient accuracy to yield a good estimate of background.

Once counts have been obtained for both the standards and the unknown for as many elements as there are detectors, and background intensities have been measured, the detectors and analyzing crystals are reset for a new group of elements, until all the desired elements have been done. The detected counts are than expressed in counts per second (or per 10 or 20 sec, if necessary to maintain precision), the backgrounds are subtracted, and a linear drift correction is applied between standards for each element. For high count-rates, a dead-time correction may also be needed; this compensates for the time that the detector is saturated and unable to detect incoming x-rays. The data are now ready to be converted to elemental concentrations by either a full matrix correction or by an empirical Bence–Albee correction procedure. We shall consider them in that order.

ZAF Corrections

If C_{unk} and C_{std} are the concentrations of some element in the unknown and in the standard, respectively, and I_{unk} and I_{std} are the intensities of generated x-rays, then

$$C_{unk} = M[C_{std}\,(I_{unk}/I_{std})].$$

M is the matrix correction factor, and it accounts for three phenomena: First, the efficiency of generation of x-rays varies with the atomic number, Z; second, elements in the unknown or the standard, or both, may preferentially absorb x-rays from the element of interest, yielding an anomalously low apparent concentration; and third, x-rays from other elements in the specimen may fluoresce the element of interest, causing an anomalously high apparent concentration. The factor M is usually called the *ZAF* correction, the letters in the acronym representing atomic number, absorption, and fluorescence. Detailed instructions for calculating the correction factors are available in texts that deal with the subject, but a conceptual review is appropriate here.

Atomic Number Correction

This correction is an "efficiency-of-generation" factor, and it is related to the atomic number in two ways. First, inelastic scattering of bombarding electrons does not produce characteristic x-rays, but only contributes to the continuum (p. 345). It turns out that the probability of inelastic scattering is higher for electrons bombarding light elements than it is for heavy elements. Second, backscattering of incident electrons from the surface of the specimen is greater for heavy elements than for light ones. These two effects counteract one another, but in general they do not cancel. The Z factor corrects for differences in the mean atomic numbers of the standard and the unknown.

The concentration, corrected for the atomic-number effect, is

$$C_{Z,\text{corrected}} = C_{\text{measured}}(S_{\text{unk}}/S_{\text{std}})(R_{\text{std}}/R_{\text{unk}}),$$

where S is the loss of characteristic x-rays due to absorption of inelastically scattered electrons, and R is the backscattering loss. With the appropriate formulas and data from published tables, the calculations of S and R are not difficult, but they rapidly become tedious.

Absorption Correction

Characteristic x-rays from atoms of one element are of the appropriate energy to dislodge electrons from the inner shells of atoms of other elements with atomic numbers just lower. For example, if a specimen contains both Cr and Fe, some of the characteristic x-rays from Fe atoms will eject inner-shell electrons from Cr atoms. This results in the absorption of the Fe x-ray by the Cr atom, and generation of a fluorescent x-ray from the Cr atom; the intensity of Fe radiation is thus reduced, while that of Cr is enhanced. In addition, absorption of x-rays occurs as they pass through matter in any case, even without generation of fluorescent radiation. If θ is the angle made with the specimen surface by generated x-rays directed toward the analyzing crystal (an angle fixed by the design of the microprobe), and μ/ρ is the linear absorption coefficient of the material for the element being sought, then χ is defined as

$$\frac{\mu}{\rho}\text{cosec }\theta = (\mu/\rho)(\text{cosec }\theta).$$

The absorption factor is a function of χ, called $f(\chi)$, and it also involves the critical excitation voltage for generating x-rays from the element of interest, the actual operating voltage, the atomic weight of the desired element, and its atomic number. The concentration of the element, corrected for absorption, is

$$C_{ZA\text{corrected}} = C_{Z\text{corrected}}[f(\chi_{\text{std}})/f(\chi_{\text{unk}})].$$

Fluorescence Correction

As pointed out in the previous section on the absorption correction, when a characteristic x-ray from another element is absorbed by an atom of the element of interest in such a way that one of its inner electrons is ejected, the result is a secondary fluorescent x-ray that increases the apparent concentration of the latter element. It is often possible (but not always practical) to choose as standards minerals in which fluorescence is not a significant effect, but this is of course not an option with the unknown specimen. Application of the correction requires calculation of the ratio, γ, of the intensity of fluorescent x-rays to the intensity of primary x-rays:

$$\gamma = \sum_{i,j} \frac{I_{ij}^f}{I^p},$$

where I_{ij}^f is the secondary fluorescent intensity from element i generated by element j, I^p is the intensity of primary radiation from element i, and the summation is carried out over all elements j that produce significant amounts of fluorescence in element i.

The correction for fluorescence is

$$C_{ZAF\text{corrected}} = C_{ZA\text{corrected}}[(1+\gamma_{\text{std}})/(1+\gamma_{\text{unk}})],$$

and it constitutes the final step in one ZAF correction cycle.

One cycle is not sufficient, however, because many of the quantities calculated during the correction procedure require an estimate of the composition of the unknown, and all we have to begin with is the ratio of the counts for the unknown to the counts for the standard, multiplied by the concentration in the standard. With the new estimate of composition for the unknown, the procedures are repeated (always multiplying the original ratio, not the results of the previous cycle, by ZAF). When the difference from one cycle to the next is as small as we wish, the correction procedure has converged and the composition of the unknown has been determined.

Bence–Albee Corrections

The ZAF correction just described is time-consuming and involves intricate calculations, any of which can result in an undetected arithmetic error. It is, of course, desirable to program a computer to carry out the calculations, but even then, either a lot of input data are required or else a lot of standardized data must be stored in memory, and the program is likely to be quite large. An empirical correction scheme, the Bence–Albee correction, has been shown to be as accurate as a ZAF correction, and it requires much less computation; it has been developed only for oxygen-based minerals, however (oxides, silicates, sulfates, phosphates, and so forth), and is not currently useful for sulfides, halides, and so on.

For a binary compound AB, calibration curves of K^A (intensity of x-rays from element A in counts/second) versus C^A (weight fraction of element A) have a form described by the equation

$$(1 - K^A)/K^A = \alpha^A(1 - C^A)/C^A,$$

where α^A depends on the relative importance of fluorescence and absorption. In a multicomponent system of n elements, the factor α^i describes the calibration curve for counts versus concentration of component i in the matrix of n-1 other specific components. It is calculated as

$$\alpha^A_{ABC\ldots n} = (C^B_{ABC\ldots n}\alpha^A_B + C^C_{ABC\ldots n}\alpha^A_C + \cdots + C^n_{ABC\ldots n}\alpha^A_n)$$
$$/(C^B_{ABC\ldots n} + C^C_{ABC\ldots n} + \cdots + C^n_{ABC\ldots n}).$$

Another factor, $\beta^A_{ABC\ldots n}$, defined as

$$\beta^A_{ABC\ldots n} = (C^A_{ABC\ldots n}\alpha^A_A + C^B_{ABC\ldots n}\alpha^A_B + \cdots + C^n_{ABC\ldots n}\alpha^A_n)$$
$$/(C^A_{ABC\ldots n} + C^B_{ABC\ldots n} + \cdots + C^n_{ABC\ldots n}),$$

is analogous to $\alpha^A_{ABC\ldots n}$, but is the correction factor of component A relative to pure end-member A (rather than relative to the particular matrix of the specimen). Note that the α factors are required to compute the β factors.

Let us assume, for the moment, that pure components (for example, pure oxides of the elements of interest) were not available to use as standards, but instead there were well-analyzed "working standards" of more than one oxide component. The so-called K ratio is the ratio of counts per second in the unknown to counts per second in the working standard: K^A_u/K^A_{WS}. The equation used to calculate concentration of component A in the unknown is then

$$C^A_u = \beta^A_u(K^A_u/K^A_{WS})(C^A_{WS}/\beta^A_{WS}).$$

If pure components were used as standards, both C^A_{WS} and β^A_{WS} would be unity, and the equation would be

$$C^A_u = \beta^A_u(K^A_u/K^A_A).$$

In practice, the procedure for computing concentrations of oxides is iterative. The first approximation to the concentration of oxide A is taken as

$$C^{*A}_u = (K^A_u/K^A_{WS})(C^A_{WS}/\beta^A_{WS}),$$

that is, by setting all the β-factors equal to one in the above formula. When first approximations for all n oxide components in the unknown have been determined in this way, the β-factor for oxide A (and B, and C, and so on to n) is calculated as

$$\beta^A_u = [C^{*A}_u(\alpha^A_A) + C^{*B}_u(\alpha^A_B) + \cdots + C^{*n}_u(\alpha^A_n)]/(C^{*A}_u + C^{*B}_u + \cdots + C^{*n}_u.)$$

These β factors are then substituted into the equation for C^A_u above and a second estimate of concentration for each oxide is computed. This is followed by another calculation of β factors, and a new estimate of concentrations, and so on, until the calculations converge. The method is rapid, as only a few iterations (generally no more than three) result in convergence. Tables of α factors for common operating conditions are to be found in the mineralogical literature, and from these, it is easy to calculate the β factors for standards being employed.

Energy-Dispersive Analysis

Detection of the x-rays generated in a microprobe by using an analyzing crystal to direct them toward a detector is called wavelength-dispersive analysis because discrimination between x-rays of the various elements is by diffraction at different angles governed by Bragg's equation. The fact that the x-rays of different elements have different wavelengths requires that they also have different energies, however, and the technique of discriminating among elements by the energies of their characteristic x-rays is called *energy dispersive analysis* (sometimes EDS—*energy-dispersive spectrometry*). The geometry and operation of the microprobe itself are the same as for wavelength-dispersive analysis, but the instrumentation for detection is different.

Instead of recording x-rays from one element at a time, as with the wavelength-dispersive method, a lithium-drifted silicon detector is used and a multichannel pulse-height analyzer records the entire energy spectrum of the specimen at once. The multichannel analyzer has 512 or 1024 channels, and each channel is sensitive to an energy band of 20 eV. Beginning from about 1 keV, then, the spectrum is recorded to at least 11 keV, which includes the energies of the $K\alpha$ emissions for elements from sodium through the first transition series (actually, as high as arsenic), as well as higher spectra of heavier elements. A typical spectrum is shown in Figure 10-13. Elements lighter than sodium are not generally analyzed by energy-dispersive spectrometry.

Background corrections are determined for all elements simultaneously by fitting a curve to the continuum, using readings from a few channels that correspond to energies of no characteristic x-ray peaks. Corrections must be made for overlap of adjacent peaks, for dead time (which is more severe than in wavelength-dispersive spectrometry), and perhaps for some other electronic effects unique to this type of detector (pulse pileup corrections and corrections for so-called escape peaks). Only a few minutes are required to collect the entire spectrum for an unknown, and with a suitably stable instrument and constant operating conditions, standard intensities need to be measured only infrequently. Because of the Si(Li) detector, the electron-beam current can be very low for energy-dispersive detection, resulting in less concern about specimen degradation. Matrix corrections described previously for reduction of data collected by wavelength-dispersive spectrometry can be used for energy-dispersive data with little modification.

Despite the obvious advantages just cited, there are also disadvantages to energy-dispersive analysis. The principal one is relatively poor energy-resolution; overlap between peaks, both $K\alpha$–$K\alpha$ and $K\beta$–$K\alpha$ overlap, must be corrected for. In addition, detection limits are somewhat higher than with wavelength-dispersive analysis, and are also adversely affected by peak overlap. Even so, concentrations of as little as 0.1 wt % are easily detectable and measurable for elements from atomic numbers 11 through 30 in silicates. It has been adequately demonstrated that analyses of rock-forming silicates by energy-dispersive spectrometry is comparable in accuracy to wavelength-dispersive microprobe analyses for elements present in concentrations higher than 0.2 wt %. Moreover, it is significantly faster, less subject to operator error, and the method can be used with any instrument that generates x-rays, such as scanning and transmission electron microscopes.

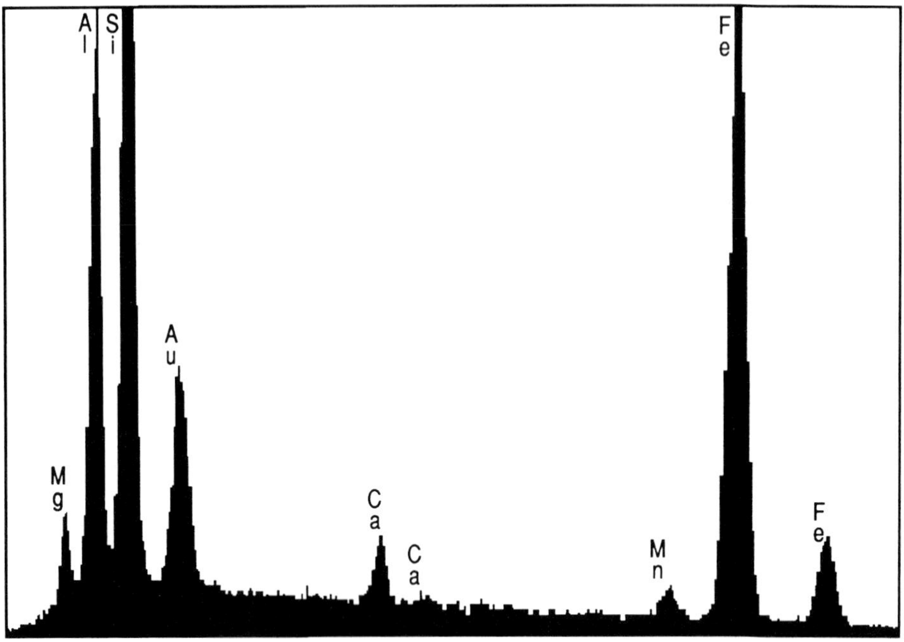

Figure 10-13. An energy-dispersive spectrum of an almandine-rich garnet that contains pyrope, grossular, and spessartine components as well. The smaller Ca and Fe peaks are $K\beta$ peaks, and the Mg, Al, Si, Mn, and larger Ca and Fe peaks are $K\alpha$ peaks. The Au peak is an $L\alpha$ peak, and is present because the specimen was coated with gold to prevent the accumulation of charge.

X-RAY FLUORESCENCE SPECTROMETRY

Figure 10-14 shows that x-ray fluorescence spectrometry (XRF, for short) has a great deal in common with electron microprobe analysis, along with some important differences. Whereas microprobes use curved analyzing crystals, XRF uses flat ones because collimators are employed rather than requiring focusing conditions. The most important distinction, however, is the use of incident x-rays instead of electrons to generate x-rays in the specimen. This results in different detection limits and sensitivities, different sample characteristics, and often in different applications. X-ray fluorescence spectrometry is more often used for whole-rock analyses than for analyses of individual minerals; it can, nonetheless, be quite successfully used for mineral analysis if care is taken to obtain a sample free of foreign material. It is most effective for elements heavier than sodium.

Instrumentation for XRF Analysis

Characteristic x-rays are produced in the atoms of an XRF specimen exactly as they are in an x-ray tube or in an electron microprobe—by outer electrons dropping in to fill vacancies created when inner-shell electrons are ejected. Instead of using electrons to generate the vacancies, however, the electron vacancies are created by x-rays in the same way that (unwanted) fluorescent x-rays are sometimes produced in microprobe specimens. The x-ray tube used to generate the incident x-rays is not different in principle from the sort used in x-ray diffractometers, except that the target metal may differ. The intent is to produce an intense continuum, for it is the energy of the continuum that ejects electrons from the specimen; if the continuum is energetic enough, then it will contain many x-rays of appropriate energies to excite fluorescence in a wide variety of elements. (Energy depends, of course, not on intensity, but on wavelength; but the number of fluorescent x-rays generated depends on the intensity of the incident continuum, given that x-rays of the appropriate energies are present.)

Because the incident x-rays cannot be focused as electrons can, a large area of the specimen is irradiated. To ensure the necessary resolution in wavelength, collimators are placed between the specimen and the analyzing crystal, and between the analyzing crystal and the detector. The latter may be a flow proportional counter or a scintillation counter. The amplified signal passes through a pulse-height selector, and thence to a recorder where the characteristic x-ray intensity is displayed graphically, or to a scaler where the counts are registered.

Pulse-height selection has been mentioned in connection with microprobe analysis, and has been outlined briefly and conceptually (p. 366). A numerical example is appropriate here, and it applies equally well to microprobe instrumentation. The wavelength of barium $L\alpha$ radiation is 2.776 Å, and that of rubidium $K\alpha$ radiation is 0.926 Å; the respective photon energies are 4.465 keV and 13.393 keV. For barium, the Bragg equation is satisfied at

$$\theta = \sin^{-1}(2.776/2d),$$

where d is the d spacing of the diffracting planes of the analyzing crystal. Note that at virtually the same diffraction angle, the third-order diffraction of Rb$K\alpha$ radiation also satisfies Bragg's law:

$$\theta = \sin^{-1}[(3)(0.926)/d] = \sin^{-1}(2.778/2d).$$

Were it not for pulse-height discrimination, the Rb$K\alpha$ intensity would be superimposed on the Ba$L\alpha$ peak, inasmuch as both arrive at the detector at the same spectrometer setting. The Rb peak is, however, a third-order peak, not a first-order one; so its energy is still 13.393 keV. The window on the pulse-height selector is set

Figure 10-14. The geometry of x-ray fluorescence spectrometry.

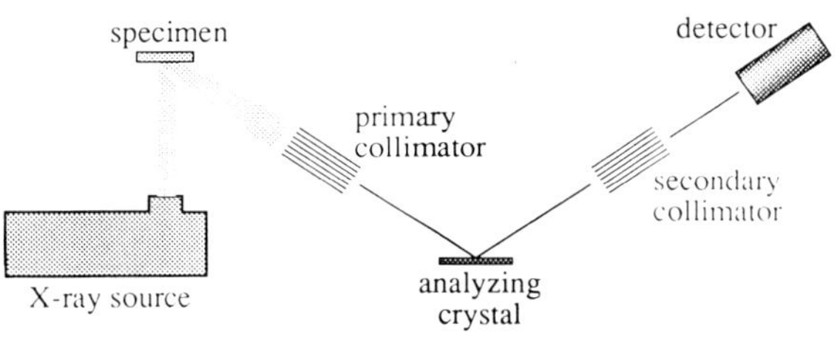

to register the arrival of photons that have only a restricted energy range above the energy at which the Ba$L\alpha$ peak begins (the "baseline" energy), and if the window is set properly, only photons with energies between the low and high sides of the Ba$L\alpha$ peak will be counted. This will obviously exclude x-rays of Rb$K\alpha$ radiation.

Like wavelength-dispersive electron microprobes, x-ray fluorescence spectrometers detect a single element at a time, and they are fitted with a variety of analyzing crystals to provide the flexibility needed to analyze for a broad range of elements with a correspondingly broad range of characteristic x-ray wavelengths.

Specimen Preparation

Because the incident x-ray beam cannot be focused, a rather wide collimated beam strikes the specimen. It is therefore not important for the specimen to be highly polished, as for microprobe specimens, because the fluorescent x-rays are generated over a large area, and small surface irregularities are, in a sense, averaged out. Samples are usually prepared as discs of glass or compressed powder 25 or 30 mm in diameter with fairly flat, but neither polished nor ground, surfaces.

Like microprobe analyses, XRF analyses suffer from matrix effects, and it is important that the irradiated part of the specimen be homogeneous. Slabs of rock, aggregates of coarse grains, and so forth are thus not suitable for quantitative XRF analysis. If the specimen is ground to a very fine powder, and if the mass absorption coefficient of the minerals in it are both relatively small and uniform, then the fluorescent x-rays pass through enough grains of powder that the specimen can be considered homogeneous. If the grains of powder differ substantially in their mass absorption coefficients, however, then the x-rays will not be uniformly attenuated, and the specimen cannot be considered homogeneous. The absorption of the fluorescent x-rays depends to a large extent on their wavelengths, as well as on the absorption characteristics of the grains through which they pass, and the absorption is more severe for longer wavelengths; thus particles must be ground much more finely—submicrometer in size—for analyses of light elements than for analyses of heavy ones, and the risk of contamination from the grinding apparatus is increased. Pellets of compressed powder are generally used for the analysis of trace elements such as rare earths and other heavy elements.

For the determination of major elements in rocks and minerals, fused specimens are used. Fusion is accomplished by mixing a small amount of ground rock or mineral with a flux made from lithium tetraborate, lithium carbonate, and a "heavy absorber" like lanthanum oxide to reduce matrix effects. Various "recipes" are recommended by different workers, but they all have in common the reduction of matrix effects by dilution of the specimen in a flux that does not interfere with the analysis of desired elements. If most of what is irradiated consists of flux of a composition that is constant from one specimen to another, then matrix corrections are small, and accuracy is likely to be enhanced.

Data Reduction

As with wavelength-dispersive microprobe analysis, one element at a time is analyzed, and standards are counted before and after (and sometimes in the middle of) analytical runs. The standard for XRF analyses is made from known concentrations of pure chemicals, generally with all desired elements in a single standard. The intensities (counts per second) for analyzed elements are corrected for dead time (the time during which, at high count rates, the detector is saturated and unable to register counts) and for electronic drift as determined from sequential readings of the standard.

The first approximation to the concentration for element i is calculated as

$$C_i^* = (I_i^{\text{unk}}/I_i^{\text{std}})C_i^{\text{std}},$$

where C_i^{std} is the concentration of element i in the standard, and I is intensity in counts per second. Actual concentration is calculated iteratively by the equation

$$C_i^{\text{unk}} = AC_i^* - B,$$

where A is the absorption factor, and B is the background, expressed as a concentration. Backgrounds are determined on blanks containing none of the relevant analyte elements, and those values are used for all standards and unknowns thereafter. The factor A is calculated from

$$A = \mu_{\text{glass}} + \sum_i X_i(\Delta\mu)_i;$$

μ_{glass} is the mass absorption coefficient of the fused flux without the unknown in it, X_i is the weight fraction of element i in the unknown, and $(\Delta\mu)_i$ is the difference between the mass absorption coefficient of the glass and that of the component i. The first approximation to A is found by inserting C_i^* for X_i, and then that value of A is used to make an initial determination of C_i^{unk}. This provides new estimates of X_i, and the process is carried out iteratively until no significant changes occur in C_i^{unk} from one iteration to the next.

Detection Limits and Sensitivity

The lower limit of detectability is proportional to $(I_b/T)^{1/2}(1/m)$, where I_b is the background intensity, T is the counting time, and m is counts per second per percent of element in the specimen. Thus the lower the background intensity, or the higher the counting time or the efficiency of fluorescent x-ray production (m), the better (lower) the detection limits. Because x-ray fluorescence spectrometry uses x-rays to excite fluorescent radiation instead of using electrons, there is virtually no continuum, and this results in a very small background compared to microprobe analysis. Whereas it is possible to detect concentrations as small as 100 ppm in microprobe specimens, 1 ppm or even less is commonly detected by XRF.

On the other hand, the sensitivity of the microprobe is much higher than the sensitivity of the XRF spectrometer; the microprobe, with its focused electron beam, is capable of detecting as little as 10^{-15} g of material, whereas the XRF instrument requires at least 10^{-8} g of material. Put concisely, the XRF technique requires much more material than the electron microprobe, but can detect much smaller concentrations of trace elements in those larger specimens. The two methods of analysis are complementary, and the choice of method depends on the amount of specimen available, the nature of inclusions or other impurities (which are easier to avoid with the microprobe), the intent of the investigation—and, of course, availability of the instruments.

MÖSSBAUER SPECTROSCOPY

Sometimes it is important to know the ratio of ferric to ferrous iron in a mineral, and none of the analytical techniques discussed to this point can shed any light on that question. Mössbauer spectroscopy provides a reliable method not only for determining ferric-to-ferrous ratios, but for revealing the coordination number of iron (both Fe^{3+} and Fe^{2+}), and often the number of different coordination polyhedra in which iron occurs in a given mineral.

Overview of the Method

Unlike the instrumental methods so far described, which depend on the emission of radiation and in which absorption is a problem to be avoided as far as possible, Mössbauer spectroscopy is an absorption phenomenon. The nucleus of an atom, like the electrons that surround it, normally exists in a ground state (a state of lowest energy), but may be excited to higher energy levels by absorption of radiation. In Mössbauer spectroscopy, atoms are irradiated with γ rays produced by transitions in radioactive nuclei, and the nuclei of the irradiated atoms absorb some fraction of those γ rays. Measurement of the transmitted γ-ray intensity reveals the absorption. In order to appreciate how the method is useful for mineralogists, as well as why its use has been essentially limited to iron, discussion of some details and complications is necessary.

If we were to draw an accurate energy-level diagram for the nuclear transitions in some particular nucleus, the energies of the excited states would not appear as sharp lines, but as somewhat fuzzy (but narrow) bands. A photon emitted as the nucleus dropped from some excited state to the ground state might then have any energy in the range $E_0 \pm \Delta E/2$, where ΔE is the width of the excited-state band and E_0 is the difference in energy between the center of that band and the ground-state energy. For transitions occurring in a large number of nuclei, a plot of photon intensity versus transition energy would be a narrow peak of Lorentzian shape (Figure 10-15). ΔE can be calculated from the Heisenberg uncertainty principle. Recall that it states that the product of the uncertainty in momentum (Δp) and the uncertainty in position (Δx) is no smaller than Planck's constant (h):

$$(\Delta p)(\Delta x) \geq h$$

$$[\Delta(mv)](\Delta x) \geq h$$

$$[\Delta(mv^2)][\Delta(x/v)] > h;$$

x/v is time, mc^2 is E, and for a photon, $v = c$ so

$$[\Delta(mc^2)](\Delta t) \geq h$$

$$(\Delta E)(\Delta t) \geq h.$$

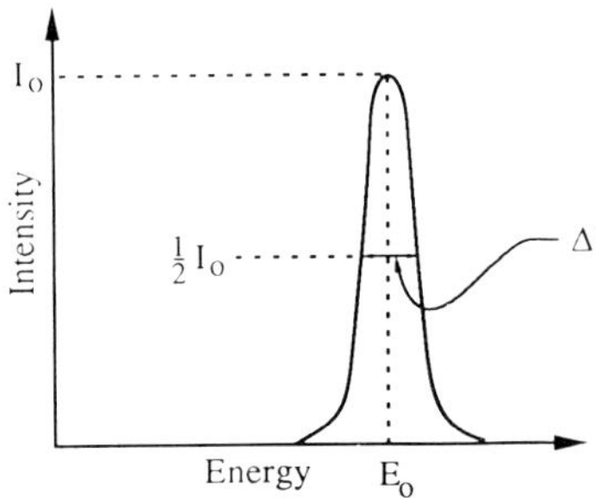

Figure 10-15. The Lorentzian peak shape of a plot of energy versus intensity for γ rays emitted during nuclear transitions.

Thus, $$\Delta E \geq h/\Delta t.$$

The time, Δt, required to measure the transition energy can be no longer than the mean half-life, τ, of the transition, or measurement would be impossible. Setting Δt to τ, recognizing ΔE as Γ, the full width of the peak at one-half the maximum height (Figure 10-15), and substituting the numerical value for h, Γ is calculated as

$$\Gamma(\text{in eV}) = 6.58 \times 10^{-16}/\tau \text{ (in sec)}.$$

Given a typical half-life of 10^{-8} sec, we see that the natural line width, Γ, is quite small indeed, especially compared to decay energies (E_0) of several to many tens of keV. This leads to a problem.

When the nucleus emits a γ ray, conservation of momentum requires that the nucleus recoil in a direction opposite to emission of the photon. Numerically, the magnitude of the momentum of the nucleus must be equal to that of the momentum of the photon. Consider the experiment described above: γ rays generated by radioactive nuclear transitions pass through a specimen and are partially absorbed as they raise some of the nuclei to excited states. (These nuclei subsequently decay, of course, emitting γ rays of the same energy, but they emit them in all directions, so that their intensity is insignificant at the detector. The absorption is due to resonance—the incident photons have just the right energy—and emission by the target nuclei is called *resonance fluorescence*.) We need the incident γ rays to have energy E_0, the energy required to excite the target nuclei, but the energy they actually have, if the emitting nuclei begin at rest, is E_0 decreased by the recoil energy R: $E = E_0 - R$. If the target nuclei also begin at rest, then the absorption of photons causes them to recoil, and the recoil energy is unavailable for excitation; the energy needed by a photon to excite the nucleus is not just E_0, then, but $E_0 + R$. Because of natural line broadening, the emission and absorption peaks have finite width, and whether absorption occurs at all will depend on the relative magnitudes of Γ and R (Figure 10-16). R may be calculated numerically by

$$R(\text{in eV}) = 5.37 \times 10^{-4} E_0^2 \text{ (in keV)}/A,$$

where A is the nuclear mass number. For all cases of interest, it turns out that R is very much larger than Γ, so there is no overlap between the peaks representing the

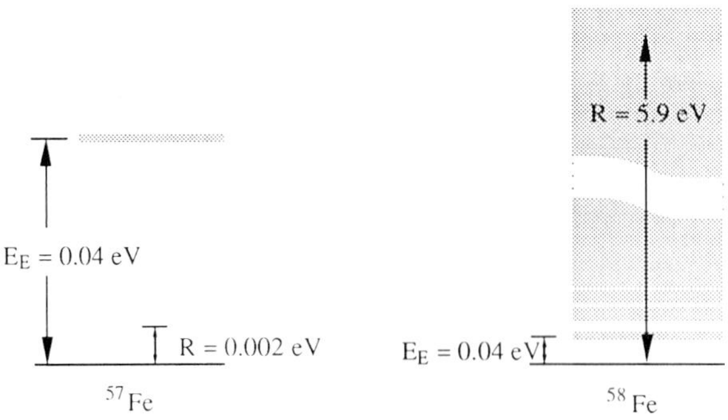

Figure 10-16. ^{57}Fe emits a recoil-free γ ray because the recoil energy R is very much less than the Einstein energy E_E. For ^{58}Fe the recoil energy is very large compared to E_E, and the emitted γ ray is affected by recoil.

energy of the emitted photons and the energy of absorbed photons, if the nuclei begin at rest. Hence no absorption and no resonance fluorescence can occur.

Before we see how a situation that appears hopeless is resolved, one other apparently negative point needs to be made. The nuclei do not begin at rest, but are always in motion. The kinetic energy they possess will be added (in a vectorial sense) to the energies of the photons they emit. This is a Doppler effect, and so it is called *Doppler broadening.* Doppler broadening smears out the emission and absorption peaks, and concomitantly reduces their heights. Because the Doppler broadening has a magnitude of the same order as the recoil energy R, the peaks may overlap a small amount, but their intensities are so low that the effect is now unobservable!

The fact is that absorption and resonance fluorescence do occur for some isotopes despite the discussion presented, and the reason is that the vibrational energy states of the crystal lattice of both the emitter and the absorber are quantized. This means that the vibrational states of the solid are separated by energies that are integral multiples of some quantum of energy. Moreover, the natural linewidths of the excited states increase with energy. An approximation to the magnitude of the quanta was proposed by Einstein, and is called the *Einstein energy* E_E. Consider two isotopes of iron, ^{57}Fe and ^{58}Fe. For the transition from the first excited state to the ground state, the energies (E_o) are 14 keV and 800 keV, respectively. The Einstein energy for both is 0.04 eV. From the equation for R, we calculate that the recoil energy for the ^{57}Fe transition is 0.002 eV, and that for the ^{58}Fe transition is 5.9 eV.

For ^{57}Fe, R is very much less than E_E, and so the nucleus, bound as it is in a crystal lattice, is unable to accept the recoil energy. It will thus not recoil either when emitting or when absorbing a γ ray; in addition, it will not experience Doppler broadening, because the photon escapes without changing the internal energy (thermal excitation) of the crystal lattice. Both the incident and the fluorescent γ-ray photon undergo recoil-free emission, and have energy E_o.

For ^{58}Fe, R is very much greater than E_E, and represents a large number of energy quanta. That many quanta above the ground state, the broadened energy levels overlap, and any amount of energy can be accepted by the lattice. Hence the nucleus *does* recoil, both on emission and absorption, Doppler broadening occurs, and the situation shown in Figure 10-16 obtains. Resonance fluorescence thus fails to occur at detectable intensities in ^{58}Fe.

It is clear, then, that the Mössbauer effect is observable in only a limited number of isotopes, and ^{57}Fe is by far the most important one for mineralogy, although it has been observed in over 100. About 2% of any natural sample of iron atoms is ^{57}Fe, and, inasmuch as the chemical reactions by which minerals form depend on electronic configurations rather than on nuclear masses, any mineral that contains iron will contain some ^{57}Fe. What is not clear to this point is how the Mössbauer effect is actually observed, and how it is useful to mineralogists, given that there are certainly simpler ways of detecting the presence of iron. We proceed along those lines.

Observation of the Mössbauer Effect

The incident photons for observation of the Mössbauer effect in iron are produced as radioactive ^{57}Co nuclei embedded in an iron matrix decay to ^{57}Fe in an excited state, which then decay to ground-state ^{57}Fe nuclei, emitting recoil-free 14 keV photons in the process. The radioactive Co in the Fe matrix is called the *source*. γ rays from the source then encounter the sample, which is positioned between the source and the detector. The sample is called the *absorber*, and two kinds of absorption occur. One is the normal absorption that would be experienced by photons of any energy, and is due to the Compton and photo effects; the other is the resonant absorption that occurs for photons of energy E_o = 14 keV, and constitutes the Mössbauer effect.

A plot of energy versus absorption would thus show some background absorption level, with a peak of width Γ (at half-maximum) centered on E_o. What we need to do, then, is to vary the energy in order to obtain such a plot. This is done by placing the source on a platform that moves at velocity, v. The energy of the incident γ rays is then $E_o + (v/c)E_o$, and their energies can be altered by varying v. In actual experiments it is transmission, rather than absorption, that is measured; so the "peaks" in absorption appear as dips in transmission. (For simplicity, we shall nonetheless refer to these as peaks.) The Mössbauer spectrometer consists essentially of a source, a mount for the absorber, and a multichannel analyzer for measuring photon intensity as a function of energy. The velocity of the source, in millimeters per second, is usually plotted instead of energy, with $v=0$ corresponding to E_o. The time required to accumulate a Mössbauer spectrum with acceptable signal-to-noise ratio is several hours to a few days.

Appearance of the Mössbauer Spectrum

The Mössbauer spectrum, which, from the discussion so far, might be expected to show a single peak at $v=0$, is modified by two factors important to its use in min-

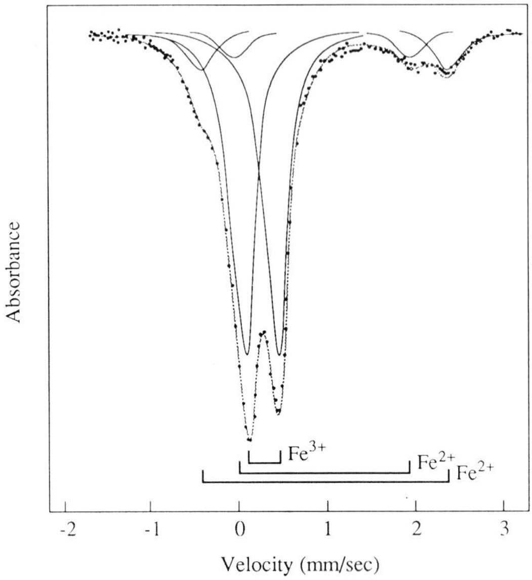

Figure 10-17. A Mössbauer spectrum showing one large Fe^{3+} doublet and two smaller Fe^{2+} doublets. The spectrum shows that there is much more Fe^{3+} than Fe^{2+} in the specimen (the concentrations are proportional to the areas under the peaks), and that Fe^{2+} occupies two different coordination polyhedra, resulting in slightly different quadrupole splittings.

eralogy. They are the *isomer shift* (sometimes called the chemical shift) and the *quadrupole splitting.*

The charge of the nucleus is distributed over the volume of the nucleus, and the nucleus also interacts with the charge-density distribution of the electrons over the same volume. If the electronic charge distribution is different for source and absorber, then the resonant peak in the spectrum will not occur at $v = 0$, but at some other value of source velocity. A typical source is ^{57}Fe (from ^{57}Co) in a stainless steel matrix, and it is obvious that the electronic configuration about an iron atom in a metal is quite different from that about an iron atom surrounded by oxygens in a mineral. Moreover, the difference will not be the same for Fe^{2+} as for Fe^{3+}. By observing the isomer shift, then, it is possible to discriminate between the two common oxidation states of iron in minerals, and the relative areas under the peaks yield the relative concentrations of the two forms. Because the isomer shift depends on the exact electronic charge distribution at the nucleus—and that will vary with coordination number, distortion of the coordination polyhedron, and perhaps other factors—Fe^{2+} and Fe^{3+} do not have single unique isomer shifts, but ranges within which their isomer shifts are expected to occur. For silicates, these ranges do not overlap, and a well-resolved Mössbauer spectrum is unambiguous in discriminating between oxidation states.

An iron nucleus in an excited state is not spherical, and thus has a nonspherical nuclear charge distribution characterized by a nuclear quadrupole moment. The interaction of this charge distribution with the asymmetric electronic charge density results in splitting of the nuclear energy levels, with a concomitant splitting of the Mössbauer spectral line into a doublet. The separation of the two peaks, measured in units of velocity, is the quadrupole splitting (and the isomer shift is measured half-way between the two peaks). The magnitude of the quadrupole splitting depends on the details of the coordination geometry (both as to number of coordinating atoms and distortion of the coordination polyhedron). In general, it is difficult to equate a specific quadrupole splitting with some particular geometry, but comparisons, using quadrupole splittings resulting from known ideal geometries as benchmarks, are useful.

Early Mössbauer spectra were resolved by eye. Nowadays it is common to subject them to a computerized curve-fitting process by which the spectra are resolved into appropriate numbers of doublets, all having a Lorentzian shape. Figure 10-17 shows a typical Mössbauer spectrum with two oxidation states of Fe represented, one of them in two different atomic sites.

NUCLEAR MAGNETIC RESONANCE SPECTROSCOPY

Another powerful instrumental technique that exploits transitions in nuclear energy levels is nuclear magnetic resonance spectroscopy (NMR). NMR has been used for decades for investigating molecular structure in liquids, but only in the past decade were methods developed that make it useful for solids and, in particular, for minerals. The reason for this will be discussed in the following. Because several geologically common nuclides are NMR-active, NMR is applicable to a larger group of mineralogically important nuclides than is Mössbauer spectroscopy, but suffers (temporarily, to be sure) from lack of a large database to aid in interpretation. As more data from NMR studies are accumulated, this technique will become increasingly more important and increasingly more common in crystal-chemical investigations.

Theoretical Overview

The nucleus of an atom has a nuclear spin I, which is quantized. The spin energy levels are normally degenerate, which means that they all have the same energy, but, in the presence of a strong magnetic field, this degeneracy is removed. The differences in these nondegenerate energy levels correspond to radio frequencies. So, if the nucleus of an atom in a strong magnetic field is irradiated with radio-frequency energy (the proper frequency of which depends on the magnitude of the external magnetic field) and absorbs an amount of energy necessary to promote it to a higher energy level, then it will radiate that energy on returning to the lower energy level. The fact that makes NMR useful for mineralogists is that the magnetic field felt at the nucleus is the external field plus a perturbation caused by the surrounding electrons, and this perturbation depends on the specific chemical environment of the nucleus. Thus, nuclei of the same cation in different coordination states, or with different next-nearest-neighbor cations, or different polyhedral geometries, yield different NMR spectra.

Like Mössbauer spectroscopy, nuclear magnetic resonance spectroscopy suffers from some experimental problems that must be overcome in order for it to be useful at all. The problem of greatest consequence is broadening of the lines in the spectrum—that is, poor resolution. One of the causes of poor resolution is an orientation-dependent line broadening that is experienced by solids at rest. In liquids, the molecules are in constant and rapid random motion, and the orientation-dependent factors are thus eliminated by averaging over time, leading to sharp spectra. For solids, the same effect can be achieved by rotating the specimen very rapidly (150,000–350,000 rpm) in the magnetic field. Moreover, the mathematical expression governing the orientation-dependent line broadening contains the factor $(3\cos^2\theta - 1)$, where θ is the angle between the magnetic-field vector and the sample-spinning axis, and this term vanishes if θ is 54.7°. This is the so-called "magic angle" for NMR spectroscopy of solids, and spinning the sample at this angle results in very small peak widths for some nuclides in solids. The technique is called *magic-angle sample-spinning NMR*, or MASS-NMR. (Sometimes the word *sample* is omitted, to yield the acronym MAS-NMR.)

Another factor affecting linewidth is the value of the quantized nuclear spin. Application of the external magnetic field results in $(2I+1)$ nondegenerate energy levels. For nuclides with $I = \frac{1}{2}$, then, there are only two spin states, and the nucleus acts as a dipole. This yields quite sharp spectral lines. ^{29}Si is one such nuclide, and is obviously a very important one for the study of silicates. Most NMR-active nuclides possess nuclear spins greater than $\frac{1}{2}$, however, and thus display a quadrupole interaction that can lead to serious line broadening. For those with half-integral spins, though, the dipolar interactions turn out to be the most important, and the broadening, although not completely eliminated, is not intolerable. Examples of such nuclides are ^{27}Al, ^{17}O, ^{31}P, ^{11}B, ^{23}Na, and ^{19}F. Greater experimental difficulties are present for some nuclides than for others; for example, the natural abundance of ^{17}O is only 0.04% of all oxygen, and its spectrum is prohibitively broadened except where its site symmetry is favorable.

Instrumentation and Specimen Preparation

In a modern NMR spectrometer, the external magnetic field is imposed with a superconducting solenoid. The specimen is 100–500 mg of finely ground powder that is packed into to sample chamber surrounded by a radio-frequency coil. This coil produces the energy that is resonantly absorbed by the specimen. The sample chamber sits in a mounting that allows the angle of the rotation axis to be varied. It is rotated by forced air, and rides on a virtually frictionless air cushion. Various amplifiers, radio frequency detectors, and data-storage and -processing systems are used.

Spectra are measured relative to some standard—a different one for each nuclide, of course—in units of parts per million. Typical values range from zero to a few hundred parts per million, depending on the nuclide and the standard, with less positive values representing greater shielding of the nucleus by the surrounding electrons. Because NMR spectra depend mostly on nearest-neighbor and next-nearest-neighbor interactions, there are as many peaks as there are significantly different sites in terms of coordination number and identity of nearest-neighbor and next-nearest-neighbor atoms. Only one nuclide is recorded on a single spectrum.

TRANSMISSION ELECTRON MICROSCOPY

General Description

A transmission electron microscope (TEM) is, in many respects, like an optical microscope. Instead of a light source, there is an electron gun like the one in an electron microprobe, and, instead of being at the bottom of the microscope, it is at the top. Electrons are generated by heating a filament, then they are accelerated down the evacuated microscope body by an electrical potential. They then encounter lenses, which are of course electromagnetic, as in the microprobe. By analogy with an optical microscope, there are condenser lenses (generally two) that cause beam convergence on the specimen, and then an objective lens beyond the specimen. Next come one or two intermediate lenses (with no analog in the optical microscope, but they are usually used as the principal magnifying lenses), and then a projector lens, analogous to the ocular. Various apertures for restricting or collimating the beam are also placed along the column of the microscope. Viewing is accomplished by inserting a fluorescent screen or a photographic film beyond the projector lens. Varying the current in the various electromagnetic lenses allows the image to be focused, and also permits use of the TEM in various modes of operation to be discussed in the following.

The Specimen

Electrons interact very strongly with matter, and so TEM specimens must be very thin (~1000 Å) in order not to absorb too much of the incident beam. Biological specimens are usually sliced with a microtome, but this is obviously not practical for minerals. Instead, minerals may be crushed and observation carried out on the thin edges of flakes, or else they may be mechanically thinned by bombardment with an ion beam. Crystallographic orientation is important for minerals, and ion thinning has the advantage that crystals with an appropriate orientation can be chosen in advance from an uncovered thin section; the section is then removed from the slide, a small (~3 mm diameter) disc is cut from or around the desired crystal, and the beam of an ion milling device is directed onto this disc. Another advantage of ion thinning is that it generally results in a larger specimen area having appropriate thickness.

Ion-thinned discs may be mounted on a small copper ring, and then covered with an evaporative coating of carbon to ensure that specimen charge does not build up. Crushed fragments are suspended in alcohol or distilled water, and a drop is placed on a small copper grid on which a thin carbon film has been deposited; the fluid, of course, evaporates, leaving the fragments on the film. Alternatively, a thin organic film containing thousands of small holes is deposited on the grid, the carbon coating is evaporated over this, and one views the parts of flakes that project out over the holes so that there is no interference from the film.

Because crystallographic orientation is important for mineralogical samples, the copper grid containing the specimen is placed in a mounting that allows it to be both rotated and tilted so that desirable orientations can be achieved. The importance of orientation will become apparent as we discuss the origin of contrast in the TEM and the modes of operation possible.

Image Formation and Modes of Operation

Formation of an image—any kind of image—depends on contrast. The image in an optical microscope results from different colors or indices of refraction of objects in the field of view. In a transmission electron microscope, the image arises from the ways in which electrons interact with matter. For biological specimens, variable absorption of electrons creates the image, with those features that absorb electrons more strongly appearing darker on the fluorescent screen or photograph. For crystalline specimens, the interaction of interest is not absorption, but diffraction.

As with any very small particles, the wave–particle duality apparent for photons is also important for the high-speed electrons that are accelerated down the tube of the transmission electron microscope. When the electrons strike a crystalline specimen, they are diffracted—just like x-rays—by planes of atoms that have the appropriate orientation according to Bragg's equation (p. 351). As in x-ray diffraction, a substantial amount of the intensity of the direct beam passes through the specimen undeflected. The direct beam then contains information from those parts of the specimen that were not in a diffracting position, but not from those that were. Thus, contrast is created by intensity differences arising from diffraction of electrons in the incident beam.

Resolution in the transmission electron microscope is limited by the wavelengths of the electrons and lens aberrations, chief among which is *spherical aberration.* Wavelength depends upon the speed of the electrons, which can be varied by changing the accelerating voltage. Spherical aberration is the inability of the magnetic lenses to focus all of the electrons onto a common point; it is a technological problem, but constitutes a limitation only when the highest resolution (around 2–5 Å) is needed. *Chromatic aberration,* caused by electrons of different wavelengths in the beam, and *astigmatism,* caused by radial asymmetry of the magnetic lenses, can also cause poor resolution, but have largely been overcome in modern instruments.

There are several different methods of extracting information contained in the electron beams (direct and diffracted) of the TEM. Because the electrons are diffracted, a diffraction pattern is formed above the image plane (in fact, above *each* image plane, inasmuch as intermediate images are formed along the electron-optical column). If an aperture is used to select a particular area of the first intermediate image (below the objective lens), then the resulting diffraction pattern that is viewed beyond the projector lens is the diffraction pattern for that small, selected area. For this reason, this technique is called *selected area diffraction* (SAD). As with x-ray diffraction, it enables the investigator to check crystal symmetry, space-group extinctions, and so forth.

Each spot in an electron diffraction pattern represents diffraction from some set of planes, *hkl*. If one or a few diffracted beams are enlarged to provide an image, they show only that information contained in them. Because the main beam no longer takes part in forming the image, the background of the image is dark, and only those features for which information is included in the selected beam(s) are bright. That is, the contrast in such an image comes only from the selected diffracted beam or beams; any features that do not contribute to the selected diffracted beam(s) are dark. Such an image is called a *dark-field image.* Resolution in a dark-field image is improved if the entire diffraction pattern is tilted so that the selected diffracted beam is parallel to the microscope body and replaces the main beam.

If the direct beam of an SAD is selected and enlarged, the image produced has a bright background due to the direct beam, and any dark features are dark because they have diffracted some electrons away from the direct beam. This sort of image is called a *bright field image.* A bright-field image resembles a dark-field image viewed in the same orientation, but bright features in one are dark in the other, and vice versa.

A final technique that has relatively recently come into use is called *high resolution transmission electron microscopy.* Very high accelerating voltages (100 kV or higher) are used, and the image is defocused by a few hundred Ångstroms. With specimens of proper thickness and orientation, the image thus yielded resembles very closely the actual atomic arrangement seen at a resolution of only a couple of Ångstroms or so. Areas of high electron density (that is, the locations of atoms) appear dark, and channels or holes in the structures appear bright. The images are "fuzzy," and one does not use this technique for measuring bond lengths or other structural parameters, but it is very useful for dealing with vacancy substitutions and other such phenomena that are obscured by the averaging effect of x-ray diffraction. A related technique, lattice imaging, allows one to view lattice repeats with a resolution of only a few Ångstroms, and therefore to study dislocations and distortions of the crystal lattice.

SCANNING ELECTRON MICROSCOPY

Like the transmission electron microscope, the electron beam in a scanning electron microscope (SEM) is produced by heating a filament and using an electrical potential to accelerate the electrons down the evacuated microscope tube. A typical SEM has two condenser lenses and an objective lens that focus the beam to a very fine point at the specimen surface. (It can be defocused to provide a wider beam if desired, for example to avoid damage to a specimen containing elements that volatilize easily.)

For most SEM work, specimen preparation is trivially simple. Sample chambers of most scanning electron microscopes can accommodate specimens up to a centimeter or so in any dimension, and it is sufficient just to glue a piece of the specimen onto a small platform, or stub, that holds it in place. The stub may be made of aluminum or any other metal, unless energy-dispersive spectrometry is to be used concurrently to determine chemistry (p. 371), in which case a carbon stub should be used in order to avoid analyzing the stub as well as the specimen. (Many SEMs, but by no means all, are equipped with energy-dispersive spectrometers; these are, of course, not necessary for the microscopy function of an SEM.) The specimen is evaporatively coated with a thin conducting deposit of gold or carbon in order to avoid the buildup of a specimen charge. In general, polishing or other preparation is not only unnecessary, but undesirable; for some applications, it may be required, however. The sample holder can translate the specimen in two perpendicular directions, can rotate it about an axis perpendicular to the surface of the stub, and can tilt it with respect to the electron beam. The specimen need not be any particular thickness, because the electrons are not intended to penetrate through it; the intent of SEM work is to view the surface of the specimen.

A major difference between the TEM and the SEM is the scanning function of the latter. The electron beam passes between scanning coils that use an electric field to move the beam back and forth across the specimen surface. The scan is a raster scan so that the beam not only moves across the specimen, but down it scan by scan. The specimen is tilted with respect to the beam, and a charged electron collector sits beside it. The electrons (either backscattered primary electrons or low-energy secondary electrons) that leave the specimen surface are attracted to the electron collector. If the surface is irregular, as it usually is for SEM specimens, then electrons from surfaces facing the collector will reach it in higher numbers than electrons from surfaces away from the collector. Thus the final image shows some parts of the specimen in relative brightness, and other parts in relative shadow, and the effect is quite three dimensional—as if viewing a surface illuminated by a light source at a shallow angle.

The next problem is how to construct an image from the electrons arriving at the collector. The collector is connected to a cathode-ray tube (essentially a television picture tube) via a photomultiplier and some amplification circuitry. A cathode-ray tube creates an image with the same sort of raster scan as that used on the specimen. The scan of the microscope beam over the specimen surface is therefore electronically correlated with the scan of the image-forming electron beam in the cathode-ray tube, and the intensity of the latter is governed by the strength of the signal from the collector. Each scan of the image-forming beam across the cathode-ray tube is thus a replica of the response of the collector to a scan across the specimen, and the image of the specimen appears on the cathode-ray tube.

11

Atomic Bonding Models in Mineralogy

In introductory chemistry courses, students are generally taught a number of elementary models of chemical bonding: ionic bonding, covalent bonding, metallic bonding, van der Waals bonding, and hydrogen bonding. Often these models are taught as if one or another applied directly and uniquely to any chemical compound, minerals included. In fact, many silicates contain two or more distinctly different sorts of bonds between different atoms, and pure ionic bonds, of the kind often taught as "correct" in undergraduate mineralogy courses, are virtually nonexistent.

IONIZATION POTENTIAL, ELECTRONEGATIVITY, AND PARTIAL COVALENCY

The amount of energy required to remove an electron completely from an atom is called the *ionization potential.* It varies not only from one type of atom to another, but also changes for a given atom as successive electrons are removed; this is because electrons tend to shield one another partially from the positive nucleus, and as this shielding is reduced by the removal of electrons, the effective nuclear charge (that is, the net charge actually "seen" by an electron) increases. The energy required to remove the first electron is called the first ionization potential, that required to remove the second is the second ionization potential, and so forth. Ionization potentials for some of the elements are given in Table 11-1.

Consider quartz, which has the chemical formula SiO_2. The oxidation numbers of silicon and oxygen are $+4$ and -2, respectively. On the presumption that quartz is ionically bonded, each silicon would give up four electrons to become Si^{4+}, and each oxygen would gain two electrons to become O^{2-}. Table 11-1 shows that the removal of the third and fourth electrons from silicon requires more energy than removal of the second electron from argon—a noble gas, highly unlikely to lose two electrons! Indeed, removal of the second electron from silicon requires about the same amount of energy as removal of the first from argon. In the face of such data, it hardly seems likely that Si is fully ionized to Si^{4+}.

Some elements are much more effective than others at attracting electrons. The

Table 11-1. Ionization Potentials of Some Elements (in Electron Volts)

	Mg	Al	Si	Ar
IP^{I}	7.65	5.99	8.15	15.76
IP^{II}	15.03	18.82	16.34	27.60
IP^{III}	80.12	28.44	33.49	
IP^{IV}			45.13	

tendency of an atom to attract electrons is called *electronegativity,* and various slightly differing tables of electronegativities have been published, based on various methods of calculating them. The values shown in Table 11-2 are based on thermochemical data.

Elements with identical electronegativities will exert identical attractions on electrons, and will therefore share them, forming covalent bonds. On the other hand, for those with very different electronegativities the bonding electrons will spend most of their time in the vicinity of the more electronegative atom, resulting in a more ionic bond. Electronegativity difference, then, is obviously related in some way to the degree of ionic (or covalent) character of a bond. An expression that has been found useful in estimating the fractional ionic character of a bond is

$$f = 1 - e^{-(1/4)(x_A - x_B)^2},$$

Table 11-2. Electronegativities of some of the chemical elements

I A	II A	III B	IV B	V B	VI B	VII B	VIII B	VIII B	VIII B	I B	II B	III A	IV A	V A	VI A	VII A
H 2.20																
Li 0.98	Be 1.57											B 2.04	C 2.55	N 3.04	O 3.44	F 3.98
Na 0.93	Mg 1.31											Al 1.61	Si 1.90	P 2.19	S 2.58	Cl 3.16
K 0.82	Ca 1.00	Sc 1.36	Ti 1.54	V 1.63	Cr 1.66	Mn 1.55	Fe* 1.83	Co 1.88	Ni 1.91	Cu 1.90	Zn 1.65	Ga 1.81	Ge 2.01	As 2.18	Se 2.55	Br 2.96
Rb 0.82	Sr 0.95	Y 1.22	Zr 1.33		Mo* 2.16			Rh 2.28	Pd 2.20	Ag 1.93	Cd 1.69	In 1.78	Sn* 1.96	Sb 2.05		I 2.66
Cs 0.79	Ba 0.89	La 1.10			W 2.36			Ir 2.20	Pt 2.28	Au 2.54	Hg 2.00	Tl 2.04	Pb* 2.33	Bi 2.02		
		Ce 1.12	Pr 1.13	Nd 1.14		Sm 1.17		Gd 1.20		Dy 1.22	Ho 1.23	Er 1.24	Tm 1.25		Lu 1.27	
					U 1.38	Np 1.36	Pu 1.28									

* Fe(III) 1.96 Mo(IV) 2.24

Sn(II) 1.80 Pb(II) 1.87

Oxidation numbers are given at the top of each group. [After A. L. Allred (1961). "Electronegativity Values from Thermochemical Data." *Journal of Inorganic and Nuclear Chemistry* 17, 215–21.]

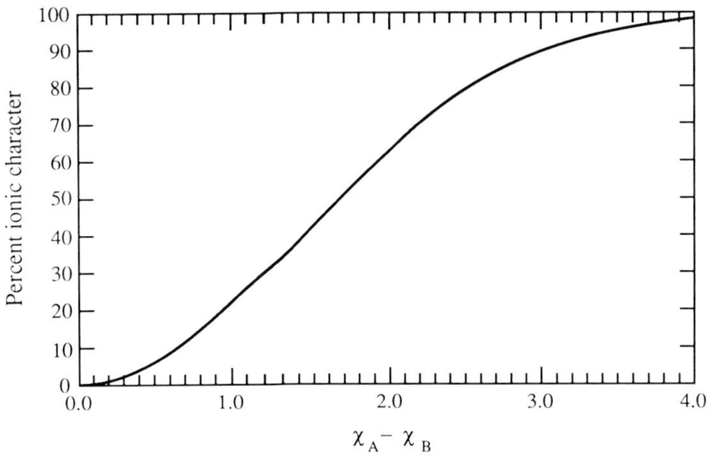

Figure 11-1. The percent ionic character as a function of the electronegativity difference between atoms *A* and *B* ($\chi_A - \chi_B$). Electronegativies may be found in Table 11-2.

where χ_A and χ_B are electronegativities of the two atoms involved. In Figure 11-1 fractional ionic character is plotted as a function of electronegativity difference, and we see that the Si–O bond of quartz is predicted to be 45% ionic and 55% covalent. On this basis, we calculate the charge of the silicon atom to be $(0.45)(+4) = +1.80$, a much more reasonable estimate than +4. Oxidation numbers are useful and practical for a number of purposes (such as balancing chemical formulas), but are not very accurate estimates of actual charges on atoms in silicates.

THE ELECTRONEUTRALITY PRINCIPLE

With Table 11-1, Figure 11-1, and some information about the arrangements of atoms in a given mineral of interest, we can estimate the actual charges on atoms. Consider, for example, the Mg end-member olivine, forsterite. Each silicon is surrounded by four oxygens, each magnesium by six oxygens, and each oxygen by one silicon and three magnesiums (see Chapter 7). The electronegativity difference between Mg and O is 2.13, corresponding to 68% ionic character. We have seen that the Si–O bond is 45% ionic. (That the ionic character of a bond between any two particular types of atoms is constant because it depends only on electronegativity difference is an approximation that is not entirely correct, but is tolerable for our present purposes.)

We begin with the electrostatic bond strength (EBS), which is defined as the formal charge of the cation (assuming a 100% ionic bond) divided by its coordination number. For magnesium the EBS is $\frac{2}{6} = \frac{1}{3}$, and for silicon it is $\frac{4}{4} = 1$. Thus, if these bonds were completely ionic, silicon would donate one electron to each of four oxygens, and Mg would distribute its two valence electrons among six surrounding oxygens (resulting in $\frac{1}{3}$ electron to each oxygen, on the average). The charges on Mg and

Si would then be 2+ and 4+, respectively. Because the Mg–O bond is only 68% ionic, however, the actual charge is (0.68)(2+) = 1.36; that is, on the average, 1.36 of the two valence electrons of magnesium are transferred to the oxygen anions, while 0.64 electrons remain in covalent bonds nearer magnesium. (Since fractional electrons have no meaning in reality, this may be interpreted as a time average: Each electron spends 68% of its time near the anion and 32% of its time in the "covalent" internuclear region.) We have seen above that, by the same procedure, Si has a charge of 1.80. To determine the charge on oxygen, recognize that each one receives 68% of $\frac{1}{3}$ of an electron from each magnesium (of which there are three), and 45% of 1 electron from the single silicon to which it is bonded. Since the oxygen atom was neutral before the addition of these electrons, its charge is $-[(0.68)(\frac{1}{3})(3) + (0.45)(1)(1)] = -1.13$. A check of the formula reveals that this charge distribution results in 4.52 positive charges and 4.52 negative charges, which balances as required.

The example just given illustrates the *electroneutrality principle,* which states that the charges on atoms are less (and often much less) than their formal oxidation numbers. Although we do not want to take the particular numerical charges just calculated too seriously—after all, the relationship between electronegativity and ionicity is approximate, and we failed to take account of such factors as variation in Mg–O and Si–O bond distances in forsterite—they certainly support the electroneutrality principle in a general way. In recent years, it has become possible to estimate atomic charges experimentally by x-ray diffraction methods (Chapter 10), and Table 11-3 shows some typical results. The three or four methods that have been

Table 11-3. Atomic Charges Determined by Experimental Means

Atom	Charge[a]	Mineral	Reference
Na	+0.51	Kernite	Cooper et al. (1973) *Am. Min.* 58, 21–31
B	+0.53		
O	−0.45		
H	+0.22		
Mg	+1.82	Orthoenstatite	Sasaki et al. (1982) *Z. Krist.* 158, 279–97
Si	+2.28		
O	−1.37		
Fe	+1.12	Orthoferrosilite	Sasaki, et al. (1982) *Z. Krist.* 158, 279–97
Si	+2.19		
O	−1.10		
Mg	+1.75	Forsterite	Fujino, et al. (1981) *Acta Cryst.* B37, 513–18
Si	+2.11		
O	−1.40		
Fe	+1.20	Fayalite	Fujino, et al. (1981) *Acta Cryst.* B37, 513–18
Si	+2.43		
O	−1.19		
Mn	+1.35	Tephroite	Fujino, et al. (1981) *Acta Cryst.* B37, 513–18
Si	+2.28		
O	−1.23		
Al	+1.43	Corundum	Iishi (1978) *Phys. Chem. Minerals* 3, 1–10
O	−0.96		

[a]Charges shown are averages for all atoms of a given element in a mineral

used to obtain atomic charges give somewhat different answers. The ranges shown in the table reflect not only results of these different methods, but also the effects of numerous factors that depend on, and vary with, the atomic arrangement of each mineral studied. Among these factors are variations in numbers of anions surrounding each cation (and vice versa), variations in individual cation–anion bond lengths, and the effects of next-nearest-neighbor atoms.

This example has demonstrated that the bonding in forsterite is neither purely ionic nor purely covalent. It can be described best by some combination of the two models, and this is true for silicates in general. While there is little point in becoming concerned with trying to designate some model as "best" for its own sake, calculations of mineral properties and behavior based on chemical bonding must make use of some mathematical procedures that are based on the sort of bonding model chosen. Such calculations form the basis for some of the more interesting approaches to mineralogical problems today.

BONDING IN MINERALS: SOME APPLICATIONS

The quantitative application of bonding models to mineralogical problems usually requires calculations that are rather tedious and sometimes very complicated. The advent of modern high-speed digital computers has made such calculations practical, however, and research spanning a broad range of problems and models has been carried out in recent years. Detailing the instructions for conducting such studies is not within the scope of this text, but a brief and necessarily somewhat qualitative introduction to three bonding models will help acquaint the student with the potential of this significant research frontier.

Ionic Bonding via the Electrostatic Model

If two ions are considered to be fully oxidized point charges separated by some distance r, then the electrostatic potential energy is

$$u = (z_1 e)(z_2 e)/r = z_1 z_2 e^2/r,$$

where z_1 and z_2 are the oxidation numbers of the ions, and e is the charge of an electron. Because a mineral is an array of atoms, the electrostatic potential energy between any two ions is perturbed by the presence of all the other ions. To take account of this, we introduce a numerical factor A, called the Madelung constant, such that

$$u = Az_1 z_2 e^2/r.$$

The Madelung constant is a function of the particular kind of arrangement of ions, and values for many arrangements have been calculated. If z_1 and z_2 are of opposite sign, so that the ions attract, then bringing them closer together (decreasing r) lowers the potential energy (that is, makes it more negative), which is a favorable situation.

Obviously, there is some limit beyond which repulsions between the electrons in the outer shells of the ions prohibit closer approach. The equation describing electrostatic interactions must, then, contain a repulsive term that becomes significant as we approach the equilibrium interionic separation, which is the value of r actually observed in nature for ions of a given mineral. Thus modified, we have

$$u = Az_1z_2e^2/r + Be^{-(r/\rho)}$$

Here B and ρ are constants, and, like A, they depend upon the particular substance under investigation.

This expression provides the electrostatic potential energy between two ions that are separated by a distance r. If we wish to know the total electrostatic potential energy of a particular ion located at a particular position in a crystal (the so-called *site energy*), then we must add the electrostatic potential energies calculated between that ion and every other ion in the crystal. This is, of course, impractical, but, inasmuch as the interionic forces are not significant over very large distances, summation over the entire crystal is unnecessary. Even so, the computational task is gargantuan, and more efficient methods than direct summation for point charges have been developed.

As mineralogists have sought computational methods of understanding the structural details of minerals, electrostatic energy calculations have provided mixed results. They have furnished insight into the locations of hydrogen atoms in humites, micas, and other hydroxyl-containing minerals, for example, but have failed to adequately predict the crystal structures of silicate minerals in general. Presumably, this is because the former task addresses only the "ionic" aspects of the crystal structure, whereas the latter involves considerable covalent character not accounted for in the electrostatic model. On the other hand, electrostatic energy calculations have done rather well in some areas that might intuitively seem to require more than the ionic model—predictions concerning the stabilities of mica polytypes, for instance (see Chapter 3).

Crystal-Field Theory: A Bridge between the Ionic and Covalent Models for Transition Metals

The bonding electrons of atoms through atomic number 20 (calcium), in their ground states, are located in s and p orbitals. Figure 11-2 shows that s orbitals are spherical, and that the "dumbbell" shaped lobes of the three p orbitals have identical symmetry properties and differ only in direction. Consequently, cations with only s and p bonding orbitals exhibit no preference for one site over another on the basis of the symmetry of the site. That is, such cations do not preferentially occupy sites surrounded by four anions at the corners of a tetrahedron, or by six anions arranged at the corners of an octahedron, or any other particular kind of site, for reasons originating in the symmetry of either the site or the bonding orbitals. The

Figure 11-2. The distribution functions (i.e., the squares of the angular portions of the atomic wave functions) for hydrogenlike atoms. In all cases, the x axis slants out of the page to the left, the positive y axis points to the right, and the positive z axis points vertically upward in the plane of the page.

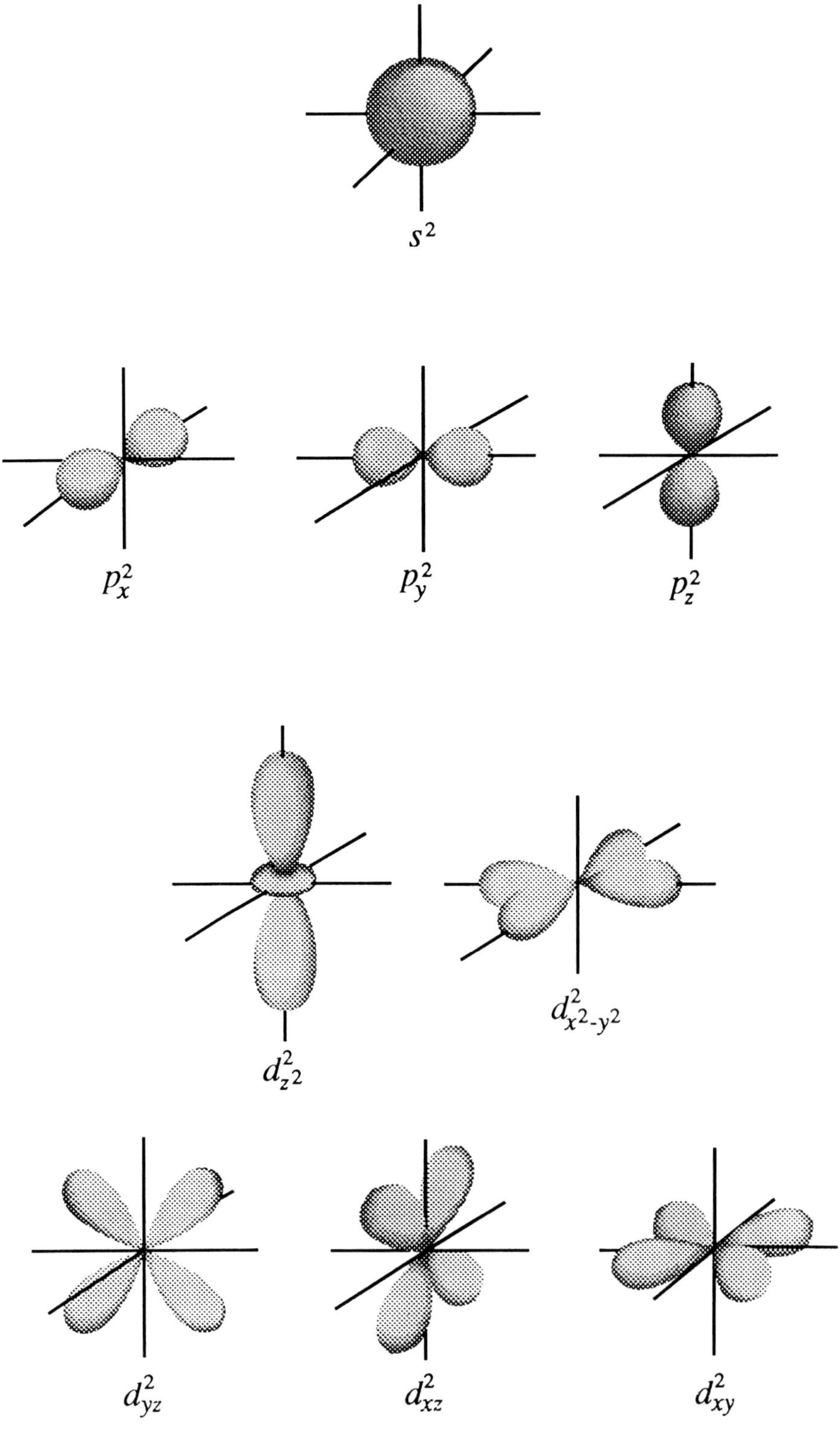
s^2
p_x^2
p_y^2
p_z^2
$d_{z^2}^2$
$d_{x^2-y^2}^2$
d_{yz}^2
d_{xz}^2
d_{xy}^2

five *d* orbitals, on the other hand, consist of two groups with different symmetry properties. The d_{z^2} and $d_{x^2-y^2}$ orbitals are directed along orthogonal axes and belong to one group, while the d_{xy}, d_{xz}, and d_{yz} orbitals are each directed midway between, and in the plane of, two axes. The fact that they have different symmetry properties means that they behave differently under the influence of electrostatic fields of different symmetries. Using some shorthand notation from mathematical group theory, which provides the theoretical basis for crystal field theory, we refer to the d_{z^2} and $d_{x^2-y^2}$ orbitals as e_g orbitals, and to the other three as t_{2g} orbitals.

Let us imagine a single atom of, say, iron. Because iron is a transition metal, some of its electrons occupy *d* orbitals. If the atom is subject to a spherical electrical field, then all five *d* orbitals are said to be *degenerate*—that is, they have the same energy (see Figure 11-3). A spherical field arising from the atomic arrangement of a mineral is obviously unlikely, so imagine now that six O^{2-} anions are brought in along orthogonal (Cartesian) axes centered on the iron atom. In crystal field theory the anions are considered to be point negative charges, and they thus exert a repulsive force on the electrons in the *d* orbitals. The anions approach the e_g orbitals more closely than they approach the t_{2g} group, and so the energy of the former set is relatively raised, while that of the latter is relatively lowered. The amount of energy change follows a simple center of gravity rule, so that the two e_g orbitals are raised $\frac{3}{5}\Delta$ and the three t_{2g} orbitals are lowered $\frac{2}{5}\Delta$, where Δ is the total *octahedral crystal field splitting*. Energy level diagrams for other arrangements of anions look different from Figure 11-3 in detail, but they are derived in the same way.

Recall that Hund's rule requires that orbitals of the same energy be singly occupied before electrons are paired. In the current example, this means that three of the six *d* electrons of Fe^{2+} will occupy the t_{2g} orbitals—one each in the d_{xy}, d_{xz}, and d_{yz} orbitals—before the possibility of electron pairing arises. The placement of the fourth electron depends on whether the energy required to pair two electrons is

Figure 11-3. A qualitative energy-level diagram for a transition-metal ion (Fe^{2+} in this case) in a fieldless space, in a spherical potential field, and in an octahedral crystal field. The anions generating the octahedral field lie along the Cartesian coordinate axes of Figure 11-2 and approach the e_g orbitals more closely than the t_{2g} orbitals, thereby raising the energy of the former and lowering that of the latter.

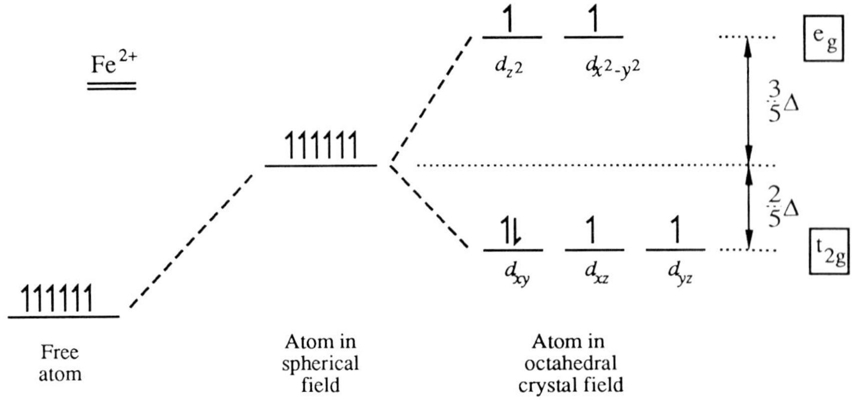

greater or lesser than Δ. If greater, then it will be energetically favorable to place the fourth and fifth electrons in d_{z^2} and $d_{x^2-y^2}$ orbitals, and then to pair the sixth one with an electron already in one of the t_{2g} orbitals. If lesser, then the fourth, fifth, and sixth electrons will all be paired with other electrons in the t_{2g} orbitals, leaving the e_g set empty. The factor that determines which of these two alternatives will occur for a given transition element is the nature of the anions (and perhaps external variables like pressure and temperature). For most anions of mineralogical interest (e.g., O^{2-}, F^-, and Cl^-), Δ is generally less than the pairing energy. Thus, the fourth and fifth electrons of Fe^{2+} occupy the relatively high-energy e_g orbitals (Fig. 11-3), and Fe^{2+} coordinated by six oxygens is said to possess a "high spin electronic configuration" (as opposed to the "low spin configuration" that obtains if Δ is greater than the pairing energy).

The Fe^{2+} atom in this example thus possesses four electrons in t_{2g} orbitals, each electron having $\frac{2}{5}\Delta$ less energy than it would have in a spherical field of equivalent strength, and two electrons in e_g orbitals, each electron having $\frac{3}{5}\Delta$ greater energy than it would have in the spherical field. The total crystal field stabilization energy (CFSE) is

$$(4)(\tfrac{2}{5}\Delta) - (2)(\tfrac{3}{5}\Delta) = \tfrac{2}{5}\Delta.$$

That is, Fe^{2+} is stabilized (relative to Fe^{2+} in a spherical field) by $\frac{2}{5}\Delta$ when placed in an octahedral crystal field. Moreover, Δ is an amount of energy that can be measured experimentally by absorption spectroscopy.

Of course, geometrically ideal coordination polyhedra are comparatively rare in minerals. As an example of distorted coordination, suppose that the octahedron is compressed along the z axis. It is apparent that the d_{z^2} orbital will experience a closer approach by anions than will $d_{x^2-y^2}$ orbital (Figure 11-2), and this will raise the energy of the former and lower that of the latter relative to the energies they had in the undistorted octahedral field (Figure 11-4). The energy difference α is smaller than Δ. In addition, the d_{xz} and d_{yz} orbitals will be raised relative to the d_{xy} orbital, resulting in a separation of δ between the two groups. With this configuration of anions, the CFSE for Fe^{2+} is

$$\{(4)(\tfrac{2}{5}\Delta) + [(2)(\tfrac{2}{3}\delta) - (2)(\tfrac{1}{3}\delta)]\} - \{(2)(\tfrac{3}{5}\Delta) - [(1)(\tfrac{1}{2}\alpha) - (1)(\tfrac{1}{2}\alpha)]\} = \tfrac{2}{5}\Delta + \tfrac{2}{3}\delta.$$

Inasmuch as $(\frac{2}{5}\Delta + \frac{2}{3}\delta) > \frac{2}{5}\Delta$, Fe^{2+} will preferentially occupy an octahedron that is compressionally distorted along z, given the choice between that and one of ideal geometry. In fact, most transition-metal atoms are stabilized in distorted octahedra relative to ideal ones, regardless of the nature of the distortion. (It is simple enough to discover which are not, by realizing that they must be those with the same number of electrons in every orbital. The center-of-gravity rule then requires that the CFSE for distorted and undistorted sites be equal.)

Because the anions of minerals provide distorted polyhedra for cations more commonly than they provide regular polyhedra, the qualitative application of crystal-field theory is useful for understanding why transition-metal cations preferentially occupy certain atomic positions. There are obvious limitations to the model, however. First, it applies only to transition-metal cations. Second, while it may provide a rationalization for assuming that a given transition-metal atom occupies a

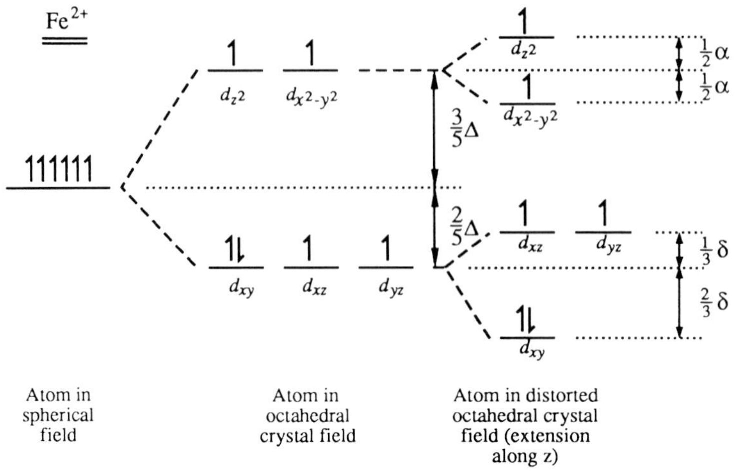

Figure 11-4. When a Fe^{2+} ion is subjected to an extensionally distorted octahedral crystal field, additional crystal-field splitting occurs. The energy of the d_{z^2} orbital is raised relative to that of the $d_{x^2-y^2}$ orbital because the anions along the z axis approach the former more closely than those along the x and y axes approach the latter. Similarly, the d_{xz} and d_{yz} orbitals are approached more closely than is the d_{xy} orbital, so additional splitting occurs in the t_{2g} orbitals.

particular site, it fails to take into account other important, and sometimes dominant, factors—such as atomic size. Third, quantitative application requires that crystal field splitting parameters be available, and this is generally not the case for any but common aqueous cations in geometrically ideal coordination. Despite these limitations, the theory has proved useful in rationalizing, in particular, Mg–Fe ordering in many ferromagnesian silicates (see Chapters 4 and 7, for instance).

Molecular Orbital Theory

An electron can be described by its spin and its position in space, and the theory of quantum mechanics requires that this description be contained in the electronic wavefunction Φ. (We use Φ here instead of the often-used ψ or ϕ in order to avoid confusion with either *molecular* orbitals or the standard notation for one of the spherical polar coordinates used in the following.) The wave function is a solution of the Schrödinger equation,

$$\left[\frac{-h^2}{4m}\left(\frac{\partial^2\Phi}{\partial x^2} + \frac{\partial^2\Phi}{\partial y^2} + \frac{\partial^2\Phi}{\partial z^2}\right) + V(x,y,z)\right]\Phi = E\Phi,$$

where h is Planck's constant, m is the mass of the electron, V is the potential energy, E is total energy, and (x,y,z) are positional coordinates. The expression $[-h^2/4m(\partial^2/\partial x^2 + \partial^2/\partial y^2 + \partial^2/\partial z^2) + V(x,y,z)]$ is called an *operator,* because it mathematically operates on some function, in this case the wave function Φ. This operator is called the *Hamiltonian operator* (or, simply, the *Hamiltonian*), and is abbreviated $\mathcal{H}$. Thus, the same equation (with a deceptively simpler appearance!) is written,

$$\mathcal{H}\Phi = E\Phi.$$

The solution of the Schrödinger equation depends upon the nature of the potential energy term $V(x,y,z)$, and a complicated potential energy can make an exact solution impossible to obtain.

Among the few systems that admit of an exact solution to the Schrödinger equation is the hydrogenlike atom—an atom consisting of a nucleus of charge $+Ze$, and one electron. (Note that the nucleus need not be a single proton, as in the hydrogen atom itself.) In Table 11-4 the solutions (Φ) for the hydrogenlike atom are presented. $R(r)$ is the radial part of the wavefunction (that is, it describes the orbital as a function of distance from the nucleus), and $Y(\theta,\phi)$ is the angular part, where r, θ, and ϕ are spherical polar coordinates centered on the nucleus. The product $R(r)Y(\theta,\phi)$ is Φ. The function Φ^2 is the *distribution function*, which is the probability of finding the electron in any particular volume of space. Unlike R (and therefore Φ), Y does not depend on the principal quantum number n, and so Y^2 gives the shape of the volume of space in which the electron is likely to be for any electron shell. The orbital drawings of Figure 11-2 represent the squares of the angular portions of the wave functions (Y^2).

The principal importance of the exact solutions for the hydrogenlike atom is that they are approximate solutions for the many-electron atom, and we can picture the angular distribution functions of Figure 11-2 as being similar to those for any atom. (Radial distribution functions would, of course, vary with the principal quantum number). This approximation forms the basis for molecular-orbital theory.

As two atoms are brought together to form a diatomic molecule, the electrons interact and the atomic orbitals are no longer adequate descriptions. Instead, overlapping atomic orbitals merge to form *molecular orbitals* (MOs), which are usually written as linear combinations of the interacting atomic orbitals (the so-called MO-LCAO approximation). Calculations of the energies of MOs made up only of inner-shell electrons have shown that these electrons can be generally considered as remaining in nonbonding (i.e., noninteracting) AOs with very little error, so MOs are often considered to be made up of valence-shell AOs only.

The degree of complexity of molecular-orbital calculations varies greatly, depending upon the level of approximation used. In general, a molecular orbital, ψ, is written as

$$\psi = \sum_{i=1}^{n} c_i\Phi_i,$$

where there are n atoms in the molecule, the Φ_i are the atomic orbitals, and the c_i are coefficients; the MOs are assumed to be solutions of the molecular Schrödinger equation,

$$\mathcal{H}\psi = E\psi.$$

Table 11-4. Wavefunctions for the hydrogenlike atom

Orbital	$R(r)$	$Y(\theta,\phi)$
$1s$	$2\left(\frac{Z}{a_0}\right)^{3/2} e^{-Zr/a_0}$	$\frac{1}{\sqrt{4\pi}}$
$2s$	$\left(\frac{Z}{2a_0}\right)^{3/2}\left(2-\frac{Zr}{a_0}\right)e^{-Zr/2a_0}$	$\frac{1}{\sqrt{4\pi}}$
$2p_z$	$\frac{1}{2\sqrt{6}}\left(\frac{Z}{a_0}\right)^{3/2}\frac{Zr}{a_0}e^{-Zr/2a_0}$	$\sqrt{\frac{3}{4\pi}}\cos\theta$
$2p_x$		$\sqrt{\frac{3}{4\pi}}\sin\theta\cos\phi$
$2p_y$		$\sqrt{\frac{3}{4\pi}}\sin\theta\sin\phi$
$3s$	$\frac{2}{81\sqrt{3}}\left(\frac{Z}{a_0}\right)^{3/2}\left[27-\frac{18Zr}{a_0}+2\left(\frac{Zr}{a_0}\right)^2\right]e^{-\frac{Zr}{3a_0}}$	$\frac{1}{\sqrt{4\pi}}$
$3p_z$	$\frac{1}{81\sqrt{3}}\left(\frac{2Z}{a_0}\right)^{3/2}\left[\frac{6Zr}{a_0}-\left(\frac{Zr}{a_0}\right)^2\right]e^{-\frac{Zr}{3a_0}}$	$\sqrt{\frac{3}{4\pi}}\cos\theta$
$3p_x$		$\sqrt{\frac{3}{4\pi}}\sin\theta\cos\phi$
$3p_y$		$\sqrt{\frac{3}{4\pi}}\sin\theta\sin\phi$
$3d_{z^2}$	$\frac{4}{81\sqrt{30}}\left(\frac{Z}{a_0}\right)^{3/2}\left(\frac{Zr}{a_0}\right)^2 e^{-\frac{Zr}{3a_0}}$	$\sqrt{\frac{5}{16\pi}}(3\cos^2\theta-1)$
$3d_{x^2-y^2}$		$\sqrt{\frac{15}{16\pi}}\sin^2\theta\cos2\phi$
$3d_{xy}$		$\sqrt{\frac{15}{16\pi}}\sin^2\theta\sin2\phi$
$3d_{xz}$		$\sqrt{\frac{15}{4\pi}}\sin\theta\cos\theta\cos\phi$
$3d_{yz}$		$\sqrt{\frac{15}{4\pi}}\sin\theta\cos\theta\sin\phi$

After C. R. Gatz (1971). *Introduction to Quantum Chemistry.* Merrill, Columbus, Ohio.

Substituting for ψ, we get

$$\mathcal{H}\left(\sum_{i=1}^{n} c_i\Phi_i\right) = E\left(\sum_{i=1}^{n} c_i\Phi_i\right),$$

whence

$$\sum_{i=1}^{n} c_i\mathcal{H}\Phi_i - \sum_{i=1}^{n} c_iE\Phi_i = 0,$$

and

$$\sum_{i=1}^{n} c_i(\mathcal{H} - E)\Phi_i = 0.$$

The solutions are found by multiplying this equation by one of the AOs, say Φ_j, and integrating over all spatial coordinates:

$$\sum_{i=1}^{n} c_i \int \Phi_j(\mathcal{H} - E)\Phi_i \, dV = 0.$$

Here a space-saving notation is introduced whereby

$$\int \Phi_j \mathcal{H} \Phi_i \, dV = H_{ji},$$

and

$$\int \Phi_j E \Phi_i \, dV = E \int \Phi_j \Phi_i \, dV = ES_{ji}.$$

Thus,

$$\sum_{i=1}^{n} c_i(H_{ji} - ES_{ji}) = 0.$$

Because there is one such equation for each AO Φ_j, in the set of n AOs used to construct ψ, we have a set of n homogeneous linear equations:

$$\begin{array}{l}
c_1(H_{11} - ES_{11}) + c_2(H_{12} - ES_{12}) + c_3(H_{13} - ES_{13}) + \ldots + c_i(H_{1i} - ES_{1i}) = 0 \\
c_1(H_{21} - ES_{21}) + c_2(H_{22} - ES_{22}) + c_3(H_{23} - ES_{23}) + \ldots + c_i(H_{2i} - ES_{2i}) = 0 \\
c_1(H_{31} - ES_{31}) + c_2(H_{32} - ES_{32}) + c_3(H_{33} - ES_{33}) + \ldots + c_i(H_{3i} - ES_{3i}) = 0 \\
\quad \vdots \qquad\qquad\qquad \vdots \qquad\qquad\qquad \vdots \qquad\qquad\qquad\qquad \vdots \\
c_1(H_{j1} - ES_{j1}) + c_2(H_{j2} - ES_{j2}) + c_3H_{j3} - ES_{j3}) + \ldots + c_i(H_{ji} - ES_{ji}) = 0.
\end{array}$$

Obviously, $c_i = 0$ for all i satisfies this set of equations, but this solution is trivial, because then the wave function ψ is zero. Nontrivial solutions may be found by setting the so-called *secular determinant* equal to zero:

$$\begin{vmatrix}
H_{11} - ES_{11} & H_{12} - ES_{12} & H_{13} - ES_{13} & \cdots & H_{1i} - ES_{1i} \\
H_{21} - ES_{21} & H_{22} - ES_{22} & H_{23} - ES_{23} & \cdots & H_{2i} - ES_{2i} \\
H_{31} - ES_{31} & H_{32} - ES_{32} & H_{33} - ES_{33} & \cdots & H_{3i} - ES_{3i} \\
\vdots & \vdots & \vdots & & \vdots \\
H_{j1} - ES_{j1} & H_{j2} - ES_{j2} & H_{j3} - ES_{j3} & \cdots & H_{ji} - ES_{ji}
\end{vmatrix} = 0.$$

At this point the various approximate approaches to the problem diverge. An approximation of some sort is necessary, because the Schrödinger equation cannot be solved exactly for any but the very simplest atomic systems, but some methods are more approximate than others. The Hückel molecular-orbital (HMO) method makes no attempt to specify the Hamiltonian $\mathcal{H}$ (beyond assuming it to be a one-electron operator, so that electron–electron interactions are ignored), but instead fits empirically determined parameters to the integrals H_{ji} and S_{ji}. Moreover, S_{ji} (called the *overlap integral* because it is a measure of the amount of overlap between

AOs Φ_j and Φ_i) is assumed to be 0 for $j \neq i$ and 1 for $j = i$, and the H_{ji} are taken to be 0 for atoms that are not adjacent. In the extended Hückel (EHMO) treatment the overlap integrals are calculated from AO wavefunctions.

At the other end of the spectrum are the ab initio methods. Ab initio is Latin for "from the beginning," and these methods calculate all necessary quantities from first principles. They employ a full and explicit Hamiltonian and evaluate all integrals. In between the ab initio and the HMO methods are several other techniques that resort to various intermediate levels of approximation. Among these are methods called zero differential overlap (ZDO), complete neglect of differential overlap (CNDO), and neglect of diatomic differential overlap (NDDO), plus some others. The names indicate the ways in which overlap integrals are handled.

Although it may seem that ab initio calculations should always be used because they are the most exact, this is not always possible. The larger the number of atoms that constitute a molecule, the larger and more complicated are the molecular orbitals. Even with the most modern computers, calculations on molecules that are moderately large take too long and require too much computer memory to be practical. Because a mineral consists of a three dimensionally continuous array of atoms, with no cluster of atoms that can be isolated as a "molecule" of the mineral, an immediate difficulty is encountered. A "molecular orbital" for a mineral, or even a single unit cell, must be very complicated indeed! The practical solution is to consider polyatomic fragments modeled on similar parts of the crystal structure of a mineral of interest. If such fragments are judiciously chosen, results can be cautiously applied to the relevant parts of the whole atomic arrangement. Good examples of applications of molecular-orbital methods are found in Chapters 1 and 3.

12
Phase Diagrams

Throughout Part I of this text from Figure 1-1 on, phase diagrams have been presented to show mineral stability as a function of some combination of pressure, temperature, and composition. Sometimes, as in Figure 2-17, all three variables are used; more often we have seen pressure–temperature or temperature–composition phase diagrams. As phase diagrams are sometimes given only passing attention in undergraduate mineralogy courses, but constitute a tool of great importance in crystal chemistry, we shall consider them a bit further here. There is no pretense that this short chapter is an adequate substitute for lengthier and more detailed works on the subject; the object here is only to enhance the reader's working acquaintance with the sorts of phase diagrams used in this text.

At the outset it is important to understand that phase diagrams depict equilibrium behavior, and often the phenomena of interest are not equilibrium phenomena. Further on in this chapter we shall consider the influence of kinetic (nonequilibrium) effects on mineral behavior.

A phase is a homogeneous part of a chemical system that is mechanically separable from the rest of the system—at least conceptually. The quartz (biotite, and so forth) in a granite could be mechanically separated from the rest of the rock, and is thus a phase. (It is true that a great deal of patience and uncommon dexterity might be required, but the task is conceptually possible!) The undissolved vapor in a magma is physically distinct from the liquid, and is therefore also a phase in that system; however, the H_2O molecules and CO_2 molecules in the vapor are not physically distinct components of the vapor because the two gases are miscible, so they are not separate phases but are constituents of a single vapor phase. The components of a phase are the chemical entities that constitute it. For example, the potassium feldspar $KAlSi_3O_8$ can be thought of as consisting of three components: K_2O, Al_2O_3, and SiO_2 combined in the proportions 0.5:0.5:3.

A phase diagram is a graphical way of showing which phases in a system are stable under various conditions. The types of phase diagrams most commonly used in mineralogy and petrology show stability fields as a function of pressure and temperature (*P–T* diagrams), or of temperature and composition (*T–X* diagrams); less frequently, phase diagrams display stability as a function of pressure and composition (*P–X* diagrams) or of all three variables (*P–T–X* diagrams). Examples of these four types of phase diagrams are Figures 1-1, 2-2, 2-16(b), and 2-17, respectively. Although such diagrams can be used with virtually no insight into their origins, some understanding of how they come about is useful.

PHASE STABILITY—THE EXPERIMENTAL APPROACH

The strategy in the experimental determination of phase stability is to place a specimen in some sort of appropriate apparatus (described below); raise it to a desired temperature, or pressure, or both; allow enough time for some reaction to take place; and then examine the sample to ascertain what changes, if any, occurred during the experimental run. By carrying out such experiments under a sufficient variety of conditions, the locations of the boundaries separating regions of stability of different minerals or combinations of minerals in P–T, T–X, P–X, or P–T–X space can be estimated. The nature of the specimen used depends on the problem being investigated and on the type of apparatus used. It may be a natural or synthetic mineral, or an assemblage of minerals, or some chemical constituents of minerals such as synthetic oxides or carbonates; any of these may or may not be accompanied by water, which is a vapor phase at high temperature, and is sometimes chemically essential and sometimes simply acts as a catalyst. We shall consider five representative kinds of experimental apparatus very briefly: the hydrothermal bomb, the piston–cylinder apparatus, the tetrahedral (or cubic) press, the diamond-anvil cell, and the single-crystal heater.

Hydrothermal Pressure Vessels

Equipment for raising the temperature of a sample while it is under gas (usually water or argon) pressure of up to 10 kbar or so falls into two general categories: externally heated pressure vessels and internally heated pressure vessels (often called *hydrothermal bombs*). Externally heated vessels are the smaller, less expensive, and generally less dangerous kinds, and they can be used for pressures up to 10 kbar at temperatures of 750°C—although they are not usually used at pressures that high for runs of long duration and high temperature. Most consist of cylinders 2.5–4 cm in diameter constructed of very strong alloy and possessing an axial bore about 0.6 cm in diameter. At one end of the cylinder is a threaded closure through which comes the high-pressure tubing that carries the water (or argon if the pressure will be too high for water to remain in the liquid state at room temperature while the vessel is being prepared for elevated temperature). The sample, consisting of whatever materials it is desired to react (including water), is sealed by welding in a nonreactive capsule made of a noble metal and placed at the other end of the vessel. A filler rod, the function of which is to reduce the risk of explosion by taking up space that would otherwise have to be filled with the pressure medium, is inserted behind the sample, and the vessel is pressurized. After sufficient time has elapsed to ensure that there are no leaks, the entire vessel is inserted up to the closure in a furnace that has been raised to the desired temperature. (In early designs, the entire vessel—including the closure—was placed in the furnace. Because with most externally heated bombs the closures are now outside the furnace, they are often called *cold-seal vessels,* although the closures are actually quite hot during the runs.)

As implied by the name, the furnace in internally heated pressure vessels is inside the vessel itself. This requires that the apparatus be larger, more complicated and expensive, and somewhat more dangerous than the externally heated models. The bulk of the apparatus is a fairly large, thick-walled, water-cooled cylinder containing a furnace in its central axial bore, and the samples are placed within the furnace. There is usually a closure on each end, through one of which is passed the tubing for the pressure medium, and through the other of which pass the electrical leads for heating the furnace. Because the vessel is water cooled, pressures of 10–12 kbar can be reached at temperatures of 1200 to about 1500°C.

Piston–Cylinder Apparatus

The piston–cylinder device is a solid-medium apparatus in which pressures of 60 kbar can routinely be reached, and 100 kbar is accessible with some designs. Simultaneously, temperatures in the neighborhood of 1500°C are attainable. A typical design consists essentially of the following elements: There are two opposed, hydraulically driven rams; the larger one loads the pressure vessel, while the smaller one drives a piston that puts pressure on the sample. The pressure vessel is a carbide cylinder surrounded by steel supporting rings, and it has an axial hole about 1.3 cm in diameter that contains the piston. Within that is a hollow cylinder of talc that surrounds another hollow cylinder of graphite that serves as the furnace. The sample consists of powdered minerals or synthetic chemical components, and may also include water. It is enclosed in a platinum capsule about 0.25 cm in diameter and the same length, welded if water is included, and this is placed inside the furnace and packed about with boron nitride powder and ground porcelain. Pressure is applied through the piston driven by the smaller ram, and temperature is raised by passing an electrical current through the graphite furnace.

Tetrahedral Press

This device, which has much is common with the piston–cylinder apparatus, consists of four pistonlike anvils that converge on a specimen from the corners of a tetrahedral frame. The anvils are hydraulically driven, and very high pressures can be generated. High temperature is obtained by passing an electrical current through a conducting graphite sample holder. Maximum pressures and temperatures depend on the construction and design of each specific instrument, and can be as high as 100 kbar and 2000°C.

The specimen, consisting of powdered minerals or synthetic chemicals, is packed into a graphite cylinder with graphite end caps. This is placed in a cylindrical hole in a pyrophyllite [$Al_4Si_4O_{10}(OH)_8$] tetrahedron. Metal tabs, one in contact with each end of the graphite cylinder, project from the tetrahedron to ensure an electrical pathway through the graphite for heating.

If the specimen is reactive with the graphite cylinder, it may be packed into an inner cylinder of boron nitride. It is also possible to use a noble-metal capsule (gold, silver, etc.) inside the graphite. Water may be included in the specimen holder, but caution is required because the system is not well sealed at low pressures, and the expansion of the fluid phase at high temperature may be explosive. The apparatus is better suited to mineral synthesis than to studies of phase stability.

A variation on the tetrahedral press is the cubic press. Here there are six anvils that approach the specimen along three mutually perpendicular directions. The specimen is prepared in the same way, but is embedded in a cube of pyrophyllite. Further advances in multianvil technology have resulted in devices capable of reaching well over 100 kbar.

Diamond-Anvil Cell

The diamond-anvil cell is designed primarily to accommodate small single crystals. It consists of two high-quality diamonds with culets opposed and separated by a metal gasket generally made of inconel or stainless steel. The gasket contains a small ($\sim$300 μm) hole in which there is a fluid—usually a mixture of ethanol and methanol that remains liquid at room temperature to about 100 kbar—and the crystal specimen. All of this is supported by a metal device that forces the diamonds toward one another either in a viselike or a "nutcracker"-like fashion. The crystal must not only have a considerably smaller diameter than the hole, but must also be quite thin so as not to be crushed when the culets ("anvils") are forced together. As long as the gasket does not rupture or become as thin as the specimen, and as long as the fluid medium does not solidify, the pressure is hydrostatic.

Diamond-anvil cells are relatively small, and can be placed inside an oven to provide external heating. Alternatively, the temperature can be raised with a resistance winding. One of the great advantages of the diamond-anvil cell also seriously limits the temperatures attainable, however. The cell, particularly the miniature varieties that can be mounted directly on x-ray diffractometers, can be used for in situ observation of crystallographic properties at high pressure; the diamonds are backed by beryllium plates, which are essentially transparent to x-rays, however, and the beryllium begins to soften by about 200°C. Thus pressure and temperature cannot be raised very high simultaneously. At ambient temperatures, though, pressures of 150 to 200 kbar can be reached in miniature cells of sturdy construction. Pressures up to 200 Mbar are attainable with more bulky cells that cannot be conveniently mounted on x-ray diffractometers. Heating can also be accomplished with a laser, although only a very small spot on the specimen attains the maximum temperature and temperature gradients are large.

With the miniature diamond-anvil cell, one can take precession photos of a single crystal, or collect x-ray intensity data for crystal- structure determination, at elevated pressures. If holes are drilled in the beryllium plates that back the diamonds, then the crystal in the cell can be observed by optical microscopy. Moreover, pressure can be increased without removing the crystal from an already pressurized cell, so that pressure-induced polymorphic transformations can be investigated in situ. Pressure is determined by measuring the frequency of a pressure-dependent laser-induced fluorescence in a small chip of ruby included with the crystal of interest in the pressure cell.

Single-Crystal Heater

Although the ability to determine crystal structures of minerals with high precision at simultaneously high pressure and high temperature would result in very significant advances in the geologic applications of crystal chemistry, it turns out that the experimental conditions required for either result in rather large errors in the estimates of atomic positions. At present the only alternative is to determine crystal structures at high pressures and at high temperatures independently and cautiously merge the results.

High temperatures in single-crystal work at atmospheric pressure can be attained with either a furnace or a flame heater. A furnace is no more than a wire resistance heater, and such devices can reach temperatures of around 900°C. A flame heater directs a jet of combustible gas onto a crystal; with appropriate gas mixtures, temperatures up to about 2000°C can be attained. The crystal is mounted on the wires of a thermocouple, and the temperature is measured directly. Heaters of either type can be used to heat crystals mounted on precession cameras for x-ray photographs, or on diffractometers for collection of intensity data.

PHASE STABILITY—THE THEORETICAL APPROACH

The discipline of thermodynamics provides a mathematical framework for the determination of mineral stability that is complementary to, rather than exclusive of, the techniques of experimental investigation described here. With accurate values of thermodynamic properties, themselves determined experimentally, boundaries of the stability fields for minerals and mineral assemblages can, in principle, be calculated, and even without those values the principles of thermodynamics impose some limitations on the qualitative geometries of phase diagrams. Our purpose here is to set forth some of the fundamental equations and definitions that bear on phase diagrams.

Fundamental Equations and Relationships

We begin with the internal energy of a system, which is the energy characteristic of the system and its particular state. (The system may be an individual mineral, an assemblage of minerals, a crystallizing magma, or any other volume of the universe that interests us; in general it will be an open system, in that chemical components can cross its boundaries. The state may be characterized, for example, by a particular pressure and temperature.) The internal energy is conventionally designated by either the letter E or the letter U, and is defined in its differential form for a closed system as

$$dE = dw + dq, \tag{12.1}$$

where dw and dq are differential changes in work and heat. For a reversible change, $dw = -P\,dV$ and $dq = T\,dS$; T, P, and V are temperature, pressure, and volume, respectively, and S is an extensive variable called entropy. Substituting we have

$$dE = T\,dS - P\,dV. \tag{12.2}$$

For an open system, which includes virtually all of interest among real geologic systems,

$$dE = T\,dS - P\,dV + \sum_{i=1}^{n} \mu_i dm_i. \tag{12.3}$$

The summation in this equation is over all n chemical components in the system; m_i is the number of moles of component i, and μ_i is the chemical potential of component i. The chemical potential of any component, say component j, is defined as the infinitesimally small change in internal energy of the system with an infinitesimally small change in concentration of that component, while entropy, volume, and concentrations of all other components are held constant:

$$\mu_j = \left(\frac{\partial E}{\partial m_j}\right)_{S,V,m_i}, \qquad j \neq i. \tag{12.4}$$

By some straightforward manipulations that we need not pursue here, dE can be shown to be an exact differential, leading to

$$E = TS - PV + \sum_{i=1}^{n} m_i \mu_i. \tag{12.5}$$

Having defined internal energy for a phase of some invariant composition in terms of TS and PV, there are three other useful relations that can also be expressed in terms of E, TS, and PV:

$$A = E - TS, \tag{12.6}$$

$$H = E + PV, \text{ and} \tag{12.7}$$

$$G = E - TS + PV. \tag{12.8}$$

A, H, and G are, respectively, the Helmholtz free energy, the enthalpy, and the Gibbs free energy. These three equations plus the one for internal energy constitute the four fundamental equations of thermodynamics. Differentiating the last three (dE has been found above) yields

$$dA = dE - T\,dS - S\,dT, \tag{12.9}$$

$$dH = dE + P\,dV + V\,dP, \tag{12.10}$$

$$dG = dE - T\,dS - S\,dT + P\,dV + V\,dP. \tag{12.11}$$

By substituting for dE from Equation (12.3) and simplifying we obtain

$$dA = -S\,dT - P\,dV + \sum_{i=1}^{n} \mu_i\,dm_i, \tag{12.12}$$

$$dH = T\,dS + V\,dP + \sum_{i=1}^{n} \mu_i\,dm_i, \tag{12.13}$$

$$dG = -S\,dT + V\,dP + \sum_{i=1}^{n} \mu_i\,dm_i. \tag{12.14}$$

The experimental systems and simple model systems used to determine phase equilibria are often closed systems, and the terms involving chemical potential are ignored when that is the case because $dm_i = 0$ for all i.

The Gibbs free energy, often referred to simply as the free energy, is particularly significant. When temperature and pressure are constant, the processes that move a system toward equilibrium always result in a decrease in the (Gibbs) free energy. Another way of stating this is that the phase (or combination of phases) exhibiting the lowest free energy at a given temperature and pressure is the stable phase (or combination of phases) under those conditions.

Free Energy Surfaces

Consider a closed system and examine how G changes with P or V when the other (V or P) is held constant:

$$\left(\frac{\partial G}{\partial P}\right)_T = V \tag{12.15}$$

$$\left(\frac{\partial G}{\partial T}\right)_P = -S. \tag{12.16}$$

V is always positive (inasmuch as a negative volume has no physical meaning); so G always increases with P. S is also a positive quantity, which makes $-S$ negative, so that G always decreases with increasing T. Furthermore,

$$\left(\frac{\partial^2 G}{\partial P^2}\right)_T = \left(\frac{\partial V}{\partial P}\right)_T \tag{12.17}$$

$$\left(\frac{\partial^2 G}{\partial T^2}\right)_T = \left(\frac{\partial S}{\partial T}\right)_P. \tag{12.18}$$

Because the change in volume with increasing pressure is always negative, the curve $(\partial G/\partial P)_T$ is concave downward. The same is true of the $(\partial G/\partial T)_P$ curve because entropy increases as temperature increases.

In schematic fashion, then, we can sketch a typical free energy surface in G–P–T space, as shown in Figure 12-1. Consider now some mineral that occurs in two polymorphs (dimorphs, technically). Each mineral has its own free energy surface in G–P–T space, and these surfaces intersect along the line common to both, as in Figure 12-2. The projection of this intersection onto the P–T plane constitutes the P–T phase diagram for the system. Along the line of intersection (that is, the phase boundary in P–T space) the values of G for each phase are equal at given values of pressure and temperature; in other words $G_{\text{phase1}} - G_{\text{phase2}} \equiv \Delta G = 0$.

In principle, then, it is possible to calculate the P–T phase diagram for a system by analytically solving for the pressures and temperatures at which $dG = 0$, assuming that one has the required values for volume and entropy. Consider a closed system so that Equation (12.14) reduces to

$$dG = -S\,dT + V\,dP.$$

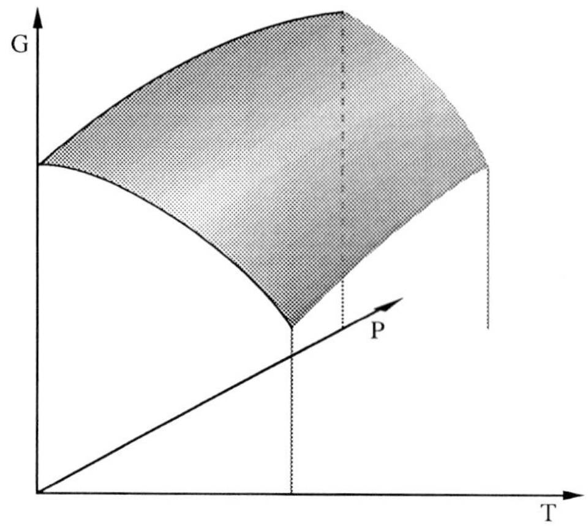

Figure 12-1. A Gibbs free-energy surface in G–P–T space, showing the decrease in G as T increases and as P decreases.

Given that under some conditions a reaction takes place in the system, both the product(s) and the reactant(s) will be present when the system is at equilibrium:

$$dG_{\text{products}} = -S_{\text{products}}\,dT + V_{\text{products}}\,dP,$$

$$dG_{\text{reactants}} = -S_{\text{reactants}}\,dT + V_{\text{reactants}}\,dP.$$

Subtracting, we obtain

$$dG_{\text{products}} - dG_{\text{reactants}} = -(S_{\text{products}} - S_{\text{reactants}})\,dT + (V_{\text{products}} - V_{\text{reactants}})\,dP.$$

Along the intersection of the two free-energy surfaces $G_{\text{products}} = G_{\text{reactants}}$ and dG can be made arbitrarily small, so that $dG_{\text{products}} - dG_{\text{reactants}} = 0$. Thus

$$0 = -(S_{\text{products}} - S_{\text{reactants}})\,dT + (V_{\text{products}} - V_{\text{reactants}})\,dP,$$

$$(S_{\text{products}} - S_{\text{reactants}})\,dT = (V_{\text{products}} - V_{\text{reactants}})\,dP, \tag{12.19}$$

$$\frac{dP}{dT} = \frac{S_{\text{products}} - S_{\text{reactants}}}{V_{\text{products}} - V_{\text{reactants}}},$$

$$\frac{dP}{dT} = \frac{\Delta S_{\text{R}}}{\Delta V_{\text{R}}}.$$

The subscript R stands for the word *reaction.* Equation (12.19) is the Clausius–Clapeyron equation, and in words, it states that the slope of the boundary on a P–T phase diagram between the region in which one phase or set of phases is stable and the region where the other is stable is the entropy change of reaction divided by the volume change of reaction (actually, molar entropy change and molar volume change). One way of constructing phase diagrams from thermochemical data, then, would be to know the molar entropies and molar volumes of the phases of interest at the temperatures and pressures of interest, and to know one point on the phase boundary in P–T space (inasmuch as the slope alone does not locate the line).

Calculation of a Phase Diagram Using Thermochemical Data

We need first to know that ΔV_R can (and almost always is) reasonably approximated as independent of both P and T *for solids* at pressures and temperatures that obtain in the earth's crust. This would obviously be a very bad approximation if it were used for reactions involving a fluid phase, but we shall restrict this brief treatment to reactions involving only solids. ΔS_R is considered independent of pressure, although it is strongly dependent on temperature.

If we wish to calculate the equilibrium pressures at several different temperatures for some reaction, and thus establish the phase boundary, we can begin again with Equation (12.14), this time omitting not only the last term (because we are dealing with a closed system) but also the $-S\,dT$ term (because we are calculating equilibrium pressures at a series of fixed temperatures; so $dT = 0$):

$$dG = V\,dP.$$

Integrating yields

$$\begin{aligned}\Delta G(T,P) &= \Delta G^0_R(T,1) + \int_1^P \Delta V_R\,dP \\ &= \Delta G^0_R(T,1) + \Delta V_R(P-1),\end{aligned} \tag{12.20}$$

where temperature is in kelvin, pressure is in bars (1 bar = 0.98692 atm), and $\Delta G^0_R(T,1)$ is a free-energy change of reaction for formation from the elements at the temperature of interest and pressure of 1 bar. Recognizing that $\Delta G(T,P)$ is zero at equilibrium and rearranging Equation (12.20) gives an equation for the equilibrium pressure at temperature T:

$$P = 1 - [\Delta G^0_R(T,1)/\Delta_R].$$

Values of $G^0_R(T,1)$ and molar volumes can be found in various tabulations of thermochemical data for many pure minerals.

Figure 12-2. Free-energy surfaces for two polymorphic phases, α and β, in G–P–T space. The projection of their intersection on the P–T plane is the P–T phase diagram.

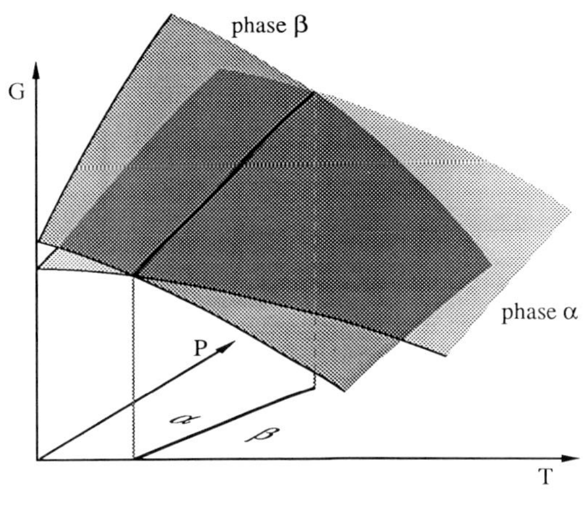

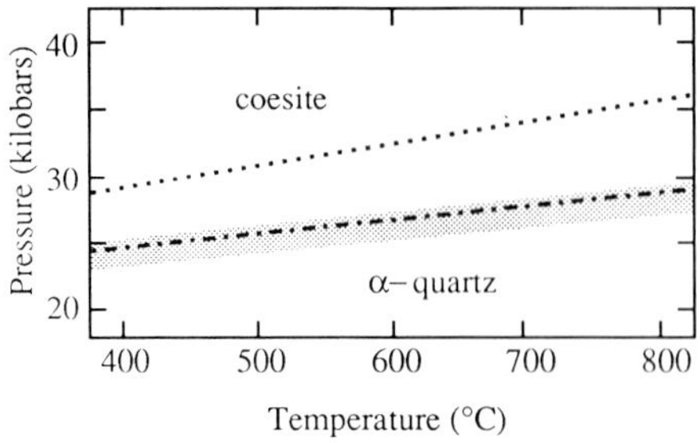

Figure 12-3. The quartz–coesite phase boundary from 400 to 800°C, calculated from thermochemical data not constrained to fit experimental data (dotted line), thermochemical data derived from experimental data (dash-dot line), and as experimentally determined (shaded band).

Consider the phase boundary between α-quartz and coesite shown in Figure 1-1. Appropriate data from one tabulation of thermochemical data [Robie, R. A., and D. R. Waldbaum (1968). "Thermodynamic Properties of Minerals and Related Substances at 298.15 K (25.0°C) and One Atmosphere (1.013 Bars) Pressure and at Higher Temperatures." United States Geological Survey Bulletin 1259, Washington, D.C.] are as follows:

Coesite

V	0.49333 cal/bar
$G^0(600\text{ K},1)$	$-190{,}106$ cal/gfw
$G^0(1000\text{ K},1)$	$-172{,}735$ cal/gfw

Quartz

V	0.54226 cal/bar
$G^0(600\text{ K},1)$	$-191{,}486$ cal/gfw
$G^0(1000\text{ K},1)$	$-174{,}494$ cal/gfw

It is not intuitively evident that calories per bar, the units of molar volume, really describe a volume, but they do, and those units are necessary for pressure to come out in bars. Calculation of the equilibrium pressures at 600 K (327°C) and 1000 K (727°C) yields 28.2 and 36.0 kbar, respectively. The line joining these points is the dotted line in Figure 12-3. The shaded band is the experimental boundary (its width indicating reasonable experimental error) for this polymorphic reaction, and it is clear that the agreement is not particularly good. The uncertainties in phase boundaries calculated from thermochemical data are typically rather large, and the experimental determinations are generally considered more accurate. Because of this there has been an effort to derive a usable set of free energies of formation from the experimental data. The calculations involved are somewhat more tedious than those just shown and we shall not consider them, but the dash-dot line in Figure 12-3 is based on the data from one such compilation [Helgeson, H. C., J. M. Delaney, W. H. Nesbitt, and D. K. Bird (1978). "Summary and Critique of the Thermodynamic Properties of Rock-Forming Minerals." *American Journal of Science* 278-A], and the agreement between calculation and experiment is obviously greatly improved.

Calculations of phase equilibria are not restricted to solid–solid reactions involving only one reactant and one product phase. Equilibria involving several phases, fluid phases, or solid solutions can also be calculated, although the computations are more complicated than in the simple example shown here, and some additional thermodynamic relationships are needed.

One might wonder why the calculations are done at all if the experiments yield better answers, and there are at least two good reasons. First, carrying out properly executed experiments is not a trivial exercise: Great care must be taken to determine the compositions of starting materials (whether they are "pure" synthetic compounds or natural minerals) and the compositions of products, and to confirm that the products derived in a reaction are actually stable minerals and not metastable products resulting because the stable reaction was kinetically hindered; equipment must be properly calibrated, which in itself is a difficult task; many experiments must be carried out in order to "zero in" on the phase boundary. Given these (and other) experimental difficulties, it is not possible for all reactions of interest to be investigated experimentally at all pressures and temperatures of interest; even though the uncertainties in calculated equilibria may be large, there may be no better answers for particular systems, and an approximation may serve better than nothing. Second, advances continue to be made in calorimetric instrumentation, thereby increasing the reliability of the thermochemical database and reducing the uncertainties in calculated equilibria.

The Phase Rule

Consider the schematic phase diagram for Al_2SiO_5 as shown in Figure 12-4(a). The smallest number of components with which the phases shown can be described is one: Al_2SiO_5. The number of phases in Figure 12-4(a) is three, and all are stable

Figure 12-4. (a) A schematic *P–T* phase diagram for the aluminum silicate polymorphs. (b) A hypothetical phase diagram involving a fourth phase stable at the same point as the other three polymorphs; this is not a possible geometry for the phase diagram.

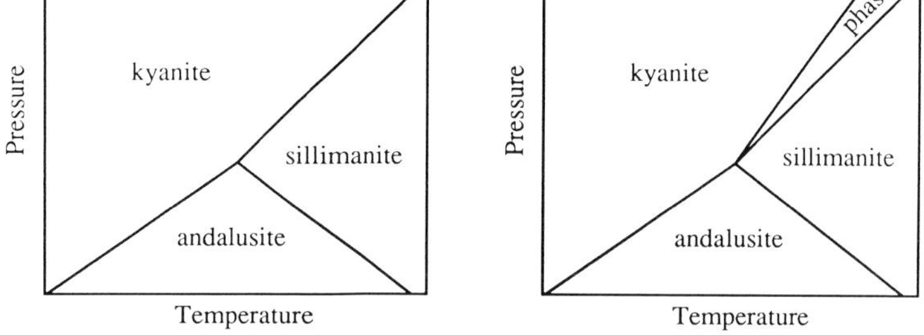

together at the pressure and temperature where the three stability fields meet. One might ask the question, "Is it possible for a fourth polymorph of Al_2SiO_5 to coexist in this system with the other three polymorphs, as shown in Figure 12-4(b)?" Figure 12-5(a) shows a single schematic boundary that separates the stability field of quartz plus kyanite plus grossular from that of anorthite; the balanced chemical reaction for ideal compositions is shown with abbreviated mineral names. This is a three-component system, because all phases may be described by some combination of SiO_2, Al_2O_3, and CaO, and three is the smallest number of components possible for the description of all of these four phases. One might ask, "Could there be a third assemblage as shown in Figure 12-5(b) that forms from quartz, kyanite, and grossular with increasing temperature and from anorthite with increasing pressure, such that it is stable along with the other four phases at some T and P?" To answer these questions we develop the *phase rule.*

Let us begin with a system, at equilibrium with temperature and pressure uniform throughout, that can be described in terms of C components that constitute p phases. All components need not be present in every phase, but all phases may be described by some combination of addition and subtraction of components. Therefore the composition of any phase may be specified by $C-1$ independent concentrations (possibly including 0 concentration for some components), the concentration of the last component being determined by difference. Thus the composition of the system of p phases may be specified by $p(C-1)$ concentrations, so that the system can be described in terms of at least that many variables. There are two additional variables, T and P, that must enter into the description of the system, so that the total number of variables is $p(C-1)+2 = pC-p+2$.

Now consider the number of equations available to describe the system. By definition, at equilibrium the chemical potential μ of every component in each phase

Figure 12-5. (a) A schematic P–T diagram for the equilibrium reaction quartz plus kyanite plus grossular ↔ anorthite. (b) A hypothetical phase diagram involving a third assemblage that intervenes between the first two assemblages at high T and P. This is a possible geometry for the phase diagram, but assemblage 3 must necessarily be a polymorphic form of anorthite.

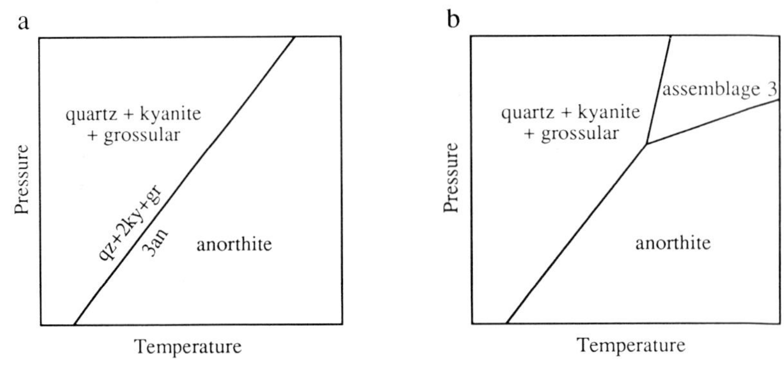

in which it occurs is the same, so that for components *i*, *j*, *k*, and so forth in phases α, β, γ, and so forth,

$$\mu_i^\alpha = \mu_i^\beta = \mu_i^\gamma = \mu_i^\delta \cdots,$$

$$\mu_j^\alpha = \mu_j^\beta = \mu_j^\gamma = \mu_j^\delta \cdots,$$

$$\mu_k^\alpha = \mu_k^\beta = \mu_k^\gamma = \mu_k^\delta \cdots,$$

$$\vdots$$

This leads to $p-1$ independent equations for each of C components, or $C(p-1) = Cp-C$ equations. Subtracting the number of equations from the number of unknowns yields the number of degrees of freedom F:

$$\begin{aligned} F &= pC - p + 2 - (Cp - C) \\ &= C - p + 2. \end{aligned}$$

In words, the number of degrees of freedom is the number of variables that can be specified arbitrarily (i.e., that are not fixed) and still maintain the specified conditions of the system.

Returning now to Figure 12-4(a), $C=1$, so that $F = 1 - p + 2 = 3 - p$. If three phases (e.g., kyanite, andalusite, and sillimanite) are in stable equilibrium together, as at the "triple point" where all three fields meet, then $F = 0$—the point is said to be an *invariant point.* This means that we can specify nothing arbitrarily and still maintain the system in the designated state (that is, all three phases coexisting in equilibrium). In other words, temperature and pressure are fixed for that condition of the system, and if either is changed all three phases cannot remain in equilibrium. The addition of a fourth stable phase would result in $F = -1$, which is physically meaningless, and so Figure 12-4(b) is an impossible phase diagram. This is not to say that there could not be a fourth polymorph of Al_2SiO_5, but only that if there were it could not coexist with all three other polymorphs in equilibrium.

Figure 12-5(a) represents a three-component system, so $F = 3 - p + 2 = 5 - p$. Along the phase boundary there are four phases stable: quartz, kyanite, grossular, and anorthite. To the right of it only anorthite is stable, and to the left only the assemblage quartz plus kyanite plus grossular is stable. On the phase boundary, therefore, $F = 5 - 4 = 1$. "One degree of freedom" means that one variable—say, either T or P—may be changed arbitrarily (within limits) while still maintaining the equilibrium qz+2ky+gr=3an—for which reason the boundary is said to be a *univariant line.* Of course the arbitrary change in either T or P specifies the change in the other required to maintain that equilibrium. Within the quartz-plus-kyanite-plus-grossular stability field, $F = 3 - 3 + 2 = 2$; so both T and P may be changed independently and the system still remain in that phase field—it is a *divariant field.*

Considering then Figure 12-5b, we see that at the triple point, $F = 3 - p + 2 = 5 - p$; because $F = 0$ at the triple point (i.e., no variable may be altered and remain at the triple point—it is an invariant point), $p = 5$. There are already four phases in the system (quartz, kyanite, grossular, and anorthite), so assemblage 3 could consist of at most a single phase; it would apparently have to be a polymorph of anorthite. Given this constraint, though, Figure 12-5(b) is a geometrically possible phase diagram. (Note that it is entirely contrived for the purpose of illustration, however, and is not intended to represent reality.)

Binary *T–X* Phase Diagrams

Binary (or two-component) temperature–composition, or *T–X*, diagrams are widely used in mineralogy, and appear often throughout this text. These represent the way in which a system changes with temperature as composition is varied between two extremes. A single *T–X* diagram is valid for only the single pressure at which it was determined, thus removing one variable from the phase rule, so that it becomes $F = C - p + 1$.

Figure 12-6(a) shows the form of a binary *T–X* diagram for two minerals that exhibit no solid solution behavior and have no stable intermediate compositions. It could represent, schematically, the diagram for diopside–albite, or anorthite–titanite, or SiO_2–fluorite, but not albite–anorthite (in which there is solid solution) or SiO_2–forsterite (in which there is an intermediate composition, enstatite). The *liquidus* is that curve above which the system is entirely liquid for a given bulk composition. The liquidus is a univariant line because a liquid and one crystalline phase are in stable coexistence on it: $F = 2 - 2 + 1 = 1$. The *solidus* is the line below which the system is entirely solid for a given bulk composition. As temperature is lowered for a system of bulk composition *A* [Figure 12-6(b)], crystals of phase β begin to crystallize at temperature T_1 on the liquidus. As the temperature continues to fall, more crystals of β are formed. At any temperature between the liquidus and the solidus, for example at temperature T_2, the proportion of liquid and solid is determined by the *lever rule:* % liquid $= (BD/CD) \times 100$, % crystals $= (BC/CD) \times 100$. At this temperature the composition of the crystals is β while the composition of the liquid is found by dropping a perpendicular from point *C* to the composition axis.

As crystallization continues on to lower temperatures the composition of the residual liquid migrates down the liquidus. When the temperature reaches T_3, the liquid is at point *E*, known as the *eutectic,* and its composition is found by dropping a perpendicular from *E* to the composition axis. The percentage of the system that is still liquid is again found by the lever rule, and it is apparent that as temperature has decreased the proportion of solid has increased, because the horizontal distance between line *AJ* and the liquidus has increased while that between line *AJ* and the right edge of the phase diagram has remained constant. The eutectic is an invariant point: $F = 2 - 3 + 1 = 0$. As more heat is removed from the system, the temperature remains constant at the eutectic temperature until all of the liquid has crystallized. At this temperature both α and β crystallize; this is shown by the fact that a horizontal line drawn through the eutectic (i.e., the solidus) intersects the edges

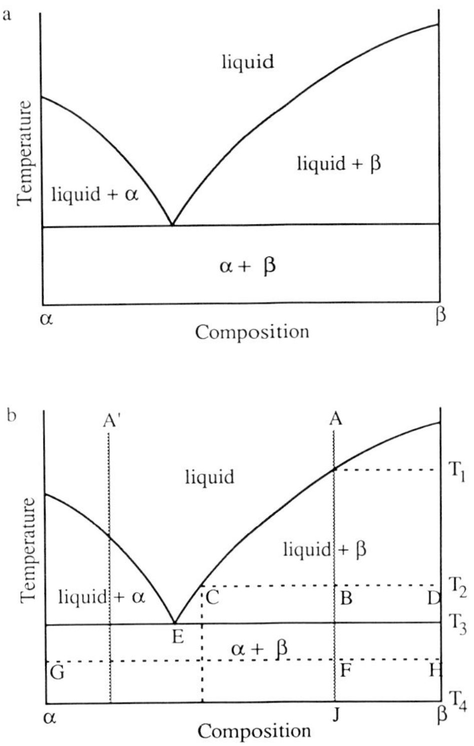

Figure 12-6. (a) A simple binary eutectic T–X phase diagram for phases α and β, which form no mutual solid solution. (b) The same phase diagram with two arbitrary bulk compositions (A and A') and equilibrium cooling paths from them indicated.

of the diagram at compositions α and β. This is, of course, the first and only temperature at which α crystallizes at all. When all of the liquid has crystallized, further removal of heat reduces the temperature. The system is now entirely solid, consisting of a mechanical mixture (as opposed to a solid solution) of phases α and β. The final proportions are again found by the lever rule: $\%\ \alpha = (FH/GH) \times 100$, and $\%\ \beta = (FG/GH) \times 100$.

Had we begun with a system at point A', the scenario would have been qualitatively identical, except that the first phase to crystallize would have been α. At the eutectic temperature the liquid composition would have been that at E, and phase β would have begun to crystallize along with phase α. When all of the liquid was crystallized the system would have consisted of solid α + solid β in proportions dictated by the lever rule, and with bulk composition identical to that at A'.

Figure 12-7 shows a schematic phase diagram wherein a stable phase of intermediate composition exists between the extreme compositions. This might, for example, represent the phase diagram for the system SiO_2–Mg_2SiO_4 (forsterite), with $Mg_2Si_2O_6$ (enstatite) as the intermediate phase. Crystallization from bulk compositions C and D is qualitatively identical to crystallization from A' and A in the

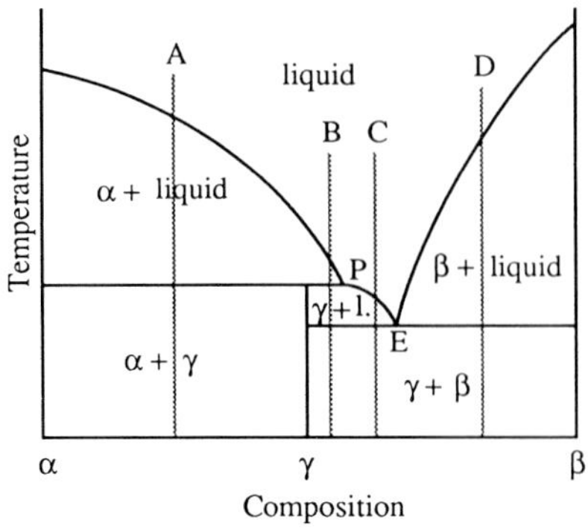

Figure 12-7. A binary *T–X* phase diagram containing both a peritectic and a eutectic. Four cooling paths are shown.

previous example, with γ and β being the extreme compositions and both starting bulk compositions resulting in solid mixtures of these two phases. The natures of crystallization from *A* and *B* are different.

As a system at *A* cools, it encounters the liquidus and begins to crystallize phase α while the residual liquid migrates down the liquidus toward point *P*. When the temperature reaches that of the solidus, only crystals of composition α have formed and the liquid has the composition shown by dropping a perpendicular from point *P* to the composition axis. Point *P* is called the *peritectic* and is an invariant point. Here, as with a eutectic, the temperature remains constant until one phase has disappeared so that some variable can change. In this case, unlike a eutectic, the reaction α + liquid = γ ensues and proceeds until the liquid is used up and the resulting solid assemblage consists of $\alpha + \gamma$. This behavior is caused by the *incongruent melting* of phase γ—that is, pure γ melts on heating to a liquid not of its own composition and a solid phase. Crystallization from *B* is nearly the same, except that now the reaction at the peritectic runs out of phase α before running out of liquid, so the remaining liquid moves down the univariant curve toward eutectic *E* crystallizing phase γ. At *E* the last liquid crystallizes both γ and β until it is used up, and the final assemblage is $\gamma + \beta$.

[It is well known that forsterite and quartz do not coexist in stable equilibrium, and this sort of phase diagram explains why. Let α = forsterite, β = quartz, and γ = enstatite. Compositions to the right of enstatite (γ) will end up as enstatite plus quartz assemblages, and those to the left will end up as forsterite plus enstatite assemblages. Melts with the exact composition of enstatite will end up as monomineralic enstatite, because in the formation of enstatite the peritectic reaction will use up all of the liquid and all of the forsterite simultaneously, and quartz will never form because point E will never be reached. Not all binary systems with stable intermediate compositions show incongruent melting—see Figure 12-8, for instance—but the presence of the intermediate phase nonetheless ensures that the two extreme compositions will not coexist in equilibrium.]

Next let us examine the *T–X* phase diagram for a binary system in which there is partial solid solution between the two extreme compositions (Figure 12-9). As

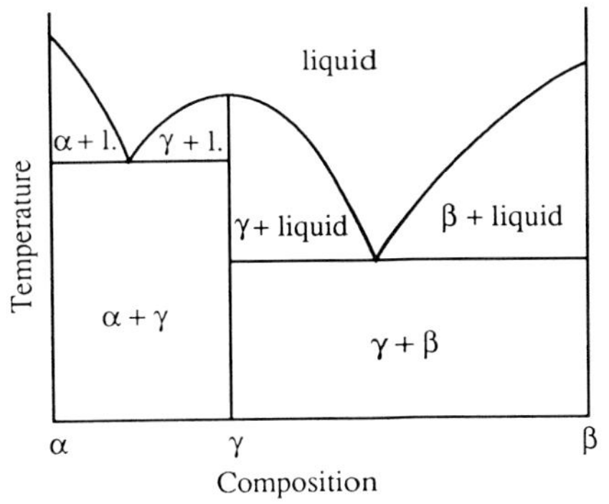

Figure 12-8. A binary *T–X* phase diagram that involves an intermediate stable composition but no peritectic.

melt of composition *A* cools and encounters the liquidus, crystals of $\alpha_{\text{solid solution}}$ begin to form. At temperature T_1 an isothermal (i.e., horizontal) line intersects the liquidus on the right and the solidus on the left, indicating the compositions of the crystals (by dropping a perpendicular from the intersection between the isothermal line and the solidus to the composition axis) and of the residual liquid (found in a similar way, using the intersection of the isotherm and the liquidus). When the liquid reaches the eutectic, crystals of the most β-rich $\alpha_{\text{solid solution}}$ and α-rich $\beta_{\text{solid solution}}$ possible are formed. Thereafter as the temperature decreases, exsolution occurs as the compositions of both phases move in the crystalline state closer to the end-member compositions. At any temperature, say T_2, an isothermal line connects the two limbs of the *solvus*, and perpendiculars dropped from those intersections indicate the equilibrium compositions of the two solid phases at that temperature. Of course, the movement of atoms in the solid state requires energy, usually thermal, and, if that energy is unavailable, then the exsolution is kinetically stranded at some stage.

Binary *T–X* diagrams showing complete solid solution appear in Figure 12-10.

Figure 12-9. A binary *T–X* phase diagram that involves partial solid solution between the end members, indicated by the subscript *ss*.

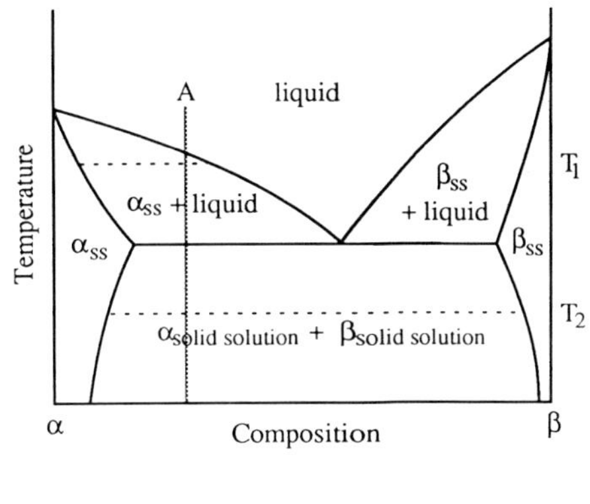

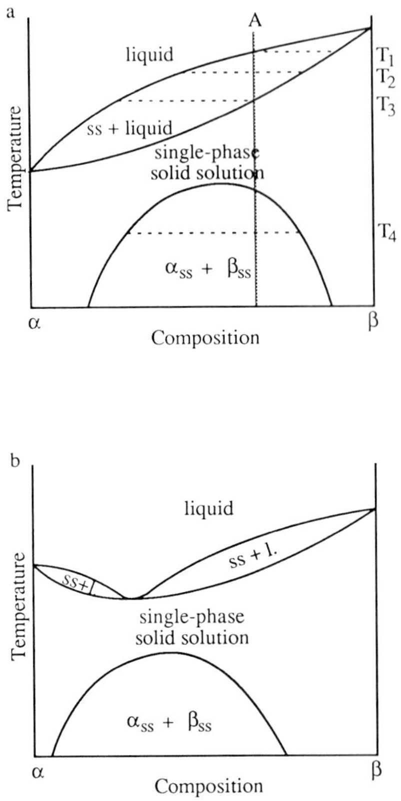

Figure 12-10. (a) A binary *T–X* phase diagram that shows complete solid solution between end members as well as a solvus at lower temperatures. This diagram bears a resemblance (greatly simplified) to the plagioclase *T–X* phase diagram shown in Figure 2-44. (b) A binary *T–X* phase diagram similar to that in Figure 12-10(a), but showing a minimum melting point in the liquidus. This system is superficially similar to the alkali feldspar system shown in Figure 2-2 (except that the latter does not extend to high enough temperatures to show the liquidus and solidus).

In Figure 12-10(a) a liquid at A begins to cool, encountering the liquidus at T_1. The compositions of crystals first produced is given by the intersection of the isotherm with the solidus, as in previous examples. As the temperature decreases the crystals that formed previously are no longer in equilibrium with the liquid, which is migrating down the liquidus. The crystals therefore reequilibrate (changing their compositions by solid-state diffusion) and migrate down the solidus. As the last liquid crystallizes at T_3, the composition of all crystals is uniformly the same composition as the starting bulk composition, A, if equilibrium crystallization has occurred. If cooling has been too fast for equilibrium to be maintained between crystals and the changing liquid (i.e., if fractional crystallization has occurred), then compositionally zoned crystals will have formed. When the temperature has decreased to that of the solvus for the given bulk composition, subsolvus exsolution can begin, governed, of course, by kinetic factors. At T_4 the compositions of the equilibrium exsolved phases is given, as before, by the intersection of an isotherm with the limbs of the solvus. Figure 12-10(b) is a variation on the theme, showing a solid solution that has a minimum melting point between the two end members. In a very simplified fashion, Figure 12-10(a) is like the plagioclase phase diagram of Figure 2-44, and Figure 12-10(b) is like the alkali feldspar phase diagram of Figure 2-2 (except that the temperature was not shown high enough there to include the solidus and liquidus on the phase diagram).

Ternary *T–X* Phase Diagrams

It is possible to represent graphically three-component *T–X* phase diagrams if the components are placed at the corners of a triangle and the temperature axis is oriented vertical to the plane of that triangle (Figure 12-11), so as to make a three-dimensional triangular prism. The liquidus is then a curved surface with thermal "valleys" leading, in the simplest case, from the binary eutectics on the faces of the triangular prism to the ternary eutectic—the point of lowest crystallization temperature, at which three phases crystallize simultaneously. Such ternary diagrams are generally drawn as if viewed vertically down the temperature axis, so that only the top surface of the liquidus is visible (Figure 12-12).

Figure 12-12 shows a simple eutectic ternary phase diagram. The corners are labeled by components (which here are also the only three phases in the system), and the *primary phase fields* for each phase are labeled within the diagram. The

Figure 12-11. A ternary *T–X* phase diagram in perspective. This is not the usual way of looking at ternary diagrams, but it illustrates that each side of the triangular prism is a binary *T–X* phase diagram and that the liquidus surface, shown here for a eutectic system, varies in temperature (elevation above the α–β–γ plane) with bulk composition.

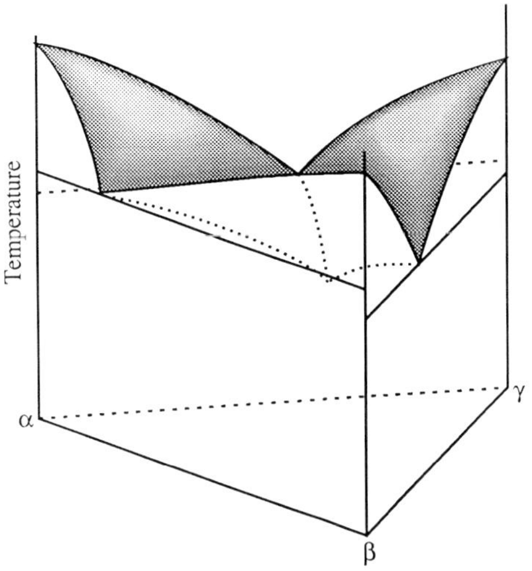

primary phase field of phase α is the compositional area on the top of the liquidus within which phase α will crystallize directly from the liquid upon cooling through the liquidus. It is, of course, possible that some combination of components α, β, and γ would yield a fourth stable phase in the system, in which case that phase would also have a primary phase field on the liquidus. Between the primary phase fields are the thermal valleys, each one with an arrowhead indicating the direction in which temperature decreases. For a starting bulk composition at A, β will be the phase to begin crystallizing as the temperature reaches that of the liquidus beneath point A, and the composition of the residual liquid will migrate down the liquidus surface directly away from the β corner (because β is being removed from the liquid). As the liquid reaches the phase boundary between the primary phase fields of β and γ, it will migrate down the boundary toward the ternary eutectic (E), crystallizing β and γ simultaneously. Upon reaching the eutectic the remaining liquid will crystallize α, β, and γ until it is completely used up. One must remember that the liquid is always a phase on the liquidus surface of a ternary T–X diagram, so the phases in equilibrium in this example are α, β, γ, and liquid. The temperature will remain constant at the eutectic until one phase is eliminated, because the eutectic is an invariant point: $F = 3 - 4 + 1 = 0$.

Crystallization paths involving phases with intermediate compositions, incongruently melting phases, and partial or complete solid solution can also be represented on ternary T–X diagrams, but are peripheral to the needs of this text, and will not be discussed. Figure 12-13 is included to show the relationship of a ternary solvus to the liquidus and solidus surfaces. This figure superficially resembles the

Figure 12-12. A ternary T–X phase diagram viewed down the temperature axis (the usual way of viewing such a diagram). From a starting composition A, the system descends toward the page until it intersects the liquidus, at which point crystals of β form and the liquid (now depleted in β) migrates away from the β corner. When it encounters the boundary between the primary phase fields of β and γ, both of these crystallize, and the liquid moves down the boundary toward the ternary eutectic. At the eutectic all three phases crystallize until the liquid is used up.

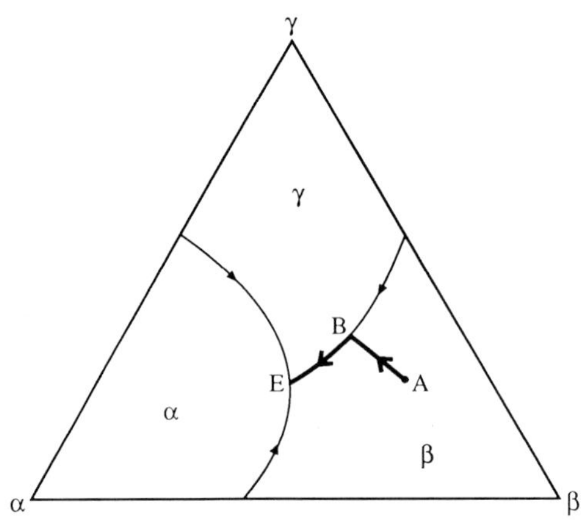

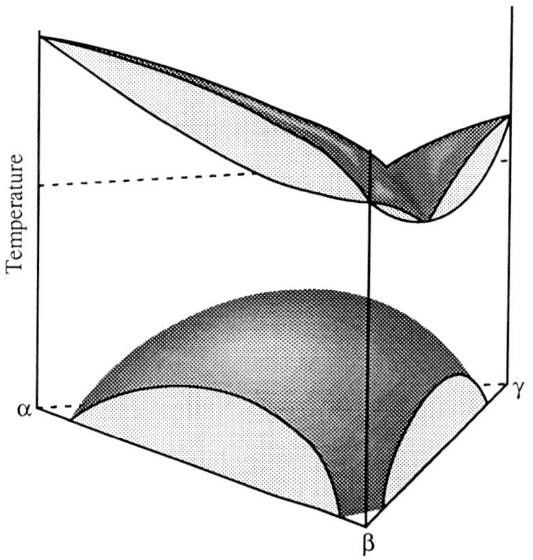

Figure 12-13. Perspective view of a ternary system that contains a solvus at low temperatures. Compare the α–β side with Figure 12-10(a) and the β–γ side with Figure 12-10(b).

ternary feldspar system, where α is anorthite, β is albite, and γ is potassium feldspar (compare with Figure 12-10).

KINETICS

Thermodynamics is concerned with equilibrium processes. Phase diagrams predict the equilibrium behavior of simple systems, and may find limited use for nonequilibrium situations such as fractional crystallization. These tools are useful for telling us what is possible, but may have little to say about what has actually happened; they reveal the initial and final states of a system under equilibrium conditions, and therefore the conditions toward which a system tends, but fail to elucidate the processes that accompanied the changes or the rates at which they occurred. Moreover, failure to attain equilibrium conditions is apparently common in geologic environments and the changes in a system are then governed instead by kinetic factors. Some discussion of these kinetic factors is thus appropriate.

Consider the simple solvus shown in Figure 12-14(a). Above the solvus the stable mineral is a single solid solution $\alpha\beta$ for any bulk composition. For bulk composition A at T_1 the stable solid solution has composition $\alpha_{x_1}\beta_{1-x_1}$. When the temperature reaches T_2 the equilibrium assemblage is a mixture of two solid solution phases: crystals of composition $\alpha_{x_2}\beta_{1-x_2}$ mixed with crystals of $\alpha_{x_3}\beta_{1-x_3}$. Continued reduction of the temperature to T_3 yields crystals of $\alpha_{x_4}\beta_{1-x_4}$ crystals of $\alpha_{x_5}\beta_{1-x_5}$, which are nearly pure α and pure β, respectively. Figure 12-14(b) shows the free-energy (G) curves associated with these temperatures. In fact, exsolution to com-

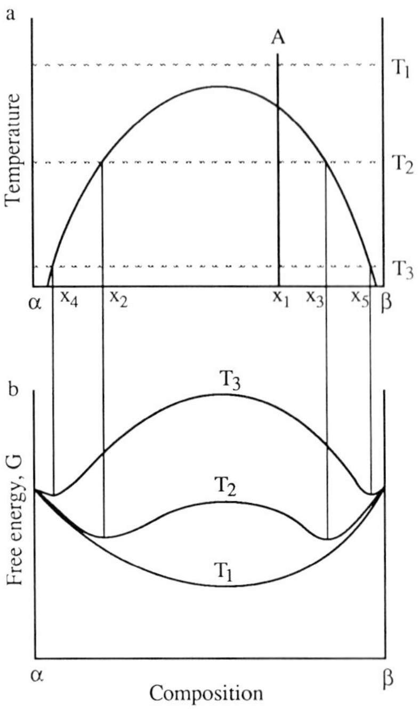

Figure 12-14. (a) A solvus in a binary system. Temperatures T_2, and T_3 on the solvus are indicated by the intersection of the solvus with the dashed lines. Compositions at those intersections are shown by the intersection of vertical lines with the composition axis. (b) The Gibbs free-energy curves for the three temperatures indicated in (a). Equilibrium exsolution compositions correspond to minima in the curves.

positions as disparate as those predicted by the equilibrium phase diagram at temperature T_3 is realistically probably unlikely, because exsolution requires the diffusion of atoms in the solid state; this process is usually slow at best, and becomes more so as the temperature is decreased. Thus the resulting assemblage is generally not the thermodynamically stable one, but the one that existed when kinetic factors—in this case, the diffusion rate—halted the transformation.

Diffusion

Diffusion is a nonlinear process described by the partial differential equation

$$\frac{\partial C}{\partial t} = D\left(\frac{\partial^2 C}{\partial x^2} + \frac{\partial^2 C}{\partial y^2} + \frac{\partial^2 C}{\partial z^2}\right)$$

Here C is concentration, t is time, D is the diffusion coefficient, and x, y, and z are positional coordinates. The form of this equation is identical to that for heat flow, and thus describes the macroscopic progress of diffusion but does not elucidate the way or ways in which atoms in a solid actually move from one place to another.

Atoms will move from regions of higher to lower chemical potential, but how the process occurs and the rate at which it occurs are not given by the equation.

D is strongly dependent on temperature, as shown by its analytical expression:

$$D = Ae^{-\Delta G_a/RT}, \tag{12.21}$$

where A is a constant, ΔG_a is the free energy of activation for diffusion, R is the gas constant, and T is temperature in kelvin. Because both the initial and final states are local energy minima in free-energy space (with the final state the lower of the two), there must be a local maximum between them; the energy required to overcome this barrier is the activation energy, and the state of the system at the intermediate local maximum is called the *activated state.* For typical values of ΔG_a the diffusion coefficient varies by many orders of magnitude over the range of reasonable geologic temperatures, increasing dramatically as temperature increases.

Diffusion may occur in several ways in crystalline materials. It may occur by the migration of atoms through crystals (so-called *volume diffusion*), or along boundaries between crystals (*grain-boundary diffusion*), or along crystal surfaces (*surface diffusion*). For phase changes in macroscopically homogeneous crystals, volume diffusion must be the dominant process, as the other two would affect only small portions of any crystal. Volume diffusion may occur in any of several possible ways, some of which seem less probable than others.

One is by simple exchange of places by two atoms, for example Si and Al in adjacent tetrahedra in a feldspar. Given that this implies simultaneous movement of atoms past each other by shoving surrounding atoms aside, this sounds like a process with a high activation energy and a low probability of occurrence. A more likely mechanism involves vacancies in the crystal structure, since every crystal has an equilibrium number of vacancies dependent on temperature.

The creation of a vacancy defect in a crystal raises the internal energy, but the mixing of vacancies among the filled lattice sites produces an entropy of mixing that is energetically favorable if the number of vacancies is small compared with the number of filled sites. If there are n vacancies among a total of N sites, then the entropy of mixing is

$$\Delta S = k \ln \{N!/[(N-n)!n!]\}, \tag{12.22}$$

where k is the Boltzmann constant. From Equation (12.6) the change in the Helmholtz free energy caused by the presence of the vacancies is

$$\Delta A = \Delta E - T\,\Delta S. \tag{12.23}$$

If E is the change in internal energy due to a single vacancy, then substituting Equation (12.22) into Equation (12.23) yields

$$\Delta A = nE - kT \ln \{N!/[(N-n)!n!]\}. \tag{12.24}$$

Expanding the logarithmic term,

$$\Delta A = nE - kT\,[\ln N! - \ln (N-n)! - \ln n!];$$

using the Sterling formula ($\ln X! = X \ln X - X$), and assuming that $n \ll N$ so that $(N - n) \approx N$, this reduces to

$$\Delta A = nE - kTn \ln N + kTn \ln n.$$

The change in A with a differential change in n is therefore

$$\Delta A/\delta n = E - kT \ln N + kT \ln n,$$

and because $\partial A/\partial n = 0$ at equilibrium,

$$0 = E - kT \ln N + kT \ln n,$$

$$-E/kT = \ln (n/N), \qquad (12.25)$$

$$n/N = e^{-E/kT}$$

Assume, for example, that $E = 1$ eV and that the temperature is 300 K (k is 8.6169×10^{-5} eV K^{-1}); then $n/N \approx 1.6 \times 10^{-17}$—that is, for every vacancy there are about 6.25×10^{16} filled sites. If the temperature is increased to 800 K, n/N is 5×10^{-7}; so that there are 2×10^{6} filled sites for every vacancy. The change from room temperature to 527°C results in an increase in vacancies present at equilibrium by a factor of about 3×10^{10} (for this hypothetical value of E). If the vacancies are paired so that charge balance is maintained, there are still more vacancies at 527 than at 27°C by a factor of over 10^{5} (again, for $E = 1$ eV). It is therefore obvious that diffusion by a vacancy mechanism must proceed at a very much faster rate at high temperature than it does at low temperature.

Point defects in crystals are of two general types: *Schottky defects* and *Frenkel defects.* The absence of an atom from the lattice constitutes a Schottky defect, whereas the relocation of an atom from its normal position to an interstitial (i.e., normally unoccupied) position is a Frenkel defect. The availability of interstitial sites at any given temperature depends, of course, on the nature of the crystal structure and on the sizes of the atoms that are migrating. Smaller atoms, in general, move through a crystal structure faster than larger ones. The formation of Schottky defects may involve coupled vacancy substitutions in order to maintain overall charge balance.

Alternative Metastable Behavior

Figure 12-15(a) shows a T–X phase diagram for a hypothetical system in which a solvus exists below the solidus. For a solid of composition X starting at point A and cooled to temperature T_1, the stable assemblage is an exsolution intergrowth of α_{ss} (α-rich solid solution) and β_{ss} (β-rich solid solution) of compositions x_1 and x_2, respectively. The G–X diagram of Figure 12-15(b) shows the free energies for α_{ss}, β_{ss}, and the single phase $\alpha\beta_{ss}$ at temperature T_1, and the heavy dotted line tangent to the α_{ss} and β_{ss} curves at x_1 and x_2, respectively, gives the assemblage of lowest free energy at that temperature. This is the assemblage that is stable under equilibrium conditions for the system of bulk composition X at temperature T_1. The dot on the G curve for $\alpha\beta_{ss}$ represents the free energy for the less stable single-phase solid solution at the same temperature, and the vertical distance from the dot to the heavy

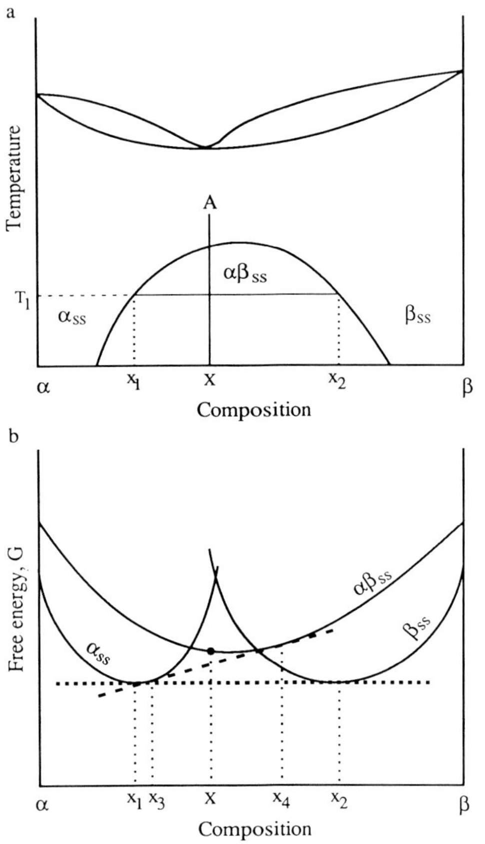

Figure 12-15. a. A binary *T–X* phase diagram containing a solvus. As the phase at *A* cools below the solvus, equilibrium thermodynamics predicts that exsolution to compositions on the solvus will occur. For example, at temperature T_1 the system should consist of an exsolution intergrowth of compositions x_1 and x_2. (b) Curves of free energy versus composition at temperature T_1. The dot on the $\alpha\beta_{ss}$ curve represents the free energy of the single unexsolved phase, whereas the heavy dotted line between compositions x_1 and x_2 indicate the decrease in free energy associated with equilibrium exsolution. Exsolution to compositions x_3 and x_4 also results in a decrease in free energy, and, while it is not as great a decrease as results from equilibrium behavior, it may be kinetically favored because the products are less different from the starting phase than are the products of the equilibrium process. Such behavior is called *alternative metastable behavior.*

dotted line is the amount by which the free energy of the exsolved assemblage is lowered relative to the single-phase solid solution.

If cooling occurs too rapidly for equilibrium to be maintained, kinetic barriers may inhibit the formation of the equilibrium assemblage because that would require large amounts of diffusion and some degree of structural rearrangement. Instead, it may be kinetically favorable to form an α_{ss} of composition x_3 and a "β-richer" $\alpha\beta_{ss}$ of composition x_4, as shown by the heavy dashed line. The free-energy difference between the single-phase solid solution and this assemblage is represented by the vertical distance from the dot to the heavy dashed line on Figure 12-15(b); the free energy of this assemblage is not as low as that of the stable assemblage, but it *is* lower than that of the single-phase solid solution (it is therefore described as *metastable*) and, unlike the stable assemblage, it is kinetically attainable. The formation of this metastable assemblage is called *alternative metastable behavior,* and it is very common in minerals.

Kinetically Controlled Processes

In this text we consider several processes that are kinetically controlled and occur in one or more of the mineral groups discussed. The equilibrium behavior of these processes is well understood in most cases, but that behavior represents only the ideal end to which the system tends. Among the important processes that are kinetically controlled are nucleation and growth of exsolved phases, spinodal decomposition, and reconstructive and order–disorder phase transformations. As these are discussed at some length where applicable in the first eight chapters of this text, they will not be reiterated here.

TTT Diagrams

As a cooling solid crosses a solvus, as in Figure 12-15(a), two factors determine the speed and extent of the predicted exsolution: the free energy of activation for the transformation and the diffusion rate. The free energy of activation is infinite at T_t but decreases with the degree of supersaturation, which means that it decreases as the temperature decreases. In terms of heterogeneous nucleation (the nucleation of a new phase at defects within a parent structure), therefore, more nuclei can form per unit time at temperatures substantially below the solvus temperature than at temperatures near the solvus temperatures. As the temperature decreases, however, the diffusion rate also decreases, which tends to counteract the decrease in activation energy, making it harder to form the nuclei that, thermodynamically, could form. The rate of nucleation, R_N, is given by the product of the two factors:

$$R_N = Ke^{-\Delta G^*/RT}e^{-\Delta G_a/RT}, \tag{12.26}$$

where K is a constant, ΔG^* is the free energy of activation for nucleation and ΔG_a is the activation energy for diffusion [see Equation (12.21)]. It is evident that at high temperatures (i.e., when the system is near the solvus) diffusion is fast but the rate of nucleation is controlled by the thermodynamic term; at low temperatures the free energy of activation for the transformation is very small, but the rate of nucleation is controlled by the kinetic term. At some temperature in between these extremes, the nucleation rate and diffusion rate combine to yield the maximum rate of transformation, and time-temperature-transformation (TTT) diagrams display these relationships between temperature and transformation rate.

Figure 12-16 shows the typical form of a TTT diagram for a cooling solid in which some sort of kinetically governed phase transformation occurs. The curve on the left indicates the onset of the transformation, and the one on the right indicates completion. T_t is the transformation temperature under equilibrium conditions—

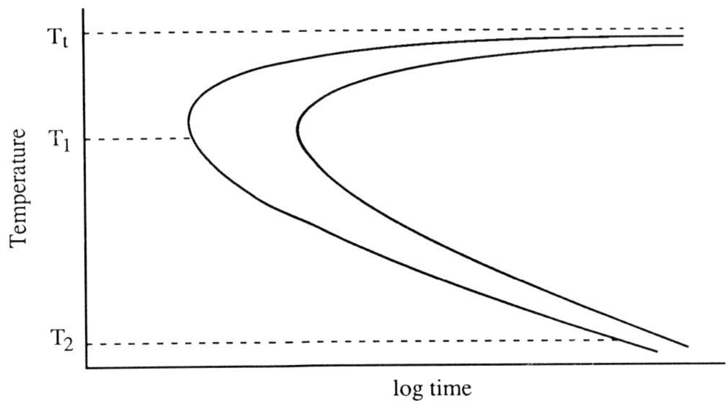

Figure 12-16. A time-temperature-transformation (TTT) diagram. The curve on the left shows the time (in terms of the logarithm of time) for the onset of a transformation at each temperature, and the curve on the right shows the time for completion of the reaction. See text for discussion of the three temperatures indicated by dashed lines.

the one that would be obtained for the system from an equilibrium phase diagram. Note that the horizontal axis is log time rather than time; so that the range of times prevalent in geological processes can be represented without an extremely long diagram. Near the temperature T_t the time required for growth of the new phase is very long, and this is the condition for equilibrium crystallization. At low temperatures (T_2, for example) the rate of formation of the new phase is also very slow. At a tem-

Figure 12-17. A hypothetical TTT diagram showing the curves for onset and completion of heterogeneous nucleation (HetNuc), homogeneous nucleation (HomNuc), and spinodal decomposition (SpinDec). The four dashed arrows show four different rates of cooling, the slowest being R_1 and the fastest being R_4. Different sorts of transformations will take place (or, in the case of R_4, no transformation at all) for different cooling rates.

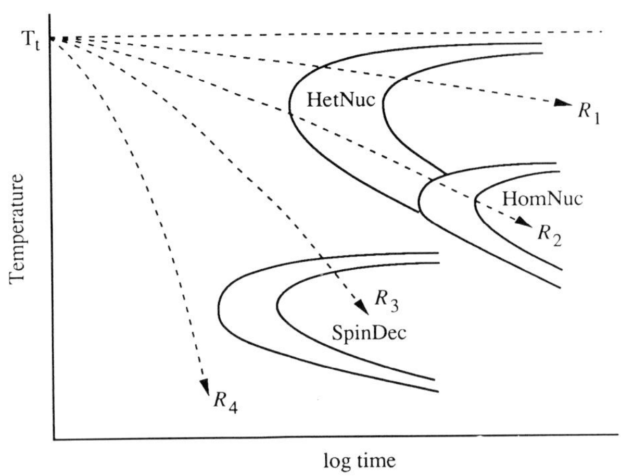

perature such as T_1, however, the time for onset of the transformation is relatively much faster, as is the time for completion. If all of the terms of Equation (12.26) are known for a system, a TTT diagram may be calculated; generally these data are unavailable, and the diagram is determined experimentally.

The arrows of Figure 12-17 schematically show four cooling rates beginning at T_t at time zero. If cooling through the solvus is slow enough (cooling rate R_1), then the only process that occurs is the heterogeneous nucleation of an exsolved phase. A faster rate of cooling (rate R_2) results in the initiation of heterogeneous nucleation, but as the temperature decreases, the system crosses the boundary marking the beginning of homogeneous nucleation and completes the transformation in that way. At a cooling rate R_3 nucleation of a discrete exsolved phase does not occur at all, but rather the system crosses the coherent spinodal, and spinodal decomposition ensues; this is alternative metastable behavior. If the system is cooled too rapidly (R_4), it persists metastably as a single phase.

Appendix
Summary of Unit-Cell Data for Ambient Conditions

(Because of variations in lattice parameters, caused largely by solid solution, only approximate values for end-member compositions are given. Lattice parameters that are fixed by symmetry are omitted.)

Mineral Name	Space Group	Unit-Cell Parameters (Å, deg) a / α	b / β	c / γ
Mineral family: Silica polymorphs				
Coesite	$C2/c$	7.17	12.33	7.18
			120	
Cristobalite (low)	$P4_12_12$ ($P4_32_12$)	4.97		6.93
Quartz (α, low)	$P3_112$ ($P3_212$)	4.91		5.40
Quartz (β, high)	$P6_222$ ($P6_422$)	5.01		5.47
Stishovite	$P4_2/mnm$	4.18		2.66
Tridymite (low)	(See Figure 1-9)			
Mineral family: Feldspars				
Albite (low)	$C\bar{1}$	8.14	12.79	7.16
		94.3	116.6	87.7
Anorthite	$P\bar{1}$	8.18	12.88	14.17
		93.2	115.9	91.2
Celsian	$I2/c$	8.62	13.08	14.41
			115.1	
Max. microcline	$C\bar{1}$	8.56	12.96	7.22
		90.6	115.8	87.7
Sanidine	$C2/m$	8.56	13.03	7.17
			116.0	
Mineral family: Micas				
Anandite-2*Or*	$Pnmn$	5.47	9.49	19.96
Annite-1*M*	$C2/m$	5.39	9.32	10.27
			100.6	
Celadonite-1*M*	$C2/m$	5.21	9.02	10.27
			100.0	
Clintonite-1*M*	$C2/m$	5.20	9.01	9.81
			100.3	

Mineral Name	Space Group	Unit-Cell Parameters (Å, deg) a α	b β	c γ
Ephesite-2M_1	$C2/c$	5.11	8.87 95.2	19.29
Ferriannite-1M	$C2/m$	5.43	9.40 100.1	10.34
Margarite-2M_1	Cc	5.1	8.83 95.5	19.15
Muscovite-2M_1	$C2/c$	5.19	9.00 95.2	20.10
Paragonite-2M_1	$C2/c$	5.13	8.99 94.4	19.29
Phengite-2M_1	$C2/c$	5.22	9.04 95.8	19.97
Phlogopite-1M	$C2/m$	5.33	9.23 100.2	10.26
Polylithionite-1M	$C2/m$	5.19	8.97 100.5	10.03
Siderophyllite-1M	$C2/m$	5.37	9.30 100.1	10.27
Zinnwaldite-1M	$C2$	5.30	9.14 100.8	10.10
Mineral family: Pyroxenes				
Aegirine (acmite)	$C2/c$	9.65	8.80 107.3	5.29
Clinoenstatite	$P2_1/c$	9.62	8.83 108.3	5.19
Diopside	$C2/c$	9.75	8.92 105.8	5.25
Enstatite	$Pbca$	18.22	8.81	5.17
Esseneite	$C2/c$	9.79	8.82 105.8	5.37
Ferrosilite	$Pbca$	18.43	9.08	5.24
Hedenbergite	$C2/c$	9.85	9.01 104.8	5.26
Jadeite	$C2/c$	9.42	8.56 107.6	5.22
Johannsenite	$C2/c$	9.98	9.16 105.5	5.29
Kanoite	$C2/c$	9.74	8.94 108.6	5.26
Omphacite	$P2/n$	9.60	8.77 106.9	5.27
Petedunnite	$C2/c$	9.82	9.00 105.6	5.27
Spodumene	$C2$	9.45	8.39 110.1	5.22
Mineral family: Amphiboles				
Anthophyllite	$Pnma$	18.56	18.01	5.28
Arfvedsonite	$C2/m$	9.77	18.03 103.7	5.33

		Unit-Cell Parameters (Å, deg)		
Mineral Name	Space Group	a α	b β	c γ
Eckermannite	$C2/m$	9.76	17.89 103.2	5.28
Edenite	$C2/m$	9.84	17.95 105.2	5.31
Ferro-actinolite	$C2/m$	9.91	18.17 105.0	5.28
Ferro-gedrite	$Pnma$	18.53	17.95	5.32
Ferro-glaucophane	$C2/m$	9.60	17.87 103.6	5.30
Ferro-hornblende	$C2/m$	9.83	18.04 105.0	5.30
Ferr. pargas. hornbl.	$C2/m$	9.96	18.19 104.9	5.32
Ferro-pargasite	$C2/m$	9.95	18.14 105.3	5.33
Ferro-tschermakite	$C2/m$	9.82	18.11 105.0	5.33
Gedrite	$Pnma$	18.60	17.84	5.28
Glaucophane	$C2/m$	9.54	17.74 103.7	5.30
Hastingsite	$C2/m$	9.95	18.24 105.0	5.34
Holmquistite	$Pnma$	18.29	17.67	5.28
Kaersutite	$C2/m$	9.83	17.99 105.2	5.30
Kozulite	$C2/m$	9.91	18.11 104.6	5.30
Magn.-arfvedsonite	$C2/m$	9.80	17.83 104.2	5.27
Magn.-hastingsite	$C2/m$	9.84	17.97 105.1	5.30
Magn.-hornblende	$C2/m$	9.82	17.97 104.9	5.30
Magn.-katophorite	$C2/m$	10.02	18.04 105.0	5.29
Magn.-riebeckite	$C2/m$	9.80	17.99 105.0	5.24
Pargasite	$C2/m$	9.87	18.01 105.3	5.30
Richterite	$C2/m$	9.91	17.98 104.3	5.27
Riebeckite	$C2/m$	9.77	18.05 103.6	5.34
Taramite	$C2/m$	.995	18.10 105.5	5.32
Tirodite	$P2_1/m$	9.55	18.01 102.7	5.30
Tremolite	$C2/m$	9.82	18.05 104.7	5.28

Mineral Name	Space Group	Unit-Cell Parameters (Å, deg) a α	b β	c γ
Mineral family: Al_2SiO_5 polymorphs and mullite				
Andalusite	*Pnnm*	7.78	7.92	5.57
Kyanite	$P\bar{1}$	7.10	7.74	5.57
		90.1	101.0	105.8
Mullite	*Pbam*	7.58	7.68	2.89
Sillimanite	*Pbnm*	7.44	7.60	5.75
Mineral family: Olivines				
Fayalite	*Pbnm*	4.82	10.47	6.10
Forsterite	*Pbnm*	4.76	10.23	5.99
Glauchochroite	*Pbnm*	4.91	11.15	6.49
Kirchsteinite	*Pbnm*	4.89	11.18	6.47
Monticellite	*Pbnm*	4.82	11.11	6.38
Tephroite	*Pbnm*	4.91	10.60	6.26
Mineral family: Garnets				
Almandine	$Ia\bar{3}d$	11.53		
Andradite	$Ia\bar{3}d$	12.05		
Blythite	$Ia\bar{3}d$	11.85		
Calderite	$Ia\bar{3}d$	11.82		
Goldmanite	$Ia\bar{3}d$	12.07		
Grossular	$Ia\bar{3}d$	11.85		
Hydrogarnet	$Ia\bar{3}d$	12.57		
Kimzeyite	$Ia\bar{3}d$	12.20		
Knorringite	$Ia\bar{3}d$	11.60		
Khoharite	$Ia\bar{3}d$	11.12		
Pyrope	$Ia\bar{3}d$	11.46		
Skiagite	$Ia\bar{3}d$	11.72		
Spessartine	$Ia\bar{3}d$	11.61		
Uvarovite	$Ia\bar{3}d$	12.00		

Index